AF695972

Springer
Proceedings in Physics 3

Springer Proceedings in Physics

Springer Proceedings in Physics is a new series dedicated to the publication of conference proceedings. Each volume is produced on the basis of camera-ready manuscripts prepared by conference contributors. In this way, publication can be achieved very soon after the conference and costs are kept low; the quality of visual presentation is, nevertheless, very high. We believe that such a series is preferable to the method of publishing conference proceedings in journals, where the typesetting requires time and considerable expense, and results in a longer publication period. Springer Proceedings in Physics can be considered as a journal in every other way: it should be cited in publications of research papers as *Springer Proc. Phys.*, followed by the respective volume number, page and year.

Nonlinear Phenomena in Physics

Proceedings of the
1984 Latin American School of Physics,
Santiago, Chile, July 16–August 3, 1984

Editor: F. Claro

With 110 Figures

Springer-Verlag
Berlin Heidelberg New York Tokyo

Professor Francisco Claro, Ph. D.
Pontificia Universidad Catolica de Chile, Facultad de Fisica,
Casilla 114-D, Santiago, Chile

ISBN-13: 978-3-642-93291-5 e-ISBN-13: 978-3-642-93289-2
DOI: 10.1007/978-3-642-93289-2

This work is subject to copyright. All rights are reserved, whether the whole or part of the material is concerned, specifically those of translation, reprinting, reuse of illustrations, broadcasting, reproduction by photocopying machine or similar means, and storage in data banks. Under § 54 of the German Copyright Law where copies are made for other than private use, a fee is payable to "Verwertungsgesellschaft Wort", Munich.

© Springer-Verlag Berlin Heidelberg 1985
Softcover reprint of the hardcover 1st edition 1985

The use of registered names, trademarks, etc. in this publication does not imply, even in the absence of a specific statement, that such names are exempt from the relevant protective laws and regulations and therefore free for general use.

2153/3130-543210

Preface

It was almost four hundred years ago that Galileo wrote in Il Saggiatore that the "Book of Nature is written in mathematical characters". This sentence, inspired at the dawn of physics has proved with the passage of time to contain a deep truth and also a warning: in order to understand Nature, first we must learn to read mathematical characters. Indeed, writing physical law in such characters has proved not as hard as unraveling the content of the resulting equations. In particular, the lack of knowledge in the field of nonlinear mathematics has been a severe limitation in the past. Thus the solution to equations such as the Navier-Stokes equation in fluid dynamics has remained elusive. The recent advent of fast computers and some important analytical and numerical results in the study of bifurcations and nonlinear waves have encouraged work both in theory and experiment involving nonlinear phenomena. An explosive growth in the specialized literature penetrating most research areas in physics in the last few years has ensued.

This book contains the most recent advances in nonlinear physics in various fields including astrophysics, gravitation, particle physics, quantum optics, fluid dynamics and the mathematics underlying the phenomena of chaos and nonlinear waves. It presents a selection from the lectures delivered at the XXI Latin American School of Physics held in Santiago, Chile in July-August 1984 (ELAF'84). This School, attended by over 100 physicists from all parts of America, covered the subject matter NONLINEAR PHENOMENA IN PHYSICS trying to balance information on the development of new mathematical methods with the recent phenomenology most relevant to the subject. The book reflects this spirit, together with the aim of publishing material that is new and not contained in available review articles or books. Authors are renowned specialists in each field, selected by the Organizing Committee of ELAF'84 considering not only their expertise but also their ability to communicate. They were asked to write an easy to read text that would be of use not only to specialists but also to newcomers to the field and to advanced students. We feel these goals were achieved and that this volume will prove to be valuable and timely.

I would like to thank all those who made ELAF'84 possible. First, I owe special gratitude to the Organizing Committee: Drs. M.C. Depassier, L. Gomberoff, D. Gottlieb, F. Lund, M. Orszag, H. Quintana, I. Schmidt, I. Schuller and E. Tirapegui, who assisted me in all important decisions and many tasks. Then, my thanks to Ms. Angela Bau who combined extreme efficiency and charm to make all practical matters work, and Ms. Liliana Pineda for her patience and valuable assistance in preparing the School and editing this book. Several colleagues and local students and friends contributed in various ways to the success of ELAF'84 including L. Arroyo, Z. Barticevic, D.R. Hofstadter, J. Krause, M. Loewe, I. Olivares, M. Pacheco, R. Ramírez, S. Rios, R. Rojas and V. Tapia. Finally, the sponsors: Organization of American States (O.A.S.), U.N.E.S.C.O., National Science Foundation, Committee on Science and Technology in Developing Countries, Facultad de Ingeniería and Instituto de Música (Catholic University of Chile), International Centre for Theoretical Physics in Trieste, Italy, C.E.R.N., European Southern Observatory (ESO), Service Culturel et de Coopération Scientifique et Technique of France, the British Council, Cerro Tololo Interamerican Observatory, Argonne National Laboratory U.S.A., CONICYT Chile, C.N.P.Q. Brasil, the Chilean Academy of Sciences, Universidad F. Santa María publishing office, DERCO S.A. Chile, CORPORA S.A. Chile, and Centro de Perfecionamiento del Magisterio, Chile. In the name of the Chilean Society of Physics, organizer of ELAF'84, I would like to thank them all for their generosity.

Santiago October 1984

Francisco Claro

Contents

Part II Quantum Optics

Part III Fluids

Part IV Astrophysics and General Relativity

Part I Mathematical Methods and General

Dynamics for Golden Mean Rotation

Mitchell J. Feigenbaum
Cornell University, Physics Department, Ithaca, NY 14853, USA

Trajectory scalings for golden mean quasiperiodic motion are determined from its renormalization group. The power spectrum is then computed in an accurate approximation.

I. Introduction

This paper, of a partially pedagogic nature, assumes some familiarity with the renormalization group methods for deterministic systems. Indeed, it represents a (not entirely trivial) reworking of the tools developed for period doubling in a somewhat more complex setting. Background material will be mentioned in allusion.

Basically, we have the following situation. A high magnification of appropriately high iterates of a mapping on the circle near a special point is an arbitrarily accurate approximation to a fixed point (function) of a transformation (the renormalization group transformation) on a space of (actually pairs of) functions. Emphatically, high iterates of the map everywhere on the circle are not well approximated. The power spectrum (as an example of a general dynamical object of study) is a transform of the entire motion, and so the question arises as to how knowledge about some particular point can suffice to determine such an object. Because (modulo a proviso to follow) indeed it is possible to accomplish just this. The proviso is that a number of measurements must be provided as initial data, and the better the desired result, the more such measurements must be provided.

Rather than a limitation that could be removed by a better method, this point of formulating a "new" initial data problem represents a significant insight into what are reasonable physical questions, and what should be the significant fundamental "variables" of a physical situation characterized by a large degree of irregularity and hence embracing large amounts of information. Thus, a "better" answer can probably require calculation approaching, if not identical to, the actual simulation of the precise process, and so a computation esthetically pleasing to only a computer. That is, it offers no "physical" insight and is "understanding" insofar as the witnessing of an event can be equated to its understanding.

The method of solution presented here is that of the "trajectory scaling function" originally implemented for the determination of the dynamical properties of period doubling systems. To my mind, it represents the deepest insight that emerged from the scenario of period doubling for reason of the content of the last paragraph, and that it offers an object of study that can be abstracted to broader contexts. Indeed, the substance of this paper is more important in this sense of learning to extend ideas, rather than in its specific results.

The idea of a trajectory scaling function is that if one knows scaling everywhere along a trajectory, approximate (coarse grained, restricted observation time, etc.) data can be refined to full information. In the dynamical processes studied to date, this scaling is only asymptotic (accurate at fine scales). Hence the refinement of coarse data entails some error--the finer the data, the smaller the error. Again, I believe that this is irremedial; that is, it is likely that no finite calculation could do better. It is important to appreciate what is feasibly computible,and then to erect an appropriate kinematical framework.

In the context of strongly scaling regimes, it is an added feature that the <u>entire</u> trajectory scaling is determined by the fixed point function that pertains to just one region of phase space, so that the renormalization group methodology can be extended to determine all global dynamical objects. To understand this, the scaling function determines the ratio of pairs of closely-spaced points on the orbit. Asympotically, (as the separation goes to zero) such a ratio is invariant under smooth transformations. However, the dynamics of a system smoothly images one point to another over short time intervals. This fails only when the orbit passes close to a critical point--that is, where the inverse loses differentiability in the present context. However, the "special" point about which the renormalization group is formulated is precisely such a point. Thus, the scaling is preserved everywhere along the orbit far from this point, and can be computed arbitrarily near to that point from the fixed point. Finally, by this semi-invariance, the trajectory scaling function can be well approximated by just a few constant (and computible) values, so that easy approximate refinement calculations can be performed. I shall do just this for the case of golden mean circle maps in this paper. Let me end this Introduction with some circle map definitions to set the context.

II. Golden Mean Circle Maps

By a circle map I mean a mapping of (0,1) 1:1 and onto itself. Thus f traverses the circle precisely once and is a monotone map on (0,1). If $\tilde{f}$ is a monotone periodic map with period 1, then $\tilde{f}$ mod 1 is a suitable f. A strictly monotone f is said to be subcritical, and one for which f' vanishes somewhere, but is never negative,is said to be critical. The <u>rotation number</u> ρ of f is defined as the average angular shift per iteration. That is

$$\rho \equiv \lim_{n\to\infty} \frac{\tilde{f}^n(x)-x}{n} \qquad x\ \varepsilon(0,1) \tag{1}$$

and by $\tilde{f}$'s monotonicity is easily proven to be independent of x. Should ρ prove to be rational, $\rho = p/q$, then there exists a point $\bar{x}$ of period q of f which is attractive:

$$\bar{f}^q(\bar{x}) = \bar{x} + p. \tag{2}$$

For ρ irrational, the limit of (1) must always be taken.

The connection of ρ to quasiperiodicity is the following. Let a system move with the two frequencies ν_1 and ν_2. Observing the system every $1/\nu_1$ seconds results in the second angle advancing by ν_2/ν_1. Regarding f as the transformation of the second angle every $1/\nu_1$ seconds, its rotation number is hence $\rho = \nu_2/\nu_1$.

Golden mean rotation means that

$$\rho = \rho^* = (\sqrt{5}-1)/2 \tag{3}$$

and is surprisingly important, since in some sense, ρ^* is the "most" irrational number. It is known (inter alia) by a result of Herman that at ρ^* a smooth map with smooth inverse is by a smooth change of variables pure rotation at ρ^*:

$$f^*(\xi) = \xi + \rho^* \mod 1. \tag{4}$$

Thus, for our purposes, a subcritical golden mean map is smoothly related to (4), while this fails to be true at criticality. We intend to study the dynamics in both cases.

The next fact to know about ρ^* is that $\{\rho_n\}$ defined as

$$\rho_n \equiv F_{n-1}/F_n \tag{5}$$

where

$$F_{n+1} = F_n + F_{n-1};\ F_1 = F_2 = 1 \tag{6}$$

is a best approximating rational sequence to ρ^*. (The F_n are the "Fibonacci" numbers.) It is easy to determine that

$$\rho^* - \rho_n \sim (-\rho^{*2})^n \quad \text{and} \tag{7}$$

$$\rho^* F_n - F_{n-1} = k(-\rho^*)^n \qquad [\text{for } k = O(1)]. \tag{8}$$

Equations (6)-(8) establish the golden mean renormalization group. First, consider the one parameter family of circle maps

$$f_\omega \equiv x + \omega \mod 1. \tag{9}$$

Defining ω_n as the parameter value for which (9) has $\rho = \rho_n$, it follows from (7) that

$$\rho^* - \rho_n \sim \delta^{-n} \quad \text{where} \tag{10}$$

$$\delta = -\rho^{*-2}. \tag{11}$$

Also, since

$$f_\omega^n(x) = x + n\omega \mod 1, \tag{12}$$

$$\begin{aligned} f_{\rho^*}^{F_n}(x) &= x + \rho^* F_n \mod 1 = x + (\rho^* F_n - F_{n-1}) \mod 1 \\ &= x + k(-\rho^*) \mod 1 \end{aligned} \tag{13}$$

Defining

$$\alpha \equiv -1/\rho^*, \tag{14}$$

for sufficiently small x (to discard the mod 1),

$$\alpha^n f^{F_n}_{\rho^*} \alpha^{-n} = x + k \equiv g. \tag{15}$$

The results (10) and (15) are highly reminiscent of period doubling, and a renormalization group follows accordingly:

$$f^{F_{n+1}} = f^{F_n + F_{n-1}} = f^{F_n} \cdot f^{F_{n-1}} \tag{16}$$

and

$$\alpha^{n+1} f^{F_{n+1}} \alpha^{-n-1} = \alpha(\alpha^n f^{F_n} \alpha^{-n}) \cdot (\alpha^n f^{F_{n-1}} \alpha^{-n}) \alpha^{-1}. \tag{17}$$

Defining

$$\begin{aligned} f_n &\equiv \alpha^n f^{F_{n-1}} \alpha^{-n} \\ g_n &\equiv \alpha^n f^{F_n} \alpha^{-n} \end{aligned} \tag{18}$$

(17) is equivalent to

$$\begin{aligned} g_{n+1} &= \alpha g_n f_n \alpha^{-1} \\ f_{n+1} &= \alpha g_n \alpha^{-1} \end{aligned} \tag{19}$$

which is the golden mean renormalization group transformation on the pair of functions (f_n, g_n). The fixed point of (19) satisfies

$$\begin{aligned} g &= \alpha g f \alpha \\ f &= \alpha g \alpha^{-1} \end{aligned} \tag{20}$$

or

$$g = \alpha g \alpha g \alpha^{-2}, \tag{21}$$

for which (15) is easily seen to be a solution. For a critical golden mean f, generically f has a cubic inflection point as then too does the object on the left hand side of (15), so that (15) is not the desired critical solution of (21). That solution does exist, but availabe only numerically. For the subcritical case, it is easy to see that the linear solution (15) is again correct: a subcritical f is smoothly conjugate (i.e., by a coordinate transformation) to (12) and in the limit as $n \to \infty$ in (15), the infinite magnification of the smooth coordinate transformation is linear.

The question we answer in this paper is what is the dynamical behavior of a golden mean map, which we now face.

III. Trajectory Scaling

Let us denote by g the fixed point of (21), which is

$$g = \lim_{n \to \infty} \alpha^n f^{F_n}_{\rho^*} \alpha^{-n}. \tag{22}$$

Since (22) entails an infinite magnification, it matters about which point we magnify. I have tacitly taken this point to be x=0, at which f_{ρ^*} has its cubic inflection in the critical case. g thus is arbitrarily well approximated by suitable iterates of f near the origin. As said in the Introduction, g suffices to compute <u>scalings</u> everywhere along the orbit under f. So, first let us define these scalings. To do so, I must first specify the objects that scale.

Thus, define first finite set of points whose spacings are to be scaled to those of a larger set. Define

$$x_t \equiv f^t(0) \tag{23}$$

$$x_t^{(N)} \equiv x_{(t-1)\mathrm{mod}F_N+1}. \tag{24}$$

That is, $x^{(N)}$ is periodic with period F_N, and is simply the first F_N images of the origin. (This is our short time observation of the system.)

Next, suitably define a set of smallest spacings among the points of $x^{(N)}$:

$$d_t^{(N)} \equiv \begin{cases} x_t^{(N)}-x_{t+F_{N-1}}^{(N)} & 0 < t \le F_{N-2} \quad (25) \\ x_t^{(N)}-x_{t+F_{N-2}}^{(N)} & F_{N-2} < t \le F_{N-1} \quad (26) \\ -d_{t-F_{N-1}}^{(N)} & F_{N-1} < t \le F_N \quad (27) \\ d_{t-F_N}^{(N)} & F_N < t \quad (28) \end{cases}$$

The good definition of $d^{(N)}$ is the essence of the calculations to follow. The intuition leading to this definition is that rotation at ρ^* is near roation at ρ_n which is periodic with period F_n, and hence the objects in $d^{(N)}$ are all very small for large n. Also, scaling relations are recursive, while the definition of the F_n in (6) require two successive F's. Relation (27) is "purely" definitional, analogous to a sign change of differences halfway through a set of points as in period doubling.

Finally, define a scaling of the $d^{(N)}$:

$$\sigma_N(t/F_{N+1}) \equiv d_t^{(N+1)}/d_t^{(N)}. \tag{29}$$

Obviously, if σ_N is known, the coarse grain data of $x^{(N)}$ will yield $d^{(N+1)}$. Then $d^{(N+2)}$, etc., allowing arbitrary refinement of the data. The point of the "good" definition of $d^{(N)}$ is that

$$\sigma = \lim_{N\to\infty} \sigma_N \tag{30}$$

is nice, is computible from g and hence universal, and finally that, as shall follow, allows an elementary recursive computation of the power spectrum of x_t.

It follows by (27) and (28) that

$$\sigma_N\left(\frac{t}{F_{N+1}} + \frac{F_N}{F_{N+1}}\right) = -\sigma_N(t/F_{N+1}) \qquad 0 \le t < F_{N-1}. \tag{31}$$

Thus,

$$\sigma(x + \rho^*) = -\sigma(x) \qquad 0 < x < 1-\rho^* = \rho^{*2} \tag{32}$$

so that we only require σ from 0 to ρ^*.

Methods now exactly analogous to those of period doubling allow us to calculate σ at those x which have a terminating expansion to the base ρ^{*-1}, i.e., for those x such that

$$x_N = \frac{F_{N-r_1}+\ldots+F_{N-r_m}}{F_N}, \quad x = \lim_{N\to\infty} x_N = \rho^{*r_1}+\ldots+\rho^{*r_m} \tag{33}$$

At such x, $f^{F_N x_N}(0)$ is arbitrarily close to x = 0, and these limits by (22) are determined by g. For other x, f smoothly and invertibly images the d's and σ is unchanged. Thus, σ has derivative zero almost everywhere with discontinuities at ρ^*-adic numbers, the discontinuities decreasing with the number of digits in x's ρ^* expansion. This leads to an approximation scheme: define σ as constant over the subintervals between x's of the same maximum digit length. We shall now compute the lowest order such approximation.

In this approximation, we keep only $r_1 \neq 0$ in (33), but notice by (25)-(27) that three distinct values need to be computed for $0 < x < \rho^*$.

i) $$\sigma_N(F_N/F_{N+1}) = d^{(N+1)}_{F_N}/d^{(N)}_{F_N} = \frac{x^{(N+1)}_{F_N} - x^{(N+1)}_{F_N+F_{N-1}}}{-d^{(N)}_{F_{N-2}}} = \frac{x^{(N+1)}_{F_N} - x^{(N+1)}_{F_{N+1}}}{x^{(N)}_{F_N} - x^{(N)}_{F_{N-2}}}$$

$$= \frac{x_{F_N} - x_{F_{N+1}}}{x_{F_N} - x_{F_{N-2}}} = \frac{f^{F_N}(0) - f^{F_{N+1}}(0)}{f^{F_N}(0) - f^{F_{N-2}}(0)}$$

$$= \frac{\alpha^N f^{F_N}(0) - \alpha^{-1}\,\alpha^{N+1} f^{F_{N+1}}(0)}{\alpha^N f^{F_N}(0) - \alpha^2\,\alpha^{N-2} f^{F_{N-2}}(0)}$$

$$\sigma_3 \equiv \sigma(\rho^*) = \lim_{N\to\infty} \sigma_N(F_N/F_{N+1}) = \frac{g(0)-\alpha^{-1}g(0)}{g(0)-\alpha^2 g(0)} = \frac{1-\alpha^{-1}}{1-\alpha^2} = \frac{-\alpha^{-2}}{1+\alpha^{-1}} \tag{34}$$

ii) Similarly, $$\sigma_2 \equiv \sigma(\rho^{*2}) = 1+\alpha^{-1} \tag{35}$$

iii) $$\sigma_1 \equiv \sigma(\rho^{*3}) = \frac{1-g(\alpha-2)}{1-\alpha^{-2}} \qquad (g(0) \quad 1) \tag{36}$$

In this approximation we take σ constant as

$$\sigma(x) \approx \begin{cases} \sigma_1 & 0 < x < \rho^{*3} \\ \sigma_2 & \rho^{*3} < x < \rho^{*2} \\ \sigma_3 & \rho^{*2} < x < \rho^{*} \end{cases} \tag{37}$$

For the subcritical case

$$\sigma_1 = \frac{1-(1+\alpha^{-2})}{1-\alpha^{-2}} = \frac{\alpha^{-2}}{\alpha^{-2}-1} = \frac{\rho^{*2}}{\rho^{*2}-1} = -\rho^{*}$$

$$\sigma_2 = 1+\alpha^{-1} = 1-\rho^{*} = \rho^{*2} \tag{38}$$

$$\sigma_3 = \frac{-\alpha^{-2}}{1+\alpha^{-1}} = \frac{-\rho^{*2}}{1-\rho^{*}} = -1.$$

In fact, by (12) it is easy to see that the approximation (37) is exact with σ_i as given by (38).

In the critical case, we need the numerical result for g:

$$\begin{aligned} \alpha &= -1.28857 \\ \sigma_1 &= -.393178 \\ \sigma_2 &= .223944 \\ \sigma_3 &= -2.68936. \end{aligned} \tag{39}$$

To summarize, a universal scaling function for golden mean dynamics exists, precisely three values in the subcritical case and in first approximation in the critical case can be so specified with renormalized values.

It is very interesting to observe that orderly scaling pertains to successive **time** values, and not to the scaling at each particular point on the circle. Indeed, successive applications of σ_N

$$\sigma_N(t/F_{N+1}) \sim \sigma(t/F_{N+1}) \tag{40}$$

at its different scales F_N produces from $d^{(N)}$ intervals of highly erratic variation along the circle. That is, rather than an ergodic analysis requiring a measure ("interval" density) at each point, only a refining time-average analysis is feasible. This presages that ergodic statistical analyses are apt to fail in the context of chaotic dynamics of dissipative systems.

In "coordinate space" (29) and (40) allow us to determine the fine scale (i.e., long time) behavior of x_t from an initial data set $x^{(N)}$. This effectively finishes the problem. However, spectral analysis is an important experimental tool, and we can ask what σ implies for a power spectrum. To anticipate the answer, quasiperiodic critical motion has a very rich and complex spectrum. It turns out, however, to be an elementary algorithmic encoding of essentially one number!

IV. The Golden Mean Power Spectrum

Define the Fourier transform of x_t as

$$\hat{x}(\omega) = \lim_{N\to\infty} \hat{x}_N(\omega) \tag{41}$$

where

$$\hat{x}_N(\omega) \equiv \frac{1}{F_N} \sum_1^{F_N} x_t e^{i\omega t}. \tag{42}$$

Our strategy is to compute $\hat{x}_N$ iteratively in N. In order to do so, again recalling the 2nd order scheme of the F's, define a pair of spectral subsums:

$$A_N(\omega) \equiv \sum_1^{F_{N-2}} d_t^{(N)} e^{i\omega t} \tag{43}$$

$$B_N(\omega) \equiv \sum_{F_{N-2}+1}^{F_{N-1}} d_t^{(N)} e^{i\omega t}. \tag{44}$$

(This is part of a general scaling strategy: If in an approximation a scaling function has n constant values, define n objects each over the support of one constant value. In this way an n dimensional recursion is constructed, which is a perturbational approach to arbitrarily accurate results.)

The spectral subsums enjoy an elementary recursion in consequence of the approximation of (34)-(36).

$$A_{N+1} = \sum_1^{F_{N-1}} d_t^{(N+1)} e^{i\omega t} = \sum_1^{F_{N-2}} d_t^{(N+1)} + \sum_{F_{N-2}+1}^{F_{N-1}} d_t^{(N+1)} e^{i\omega t}$$

$$\approx \sigma_1 \sum_1^{F_{N-2}} d_t^{(N)} e^{i\omega t} + \sigma_2 \sum_{F_{N-2}+1}^{F_{N-1}} d_t^{(N)} e^{i\omega t}$$

i.e.,

$$A_{N+1}(\omega) \approx \sigma_1 A_N(\omega) + \sigma_2 B_N(\omega). \tag{45}$$

$$B_{N+1} = \sum_{F_{N-1}+1}^{F_N} d_t^{(N+1)} e^{i\omega t} \approx \sigma_3 \sum_{F_{N-1}+1}^{F_N} d_t^{(N)} e^{i\omega t} = \sigma_3 e^{i\omega t F_{N-1}} \sum_1^{F_{N-2}} d_{t+F_{N-1}}^{(N)} e^{i\omega t}$$

i.e.,

$$B_{N+1}(\omega) \approx -\sigma_3 e^{i\omega F_{N-1}} A_N(\omega). \tag{46}$$

Together, we have

$$\begin{pmatrix} A_{N+1}(\omega) \\ B_{N+1}(\omega) \end{pmatrix} \approx \begin{bmatrix} \sigma_1 & \sigma_2 \\ -\sigma_3 e^{i\omega F_{N-1}} & 0 \end{bmatrix} \begin{pmatrix} A_N(\omega) \\ B_N(\omega) \end{pmatrix} \equiv \Sigma_N(\omega) \begin{pmatrix} A_N(\omega) \\ B_N(\omega) \end{pmatrix}. \tag{47}$$

Consider (47) for a fixed ω such that

$$\omega = 2\pi\nu, \quad \lim_{N\to\infty} \nu F_N \bmod 1 = 0. \tag{48}$$

(Such ν's have the form (33).) For large N, we then have

$$\Sigma_N(\omega) \sim \Sigma_N(0) \equiv \Sigma, \tag{49}$$

and

$$\begin{pmatrix} A_{n+r} \\ B_{n+r} \end{pmatrix} \sim \Sigma^n \begin{pmatrix} A_r \\ B_r \end{pmatrix} \qquad r>>1 \tag{50}$$

Accordingly, we simply need the spectrum of Σ to determine how $n \to \infty$. Such λ satisfy

$$\lambda^2 - \sigma_1 \lambda + \sigma_2 \sigma_3 = 0. \tag{51}$$

Consider first subcriticality, for which we have by (38)

$$\lambda^2 + \rho^* \lambda - \rho^{*2} = 0$$

or

$$\lambda^2 + (1-\rho^{*2})\lambda - \rho^{*2} = 0$$

so that

$$\lambda = -1,\ \rho^{*2}. \tag{52}$$

It is easy to see that for general ω the complex spectrum of $\Sigma_N(\omega)$ is strictly within the unit circle. Thus, under (47), A_N and B_N decay to zero for all ω not obeying (48), while at the ω's of (48), there is an initial transient decay until N is asymptotically large, after which the magnitudes of A and B remain constant, simply alternating in sign.

Next consider criticality. Observe that an eigenvalue of -1 requires by (51) that

$$1 + \sigma_1 + \sigma_2 \sigma_3 = 0 \tag{53}$$

which by (34)-(36) requires

$$\sigma_1 = \alpha^{-2} - 1 = -.39774. \tag{54}$$

which is satisfied by (39) to 1/2%. Since in criticality, that σ has just three values is an approximation, and so the choice of a particular set of three values is somewhat arbitrary, we shall take the surprisingly close agreement of (34) and (39) to motivate redefining σ_1 by (54), and so

$$\lambda = -1,\ \alpha^{-2}. \tag{55}$$

This is, of course, very important, otherwise the type of spectrum corresponding to the amplitudes of A and B converging to non-zero values should be forbidden, which we shall later argue is unacceptable. Thus, critical spectra will differ from subcritical ones simply by the replacement

$$-\rho^* \to \alpha^{-1}. \tag{56}$$

Also, this means that to a very good approximation, the iterative determination of the power spectrum depends only the one number α: spectra appear complicated in consequence of a repeated very elementary "genentic" rule.

Observe, in summary of the above considerations, that there are two very different spectra possible:

i) $$\begin{pmatrix} A_N(\omega) \\ B_N(\omega) \end{pmatrix} \sim (-1)^N \begin{pmatrix} a_1(\omega) \\ b_1(\omega) \end{pmatrix} . \tag{57}$$

ii) $$\begin{pmatrix} A_N(\omega) \\ B_N(\omega) \end{pmatrix} \sim \alpha^{-2N} \begin{pmatrix} a_2(\omega) \\ b_2(\omega) \end{pmatrix} , \tag{58}$$

where both hold for ω of the type (48), and for ii) to be possible, the A,B vector must be orthogonal to the type i) space. While this appears as generically impossible, we shall argue later that a type i) spectrum applies to a "modular" angle variable, while a type ii) applies to the (physically natural) periodic functions of this angle.

Let us end this discussion of the A, B spectra with a typical renormalization argument to determine scaling power laws.

Consider an ω which satisfies (40). Since the scaling formula (47) is an asymptotic (in N) property, take

$$r >> 1 \tag{59}$$

as the lowest value of N for which (47) is acceptably accurate,and such that $\rho^* F_r \approx F_{r-1}$. Finally, consider only those ω satisfying

$$\omega F_{r-1} << 2\pi. \tag{60}$$

Since A_r and B_r are synthesized with $t < F_{r-1}$, we have

$$(A_r,B_r)(\omega) \approx (A_r,B_r)(0). \tag{61}$$

We consider here only a type ii) spectrum, so that $(A_r,B_r)(0)$ is in the α^{-2} subspace:

$$\Sigma \begin{pmatrix} A_r(0) \\ B_r(0) \end{pmatrix} = \alpha^{-2} \begin{pmatrix} A_r(0) \\ B_r(0) \end{pmatrix} \tag{62}$$

Observe now

$$\begin{pmatrix} A_N(\omega) \\ B_N(\omega) \end{pmatrix} = \Sigma_{N-1}(\omega)\dots\Sigma_r(\omega) \begin{pmatrix} A_r(0) \\ B_r(\omega) \end{pmatrix} \tag{63}$$

$$\begin{pmatrix} A_N(\rho^*\omega) \\ B_N(\rho^*\omega) \end{pmatrix} = \Sigma_{N-1}(\rho^*\omega)\dots\Sigma_r(\rho^*\omega) \begin{pmatrix} A_r(0) \\ B_r(0 \end{pmatrix} \tag{64}$$

$$\approx \Sigma_{N-2}(\omega)\dots\Sigma_r(\omega)\Sigma_{r-1}(\omega) \begin{pmatrix} A_r(0) \\ B_r(0) \end{pmatrix} \tag{65}$$

$$\approx \Sigma_{N-2}(\omega)\dots\Sigma_r(\omega)\ \Sigma \begin{pmatrix} A_r(0) \\ B_r(0) \end{pmatrix} \tag{66}$$

$$\approx \alpha^{-2}\Sigma_{N-2}(\omega)\dots\Sigma_r \begin{pmatrix} A_r(0) \\ B_r(0) \end{pmatrix} \tag{67}$$

where $\rho^{*}F_n = F_{n-1}$ is used to obtain (65), (60) is used to replace $\Sigma_{r-1}(\omega)$ by Σ and finally (62) used to obtain (67). Comparing with (63), we have

$$\begin{pmatrix} A_N(\omega) \\ B_N(\omega) \end{pmatrix} \approx \alpha^2 \Sigma_{N-1}(\omega) \begin{pmatrix} A_N(\rho^{*}\omega) \\ B_N(\rho^{*}\omega) \end{pmatrix} \approx \alpha^2 \Sigma \begin{pmatrix} A_N(\rho^{*}\omega) \\ B_N(\rho^{*}\omega) \end{pmatrix} \tag{68}$$

since ω satisfies (48). While $(A_r, B_r)(0)$ was the eigenvector of α^{-2}, after a transient of Σ_n's, the vector on the right hand side is no longer orthogonal to the -1 subspace, and for large N the relaxed transient now has the vector as the -1 eigenvector. Thus,

$$\begin{pmatrix} A_N(\omega) \\ B_N(\omega) \end{pmatrix} \sim \alpha^2 \begin{pmatrix} A_N(\rho^{*}\omega) \\ B_N(\rho^{*}\omega) \end{pmatrix} \tag{69}$$

or,

$$A_N(\rho^{*}\omega) \sim -\alpha^{-2} A_N(\omega) \tag{70}$$

with an identical formula for B_N. We thus have the scaling law

$$A_N(\omega) \sim (-1)^N \omega^{\gamma} \tag{71}$$

$$\gamma = \frac{\ell n \alpha^{-2}}{\ell n \rho^{*}} . \tag{72}$$

In the subcritical case $|\alpha| = \rho^{*-1}$ and

$$\gamma_{sub} = 2. \tag{73}$$

For criticality, using α of (39),

$$\gamma_{crit} = 1.054. \tag{74}$$

Thus type ii) subcritical and critical A, B spectra are characterized by series of lines, a given series of the form

$$\omega_n = \rho^{*n} \omega_0 \tag{75}$$

falling approximately linearly to zero in the critical case and quadratically in the subcritical case.

We are now ready for the final step of the calculation: namely determining $\hat{x}_N$ from A_N and B_N. This will be exact for subcriticality and approximate for criticality.

It will turn out that one ingredient is missing to complete this exercise, which I'll provide now: Namely, the form of $d^{(N)}$'s definition in (25)-(27) implies extra internal informationn. By (25) and (37),

$$x_t - x_{t+F_N} \approx \sigma_1 (x_t - x_{t+F_{N-1}}) \qquad 0 < t \le F_{N-2} \tag{76}$$

$$x_t - x_{t+F_N} \approx \sigma_2 (x_t - x_{t+F_{N-2}}) \qquad F_{N-2} < t \le F_{N-1} \tag{77}$$

$$x_t - x_{t+F_{N-1}} \approx \sigma_3 (x_t - x_{t-F_{N-1}}) \qquad F_{N-1} < t \le F_N . \tag{78}$$

However, $t+F_N$ in the range of (76) overlaps with $t+F_{N-1}$ in the range of (78):

$$t + F_N \equiv t'+F_{N-1}$$

$$\Rightarrow t' = t+F_{N-2}; \tag{79}$$

$$F_{N-1} < t' \leq F_N$$

$$\Rightarrow F_{N-3} < t \leq F_{N-1}. \tag{80}$$

Thus for

$$F_{N-3} < t \leq F_{N-2} \tag{81}$$

x_{t+F_N} for t in range (76) is identical to $x'_{t+F_{N-1}}$ for t' in the range (78). and we obtain

$$x_t-x_{t+F_{N-2}} = \sigma_1(x_t-x_{t+F_{N-1}}) - \sigma_3(x_{t+F_{N-2}}-x_{t-F_{N-3}}) \tag{82}$$

for t satisfying (81). (For the three-valued approximation to be exact, it is possible to show, with more of the internal consistency relations like (82), that σ must be the subcritical (38). In the critical case, (82) is simply approximate.)

With (82) appended to (76)-(78), everything is now at hand to compute $\hat{x}$. The idea is to rewrite (42) as a transform of trajectory differences, and then to employ the scalings and definitions of A_N and B_N. By (24),

$$\sum_1^{F_N} x_t e^{i\omega t} = \sum_1^{F_N} x_t^{(N)} e^{i\omega t}$$

$$= e^{i\omega r} \sum_1^{F_N} x_{t+r}^{(N)} e^{i\omega t}$$

or

$$\sum_1^{F_N} x_{t+r}^{(N)} e^{i\omega t} = e^{-i\omega r} \sum_1^{F_N} x_t e^{i\omega t} \tag{83}$$

for

$$\omega = 2\pi k/F_N \qquad k = 0,\dots,F_N-1. \tag{84}$$

Thus, we can rewrite (42) at such ω as

$$F_N(1-e^{-i\omega F_{N-1}})\hat{x}_N(\omega) = \sum_1^{F_N} (x_t^{(N)}-x_{t+F_{N-1}}^{(N)})e^{i\omega t}. \tag{85}$$

We now regard the right hand side of (85) as the definition of the analytic continuation of $\hat{x}_N$ for all ω, and proceed to split the sum into parts given by A_N and B_N and other parts which by the scaling formulae (76)-(78) and (82) can be expressed in terms of A_n, B_n for n=N-1, N-2. Finally, the lower indexed A_n, B_n are expressed in terms of A_N, B_N by the relations (47). Omitting details, the result is

$$\hat{x}_N(\omega) \approx \frac{(1+\beta e^{i\omega F_{N-2}})A_N(\omega)+(\sigma_3^{-1}(1+\sigma_1\beta)-e^{i\omega F_{N-2}})B_N(\omega)}{F_N(1-e^{-i\omega F_{N-1}})} \tag{86}$$

where

$$\beta \equiv \frac{\sigma_1}{\sigma_1}(\sigma_1 - \sigma_2). \tag{87}$$

(For subcritical spectra, the formula simplifies since then $1+\sigma_1\beta=0$.)

Formula (86) finishes the computation. To employ it, A_r and B_r are computed from the measured initial data $x_t^{(r)}$; (47) is used repeatedly to reach asymptotic A_N, B_N and (86) then employed to compute the spectral lines at ω satisfying (84). Let us comment on the spectra so generated.

The principal difference between the A, B spectra and that of $\hat{x}$ is determined by the denominator of (86): for ω of the form (48) for which A_N, B_N don't vanish as $N\to\infty$, the numerator factors are simply constant. Using (71), the spectrum of $\hat{x}_N$ for N asymptotically large behaves as

$$\hat{x}_n(\omega) \sim \frac{(-1)^N}{F_N(1-e^{i\omega F_{N-1}})} c(\omega) \equiv \hat{x}_N(\omega)c(\omega) \tag{88}$$

where $c(\omega)$ scales as ω^γ with γ of (72). The first factor is essentially the Fourier transform of pure rotation at ρ^*. To evaluate it, set

$$\frac{\omega}{2\pi} = \frac{pF_{N-1}}{F_N} \bmod 1 \sim \rho^* p \bmod 1 \tag{89}$$

and noting the identity

$$F_{N-1}^2 = F_N F_{N-2} + (-1)^N, \tag{90}$$

$$\hat{x}_N(\omega) = \frac{1}{F_N}\,\frac{(-1)^N}{1-e^{-2\pi i p F_{N-1}^2/F_N}} = \frac{1}{F_N}\,\frac{(-1)^N}{1-e^{-2\pi i p(-1)^N/F_N}}$$

Thus,

$$\hat{x}(2\pi\rho^* p \bmod 1) = \frac{1}{2\pi i p} \tag{91}$$

the characteristic "rearrangement" spectrum of golden mean rotation. It is easy to see that this spectrum rises linearly with ω: $\rho^* p$ is close to integer for p the sum of F_n's with large n, with fractional part $O(1/p)$ which is also the amplitude of $\hat{x}$. Thus $\hat{x}(\omega)$ is of the same form as the A, B spectra with the power law exponent of each set of lines (75) promoted by one above γ of (72). Thus,

$$\hat{x}_{crit}(\omega) \sim \omega^{\gamma_{crit}} = \omega^{2.054} \tag{92}$$

$$\hat{x}_{sub}(\omega) \sim \omega^3. \tag{93}$$

Had $A_N(0)$, $B_n(0)$ satisfied (57), $c(\omega)$ would be constant for small ω, and the pure rotation spectrum of (91) would have emerged. This is simply the quasiperiodic rearrangement of the transform of the "ramp"

$$x_t^{(N)} = \frac{t}{F_N} \bmod 1 \tag{94}$$

whose discontinuity produces the characteristic $1/p$ spectrum. Thus (91) is the general spectrum for any <u>angle</u> variable. This is the reason that we demanded the -1 eigenvalue of Σ. Had we considered the transform of any continuous variable--namely a periodic function of an angle variable--then its rearranged spectrum would scale at a higher power than linear. Thus the type ii) spectra, which obtain if $(A_N(0),\ B_N(0))$ is in the eigenspace of $\lambda = \alpha^{-2}$ correspond to the spectrum of smooth periodic functions of an angle variable. To better understand this, recall that the scaling function was constructed with the idea that scalings are invariant under smooth transformations:

$$D_t^{(N)} \equiv P(x_t^{(N)}) - P(x_{t+F}^{(N)}) \approx P'(x_t^{(N)})d_t^{(N)} \tag{95}$$

and in the ratio of (95) with $D_t^{(N+1)}$ the P' factors cancel since $x_t^{(N)} \approx x_t^{(N+1)}$. Thus σ_N and $\Sigma_N(\omega)$ for the variable $P(x_t)$ are just those we have considered, and the recursion formulae and (86) are unaltered: whether we are considering a modular variable or a periodic function of it enters only in the initial data $x_t^{(r)}$.

There is, however, a difficulty. For P periodic, it has points of zero derivative. Since the angle x_t ergodically moves over the circle, there will be t's for very long orbits for which the cancellation of the P' factors in (95) could fail: the σ_N function for $D_t^{(N)}$ can become "noisy" for very large N. For the spectral recursion, however, σ appears within the "integrals" (43)-(44) and the "noise" can be unimportant. I am unsure of the outcome--numerically the noise appears for $F_n \sim 10000$ and if the spectrum is affected, it is of spectral lines well below any physical measurement.

V. Conclusion

In summary, we have extended period doubling trajectory scaling to golden mean rotation. It is clear that Fourier analysis is an exceedingly poor procedure for analyzing quasiperiodic motion: extracting σ from the data according to the definitions (23)-(29) produces a much cleaner analysis of the data. The idea of a difference defined differently over the orbit is necessary, for example, to extend scaling ideas to truly higher dimensional problems (Hamiltonian period doubling, for example). The idea of subsums, as in (43)-(44) is the germ of a perturbation scheme for accurate spectral and fractal dimensional analyses constructed from time series scalings, rather than ergodically. Irreducibly many-valued scaling functions have appeared in the analysis of the scaling of devil's staircases and, I suspect, will emerge in the analyses of more fully developed chaos. This paper represents, basically, a technical exercise to gain necessary experience and expertise.

Acknowledgement

This work has been partially supported by U. S. Department of Energy, DE-AC02-83ER13044. I also thank NORDITA for hospitality during research and The Rockefeller University and The Institute for Advanced Study for hospitality during its writing.

Nonlinear Phenomena in Dissipative Systems

Pierre Coullet

Laboratoire de Physique de la Matière Condensée, Parc Valrose
F-060 Nice Cedex, France

Observatoire de Nice, B.P. 139, F-06003 Nice Cedex, France

We use the Rayleigh-Bénard convection and a Chemical Reaction problem in order to introduce various concepts in Nonlinear Physics. The nonlinear phenomena described occur in many other fields, as for example Optics, Acoustics, etc.

1. Stability analysis of equilibria

1.1. Rayleigh-Bénard convection

Heating a fluid from below is the most natural way to produce convection [1]. The tendency of the hot and light fluid at the bottom of the layer is to rise, under the effect of the archimedian force, while the cold fluid at the top tends to fall. This tendency is opposed by two effects: the viscous force tends to inhibit fluid motions, and its thermal conductivity tends to homogenize the temperature across the layer. As a consequence, the onset of motion in the fluid only occurs when the destabilizing constraint is above a certain threshold. If the gradient of temperature applied on the layer is less than some critical value, the system is then in the conduction regime. In this state the fluid is at rest. The temperature and pressure fields have static values. We are first interested in describing the experimentally observed stability of this regime [2].

The mathematical description of the Rayleigh-Bénard system of a thin layer is given by the Boussinesq approximation [3,4] of the Navier-Stokes equation coupled with the heat equation

$$\partial_t \vec{V} + (\vec{V} \cdot \vec{\nabla})\vec{V} = -\vec{\nabla}P + \sigma\nabla^2\vec{V} + R\sigma T\hat{z} \tag{1.a}$$

$$\partial_t T + (\vec{V} \cdot \vec{\nabla})T = \nabla^2 T \tag{1.b}$$

$$0 = \vec{\nabla} \cdot \vec{V} . \tag{1.c}$$

The unit of distance, time, temperature have been respectively chosen to be d, d^2/κ, ΔT where

d is the thickness of the layer

$\Delta T = T_1 - T_2$, the gradient of temperature

T_1, the temperature at the bottom $(z = 0)$

T_2, the temperature at the top $(z = 1)$

κ, the thermal conductivity of the fluid

$\sigma = \nu/\kappa$, its Prandlt number

ν, its kinematic viscosity

$R = \dfrac{\alpha\Delta T d^3}{\nu\kappa}$, the Rayleigh number

α, its dilatation coefficient.

This system of equations has a simple steady solution $\vec{V}=0$, P_0, T_0 which describes the conductive regime. P_0, T_0 satisfy the equations

$$\nabla^2 T_0 = 0 \tag{2}$$

$$\partial_z P_0 = R\sigma T_0 . \tag{3}$$

In order to study the stability of the conduction solution, one introduces the deviations θ and π from this state,

$$T = T_0 + \theta \tag{4}$$

$$P = P_0 + \pi \tag{5}$$

Then (1) reads

$$\partial_t \vec{V} + (\vec{V} \cdot \vec{\nabla})\vec{V} = \sigma\nabla^2\vec{V} - \vec{\nabla}\pi + R\sigma\theta\hat{z} \tag{6.a}$$

$$\partial_t \theta + (V \cdot \vec{\nabla})\theta = \vec{V} \cdot \hat{z} + \nabla^2\theta \tag{6.b}$$

$$0 = \vec{\nabla} \cdot \vec{V}. \tag{6.c}$$

For small perturbations,one first considers the linear version of (6)

$$\partial_t \vec{V} = \sigma\nabla^2\vec{V} - \vec{\nabla}\pi + R\sigma\theta\hat{z} \tag{7.a}$$

$$\partial_t \theta = \vec{V} \cdot \hat{z} + \nabla^2\theta \tag{7.b}$$

$$0 = \vec{\nabla} \cdot \vec{V} . \tag{7.c}$$

One gets a relation between the pressure and the temperature field by taking the divergence of (7.a)

$$\nabla^2\pi = R\sigma\partial_z\theta \tag{8}$$

Using this relation, one can suppress the pressure in (7.a)

$$\begin{aligned} \partial_t \nabla^2\vec{V} &= \sigma\nabla^4\vec{V} + R\sigma(\nabla^2\theta\hat{z} - \partial_z\vec{\nabla}\theta) \\ \partial_t\theta &= \vec{V} \cdot \vec{z} + \nabla^2\theta \end{aligned} \tag{9}$$

Introducing the components of the velocity field

$$\vec{V} = (u, v, w), \tag{10}$$

(9) reads

$$\partial_t \nabla^2 u = \sigma\nabla^4 u - R\sigma\partial_x\partial_z\theta$$

$$\partial_t \nabla^2 v = \sigma\nabla^4 v - R\sigma\partial_y\partial_z\theta$$

$$\partial_t \nabla^2 W = \sigma\nabla^4 W + R\sigma\nabla_{\parallel}^2\theta$$

$$\partial_t\theta = W + \nabla^2\theta$$

where $\nabla_{\parallel}^2 = \frac{\partial^2}{\partial x^2} + \frac{\partial^2}{\partial y^2}$. The boundary conditions chosen to solve (1) are the following:

Vertical boundary conditions

The plates are supposed to be perfect heat conductors:

$$\theta(0) = \theta(1) = 0$$

The top and bottom boundary are supposed to be the surfaces without deformations:

$$W(0) = W(1) = 0, \quad \partial_z u(0) = \partial_z u(1) = \partial_z v(0) = \partial_z v(1) = 0 \tag{13}$$

Horizontal boundary conditions

The box is supposed to be periodic in both x and y directions:

$$\begin{aligned} \theta(\vec{x} + \vec{\Gamma}, z) &= \theta(\vec{x} + \vec{\Gamma}, z) \\ \vec{v}(\vec{x} + \vec{\Gamma}, z) &= \vec{v}(\vec{x} + \vec{\Gamma}, z) \end{aligned} \tag{14}$$

where $\vec{\Gamma} = (\Gamma_x, \Gamma_y)$ and $\vec{x} = (x,y)$. Γ_x, Γ_y are called the aspect ratios of the convective box.

We now expand u, v, w, θ, in Fourier series

$$\begin{aligned} u &= \sum_{n=0}^{\infty} \sum_{\vec{n}} u_{n,\vec{k}} \cos(n\pi z) e^{i\vec{k}\cdot\vec{x}} \\ v &= \sum_{n=0}^{\infty} \sum_{\vec{k}} v_{n,\vec{k}} \cos(n\pi z) e^{i\vec{k}\cdot\vec{x}} \\ w &= \sum_{n=1}^{\infty} \sum_{\vec{k}} w_{n,\vec{k}} \sin(n\pi z) e^{i\vec{k}\cdot\vec{x}} \\ \theta &= \sum_{n=1}^{\infty} \sum_{\vec{k}} \theta_{n,\vec{k}} \sin(n\pi z) e^{i\vec{k}\cdot\vec{x}} \end{aligned} \tag{15}$$

where

$$\sum_{\vec{k}} \equiv \sum_{n_x=-\infty}^{+\infty} \sum_{n_y=-\infty}^{+\infty} \quad \text{and} \tag{16}$$

$$\vec{k} = 2\pi\left(\frac{n_x}{\Gamma_x}\vec{i} + \frac{n_y}{\Gamma_y}\vec{j}\right) \tag{17}$$

where $\vec{i}$ and $\vec{j}$ are the unit vectors in the x and y directions respectively. One has

$$\overline{\alpha}_{n,\vec{k}} = \alpha_{n,-\vec{k}} \tag{18}$$

where α stands for (u,v,w,θ), and $\overline{\alpha}$ denotes the complex conjugate of α. Defining the quantity $a^2_{n,\vec{k}}$ by

$$a^2_{n,k} = \vec{k}^2 + n^2\pi^2 \tag{19}$$

one gets from (11) and (15) the set of equations

$$
\begin{aligned}
-a^2_{n,\vec{k}}\partial_t u_{n,\vec{k}} &= a^4_{n,\vec{k}}\sigma u_{n,\vec{k}} - iR\sigma n\pi k_x\theta_{n,\vec{k}} \\
-a^2_{n,\vec{k}}\partial_t u_{n,\vec{k}} &= a^4_{n,\vec{k}}\sigma u_{n,\vec{k}} - iR\sigma n\pi k_z\theta_{n,\vec{k}} \\
-a^2_{n,\vec{k}}\partial_t w_{n,\vec{k}} &= a^4_{n,\vec{k}}\sigma w_{n,\vec{k}} - R\sigma k^2\theta_{n,\vec{k}} \\
\partial_t\theta_{n,\vec{k}} &= -a^2_{n,\vec{k}}\theta_{n,\vec{k}} + w_{n,\vec{k}}\ .
\end{aligned}
\tag{20}
$$

(20) has solutions of the form

$$
\begin{aligned}
u_{n,\vec{k}} &= u_{n,\vec{k}}(0)e^{st}\ , \quad u_{n,\vec{k}} = u_{n,\vec{k}}(0)e^{st}\ , \\
w_{n,\vec{k}} &= w_{n,\vec{k}}(0)e^{st}\ , \quad \theta_{n,\vec{k}} = \theta_{n,\vec{k}}(0)e^{st}.
\end{aligned}
\tag{21}
$$

With the help of (21), (20) reads

$$
\Lambda \begin{pmatrix} u_{n,\vec{k}}(0) \\ v_{n,\vec{k}}(0) \\ w_{n,\vec{k}}(0) \\ \theta_{n,\vec{k}}(0) \end{pmatrix} = 0 \quad \text{where} \tag{22}
$$

$$
\Lambda = \begin{pmatrix}
-a^2_{n,\vec{k}}(s+\sigma a^2_{n,\vec{k}}) & 0 & 0 & iR\sigma k_x n\pi \\
0 & -a^2_{n,\vec{k}}(s+\sigma a^2_{n,\vec{k}}) & 0 & iR\sigma k_y n\pi \\
0 & 0 & -a^2_{n,k}(s+\sigma a^2_{n,\vec{k}}) & R\sigma k^2 \\
0 & 0 & -1 & s+a^2_{n,\vec{k}}
\end{pmatrix} \tag{23}
$$

(22) has non trivial solutions only if

$$\det\Lambda = 0 \tag{24}$$

which leads to the characteristic equation

$$a^4_{n,\vec{k}}(s+\sigma a^2_{n,\vec{k}})^2\ (a^2_{n,\vec{k}}s^2 + a^4_{n,\vec{k}}s(\sigma+1) + \sigma(a^6_{n,\vec{k}} - Rk^2)) = 0. \tag{25}$$

This equation has 4 roots

$$s_x = -\sigma a^2_{n,\vec{k}} \tag{26.a}$$

$$s_y = -\sigma a^2_{n,\vec{k}} \tag{26.b}$$

$$s_\pm = \frac{1}{2a^2_{n,\vec{k}}}\,(-a^4_{n,\vec{k}}(\sigma+1) \pm (a^8_{n,\vec{k}}(\sigma+1)^2 + 4a^2\ \ (Rk^2 - a^6_{n,\vec{k}}))^{1/2}). \tag{27}$$

The convective modes, associated with (27) are

$$\phi_{\pm n,\vec{k}}(\vec{x}, z) = (u_{\pm}\cos n\pi z,\ v_{\pm}\cos n\pi z,\ \sin n\pi z,\ \theta_{\pm}\sin n\pi z)e^{i\vec{k}\cdot\vec{x}} \tag{28}$$

where

$$u_{\pm} = i\frac{R\sigma k_x n\pi}{a^2_{n,\vec{k}}(s_{\pm} + \sigma a^2_{n,\vec{k}})(s_{\pm} + a^2_{n,\vec{k}})}, \quad v_{\pm} = i\frac{R\sigma k_y n\pi}{a^2_{n,\vec{k}}(s_{\pm} + \sigma a^2_{n,\vec{k}})(s_{\pm} + a^2_{n,\vec{k}}} \tag{29}$$

$$\theta_{\pm} = \frac{1}{(s_{+} + a^2)} . \tag{30}$$

The conductive regime is unstable when, for some values of σ, R, $\vec{k}$, n, s_+ become positive. The marginal situation is then defined by $s_+(\sigma, R, \vec{k}, n) = 0$. The marginal curves $R(\vec{k},n) = a^6{}_{n,\vec{k}}/k^2$ are represented in Fig.1. In finite boxes, thanks to the discretization in the k-space, one single mode becomes unstable first.

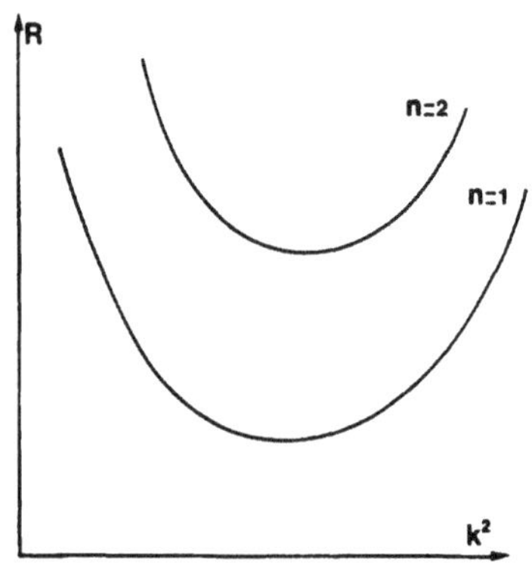

Figure 1

What we observe here is the basic mechanism for pattern formation. For a given value of an external constraint, the homogeneous solution of a translationally-invariant system loses its stability with a mode characterized by a spatial structure with a finite wavelength. In general, the system under scrutiny possesses rotational symmetry in the plane in such a way that the critical wave number is only given by its modulus k_c^2. This implies that in large systems many "planforms" are possible. One can have unidimensional periodic patterns (strips) or two-dimensional periodic patterns (rectangles, squares or hexagons). The selection of a given patern is a nonlinear dynamical problem [4]. In the following, we will only consider the case of unidimensional periodic patterns. Actually, in the case of Rayleigh-Bénard convection, under appropriate conditions, the corresponding fluid motion (straight rolls) is shown to be stable [4].

1.2. Linear stability analysis of equilibria

Let us now consider other examples for which a similar analysis can be performed. The following abstract (partial) differential equation is supposed to describe some physical system.

$$\partial_t \underline{U} = \underline{\underline{L}}(\vec{\nabla}, \lambda)\underline{U} + \underline{N}(\vec{\nabla}, \lambda, \underline{U}) \quad \text{where} \tag{31}$$

$U = (U_1 \ldots U_N)$

$\underline{\underline{L}}$ is a (differential) linear operator

$\underline{N}$ is a (differential) nonlinear operator

$\vec{\nabla}$ represents the gradient operator in d dimensions and

$\lambda = (\lambda_1 \ldots \lambda_k)$ is a set of external parameters.

A special example of systems like (31) is given by the reaction-diffusion equations

$$\partial_t \underline{C} = \underline{f}(\underline{C}) + \underline{\underline{D}}\nabla^2 \underline{C}. \tag{32}$$

These equations provide models to describe spatial effects in chemical reactors. $\underline{C} = (C_1 \ldots C_N)$ represents the chemical species. The function $\underline{f}(\underline{C})$ is given by the chemical kinetics, $\underline{\underline{D}}$ is a diffusion matrix. Let $\underline{C}^*$ be a spatially homogeneous stationary solution of (32)

$$\underline{f}(\underline{C}^*) = 0. \tag{33}$$

We introduce the deviation $\underline{U}$ from this solution

$$\underline{C}(\vec{x}, t) = \underline{C}^* + U(\vec{x}, t) \tag{34}$$

and get for $\underline{U}$ equations of the form (31)

$$\partial_t \underline{U} = (\underline{\underline{L}} + \underline{\underline{D}}\nabla^2)\underline{U} + \underline{N}(\underline{U}), \quad \text{where} \tag{35}$$

$$\underline{\underline{L}} = \left.\frac{\partial \underline{f}}{\partial \underline{C}}\right|_{\underline{C}^*} \quad \text{and} \tag{36}$$

$$\underline{N}(\underline{U}) = \underline{f}(\underline{U} + \underline{C}^*) - \underline{\underline{L}}\underline{U}. \tag{37}$$

We now illustrate the linear analysis in the case of two chemical species C_1, C_2 in a linear reactor (d = 1). The stability of the homogeneous state is again studied through the linear version of (37)

$$\partial_t \underline{U} = (\underline{\underline{L}} + D\nabla^2)\underline{U} \quad \text{where} \tag{38}$$

$$\underline{\underline{L}} = \begin{pmatrix} L_{11} & L_{12} \\ L_{21} & L_{22} \end{pmatrix}, \quad \text{and} \quad \underline{\underline{D}} = \begin{pmatrix} D_1 & 0 \\ 0 & D_2 \end{pmatrix} \tag{39}$$

Choosing the same kind of lateral boundary conditions as in the convective problem (periodic boundary conditions), one can expand $\underline{U}$ in Fourier series e^{ikx}, where $k = 2\pi n/\Gamma$, and Γ is the length of the reactor. Then (38) reads

$$\partial_t \underline{U}_k = (\underline{\underline{L}} - \underline{\underline{D}}k^2)\underline{U}_k. \tag{40}$$

Introducing the following notations

$$\begin{aligned} T &= L_{11} + L_{22} \\ R &= -L_{11} L_{22} + L_{12} L_{21} \\ P &= L_{11} D_2 + L_{22} D_1 . \end{aligned} \tag{41}$$

The characteristic equation reads

$$s^2 - s(T - (D_1 + D_2)k^2) + (D_1 D_2 k^4 - Pk^2 - R) = 0. \tag{42}$$

We choose $T < 0$. The type of marginal curve $R(k^2) = -Pk^2 + D_1 D_2 k^4$ depends on the sign of P. The different situations are sketched in the following figures. In Fig. 2 ($P < 0$) the instability occurs first for $k = 0$ when R crosses 0. One has

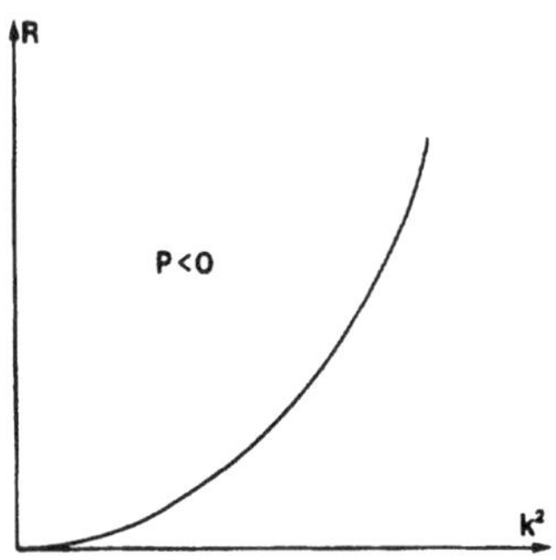

Figure 2

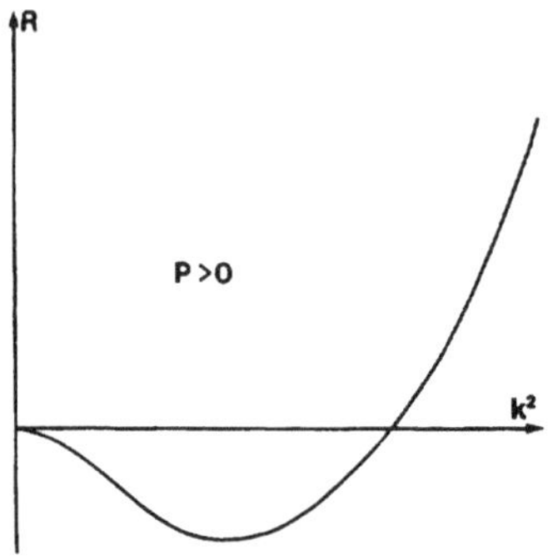

Figure 3

a long wavelength instability. In Fig.3 $(P > 0)$ the instability occurs first at finite wave number $k_c = \frac{P}{2D_1D_2}$ when R crosses the value $R_c = -\frac{1}{4}\frac{P^2}{D_1D_2}$. A periodic pattern is expected to occur. This instability is only possible if $D_1 \neq D_2$ since $T = L_{11} + L_{22}$ is negative. At the onset of instability, for a reactor of length $\Gamma = 2\pi/k_c$, the marginal mode is given by

$$U = \begin{pmatrix} 1 \\ \dfrac{D_1 k_c^2 x - L_{11}}{L_{12}} \end{pmatrix} . \qquad (43)$$

The next step in the understanding of the instability phenomena is to study its nonlinear development. The linear theory tells us that a given mode grows exponentially in time. Numerical and experimental observations are not consistent with this prediction.

2. Steady non equilibrium behavior and pattern formation

2.1. Adiabatic elimination method

When the Rayleigh number crosses its critical value ($R_c = 27\pi^4/4$ in the case of free-free vertical boundary conditions) one observes fluid motions under the form of rolls. Near the onset of convection,the roll structure is quite well descibed by the linear theory, but its amplitude is finite and constant in time, contrary to the predictions of the linear theory. What we have to understand is the basic nonlinear mechanism for the saturation of the linearly unstable mode.

We consider the case of the chemical reactor with $P > 0$. One can always expand the solution of (35) on the basis of the eigenfunctions of the linear operator

$$\underline{U} = \sum_k B_k^+ \underline{\phi}_k^+ + \sum_k B_k^- \underline{\phi}_k^- + \text{c.c.} \qquad (44)$$

where c.c. stands for complex conjugate. The $\underline{\phi}_k^\pm$ are defined by

$$(\underline{L} - \underline{\underline{D}}\nabla^2)\underline{\phi}_k^\pm = s_k^\pm \underline{\phi}_k \qquad (45)$$

$$\underline{\phi}_k(x) = \tilde{\underline{\phi}}_k e^{ikx} \qquad (46)$$

and $s_k^\pm$ are the two roots of the characteristic equation (42). Let us single out in (44) the almost marginal mode $B_{k_c}^+ \equiv A_{k_c}$

$$(\underline{L} - \underline{\underline{D}}\nabla^2)\underline{\phi}^+_{k_c} = \mu\underline{\phi}^+_{k_c} \tag{47}$$

where $s^+(k_c) \equiv \mu$. Equation (44) then reads

$$\underline{U} = A_{k_c}\underline{\phi}_{k_c} + \sum_{k\neq k_c} B^+_k \underline{\phi}^+_k + \sum_k B^-_k\underline{\phi}^-_k + \text{c.c.} \tag{48}$$

Using (48), (37) takes the form of an infinite number of coupled differential equations

$$\partial_t A_{k_c} = \mu A_{k_c} + F_{k_c}(A_{k_c}, \{B^-_k\}, \{B^+_k\}; \text{c.c.}) \tag{49.a}$$

$$\partial_t B^-_k = s^-_k B_k + G^-_k(A_{k_c}, \{B^-_k\}, \{B^+_k\}; \text{c.c.}) \tag{49.b}$$

$$\partial_t B^+_k = s^+_k B^+_k + G^+_k(A_{k_c}, \{B^-_k\}, \{B^+_k\}; \text{c.c}) \tag{49.c}$$

which, together with the corresponding complex conjugate equations, are completely equivalent to (37). In (37) we assume $\underline{N}(\underline{U})$ is quadratic in $\underline{U}$ (as in the Boussinesq equations). Then F_{k_c}, $G^\pm_k$ are also quadratic functions of the amplitudes. The translational invariance of the initial equations induces a very simple selection rule for the possible nonlinear coupling terms in (49) (conservation of the impulsions): in the equation for C_k where C_k stands for $A_{\pm k_c}$ or $B^\pm_{\pm k}$, the nonlinear terms have the form $C_{k_1}C_{k_2}$ with $k_1 + k_2 = k$. This implies in particular that there is no quadratic term in A_{k_c} and/or $\overline{A}_{k_c}$ in the equation for A_{k_c}. The quadratic coupling terms contain at least a stable mode $B^\pm_{\pm k}$. This indicates clearly that the nonlinear saturation mechanism, if there is one, involves stable modes. Let us also remark that it is not necessarily the case if $\underline{N}(\underline{U})$ contains cubic nonlinearities, since then, a term of the form $|A_{k_c}|^2 A_{k_c}$ is allowed. We now truncate (49) using the following prescriptions.

1) One writes down the equation for A_{k_c} and retains only the terms which contain A_{k_c} and $\overline{A}_{k_c}$. This singles out some modes $B^\pm_k$

2) One considers the equations for these stable modes, but one retains the coupling terms which only contain A_{k_c} and $\overline{A}_{k_c}$.

This procedure will become clear in the following. From (49) one gets the equation for A_{k_c}

$$\begin{aligned}\partial_t A_{k_c} = {} & \mu A_{k_c} + \alpha_1 A_{k_c} B^-_0 + \alpha_2 A_{k_c} B^+_0 + \alpha_3 A_{k_c}\overline{B}^-_0 + \alpha_4 A_{k_c}\overline{B}^+_0 \\ & + \alpha_5 \overline{A}_{k_c} B^-_{2k_c} + \alpha_6 \overline{A}_{k_c} B^+_{2k_c} .\end{aligned} \tag{50}$$

This singles out the modes $B^\pm_0$ and $B^\pm_{2k_c}$ whose equations read

$$\begin{aligned}\partial_t B^\pm_0 &= s^\pm_0 B^\pm_0 + \beta^\pm |A_{k_c}|^2 \\ \partial_t B^\pm_{2k_c} &= s^\pm_{2k_c} B^\pm_{2k_c} + \gamma^\pm A^2_{k_c} .\end{aligned} \tag{51}$$

These equations are sufficient enough to analyze the nonlinear behavior of the unstable mode A_{k_c} near the threshold of instability. For the sake of simplicity, we first consider a simpler mode-mode coupling which contains all the non trivial features of the previous problems (chemical reactor, Rayleigh-Bénard convection). Let A and B be the amplitudes of two real modes which obey the following equations:

$$\partial_t A = \mu A + \alpha AB \tag{52.a}$$

$$\partial_t B = sB + \beta A^2 \tag{52.b}$$

with $s < 0$ and $0 < \mu \ll \gamma$. When A and B are small enough, the linear-associated problem is characterized by two time scales $\tau_1 = \mu^{-1}$ and $\tau_2 = (-s)^{-1}$, with $\tau_2 \ll \tau_1$. On the rapid time scale τ_2, the amplitude B is damped and (52.a) predicts an exponentially growing behavior for A. Then, on the slow time scale τ_1 the linear approximation is no longer valid, and the presence of A^2 in (52.b) cannot be neglected. A formal integration of (52.b) gives

$$B = B_0 e^{st} + \beta e^{st} \int_0^t e^{-s\tau} A^2(\tau) d\tau . \tag{53}$$

In first approximation $A(\tau)$ can be taken as a constant in the integral in such a way that

$$B(t) \simeq -\frac{\beta}{s} A^2(t). \tag{54}$$

This method [5] called adiabatic elimination procedure of the stable modes can be performed at any order in μ and A. At the leading order, the nonlinear development of the instability is then given by the amplitude equation

$$\partial_t A = \mu A - \frac{\alpha\beta}{s} A^3 . \tag{55}$$

This equation, known as the Landau equation [6] , describes the so called pitchfork bifurcation. Its solution for $\mu > 0$ is given by

$$A(t) = \operatorname{sgn}(A(0)) \frac{\sqrt{\mu}\,\sqrt{C}\, e^{\mu t}}{\left(1 + \frac{\alpha\beta}{s} C\, e^{2\mu t}\right)^{1/2}} \tag{56}$$

where $C = \left| \frac{A(0)}{\left(\mu - \frac{\alpha\beta}{s} A^2(0)\right)^{1/2}} \right|^{1/2}$. According to the relative sign between μ and $\frac{\mu\beta}{s}$,

(55) has one or three stationary solutions, $A^* = A$ and $A^* = \pm(\mu s/\alpha\beta)^{1/2}$. The corresponding bifurcation diagrams are sketched in Fig. 4.

The most interesting case corresponds to $\alpha\beta/\gamma > 0$ when the bifurcation is supercritical (subcritical otherwise). The instability saturates at a finite value of the amplitude $\pm \frac{\mu s}{\alpha\beta}$ depending on the initial condition. The equations (52) are invariant under the change $A \to -A$. When μ crosses 0 the instability breaks this symmetry. For this reason, the pitchfork bifurcation is often called symmetry-breaking

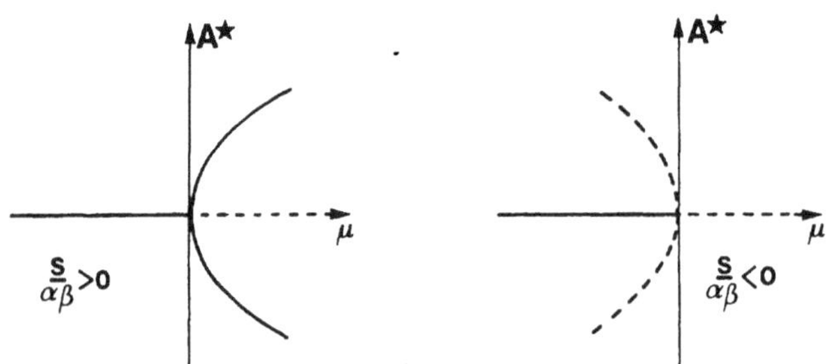

Figure 4

bifurcation. The adiabatic elimination can be performed in a more systematic way. The equation $B = -\frac{\beta}{s}A^2$ can be seen as the equation for an invariant manifold in the phase space A, B of (52) [7,8]. One can look a priori for such a manifold under the form

$$B = \psi(A, \mu). \tag{57}$$

Using (52) one gets the equation to be satisfied by ψ

$$(\mu A + \alpha A\psi(A, \mu)) \frac{\partial\psi(A, \mu)}{\partial A} = s\psi(A, \mu) + \beta A^2. \tag{58}$$

In order to solve this equation, one expands ψ in a Taylor serie in A and μ,

$$\psi(A) = \sum_{n\geqslant 2} \psi_n(\mu)A^n, \quad \text{and} \tag{59}$$

$$\psi_n(\mu) = \sum_{k\geqslant 0} \psi_{n,k}\mu^k. \tag{60}$$

Using (59), (60) and (58) one gets

$$-s\psi_{n,k} = I_{n,k}, \tag{61}$$

where $I_{n,k}$ is an explicitly known quantity at the "order"(n, k). One is then able, using (61), to compute $\psi(A, \mu)$ at each order in A and μ. For example, the leading term ψ_{20} is given by

$$\psi_{20} = \frac{\beta}{-s}. \tag{62}$$

Equations (59) and (60) are asymptotic series only valid close enough from the instability threshold. The existence of such a manifold allows to reduce, near a bifurcation point, the dynamics of a system with infinitely many degrees of freedom, to a few degrees of freedom: in our case, the Landau equation. This procedure, as well as the adiabatic elimination procedure, can be used in the chemical reactor problem. One gets

$$B_0^{\pm} = -\frac{\beta^{\pm}}{s_0^{\pm}}|A_k|^2, \quad B_{2k_c}^{\pm} = -\frac{\gamma^{\pm}}{s^{\pm}}A_{k_c}^2. \tag{63}$$

With the help of (63), (50) reads

$$\partial_t A_{k_c} = \mu A_{k_c} - \alpha|A_{k_c}|^2 A_{k_c}, \tag{64}$$

where α is a real quantity given by

$$\alpha = (\alpha_1 + \alpha_3)\frac{\beta_-}{s_0^-} + (\alpha_2 + \alpha_4)\frac{\beta_+}{s_0^+} + \alpha_5\frac{\gamma_-}{s_{2k_c}^-} + \alpha_6\frac{\gamma_+}{s_{2k_c}^+}. \tag{65}$$

One can easily check that the coupling terms retained in (50) and (51) are the only ones which contribute to α. Other terms allow to compute corrections in powers of μ of this coefficient and also the coefficients of higher order terms, as for example $|A_{k_c}|^4 A_{k_c}$. The content of (64) is slightly different from that of (55) since A is a complex amplitude. Let $A = \rho e^{i\phi}$; one then has the equations

$$\partial_t\rho = \mu\rho - \alpha\rho^3 \tag{66.a}$$

$$\partial_t\phi = 0. \tag{66.b}$$

(66.a) is identical to (55). (66.b) implies that the phase of the periodic pattern is arbitrary. This is clearly due to our choice of incomplete lateral boundary conditions. This marginal phase mode has no real meaning in a finite system, but it will become a potentially dangerous mode in an infinite system, as we will see later.

2.2. Normal forms and asymptotic techniques

We now consider more direct ways to get the Landau equation from (31). We want to solve (34) with the same boundary conditions as with the convection or the chemical reactor problems, namely the periodic boundary conditions

$$\underline{U}(\vec{x} + \vec{\Gamma}) = \underline{U}(\vec{x}), \tag{67}$$

with $\vec{\Gamma} = (\Gamma_1 \ldots \Gamma_d)$. The normal modes associated with (31) are defined by

$$\underline{\underline{L}}(\vec{\nabla}, \lambda)\, \underline{\phi}^i_{\vec{k}}(\vec{x}) = \underline{s}^i_{\vec{k}}\, \underline{\phi}^i_{\vec{k}}(\vec{x}) \tag{68}$$

with $\underline{\phi}^i_{\vec{k}}(\vec{x}) = \underline{\phi}^i_{\vec{k}} e^{i\vec{k}\cdot\vec{x}}$ where $s^i_{\vec{k}}$ are the roots of the characteristic equation, $i = 0, N-1$,

$$\det(\underline{\underline{L}}(i\vec{k},\lambda) - s\,\mathbb{1}) = 0. \tag{69}$$

This equation reads

$$s^N + C_{N-1}(\vec{k}, \lambda)s^{N-1} \ldots + C_0(\vec{k}, \lambda) = 0. \tag{70}$$

Surfaces of marginal stability in the parameter space λ are given by

$$C_0(\vec{k}, \lambda) = 0. \tag{71}$$

Let us assume that, when $\lambda = \lambda_c$, a given mode characterized by a critical wave number $\vec{k}_c$ becomes unstable first. For these parameter values one expands $\underline{U}$ in the normal mode basis

$$\underline{U} = \sum_i \sum_k B^i_{\vec{k}}\, \underline{\phi}^i_{\vec{k}}(\vec{x}) + \text{c.c.} \;. \tag{72}$$

Then the $B^i_{\vec{k}}$ obeys coupled ordinary differential equations. Using the adiabatic elimination procedure or the central manifold approach already sketched, one can express the stable modes as nonlinear functions of the critical one A_{k_c} and its complex conjugate $\bar{A}_{k_c}$ ($A_{k_c} \equiv B^0_{k_c}$, $\underline{\phi}_{k_c} \equiv \underline{\phi}^0_{k_c}$)

$$B^i_{\vec{k}} = \phi^i_{\vec{k}}(A_{k_c}, \bar{A}_{k_c}) \tag{73}$$

for $(i, \vec{k}) \neq (0, \vec{k}_c)$. Then (72) reads

$$\underline{U} = A_{k_c}\underline{\phi}_{k_c} + \bar{A}_{k_c}\underline{\bar{\phi}}_{k_c} + \underline{W}(A_{k_c}, \bar{A}_{k_c}) \tag{74}$$

where $\underline{W}$ is a nonlinear function which can be computed from (72) and (73). The set of coupled differential equations for the modes then reduces to a single equation

$$\partial_t A_{k_c} = F(A_{k_c}, \bar{A}_{k_c}) \tag{75}$$

which only contains nonlinear terms since $\lambda = \lambda_c$. Near $\lambda = \lambda_c$ all the quantities can be expanded in the small parameters $\delta = \lambda - \lambda_c$. Equations (74) and (75) become

$$\underline{U} = A_{k_c}\underline{\phi}_{k_c} + \bar{A}_{k_c}\bar{\underline{\phi}}_{k_c} + \underline{W}(A_{k_c}, \bar{A}_{k_c}, \delta) \tag{76}$$

$$\partial_t A_{k_c} = F(A_{k_c}, \bar{A}_{k_c}, \delta). \tag{77}$$

We can now take this form of solution as an ansatz [9, 10] to solve (31). It is actually not possible to determine $\underline{W}$ and $\underline{F}$ in a global way. One needs to expand them in a Taylor series in A_{k_c}, $\bar{A}_{k_c}$, and $\bar{\delta} = (\delta_1 \ldots \delta_k)$

$$\underline{W} = \sum_n \sum_{\bar{n}} \underline{W}_{n,\bar{n}}(\delta) A_{k_c}^n \bar{A}_{k_c}^{\bar{n}} \tag{78}$$

$$\underline{W}_{n,\bar{n}}(\delta) = \sum_{P_1} \cdots \sum_{P_k} \delta_1^{P_1} \cdots \delta_k^{P_k} \underline{W}_{n,\bar{n},\{P_i\}}. \tag{79}$$

Using (31), (75), (77), (78), (79) and an analogous expansion for F one gets, at the order n, $\bar{n}$, $\{P_i\}$, the following equation to solve

$$-\underline{\underline{L}}(\vec{\nabla}, \lambda_c)\underline{W}_{n,\bar{n},\{P_i\}} = \underline{I}_{n,\bar{n}\{P_i\}} - F_{n,\bar{n},\{P_i\}}\underline{\phi}_{k_c}(\vec{x}) - \bar{F}_{n,\bar{n},\{P_i\}}\bar{\underline{\phi}}_{k_c}(\vec{x}) \tag{80}$$

where $\underline{I}_{n,\bar{n},\{P_i\}}$ is a known quantity "at this order".

Since $\underline{\underline{L}}(\vec{\nabla}, \lambda_c)$ has zero eigenvalues, (80) is only solvable when the right hand side belongs to its image. Since the $F_{n,\bar{n},\{P_i\}}$ are so far arbitrary, we can choose them in order to satisfy this condition. This procedure, known as the Fredholm alternative, is actually the main tool of the bifurcation theory. Technically one defines a scalar product in the Hilbert space

$$< \underline{f}(\vec{x}), \underline{g}(\vec{x})> = \frac{1}{V} \int_0^{\Gamma_1} \cdots \int_0^{\Gamma_d} (\underline{f}(\vec{x}), \underline{g}(\vec{x}))\, d^d\vec{x} \tag{81}$$

where V represents the volume of the box and

$$(\underline{f}, \underline{g}) = \sum_{i=1}^{N} f_i(\vec{x}) \bar{g}_i(\vec{x}) . \tag{82}$$

One is then able to compute the adjoint $\underline{\underline{L}}^+(\vec{\nabla}, \lambda_c)$ of $\underline{\underline{L}}(\vec{\nabla}, \lambda_c)$ with respect to this scalar product. Its kernel is generated by $\underline{\phi}_{k_c}^+$ and $\bar{\underline{\phi}}_{k_c}^+$

$$\begin{aligned} \underline{\underline{L}}^+(\vec{\nabla}, \lambda_c)\underline{\phi}_{k_c}^+ &= 0, & \underline{\phi}_{k_c}^+ &= \tilde{\underline{\phi}}_{k_c} e^{i\vec{k}_c\cdot\vec{x}} \\ \underline{\underline{L}}^+(\vec{\nabla}, \lambda_c)\bar{\underline{\phi}}_{k_c}^+ &= 0, & \bar{\underline{\phi}}_{k_c}^+ &= \bar{\tilde{\underline{\phi}}}_{k_c} e^{-i\vec{k}_c\cdot\vec{x}} \end{aligned} \tag{83}$$

and one has the following orthogonality relations

$$\begin{aligned} &< \underline{\phi}_{k_c}(\vec{x}), \bar{\underline{\phi}}_{k_c}^+(\vec{x})> = 0 \\ &< \bar{\underline{\phi}}_{k_c}(\vec{x}), \underline{\phi}_{k_c}^+(\vec{x})> = 0. \end{aligned} \tag{84}$$

One then guarantees the solvability of (80) by demanding its right hand side to be orthogonal to the kernel of $\underline{\underline{L}}^+$. This solvability condition allows then to compute the $F_{n,\bar{n},\{P_i\}}$

$$F_{n,\bar{n},\{P_i\}} = \frac{<\underline{I}_{n,\bar{n},\{P_i\}} , \underline{\phi}^{+}_{k_c}>}{<\phi_{k_c} , \phi^{+}_{k_c}>} \quad . \tag{85}$$

The $F_{n,\bar{n},\{P_i\}}$ not involved in these solvability conditions can be chosen equal to zero in order to simplify as much as possible (77). It is then straightforward to check that the general form of the amplitude equation reads

$$\partial_t A_{k_c} = A_{k_c} g(|A_{k_c}|^2, \delta) \tag{86}$$

which becomes, at the leading order,

$$\partial_t A_{k_c} = \mu A_{k_c} - \alpha |A_{k_c}|^2 A_{k_c} \tag{87}$$

where $\mu = \sum_{i=1}^{k} \delta_i \left.\frac{\partial g(0,\delta)}{\partial \delta_i}\right|_{\delta=0}$ and $\alpha = - \left.\frac{\partial g(|A_{k_c}|^2,0)}{\partial |A_{k_c}|^2}\right|_{A_{k_c}=0}$. One of the advantages of this method on the adiabatic elimination procedure is its systematic character. The price paid for that is a loss of the physical intuition of the role played by the stable modes in the nonlinear selfinteraction of the critical mode. Its main interest, as we will see later in more detail for the case of multiple instabilities, lies in the fact that one is able to write down the general form of the amplitude equations without calculations. In practical computations, the main difficulty arises from the multiple Taylor expansions involved in this procedure.

There is a classical way, inspired from the Poincaré-Lindstedt Method [11,12] in celestial mechanics, to get the Landau equation from (31). The fact that (87) represents the correct amplitude equation implies some relation between the small parameter μ, the time scale and the magnitude of the amplitude

$$t \sim \mu \sim A_k^2 \ . \tag{88}$$

One can use these relations to reduce the multiple expansion to a single one. Introducing the following scaled variables

$$\begin{aligned} \underline{U} &= \varepsilon \underline{\tilde{U}} \\ \lambda &= \lambda_c + \varepsilon^2 \tilde{\delta} \\ t &= \varepsilon^{-2} \tilde{t} \end{aligned} \tag{89}$$

(31) then reads

$$\begin{aligned} -\underline{\underline{L}}(\vec{\nabla}, \lambda_c)\underline{U} = {} & \varepsilon \underline{N}_2(\vec{\nabla},\lambda_c;\underline{U},\underline{U}) + \varepsilon^2 \Big(\sum_{i-1}^{k} \delta_i \left.\frac{\partial \underline{\underline{L}}}{\partial \lambda_i}(\vec{\nabla},\lambda)\right|_{\lambda_c} \underline{U} \ + \\ & \underline{N}_3(\vec{\nabla}, \lambda_c;\ \underline{U},\underline{U},\underline{U}) - \partial_t U\Big) + O(\varepsilon^3) \end{aligned} \tag{90}$$

where

$$\underline{N}(\vec{\nabla}, \lambda_c;U) = \underline{N}_2(\vec{\nabla}, \lambda_c;\ \underline{U},\underline{U}) + \underline{N}_3(\vec{\nabla}, \lambda_c;\ \underline{U},\underline{U},\underline{U}) + \dots \quad . \tag{91}$$

One then expands $\tilde{U}$ in powers of $\in$

$$\underline{\tilde{U}} = \underline{\tilde{U}}_0 + \in \underline{\tilde{U}}_1 + \dots \quad . \tag{92}$$

At the order 0 one has

$$-\underline{\underline{L}}(\vec{\nabla}, \lambda_c)\underline{U}_0 = 0 \tag{93}$$

which gives

$$\underline{U}_0 = A_{k_c}\underline{\phi}_{k_c} + \bar{A}_{k_c}\bar{\underline{\phi}}_{k_c}, \quad \underline{\phi}_{k_c} = \tilde{\underline{\phi}}_{k_c}\, e^{i\vec{k}_c\cdot\vec{x}} . \tag{94}$$

At the first order has

$$-\underline{\underline{L}}(\vec{\nabla}, \lambda_c)\underline{U}_1 = \underline{N}_2(\vec{\nabla}, \lambda_c); \underline{U}_0, \underline{U}_0). \tag{95}$$

Which is always solvable since its right hand side only contains spatial frequencies 0 and $\pm 2k_c$ and hence belongs to the range of $\underline{\underline{L}}(\vec{\nabla}, \lambda_c)$

$$\underline{U}_1 = \underline{U}_{1,2k_c}\, e^{2i\vec{k}_c\cdot\vec{x}} + \underline{U}_{1,0} + \underline{U}_{1,2k_c}\, e^{-2i\vec{k}_c\cdot\vec{x}} . \tag{96}$$

At the second order one has

$$\begin{aligned} -\underline{\underline{L}}(\vec{\nabla}, \lambda_c)\underline{U}_2 = & N_2(\vec{\nabla}, \lambda_c; \underline{U}_0, \underline{U}_1) + N_2(\vec{\nabla}, \lambda_c; \underline{U}_1, \underline{U}_0) \\ & + N_3(\vec{\nabla}, \lambda_c; \underline{U}_0, \underline{U}_0, \underline{U}_0) + \sum_{i=1}^{k} \delta_i \frac{\partial \underline{\underline{L}}}{\partial \lambda_i} \underline{U}_0 \\ & - \partial_t A_{k_c}\phi_{k_c} - \partial_t \bar{A}_{k_c}\bar{\phi}_{k_c} \end{aligned} \tag{97}$$

whose solvability conditions give the Landau equation

$$\partial_t A_{k_c} = \mu A_{k_c} - \alpha |A_{k_c}|^2 A_{k_c} . \tag{98}$$

In equations (90) to (97) $\underline{U}$, δ, t stand for $\tilde{U}$, $\tilde{\delta}$, $\tilde{t}$. At the end of the calculation one can put $\epsilon = 1$ without inconvenience.

If one continues the procedure, one encounters a difficulty at the fourth order since there,one has no degree of freedom to apply the solvability condition,and one finds that the right hand side of the equation to solve does not belong to the range of $\underline{\underline{L}}(\vec{\nabla}, \lambda_c)$. The origin of the difficulty is that the amplitude equation contains higher order terms,which can be interpreted as nonlinear effects which occur on slower time scales. One is then naturally led to introduce several time scales

$$\tilde{\underline{U}}(t) = \tilde{\underline{U}}(t_0, t_1, \ldots t_n \ldots) \tag{99}$$

where $t_n = \varepsilon^n t$. Then

$$\frac{d}{dt} = \partial_t + \varepsilon\partial_{t_1} + \varepsilon^2\partial_{t_2} . \tag{100}$$

At the third order one finds $\partial_{t_1} A_{k_c} = 0$ and uses the solvability condition at the next order to determine $\partial_{t_2} A_{k_c}$

$$\partial_{t_2} A_{k_c} = \beta\mu^2 A_{k_c} + \gamma\mu |A_{k_c}|^2 A_{k_c} + \eta |A_{k_c}|^4 A_{k_c} . \tag{101}$$

In order to illustrate this method,we now consider a simple example. The motion of a ball in a rotating hoop (Fig. 5) is described by the equation

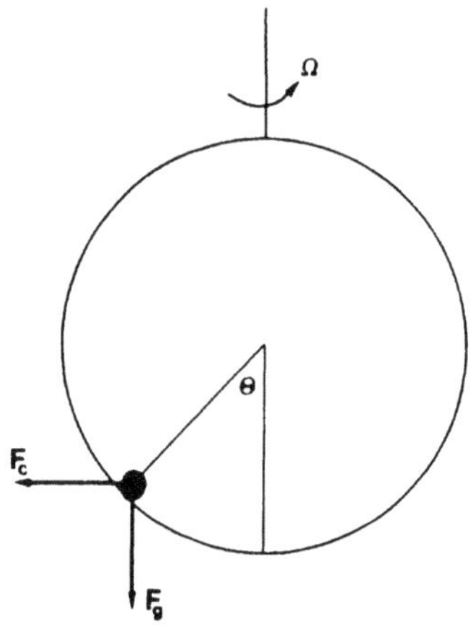

Figure 5

$$\ddot{\theta} + \frac{\nu}{m}\dot{\theta} = -g\sin\theta + R\Omega^2\sin\theta\cos\theta. \qquad (102)$$

When $\Omega = 0$, the stationary stable solution is given by $\theta = 0$. The rotation is a destabilizing effect which is opposed by the gravity. Above a given threshold Ω_c instability occurs. In the appropriate unit of time ($t = \Omega\sqrt{R}\,\tilde{t}$), (102) reads

$$\ddot{\theta} + \tilde{\nu}\dot{\theta} = \sin\theta(\cos\theta - \gamma) \qquad (103)$$

where $\tilde{\nu} = \frac{\sqrt{R}}{\Omega R}\frac{\nu}{n}$ and $\gamma = \frac{g}{\Omega^2 R}$. Introducing the following scaled variables

$$\begin{aligned} \theta &= \varepsilon\tilde{\theta} \\ \gamma &= \gamma_c + \varepsilon^2\mu \\ t &= \varepsilon^{-2}\tilde{t} \end{aligned} \qquad (104)$$

(103) becomes

$$(\gamma_c - 1)\theta = \varepsilon^2\left(-\nu\dot{\theta} - \mu\theta - \frac{\theta^3}{2}\left(1 + \frac{1-\gamma_c}{3!}\right)\right) + O(\varepsilon^4). \qquad (105)$$

One then expands θ in power of ε. At the order 0 one has

$$(\gamma_c - 1)\theta_0 = 0. \qquad (106)$$

Instability occurs when $\gamma_c = 1$ and $\theta_0 = A(t)$. The first order gives no equation and the second gives the amplitude equation

$$0 = -\nu\partial_t A - \mu A - A^3/2\ . \qquad (107)$$

This kind of method has been first used to compute the convective solution of the Rayleigh-Bénard problem [13, 14]. One finds $\alpha > 0$ in such a way that the convection is supercritical and we get near the onset of instability finite amplitude fluid motion in the form of rolls (under appropriate conditions or in bidimensional convection).

2.3. Bifurcation Theory

We have seen several examples of different systems (convection, chemical reaction, a simple mode-mode coupling model, a ball in a rotating hoop) which at the onset of instability behave in the same way. They are described by the so-called Landau equation. The universal feature involved here has a simple explanation. First, all these problems have in common the fact that a simple real eigenvalue of the linear problem becomes positive for some value of an external parameter, the others having negative real parts. In that case, the existence of the so called central manifold, which expresses the stable modes in terms of the marginal ones, allows to reduce the dynamics to a single equation, whatever the number of potentially present degrees of freedom. This equation is called an amplitude equation or normal form of the bifurcation, for reasons which will appear clear in the following. At the marginal situation it generally reads

$$\partial_t A = f(A) \tag{108}$$

with $\left.\frac{\partial f}{\partial A}\right|_{A=0} = 0$ (no linear term). Expanding f(A) in a Taylor series one has

$$\partial_t A = \alpha_2 A^2 + \alpha_3 A^3 + \alpha_4 A^4 + \dots \quad . \tag{109}$$

Near the marginal situation, the description in terms of amplitude equations holds, but (109) has to be slightly modified. Let us introduce μ_1, μ_2,... some small quantities. Then (109) becomes

$$\partial_t A = \mu_0 + \mu_1 A + (\alpha_2 + \mu_2)A^2 + (\alpha_3 + \mu_3)A^3 + \dots \tag{110}$$

Several cases have to be considered.

1) $\alpha_2 \neq 0$. Equation (110) reads

$$\partial_t A = \mu_0 + \mu_1 A + (\alpha_2 + \mu_2)A^2 + \text{h.o.t.} \tag{111}$$

If one introduces appropriate near-identity scaling and infinitesimal translation of the variable A, (111) reduces to

$$\partial_t A = \mu + \alpha_2 A^2 \tag{112}$$

which describes the so called saddle-mode bifurcation (Fig. 6). The scalings needed to get this amplitude equation from (31) are the following

$$U = \varepsilon\tilde{U}, \quad \mu = \varepsilon^2\tilde{\mu}, \quad t = \varepsilon^{-1}\tilde{t} \; . \tag{113}$$

2) $\alpha_2 = 0$. In the case of the existence of a symmetry property ($A \to -A$ in the amplitude equation) (110) reads

$$\partial_t A = \mu_1 A + (\alpha_3 + \mu_3)A^3 + \text{h.o.t.} \tag{114}$$

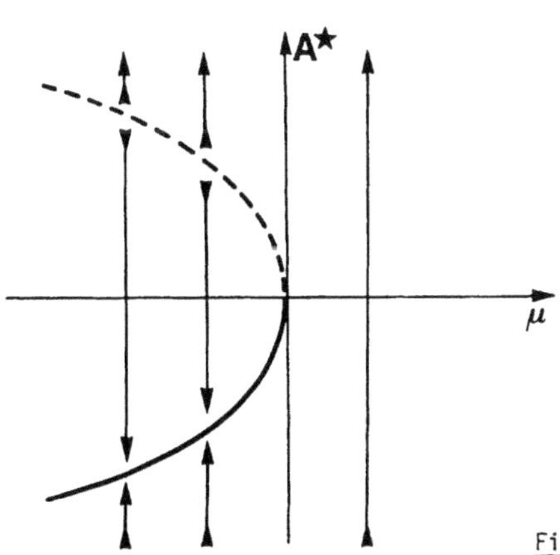

Figure 6

A near-identity scaling allows to reduce (114) to

$$\partial_t A = \mu A + \alpha_3 A^3 \qquad (115)$$

which describes the pitchfork bifurcation already discussed.

From (110) all the possible bifurcation diagrams [16] can be obtained. For example, in the first case ($\alpha_2 \neq 0$) if one only considers perturbations which preserve the origin as a stationary solution, then the constant term μ_0 is not present in (110) and the amplitude equation reads in that case

$$\partial_t A = \mu A + \alpha A^2 . \qquad (116)$$

It describes a transcritical bifurcation with exchange of symmetry (Fig. 7). One also can combine (114) and (116). When one is in a situation where the stationary solution is known to persist, a small symmetry-breaking term can lead to bifurcation diagrams as shown in Fig. 8 corresponding to the following amplitude equation

$$\partial_t A = \mu A + \nu A^2 + \alpha_3 A^3 . \qquad (117)$$

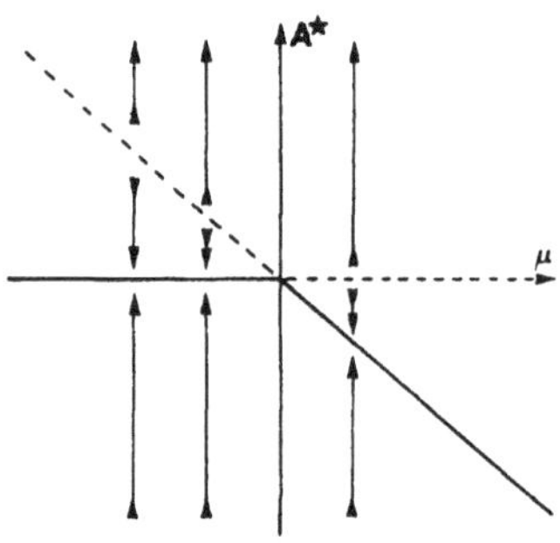

Figure 7

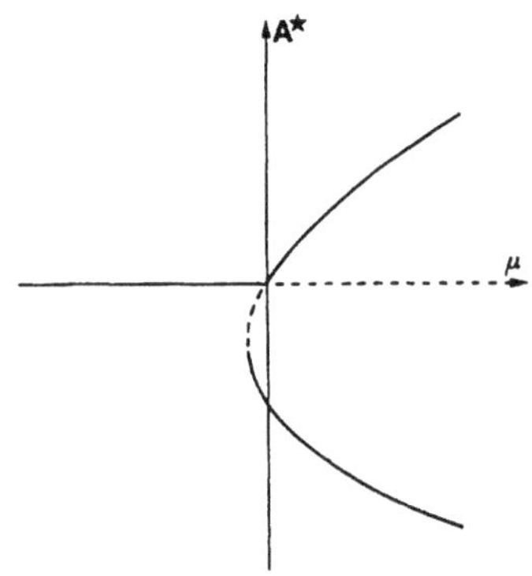

Figure 8

Since all the problems considered so far have the desired symmetry ($A \to -A$), we then understand why they are described by the same equation. The amplitude equations written below describe in some sense a nonlinear phenomena in its prototypical form. Here, the phenomena is simply the nonlinear development of a stationary instability. We will see later more complex effects, like, for example, nonlinear oscillations and even chaotic bahavior in such equations. These equations are also called normal forms for that reason.

2.4. Interaction between patterns

Up to now, we only considered situations in which one mode becomes unstable first. A simple case of multiplicity greater than one occurs when one chooses the lateral size of the "box" (in the Rayleigh-Bénard convection or in the chemical reactor) so that two modes become marginal at once [16] (Fig. 9). This double instability occurs for a precise value of the parameters Γ_x and R (one assumes $\Gamma_y = 0$), $\Gamma_x \equiv \Gamma = \Gamma_c$ and $R = R_c$. Near these parameter values, and thanks to the symmetries of the problem, the amplitude equations read

$$\partial_t A_1 = \mu_1 A_1 + \alpha_{11} A_1^3 + \alpha_{12} A_1 A_2^2 , \quad \partial_t A_2 = \mu_2 A_2 + \alpha_{21} A_2 A_1^2 + \alpha_{22} A_2^3 \qquad (118)$$

where $\mu_i = (\Gamma - \Gamma_c)\mu_\Gamma^{(i)} + (R - R_c)\mu_R^{(i)}$ for $i = 1,2$. Equation (118) describes the nonlinear interaction between two structures of different wave number. According to the values of α_{ij} one can have various phenomena, like the nonlinear extinction of one mode, or the nonlinear superposition of them and even a possible oscillation

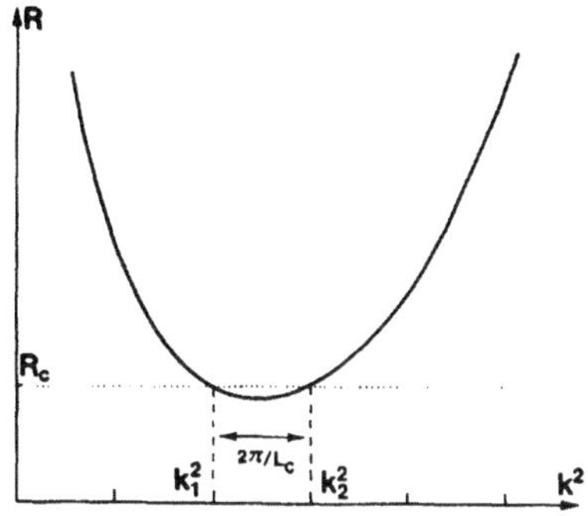

Figure 9

between them. It is a typical example of interaction between instabilities. If we look at a simple instability as a non equilibrium phase-transition, the problem of multiple instabilities appears as the equivalent of phase-transitions with several order parameters. More than one external parameter is needed to observe a situation of this type. In the parameter space (Γ, R) we have two lines. One represents the values of Γ and R for which the mode labelled by 1 becomes unstable first. The other line is the equivalent line for the second mode. At their intersection they become unstable at once. One also says that we have a codimension two bifurcation, since the intersection of two hypersurfaces of codimension one where a single instability occurs is a hypersurface of codimension two in the parameter space $(\lambda_1 \ldots \lambda_k)$: In our case, two curves which cross at a point in the two-dimensional parameter space Γ, R. In the neighborhood of such a point, any system with the appropriate symmetry is described by equations like (118).

2.5. Phase instabilities in extended systems

We now turn to the problem of removing the lateral boundary condition to infinity. When the system becomes large ($\Gamma \to \infty$) all this description in terms of ordinary differential equations, the amplitude equations, breaks down, since many modes become unstable almost at the same time. The previous approach has to be corrected by introducing amplitudes $A(\vec{x}, t)$ which slowly vary in space and time. One then looks, instead of (76), for a solution of (31) under the form

$$U = \vec{A}(\vec{x}, t)\, \underline{\phi}_{\vec{k}_c}(\vec{x}) + \overline{A}(\vec{x}, t)\, \overline{\underline{\phi}}_{\vec{k}_c}(\vec{x}) + \text{h.o.t.} \qquad (119)$$

where $A(\vec{x}, t)$ varies in space on a scale much larger than $|k_c|^{-1}$. The amplitude equation then becomes a partial derivative equation: the envelope amplitude equation [17, 18]. One assumes in the following that $\vec{k}_c$ is uniquely determined which means that one considers only one-dimensional periodic structure. In realistic cases we also have an orientational degeneracy which can give birth to multidimensional periodic patterns like squares or hexagons in the plane. The selection of a given pattern is a nonlinear interaction problem which can be approached by the methods described below. In the Rayleigh-Bénard convection, under appropriate conditions unidimensional periodic patterns (rolls) are shown to be stable [4]. We only consider this situation here. Several methods are available to get the partial derivative equation for $A(\vec{x}, t)$. One can, for example, expand $\underline{U}$ both in A, $\vec{\nabla}$, δ and get at each order an equation of the type (80) to solve. The shortest way is again to use the asymptotic technics. Let us consider the case of a two dimensional problem in which a periodic pattern in the x coordinate appears ($\vec{k}_c = k_c \vec{i}$). One introduces the following scaled quantities

$$\underline{U} = \varepsilon \underline{\tilde{U}}, \quad \lambda = \lambda_c + \varepsilon^2 \delta, \quad t = \varepsilon^{-2} \tilde{t}\,. \qquad (120)$$

$\tilde{U}$ is assumed to vary in space on different scales

$$1 \text{ and } \varepsilon^{-1} \text{ in } x, \quad \varepsilon^{-1/2} \text{ in } y. \tag{121}$$

One then as usual expands $\underline{\widetilde{U}}$ in powers of ϵ. At the zeroth order one has

$$-\underline{\underline{L}}(\vec{\nabla}, \lambda_c)\underline{\widetilde{U}}_0 = 0 \tag{122}$$

whose solution is taken in the form

$$\underline{\widetilde{U}}_0 = A(t, \varepsilon^{-1}x, \varepsilon^{-1/2}y)\, e^{ik_c x}\widetilde{\phi}_{k_c} + \text{c.c.} \tag{123}$$

At the first order, one finds that the right hand side belongs to the image of $\underline{\underline{L}}$ as previously stated. It allows us to compute $\underline{\widetilde{U}}_1$ without using solvability conditions. At the second order, the solvability condition gives the evolution equation for $A(t, x, y)$. It reads

$$\partial_t A = \mu A + \beta(\partial_x - i\partial_y^2)^2 A - \alpha|A|^2 A \tag{124}$$

with α and β real. If one changes once more the scales of the problem, (124) can be put into the form

$$\partial_t A = A + (\partial_x - i\partial_y^2)^2 A - |A|^2 A. \tag{125}$$

This equation is known as the Newell-Whitehead-Segel equation.

We first remark that (125) can be written in a potential form

$$\partial_t A = -\frac{\delta V(\{A\})}{\delta A} \tag{126}$$

where $\delta/\delta A$ represents the functional derivatives and

$$V(\{A\}) = \iint \left(-|A|^2 + \left|\frac{\partial A}{\partial x} - i\frac{\partial^2 A}{\partial y^2}\right|^2 + \frac{1}{2}|A|^4\right) dx\,dy. \tag{127}$$

As a consequence (125) only has asymptotic stationary solutions. No time dependence is expected to occur near the onset of one dimensional periodic patterns on this time scale. Nevertheless since the relaxation time diverges as one approaches the instability threshold one can observe slow variations of the pattern before it eventually reaches a stationary state. We now discuss the content of (125) in the case of one dimensional systems (bidimensional convection, linear chemical reactor).

Equation (125) then reads

$$\partial_t A = A + \frac{\delta^2}{\partial x^2} A - |A|^2 A. \tag{128}$$

This equation admits a class of spatially periodic solutions

$$A_q(x) = Qe^{iqx} \tag{129}$$

where

$$Q = (1 - q^2)^{1/2}. \tag{130}$$

The stability of these solutions is investigated through a perturbation of the form

$$A(x, t) = A_q(x) + B(x, t). \tag{131}$$

It is then convenient to choose

$$B = e^{iqx}\widetilde{B}. \tag{132}$$

In this variable the equation for the perturbation reads

$$\partial_t \widetilde{B} = -\mathcal{R}e\widetilde{B} + i\alpha\widetilde{B}_x + \widetilde{B}_{xx} - (\widetilde{B}^2 + 2|\widetilde{B}|^2) - 2|\widetilde{B}|^2\widetilde{B} \tag{133}$$

where $\alpha = \sqrt{2}q/Q$. $\widetilde{B} = i\phi$, where ϕ is constant, is a magnitude mode for the linear problem associated with (133). This mode has an important meaning. It corresponds to the translational invariance of the initial equations. If $A_q(x)$ is a solution of the problem, $A_q(x+\phi)$ is also a solution of the problem. We first look more in detail to the linear problem. Let $\widetilde{B} = R + i\phi$. The linear part of (133) then reads

$$\partial_t\phi = \phi_{xx} + \alpha R_x \tag{134.a}$$

$$\partial_t R = -R + R_{xx} - \alpha\phi_x \ . \tag{134.b}$$

Since the real part of B (amplitude mode) is damped, one can use the adiabatic elimination procedure, for example, to eliminate it. At the leading order one has

$$R = -\alpha\phi_x$$

and (143.a) reads now

$$\partial_t\phi = (1-\alpha^2)\phi_{xx} = \frac{1-3q^2}{1-q^2}\phi_{xx} \ . \tag{136}$$

This equation describes a diffusion of the phase [19] of the periodic pattern. Phase instability occurs when $(1-\alpha^2) < 0$. This instability corresponds to the so-called Eckauss instability [20]. Its marginal curve of stability lies inside the original curve of stability as sketched in Fig. 10 . As a consequence the instability cannot be naturally reached in increasing the constraint R (Rayleigh-number in convection, for example). To observe it one has to impose a periodic structure with an appropriate wavelength for a value of R above threshold, and then, after having suppressed the spatially periodic forcing decrease R [21]. When one crosses the marginal stability curve associated with the phase (see Fig. 10), the phase diffusion coefficient becomes negative and one has to look for higher order terms in the phase equation. Again several methods exist to derive this equation. The most direct way is to use an asymptotic expansion.

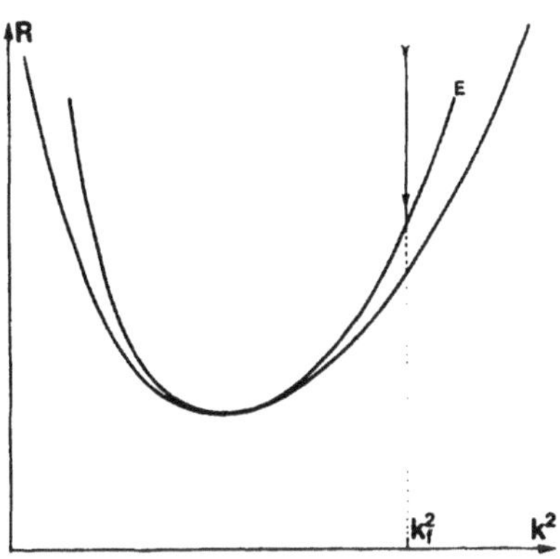

Figure 10

One introduces the following scaled quantities

$$B = \varepsilon\widetilde{B}, \quad t = \varepsilon^{-4}\widetilde{t}, \quad x = \varepsilon^{-1}\widetilde{x}, \quad \alpha = \alpha_c + \mu\varepsilon^2 \tag{137}$$

and expands $\widetilde{B}$ in powers of ε

$$B = B_0 + \varepsilon\widetilde{B}_1 + \dots$$

At the fourth order the solvability conditions give the phase equation [22,23,24]

$$\partial_t\phi = \mu\phi_{xx} - \phi_{xxxx} - 12\alpha_c\phi_x\phi_{xx} \tag{138}$$

with $\alpha_c^2 = 1$. This equation has an important meaning. It universally describes the phase instability of a one dimensional periodic pattern. It is a "generic" instability in the sense that it does not depend on the detail of the initial equation but only on its translational invariance. When one considers this instability in a large but finite "box" a long wavelength mode becomes unstable first. This describes a modulational instability of the structure.

When the one dimensional periodic pattern arises in two dimensional space the linear phase equation reads [19]

$$\partial_t \phi = D_{\parallel} \phi_{xx} + D_{\perp} \phi_{yy}. \tag{139}$$

When $D_{\perp}$ becomes negative we have a new "generic" phase instability: the so called zig-zag instability. The corresponding phase equation reads [22]

$$\partial_t \phi = \mu \phi_{yy} - \phi_{yyyy} + \phi_{xx} + (C_1 \phi_x + C_2 \phi_y^2) \phi_{yy}. \tag{140}$$

This instability describes a transversal modulation of the structure. In the convection problem the situation is more complex because of the presence in large boxes of horizontal velocity modes (26). The corresponding amplitude equations are [25] given by

$$\begin{aligned} \partial_t A &= \mu A + (\partial_x - i\partial_y^2)^2 - |A|^2 A - iCA \\ \gamma \partial_t B &= \partial_y^2 B + g\partial_y [\bar{A}(\partial_x - i\partial_y^2)A + c.c.] \end{aligned} \tag{141}$$

where $B = -\partial_y C$ represents the amplitude of the vertical vorticity mode. The vertical vorticity is defined as $(\vec{\nabla} \times \vec{V})_z = \partial_x v - \partial_y u$ and $\gamma = \sqrt{3}/(1 + \sigma)$, $g = 2\left(\frac{1+\sigma}{\sigma^2}\right)$. The second equation describes the generation of vertical vorticity from the roll curvature. New instabilities can be predicted from (141).

3. Time dependent periodic phenomena

3.1. Hopf bifurcation

We now come back to the convection experiment in a small box. When one increases the Rayleigh number several types of instabilities can occur [26]. At high Prandtl numbers, transitions towards differents cellular patterns are observed but eventually an oscillatory instability takes place. At small Prandtl numbers one observes a transition from straight roll pattern to a time dependent periodic convection under the form of stationary waves along the roll axis [27]. Since at these values of the Rayleigh number the stationary convective solution can only be gotten by means of approximations it is very difficult to follow, in that case, the nonlinear development of the oscillatory instability. For that reason we study the universal aspect of this time-dependent transition in the context of the reaction-diffusion equations (34), (37). Recalling that the characteristic equation for this problem reads

$$s^2 - sT(k^2) - R(k^2) = 0 \tag{142}$$

one remarks that, when $R(k^2) < 0$ and $T(k^2) = 0$, one has two pure imaginary roots for (142). When $T(k^2)$ becomes negative, one finds a complex mode with an oscillatory growing amplitude. The corresponding marginal curve is sketched in Fig. 11.

We first remark that the instability arises with a mode without spatial structure ($k = 0$). No pattern formation is then expected to occur. The case where such an instability arises at finite wavelength will be discussed later. The marginal mode is given by

Figure 11

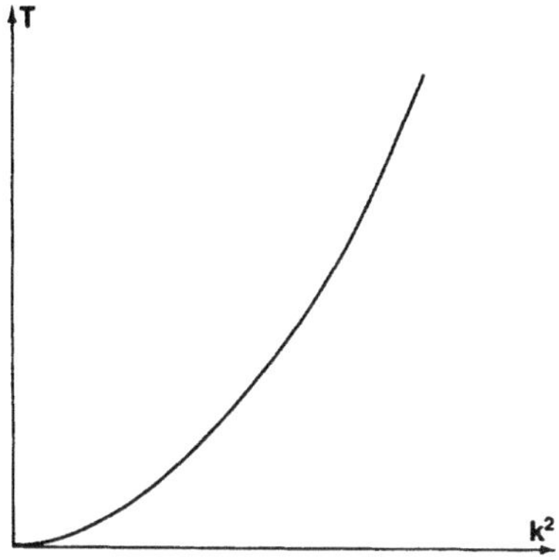

$$\underline{\phi} = \begin{pmatrix} 1 \\ \frac{i\omega + L_{22}}{L_{12}} \end{pmatrix} \tag{143}$$

where $\omega^2 = -R(0) = L_{11}L_{22} - L_{12}L_{21}$. One expands the solution on the basis of the eigenmodes of $\underline{\underline{L}}_0 - \underline{\underline{D}}\nabla^2$, where $\underline{\underline{L}}_0 = \begin{pmatrix} -L_{22} L_{12} \\ L_{21} L_{22} \end{pmatrix}$. Equation (37) then reads

$$\partial_t Z = i\omega Z + f(Z, \overline{Z}, \{B_i\}) \tag{144.a}$$

$$\partial_t B_i = s_i B_i + g_i(Z, \overline{Z}, \{B_i\}) \tag{144.b}$$

where the B_i represent the amplitudes of the stable modes, and s_i the corresponding roots of (142). Again one can define a central manifold which expresses that the stable modes follow the dynamics of Z and $\overline{Z}$ through a relation of the type

$$B_i = \psi_i(Z, \overline{Z}). \tag{145}$$

Using (145) (144a) becomes

$$\partial_t Z = i\omega Z + f(Z, \overline{Z}) \tag{146}$$

where $f(Z,\overline{Z}) \equiv f(Z,\overline{Z}, \{\psi_i(Z,\overline{Z})\}$. One is then tempted to conclude as before that the amplitude equation which describes the nonlinear development of the oscillatory instability can be obtained from (146) by keeping the first nonzero nonlinear term in the Taylor expansion of $f(Z,\overline{Z})$

$$\partial_t Z = i\omega Z + f_{20}Z^2 + f_{11}|Z|^2 + f_{02}\overline{Z}^2 \,. \tag{147}$$

Let us remark that for $T \lesssim 0$ the same kind of arguments would give

$$\partial_t Z = (i\omega(\mu) + \mu)Z + f_{20}(\mu)Z^2 + f_{11}(\mu)|Z|^2 + f_{02}(\mu)\overline{Z}^2 \tag{148}$$

where μ and $\omega(\mu)$ are given by

$$\mu = \frac{T}{2} \,, \quad \omega(\mu) = \omega\left[1 + \frac{L_{22}}{\omega^2}\mu\right] + O(\mu^2). \tag{149}$$

Actually it turns out that (147) and (148) are not the correct amplitude equations. We now jump directly to the asymptotic analysis of the oscillatory instability. We will come back later to this point in order to analyze what was wrong in our reasoning. Let us introduce the following scaled quantities

$$\underline{U} = \varepsilon\underline{\widetilde{U}} \,, \quad T = \varepsilon^2\mu. \tag{150}$$

The time is not scaled since the oscillatory mode has a frequency of order unity. Nevertheless in order to study the nonlinear saturation of its amplitude one is lead to introduce a second time which varies on a slow scale,

$$\underline{\tilde{U}} = \underline{\tilde{U}}(t, \varepsilon^2 t). \tag{151}$$

$\underline{\tilde{U}}$ is then expanded in powers of ε

$$\underline{\tilde{U}} = \underline{\tilde{U}}_0 + \varepsilon \underline{\tilde{U}}_1 + \varepsilon^2 \underline{\tilde{U}}_2 . \tag{152}$$

At the order 0 one has

$$(\partial_t - \underline{\underline{L}}_0 - \underline{\underline{D}}\nabla^2)\underline{\tilde{U}}_0 = 0 \tag{153}$$

whose solution is

$$\underline{\tilde{U}}_0 = \tilde{Z}(t_1)\ \underline{\Phi}(t) + \text{c.c.}, \tag{154}$$

where $\underline{\Phi}(t) = \underline{\Phi}(0)e^{i\omega t}$ and $t_1 = \varepsilon^2 t$. At the first order one has

$$(\partial_t - \underline{\underline{L}}_0 - \underline{\underline{D}}\nabla^2)\underline{\tilde{U}}_1 = \underline{N}_2(\underline{\tilde{U}}_0, \underline{\tilde{U}}_0). \tag{155}$$

This equation is always solvable since the right hand side of (155) only contains frequencies 0 and $\pm 2i\omega$. Otherwise speaking $\underline{N}_2(\underline{\tilde{U}}_0, \underline{\tilde{U}}_0)$ belongs to the image of the linear operator $(\partial_t - \underline{\underline{L}}_0 - \underline{\underline{D}}\nabla^2) \equiv \mathcal{L}$. One can then compute $\underline{\tilde{U}}_1$

$$\underline{\tilde{U}}_1(t) = \underline{\tilde{U}}_{1,2}\, e^{2i\omega t} + \text{c.c.} + \underline{\tilde{U}}_{1,0} . \tag{156}$$

At the second order one has the following equation

$$(\partial_t - \underline{\underline{L}}_0 - \underline{\underline{D}}\nabla^2)\underline{\tilde{U}}_1 = \underline{N}_2(\underline{\tilde{U}}_0, \underline{\tilde{U}}_1) + \underline{N}_2(\underline{\tilde{U}}_1, \underline{\tilde{U}}_0) + \underline{N}_3(\underline{\tilde{U}}_0, \underline{\tilde{U}}_0, \underline{\tilde{U}}_0)$$
$$+ \mu \underline{\underline{L}}_1 \underline{\tilde{U}}_0 - \partial_{t_1}\underline{\tilde{U}}_0 , \tag{157}$$

where $L_1 = \begin{pmatrix} 1 & 0 \\ 0 & 0 \end{pmatrix}$. The presence of frequencies $\pm i\omega$ in the right hand side of (157) does not allow in general to solve this equation. One has to introduce a solvability condition. The scalar product is defined as

$$< \underline{f} \cdot \underline{g} > = \frac{1}{T}\frac{1}{\Gamma}\int_0^T \int_0^\Gamma (f_1 \bar{g}_1 + f_2 \bar{g}_2)\,dx\,dt \tag{158}$$

where $T = 2\pi/\omega$. The adjoint of $\mathcal{L}$ is given by

$$\mathcal{L}^+ = -\frac{d}{dt} - \underline{\underline{L}}_0^+ - \underline{\underline{D}}\nabla^2 . \tag{159}$$

The general solution of

$$\mathcal{L}^+ \underline{y}(t, x) = 0 \tag{160}$$

in the space of periodic function with period T, is given by

$$\underline{y}(t, x) = y\underline{\Phi}^+(t) + \text{c.c.} \tag{161}$$

where $\underline{\Phi}^+$ is defined by

$$\underline{\Phi}^+(t) = \underline{\phi}^+ e^{i\omega t} , \quad \underline{\underline{L}}^+ \underline{\phi}^+ = -i\omega\underline{\phi}^+ . \tag{162}$$

$\underline{\Phi}(t)$ and $\underline{\Phi}^+(t)$ satisfy the following orthogonality condition

$$< \underline{\Phi}(t), \overline{\underline{\Phi}}^+(t) > = 0 . \tag{163}$$

One then imposes the orthogonality of the right hand side with $\underline{\Phi}^+$ and $\overline{\underline{\Phi}}^+$. This leads to the equation

$$\partial_{t_1}\tilde{Z} = \mu\tilde{Z} + \alpha|\tilde{Z}|^2\tilde{Z} \tag{164}$$

where α is a complex number. Back to old variables $\tilde{Z}e^{i\omega t} = Z$ one reconstitutes an equation which involves the rapid time scale $(2\pi/\omega)$ and the slow one

$$\partial_t Z = (i\omega + \mu)Z + \alpha|Z|^2 Z. \tag{165}$$

The same techniques applied to more general equations like (31) give the same amplitude equation: the complex Landau equation.

In order to illustrate this method let us consider a simple example; the van der Pol oscillator. It is described by the equation

$$\ddot{X} - (\mu - X^2)\dot{X} + X = 0. \tag{166}$$

Oscillatory behavior occurs when μ crosses 0. Let $X = \varepsilon\tilde{X}$, $\mu = \varepsilon^2\tilde{\mu}$ and $\tilde{X} = \tilde{X}_0 + \varepsilon\tilde{X}_1 + \dots$. At the zeroth order one finds the equation

$$\ddot{\tilde{X}}_0 + \tilde{X}_0 = 0 \tag{167}$$

whose solution

$$\tilde{X}_0 = \tilde{Z}_0(t_1)e^{it} + \text{c.c.} \ . \tag{168}$$

At the first order one gets nothing. The cancellation of the terms $e^{\pm it}$ at the second order (solvability condition) gives the amplitude equation

$$\partial_t Z = (1 + \mu)Z + |Z|^2 Z \ . \tag{169}$$

We now come back to the beginning of this chapter to compare the normal form approach and the asymptotic analysis. Technically the absence of quadratic terms in (165) comes from the fact that at the second order the solvability condition is always satisfied. Physically it means that there are no resonances at this order since the right hand side of (155) does not contain terms with frequencies $\pm i\omega$. When one solves (155), one gets terms like $e^{2i\omega t}$ which describe the generation of harmonics in the solution. Actually this is the key of the incorrection done in writing (148). The quadratic terms present in this equation describe both harmonics production and nonlinear saturation or amplification effects. Let us rewrite (148)

$$\partial_t Z = i\omega Z + f_{20}Z^2 + f_{11}|Z|^2 + f_{02}\overline{Z}^2 + f_{30}Z^3 + f_{21}\tilde{Z}^2\overline{Z} + f_{12}\overline{Z}^2 Z + f_{03}\overline{Z}^3. \tag{170}$$

One then introduces the following near identity nonlinear change of coordinates

$$Z = \tilde{Z} + \alpha_{20}\tilde{Z}^2 + \alpha_{11}|\tilde{Z}|^2 + \alpha_{02}\overline{\tilde{Z}} + \alpha_{30}\tilde{Z}^3 + \alpha_{21}Z^2\overline{Z} + \alpha_{12}\overline{\tilde{Z}}^2\tilde{Z}$$
$$+ f_{03}\overline{\tilde{Z}}^3 + \text{h.o.t.} \ . \tag{171}$$

It is straightforward to show that by choosing conveniently the α_{ij} one can remove all the terms but $|\tilde{Z}|^2\tilde{Z}$ in (170). Let us illustrate it in the case of the equation

$$\partial_t Z = i\omega Z + f_{pq}Z^p\overline{Z}^q + \text{h.o.t.} \ . \tag{172}$$

Let $\tilde{Z}$ be a new variable defined by

$$Z = \tilde{Z} + \alpha_{pq}\tilde{Z}^p\overline{\tilde{Z}}^q + \text{h.o.t.} \ . \tag{173}$$

Equation (172) then reads

$$\partial_t \tilde{Z} + (p-q)i\omega\alpha_{pq}\tilde{Z}^p\bar{\tilde{Z}}^q = i\omega\tilde{Z} + i\omega\alpha_{pq}\tilde{Z}^p\bar{\tilde{Z}}^q + f_{pq}\tilde{Z}^p\bar{\tilde{Z}}^q + \text{h.o.t.} \quad . \tag{174}$$

If we choose α_{pq} such that

$$\alpha_{pq} = \frac{f_{pq}}{(p-q-1)i\omega}, \tag{173}$$

(174) becomes

$$\partial_t \tilde{Z} = i\omega\tilde{Z} + \text{h.o.t.} \quad . \tag{176}$$

We then have succeeded to linearize (172) at the order $p+q$. Obviously the procedure fails when $p-q-1=0$. The corresponding term $|Z|^q Z$ is called a resonant term. When one considers it as a perturbation, it behaves at the same frequency as the linear term. At the leading order the amplitude equation then reads

$$\partial_t \tilde{Z} = i\omega\tilde{Z} + \tilde{f}_{21}|\tilde{Z}|^2\tilde{Z} \ . \tag{177}$$

This complex Landau equation is also called the normal form for the Hopf bifurcation [28]. Before analyzing its content let us summarize the whole procedure, if one wants to use the modal approach to derive (177). One first computes the central manifold, that is, the relation existing between the stable modes and the complex marginal modes. One is then left with an equation for Z only. One has then to introduce near identity nonlinear change of coordinates (171) in order to remove as much as possible nonlinear terms in (170). The resulting equation (177) universally describes the nonlinear development of an oscillatory instability.

A way somewhat in between this normal form procedure and the asymptotic method consists of a solution of (31) under the form [10]

$$\underline{U} = Z\underline{\Phi} + \bar{Z}\bar{\underline{\Phi}} + \underline{W}(Z, \bar{Z}, \mu) \tag{178}$$

together with an amplitude equation for Z.

$$\partial_t Z = i\omega Z + F(Z, \bar{Z}, \mu). \tag{179}$$

We then expand $\underline{W}$ and F in Z, $\bar{Z}$, μ. At a given order n, $\bar{n}$, k in Z, $\bar{Z}$, μ, the fact that $\underline{U}$ is a solution of (31) is expressed by the equation

$$\left(i\omega Z\frac{\partial}{\partial Z} - i\omega\bar{Z}\frac{\partial}{\partial\bar{Z}} - \underline{\underline{L}}_0\right)\underline{W}_{n,\bar{n},k} = \underline{I}_{n,\bar{n},k} - G_{n,\bar{n},k}\,\underline{\Phi} - G_{n,\bar{n},k}\bar{\underline{\Phi}} \tag{180}$$

where $\underline{I}_{n,\bar{n},k}$ is a known function at this order. The linear operator $\mathcal{L} = i\omega\left(Z\frac{\partial}{\partial Z} - \bar{Z}\frac{\partial}{\partial\bar{Z}}\right) - \underline{\underline{L}}_0$ plays an important role [7, 30]. The structure of its kernel determines the so called resonant terms and hence the form of the amplitude equation. On the basis of the eigenvector of $\underline{\underline{L}}_0$ associated with the eigenvalues $\pm i\omega$ it reads

$$\mathcal{L}_\pm = i\omega\left(Z\frac{\partial}{\partial Z} - \bar{Z}\frac{\partial}{\partial\bar{Z}} \mp 1\right) . \tag{181}$$

Let us consider, for a while, the linear equation

$$\partial_t Z = i\omega Z \tag{182}$$

and introduce a near identity nonlinear change of coordinate of constant degree n

$$Z = \tilde{Z} + \phi_n(\tilde{Z}, \bar{\tilde{Z}}). \tag{183}$$

we get for $\tilde{Z}$ the following equation

$$\partial_t \tilde{Z} = i\omega\tilde{Z} - \mathcal{L}_+\Phi_n(\tilde{Z}, \tilde{\bar{Z}}) \qquad (184)$$

where $\mathcal{L}_+$ is given by (181). We then have transformed a linear equation (182) into a nonlinear one. The nonlinear terms produced by this transformation are the non resonant terms. They belong to the range of $\mathcal{L}_\pm$. The structure of the resonant terms are determined by the kernel of the adjoint of this operator. Let us consider the following auxiliary operators

$$A = i\omega Z\partial_Z - i\omega\bar{Z}\partial_{\bar{Z}} \qquad (185)$$

$$B = \Phi_n\partial_Z + \bar{\Phi}\partial_{\bar{Z}} \qquad (186)$$

respectively associated with the following linear and nonlinear vector fields:

$$\partial_t Z = i\omega Z \quad \text{and} \qquad (187)$$

$$\partial_t Z = \bar{\Phi}\,(Z, \bar{Z}). \qquad (188)$$

The commutator $[A, B]$ of these two operators is a Lie bracket operation

$$[A, B] = C, \quad \text{where} \qquad (189)$$

$$C = \mathcal{L}_+\bar{\Phi}_n\partial_Z + \mathcal{L}_-\Phi_n\partial_{\bar{Z}} \qquad (190)$$

is associated with the vector field

$$\partial_t Z = \mathcal{L}_+\Phi_n. \qquad (191)$$

There is a deep underlying mathematical structure beyond all these perturbation calculations which is common to all problems of singularities [31]. Marginal situations are associated with singularities for vector fields. Normal forms of such singularities and their perturbations are nothing else than the dynamical analogs of catastrophes [31,32].

In practical computations (180) is solved by first applying some solvability conditions (orthoganality of the right hand side with the kernel of the adjoint of $\mathcal{L}$) which allows to compute the coefficients $F_{n,\bar{n},k}$ of the normal form and then Taylor coefficients of the solution $\underline{W}_{n,\bar{n},k}$. We now come back to the amplitude equation in order to describe its content. Let $Z = \rho e^{i\phi}$. Then (165) reads

$$\partial_t\rho = \mu\rho + \mathcal{R}e\alpha\rho^3, \quad \partial_t\phi = \omega + \mathrm{Im}\alpha\rho^2. \qquad (192)$$

The first equation describes a pitchfork bifurcation for the real amplitude of the oscillatory mode. Only the positive branch has to be taken into account. It is given by

$$\rho^* = \left(\frac{-\mu}{\mathcal{R}e\alpha}\right)^{1/2} . \qquad (193)$$

The equation for the phase on this branch reads

$$\partial_t\phi = \omega - \frac{\mathrm{Im}\alpha}{\mathcal{R}e\alpha}\mu . \qquad (194)$$

Hence we have a renormalization of the effective frequency because of the finite amplitude of the oscillatory behavior

$$\omega_{eff} = \omega - \frac{\mathrm{Im}\alpha}{\mathcal{R}e\alpha}\mu . \qquad (195)$$

3.2. Phase turbulence

Again if one considers the fate of this instability in extended systems (165) has to be replaced by a partial derivative equation. An asymptotic expansion provides the quickest way to get this equation. Equations (151) and (154) are replaced by

$$\underline{\tilde{U}}(t,\ \varepsilon^{-2}t,\ \varepsilon^{-1}x), \tag{196}$$

$$\underline{\tilde{U}}_0 = \tilde{Z}(t_1,\ \varepsilon^{-1}x)\Phi(t) + \text{c.c.} \tag{197}$$

At the second order, one gets the amplitude equation

$$\partial_t Z = (i\omega + \mu)Z + D\nabla^2 Z + \alpha|Z|^2 Z \tag{198}$$

where D is a complex quantity. Equation (198) is the complex Ginzburg-Landau equation. It is numerically known to possess chaotic behaviors as close as one wants from $\mu = 0$[33-35]. These chaotic behaviors arised as the nonlinear development of the phase instability. This phenomena, named phase turbulence [36], is universally described by Kuramoto's equation

$$\partial_t \phi = \nu\phi_{xx} - \phi_{xxxx} - (\phi_x)^2 . \tag{199}$$

Let $Z_0(t) = \rho^* e^{i\omega_{eff}t}$ be the homogeneous solution of (198). $Z_0(t + \phi)$ where ϕ is a constant, is also a solution of (198). Then $Z'_0(t)$ is a marginal mode of the equation for a perturbation to $Z_0(t)$. When μ is positive one gets a stable amplitude branch which can be adiabatically eliminated and then (198) can be reduced to (199). A way to do this reduction consists to look for a solution of (198) under the form

$$Z(t,\ x) = Z_0(t) + \phi(x,\ t)Z_0'(t) + \xi(t,\ \phi,\ \nabla,\ \nu) \tag{200}$$

together with an evolution equation for ϕ

$$\partial_t \phi = F(\phi,\ \nabla,\ \nu) \tag{201}$$

where ∇ represents the x derivative. One then formally expands ξ and F in ϕ, ∇, ν. At a given order one has singular linear non homogenous equations to solve of the type (180) or (80). One can also use the asymptotic techniques. In any case (199) is found as a solvability condition.

When ν becomes negative (phase instability) one numerically observes [36, 37] a chaotic loss of synchronization of the continuous set of oscillators described by (198).

3.3. Spatio - temporal patterns

One considers now the problem of an oscillatory instability which arises with a finite spatial wavelength. Thermosolutal convection [2] or a chemical reactor with three chemical species provide examples of systems in which such an instability can occur. Since one restricts our analysis to the case of one dimensional periodic patterns it is convenient to use the stream function ψ to write the equations for thermosolutal bidimensional convection

$$u = \frac{\partial\psi}{\partial z} \qquad w = -\frac{\partial\psi}{\partial x}, \tag{202}$$

where u, w represent the bidimensional velocity field

$$\vec{V} = (u, 0, w). \tag{203}$$

The incompressible two-dimensional Navier-Stokes equation then reads

$$\partial_t \nabla^2 \psi + J(\nabla^2\psi, \psi) = \sigma\nabla^4\psi \quad \text{where} \tag{204}$$

$$J(\alpha, \beta) = \partial_x\alpha\partial_z\beta - \partial_z\alpha\partial_x\beta . \tag{205}$$

The appropriate equations for thermosolutal convection in the Boussinesq approximation reads

$$\begin{aligned} \partial_t\nabla^2\psi + J(\nabla^2\psi, \psi) &= \sigma\nabla^4\psi + \sigma R\partial_x\theta + \sigma\tau S\partial_x\Sigma \\ \partial_t\theta + J(\theta, \psi) &= -\partial_x\psi + \nabla^2\theta \\ \partial_t\Sigma + J(\Sigma, \psi) &= \partial_x\psi + \tau\nabla^2\Sigma \end{aligned} \tag{206}$$

where Σ represents the deviation of the salinity field from its linear static value. S is the saline Rayleigh number which is proportional to the salinity gradient applied on the layer

$$S = \frac{gd^3}{K_s\nu} \nabla S. \tag{207}$$

The Lewis number τ represents the ratio between the thermal and the saline diffusivity,

$$\tau = K_s/K_T. \tag{208}$$

The modes associated with the linear equations, compatible with the boundary conditions ($\Sigma(0) = \Sigma(1) = 0$) have the form

$$\underline{U} = \begin{pmatrix} \psi \\ \theta \\ \Sigma \end{pmatrix} = \begin{pmatrix} \alpha \\ \beta \\ \gamma \end{pmatrix} e^{ikx} \sin n\pi z. \tag{209}$$

The characteristic equation of this problem reads

$$s^3 + A_{k,n}s^2 + B_{k,n}s + C_{kn} = 0, \tag{210}$$

where

$$\begin{aligned} A_{k,n} &= a_{k,n}^2(\sigma + \tau + 1) \\ B_{k,n} &= a_{k,n}^4(\tau + \sigma + \sigma\tau) - \frac{\sigma k^2}{a_{k,n}^2}(R + \tau S) \\ C_{k,n} &= k^2\sigma\tau \left\{ \frac{a^6{}_{k,n}}{k^2} - (R + S) \right\} \end{aligned} \tag{211}$$

with

$$a_{k,n}^2 = k^2 + n^2\pi^2 . \tag{212}$$

For $B_{k,n} < 0$, when $C_{k,n} = A_{k,n} B_{k,n}$, one has an oscillatory instability which comes first at $k_c \neq 0$ and $n = 1$. One then looks for a solution of (206) under the form

$$\underline{U} = A_{k_c}\phi_{k_c} + A_{-k_c}\underline{\phi}_{-k_c} + \bar{A}_{k_c}\bar{\underline{\phi}}_{k_c} + \bar{A}_{-k_c}\bar{\underline{\phi}}_{-k_c} + \underline{W}(A_{\pm k_c}, \bar{A}_{\pm k_c}, \mu) \tag{213}$$

where $\underline{\phi}_{+k_c}$, $\underline{\phi}_{-k_c}$ are the marginal modes associated with the wavenumbers k_c and $-k_c$ respectively and the frequency $i\omega$. μ measures a small deviation from the marginal situation and $A_{\pm k_c}$ are assumed to obey the following amplitude equations

$$\partial_t A_{k_c} = i\omega A_{k_c} + F_{k_c}(A_{\pm k_c}, \bar{A}_{\pm k_c}, \mu)$$
$$\partial_t A_{-k_c} = i\omega A_{-k_c} + F_{-k_c}(A_{\pm k_c}, \bar{A}_{\pm k_c}, \mu). \tag{214}$$

The only coupling terms found in these equations are of the form $|A_{k_c}|^2 A_{k_c}$ and $|A_{-k_c}|^2 A_{k_c}$ in the first equation, and $|A_{k_c}|^2 A_{-k_c}$ and $|A_{-k_c}|^2 A_{-k_c}$ in the second. When one selects the spatial phase of the system of oscillating rolls, in prescribing the fields on the lateral boundaries, these equations reduce to the complex Landau equation already discussed. One describes here a wave phenomenon. In finite "boxes" we only have stationary waves.

In extended systems phase instability can occur. The lateral (Eckauss) phase gradient is described at the leading order by a perturbation of the Korteweg-de Vries equation [22].

3.4. Interactions between instabilities

Thermosolutal convection also provides a typical example of interaction between instabilities [10, 38]. In the case $C_{k,n} = 0$ one has stationary convection. We now look for parameter values for which oscillatory and stationary convection both occur with the same wave number (k,n). One gets the following conditions

$$R = R_c(k,n) = -\frac{a_{k,n}^6}{k^2}\,\frac{\sigma+\tau}{\sigma(\tau-1)}$$
$$S = S_c(k,n) = \frac{a_{k,n}^6}{k^2}\,\frac{\tau(\sigma+1)}{\sigma(\tau-1)}. \tag{215}$$

In the parameter space R, T this point corresponds to the intersection of the line where the stationary convection comes first and the line where oscillatory convection first appears. We have already seen an example of multiple instability associated with two zero eigenvalues. In the present case the nature of the problem is quite different.

In order to analyze the typical nonlinear behavior associated with this competition between instabilities let us consider a simple example. A differential equation which combines both Hopf bifurcation and stationary bifurcation is given by

$$\ddot{x} + (x-\nu)\dot{x} - (x+\mu)x = 0. \tag{216}$$

It is convenient to rewrite (216) as a first order system

$$\dot{x} = y, \quad \dot{y} = -(x-\nu)y + (x+\mu)x. \tag{217}$$

When μ, $\nu = 0$ its associated linear operator L has two zero eigenvalues,

$$L = \begin{pmatrix} 0 & 1 \\ \mu & \nu \end{pmatrix}. \tag{218}$$

When $\nu = 0$ (for $\mu < 0$) one has an oscillatory instability with the frequency $\omega = i\sqrt{-\mu}$. When $\mu = 0$ one has a stationary pitchfork bifurcation. In the parameter space (μ, ν) the two bifurcation sets intersect at $\mu = \nu = 0$ (Fig. 12). At this point the linear operator reads

$$L = \begin{pmatrix} 0 & 1 \\ 0 & 0 \end{pmatrix}. \tag{219}$$

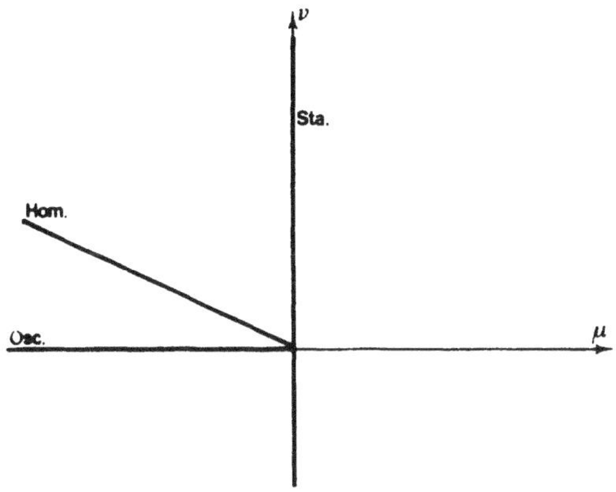

Figure 12

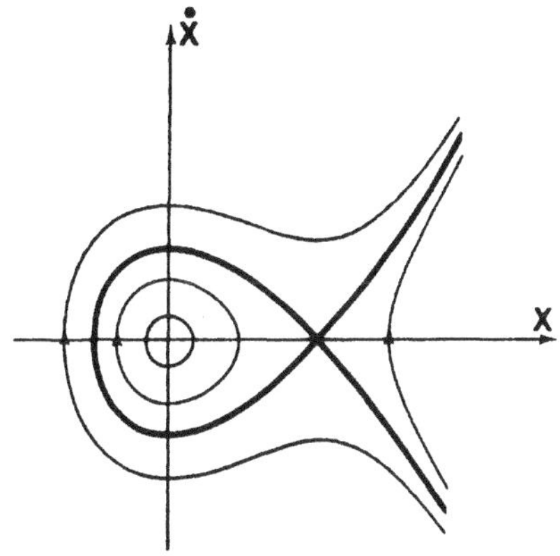

Figure 13

To study (216) it is convenient to introduce the following scaled quantities

$$x = \varepsilon^2\tilde{x}, \quad t = \varepsilon^{-1}\tilde{t}, \quad \nu = \varepsilon^2\tilde{\nu}, \quad \mu = \varepsilon^2\tilde{\mu} \,. \tag{220}$$

Equation (216) then reads

$$\ddot{x} - \mu x - x^2 + \varepsilon(x - \nu)\dot{x} = 0. \tag{221}$$

When $\varepsilon \to 0$ (221) reduces itself to a simple Hamiltonian system

$$\ddot{x} - \mu x - x^2 = 0$$

whose phase portrait is sketched in Fig 12. The dissipative term $\varepsilon(x - \nu)\dot{x}$ in (221) selects some orbits among an infinite number present in the Hamiltonian system [7, 30]. The most typical phenomena in this interaction between instabilities is the existence in the $\mu - \nu$ plane of a line which starts from the origin on which the selected orbit corresponds to the separatrix of the Hamiltonian system. On this codimension 1 set the system (216) experiences a global bifurcation for a periodic solution. This kind of global bifurcation called homoclinic bifurcation is not expected to occur near the appearance of a periodic behavior. It is a higher nonlinear effect. Nevertheless, near a codimension two bifurcation of the type (219), it naturally arises. When one crosses this bifurcation-line the period of the periodic solution diverges logarithmically.

We now comment on the difference between the two problems of competition between instabilities considered up to now. They both correspond to a double zero eigenvalue. In the first case (118) one easily identifies two independent modes with zero growth rate and the corresponding restriction of the linear operator reads

$$J = \begin{pmatrix} 0 & 0 \\ 0 & 0 \end{pmatrix} \quad . \tag{223}$$

This leads to the two coupled first order differential equations already given (118). In the second case the linear matrix reads

$$J = \begin{pmatrix} 0 & 1 \\ 0 & 0 \end{pmatrix} \tag{224}$$

which is an irreducible Jordan block. One has then only one marginal mode whose linear dynamics is second order in time. More generally [39, 40] it is convenient to represent a multiple instability by a Jordan matrix. For example in the case of three zero eigenvalues one can have the three following different linear operators

$$\begin{pmatrix} 0 & 0 & 0 \\ 0 & 0 & 0 \\ 0 & 0 & 0 \end{pmatrix} = 3\xi \quad \begin{pmatrix} 0 & 1 & 0 \\ 0 & 0 & 0 \\ 0 & 0 & 0 \end{pmatrix} = \xi^2\xi \quad \begin{pmatrix} 0 & 1 & 0 \\ 0 & 0 & 1 \\ 0 & 0 & 0 \end{pmatrix} = \xi^3 . \qquad (225)$$

Different type of situations can be considered. The singularity $\xi\omega$ for example corresponds to the matrix

$$\begin{pmatrix} 0 & 0 & 0 \\ 0 & i\omega & 0 \\ 0 & 0 & -i\omega \end{pmatrix} . \qquad (226)$$

The notation adopted follows at distance Arnold's notation [30] for singular matrices.

Each critical matrix is associated with a nonlinear equation which describes the behavior of a system near the corresponding polycritical point in the parameter space. Normal form approach [7] provides the best way to write down these equations. Let J represent a given matrix associated with such a multiple instability. The corresponding linear amplitude equation reads

$$\partial_t A = JA \qquad (227)$$

where A symbolically represents $A_1, A_2 \ldots A_p$. Let us define a new variable $\tilde{A}$ through a nonlinear near identity transformation

$$A = \tilde{A} + \phi(\tilde{A}) \qquad (228)$$

where ϕ is a homogeneous function of given degree in $\tilde{A}$. Then (227) becomes

$$\partial_t \tilde{A} = J\tilde{A} - \mathcal{L}\phi(\tilde{A}) + \text{h.o.t.} \qquad (229)$$

where $\mathcal{L} = \sum_{i,j} J_{ij} A_j \frac{\partial}{\partial A_i} - J$, called homological operator [30] , plays an important role in determining the structure of the nonlinear terms present in the normal form associated with J. Equation (228) transforms a linear system (227) into a nonlinear one (229). The nonlinear terms produced by such a change of variables represents nonlinear effects which can be removed in order to give a prototypical description of the competition between the instabilities. A procedure to compute the corresponding amplitude equation consists of looking for a solution of (31) under the form [10]

$$\underline{U} = \sum_{i=1}^{p} A_i \underline{\Phi}_i + \underline{W}(\{A_i\},\{\mu_i\}) \qquad (230)$$

together with an amplitude equation

$$\partial_t A_i = J_{ij} A_j + F_i(\{A_i\}, \{\mu_i\}) \qquad (231)$$

where $\mu = \mu_1 \ldots \mu_k$ represents small deviations from the polycritical point in the parameter space and $A = A_1 \ldots A_p$ are the amplitudes of the critical modes, $\underline{\Phi}_i$ are the critical modes defined by

$$L(\vec{\nabla}, \lambda_c)\underline{\Phi}_i = J_{ji}\underline{\Phi}_j , \qquad (232)$$

and λ_c is the polycritical point in the parameter space. We now expand $\underline{W}$ and F in the A_i and μ_i. At a given "order" $\{\alpha\} = \{\alpha_1, \ldots \alpha_p ; m_1, \ldots m_k)$ the equation (31) together with (230) and (231) reads

$$\mathcal{L}\underline{W}_{\{\alpha\}} = \underline{I}_{\{\alpha\}} - \sum_{i=1}^{p} F_{i,\{\alpha\}}\underline{\Phi}_i \qquad (233)$$

where

$$\mathcal{L} = \sum J_{ij} A_j \frac{\partial}{\partial A_i} - L(\vec{\nabla}, \lambda_c) \tag{234}$$

and $\underline{I}_{\{\alpha\}}$ is a known quantity at the order α. The orthogonality of the right hand side of (233) to the kernel of the adjoint of $\mathcal{L}$ determines, at any order $\{\alpha\}$ the normal form. The prescription adopted is that the $F_{i,\{\alpha\}}$ not involved in these solvability conditions have to be chosen zero.

In the case of the thermosolutal convection $J = \begin{pmatrix} 0 & 1 \\ 0 & 0 \end{pmatrix}$. Introducing two amplitudes A, B for the critical modes and using the procedure sketched below one gets the normal form

$$\partial_t A = B, \quad \partial_t B = -\nu B - \mu A - \alpha A^2 B - \beta A^3 \tag{235}$$

where μ, ν are linear quantities in $R-R_c$, $S-S_c$. This equation describes the nonlinear behavior of an arbitrary system near a polycritical point of ξ^2 type. The appearance of the symmetry $A \to -A$ in these equations comes from the fact that the critical modes have a finite wavelength k_c.

The chemical reactor also provides an example of critical point of the ξ^2 type. In that case the critcal modes have a null wave number. If (37) contains quadratic nonlinearities then the corresponding normal form reads

$$\partial_t A = B, \quad \partial_t B = -\nu B - \mu A - \alpha AB - \beta A^2 \tag{236}$$

where μ, ν depend linearly on R and T. We remark that (236) is nothing more than the simple model (216) already discussed.

The approach in terms of a homological operator allows us to write down the general form of the amplitude equation without tedious computations. Time periodic perturbations and stochastic perturbations of such situations can be studied in the same way [41].

4. Quasiperiodic phenomena

4.1. Poincaré Map

Coming back to our experiment in a small box in which the convection is now oscillatory, we increase the Rayleigh number. The first observation is a growth of harmonics. For a given value of the Rayleigh number a new frequency appears [42-45]. This phenomena has to be interpreted as an instability for the oscillatory regime. What we observe here is the beginning of the so-called Landau scenario for turbulence. Landau [6] expected an infinite sequence of such instabilities for higher values of the Rayleigh number leading to more and more independent frequencies. Actually, as predicted by Ruelle and Takens [46] such a sequence is not likely to be observed. At this Rayleigh number value it is very difficult to give a physical interpretation of the second oscillator. In high Prandtl number experiments it turns out to be an oscillation of the boundary layer [44]. In small Prandtl number cases this second oscillation has not been yet interpreted.

We now analyze this phenomena as a generic instability for periodic behavior. We first write the set of equations which describes the Rayleigh-Bénard convection under the following abstract form

$$\partial_t \underline{U} = \underline{\underline{L}}\,\underline{U} + \underline{N}(\underline{U},\underline{U}) \tag{237}$$

where $U = (\vec{V}, \theta)$. The velocity field $\vec{V}$ satisfied the incompressibility condition

$$\vec{\nabla} \cdot \vec{V} = 0. \tag{238}$$

The pressure field can be seen as a way to insure this constraint. For a certain range of the Rayleigh number, one knows from experiments that (237) has a stable periodic solution

$$\underline{U} = \underline{\Phi}(t) \quad \text{where} \quad \underline{\Phi}(t + T) = \underline{\Phi}(t). \tag{239}$$

The stability analysis of $\underline{\Phi}(t)$ proceeds as follows. One first writes the equation for an arbitrary perturbation $\underline{V}(t) = \underline{U}(t) - \Phi(t)$

$$\partial_t \underline{V} = \underline{\underline{M}}(t)V(t) + \underline{P}(t, \underline{V}, \underline{V}), \tag{240}$$

where

$$\underline{\underline{M}}(t + T) = \underline{\underline{M}}(t) = \underline{\underline{L}} + \frac{\partial \underline{N}}{\partial \underline{U}}(\underline{U}, \underline{U})\Big|_{\underline{U} = \underline{\Phi}(t)} \tag{241}$$

$$\underline{P}(t; \underline{V}, \underline{V}) = \underline{N}(\underline{\Phi} + \underline{V}, \underline{\Phi} + \underline{V}) - N(\underline{\Phi}, \underline{\Phi}) - \frac{\partial \underline{N}}{\partial \underline{U}}(\underline{U}, \underline{U})\Big|_{\underline{U} = \underline{\Phi}(t)} \quad . \tag{242}$$

For small perturbations one only considers the linear version of this equation

$$\partial_t V(t) = M(t)V(t) \tag{243}$$

whose formal solution is given in terms of an evolution operator $S(t)$

$$V(t) = S(t)V(0). \tag{244}$$

We are interested in the eigenvalues of $S(T)$. If $S(T)$ has all its eigenvalues inside the unit circle the periodic solution $\underline{\Phi}(t)$ is linearly (asymptotically) stable. Instability occurs when, varying an external parameter in (237) (Rayleigh number for example) some of its eigenvalues cross the unit circle. The eigenvalues of $S(T)$ are called the Floquet multipliers. They represent the quantity by which is multiplied a small perturbation of the periodic solution after one period T. It is convenient to define the logarithm of $S(T)$

$$S(T) = e^{\Lambda T} \tag{245}$$

whose eigenvalues are the Floquet exponents. The general solution of (243) has then the following Floquet form [47]

$$V(t) = Q(t)e^{\Lambda t}V(0) \tag{246}$$

where

$$Q(t + T) = Q(t) \equiv S(t)e^{-\Lambda t}. \tag{247}$$

If one introduces a new variable W by

$$V = Q(t)W, \tag{248}$$

(233) becomes an autonomous equation

$$\partial_t W = \Lambda W. \tag{249}$$

We are now able in principle to analyze the various instabilities of a periodic solution since we are back to an autonomous system. Actually we do not follow this line since in most cases one is not able to get analytically $Q(t)$ and Λ. The best way from a qualitative point of view to study the stability and the bifurcations of a periodic solution consists of introducing the notion of Poincaré map. In the simplest case it is the so-called stroboscopic map [47]. The formal integration of (240) leads to

$$\underline{V}(t) = \underline{\Phi}(t, \underline{V}(0)). \tag{250}$$

One defines the stroboscopic map as

$$\underline{V}(T) = \underline{\Phi}(T, \underline{V}(0)) \equiv \underline{F}(V(0)) \tag{251}$$

whose successive iterates give the perturbation to $\underline{\Phi}$ at successive times $\tau_n = nT$

$$\underline{V}_{i+1} = \underline{F}(\underline{V}_i), \tag{252}$$

where $\underline{V}_i = \underline{V}(iT)$. The inconvenience in studying this mapping is that $\Phi(t)$ becomes a circle of fixed points, each fixed point corresponding to a different phase of the periodic solution. One is always a Floquet multiplier of the linear part $S(T)$ of F associated with the marginal phase mode $\Phi'(t)$. One avoids this difficulty by defining a Poincaré map directly on the initial equations (237). One first defines the phase space associated with (237). In the case of a partial differential equation describing a given physical system in a small box, one can think of it as the space generated by the coefficients of the expansion of $\underline{U}$ in terms of the eigenvectors of the linear operator $\underline{L}$. In the case of an experiment where one measures only one quantity A(t) a phase space can be reconstructed from it. Introducing a given delay τ, the space generated by A(t), A(t-τ) A(t-2τ)... is a good candidate to represent such a space [48]. In the phase space the differential equation can be seen as a fluid flow. The flow lines are everywhere tangent to the vector field given by the right hand side of the differential equation. In the Poincaré method one selects a given codimension surface Σ (generally a hypersurface) transverse to the flow. One then defines a mapping on this surface by taking an initial condition on it and following the flow (integration of the equations when they are known) until the next intersection, with the same sign of the velocity with respect to the vector normal to the surface. Figure 14 represents a sketch of such a condition. This defines a local map on S

$$\underline{X}' = \underline{P}(\underline{X}). \tag{253}$$

An important property of this map is its invertibility. A periodic solution $\underline{\Phi}(t)$ of (237) now corresponds to a unique fixed point $\underline{X}^*$ of (253)

$$\underline{X}^* = \underline{P}(X^*). \tag{254}$$

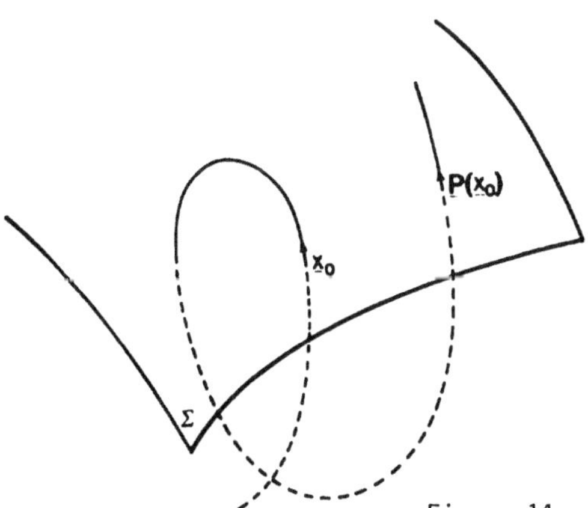

Figure 14

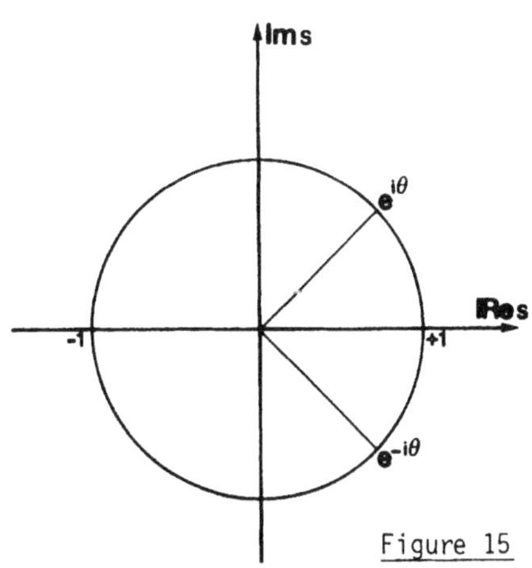

Figure 15

In order to study the stability of this fixed point, one introduces a perturbation $\underline{x}$ to the fixed point X^*,

$$\underline{X} = \underline{X}^* + \underline{x}. \tag{255}$$

Equation (254) then reads

$$\underline{x}' = \underline{\underline{\ell}}\underline{x} + \underline{n}(\underline{x}, \underline{x}). \tag{256}$$

For small perturbations one studies the linear map

$$\underline{x}' = \underline{\underline{\ell}}\underline{x}. \tag{257}$$

The eigenvalues s of $\underline{\underline{\ell}}$ are the Floquet multipliers of (243) except the multiplier one. When some of these multipliers crose the unit circle instability occurs. Typically there are three generic bifurcations for periodic solutions [49] as sketched on Fig. 15. A crossing through +1 corresponds, in the generic case, to the disappearance of the periodic solution (saddle-mode bifurcation). A crossing through -1 corresponds in the supercritical case to the appearance of a stable periodic solution with double frequency (period doubling bifurcation). A crossing through $\exp(\pm i\theta)$ ($\theta \neq 0, \frac{\pi}{2}, \frac{2\pi}{3}, \pi$) corresponds in the supercritical case to the appearance of a quasiperiodic solution with two frequencies $\tilde{\omega}_1 = \frac{2\pi}{T}$, $\omega_2 = \frac{\theta}{2\pi}\omega_1$. The cases $\theta = 0, \frac{\pi}{2}, \frac{2\pi}{3}, \pi$ have to be considered separately [49, 30] (strong resonances).

4.2. Normal form analysis

In order to study these bifurcations it is convenient to develop the same kind of approach as in the differential equation case. Near a bifurcation point the central manifold techniques allow as to reduce (256) to a finite dimensional mapping. Let J be one of the three Jordan matrices (1), (-1), $\begin{pmatrix} e^{i\theta} & 0 \\ 0 & e^{-i\theta} \end{pmatrix}$ and introduce the amplitudes A of the corresponding critical modes. On the central manifold, the linear map reads

$$\underline{A}' = JA. \tag{258}$$

If we perform a nonlinear near-identity change of variable defined by

$$A = \tilde{A} + \Phi(\tilde{A}), \tag{259}$$

where $\Phi(\tilde{A})$ is a homogeneous polynomial of given degree in $\tilde{A}$, (258) is transformed into a nonlinear map

$$\tilde{A}' = J\tilde{A} - \mathcal{L}\Phi(\tilde{A}) + \text{h.o.t.} \tag{260}$$

where $\mathcal{L}$ is the homological operator associated with maps

$$\mathcal{L}\Phi(A) = \Phi(JA) - J\Phi(A). \tag{261}$$

Again its structure determines the resonant terms and one can define a procedure quite analogous to the one for the differential equation in order to reduce (256) to its normal form. The resulting normal forms are, in the case of a +1 crossing

$$A' = \mu + \alpha A^2, \tag{262}$$

in the case of a -1 crossing,

$$A' = (-1 + \mu)A + \alpha A^3 \tag{263}$$

and in the case of a $e^{\pm i\theta}$ crossing ($\theta \neq 0$, $\pi/2$, $2\pi/3$, π),

$$A' = (1+\mu)e^{i\theta}A + \alpha|A|^2A, \qquad (264)$$

where A is a complex amplitude and α a complex number. The analysis of these maps is straightforward. In the -1 case one has to find the fixed points of the second iterate of (263) (period two orbits for (263)). In the case $e^{\pm i\theta}$ one introduces as usual the polar coordinates

$$A = \rho e^{i\phi} . \qquad (265)$$

At the leading order (264) reads

$$\rho' = (1+\mu)\rho + \mathcal{R}e(\alpha e^{-i\theta})\rho^3 \qquad (266.a)$$

$$\phi' = \phi + \theta + \mathrm{Im}(\alpha e^{-i\theta})\rho^2. \qquad (266.b)$$

Equation (266.a) has a fixed point

$$\rho^* = \left(\frac{-\mu}{\mathcal{R}e(\alpha e^{-i\theta})}\right)^{1/2} \qquad (267)$$

which is stable in the supercritical case ($\mathcal{R}e(\alpha e^{-i\theta}) < 0$). One then has an invariant circle for the Poincaré map, which has to be interpreted as an invariant torus for the corresponding flow. The dynamics on this circle is given by

$$\phi' = \phi + \theta_{eff} \quad \text{where} \qquad (268)$$

$$\theta_{eff} = \theta + \frac{\mathrm{Im}(\alpha e^{-i\theta})}{\mathcal{R}e(\alpha e^{-i\theta})}\mu . \qquad (269)$$

Equation (269) describes a pure rotation on the circle. The corresponding behavior on the torus is a winding motion with two independent frequencies $\omega_1 = 2\pi/T$, $\omega_2 = \omega_1(\theta_{eff}/2\pi)$. We then have described the following phenomena. For a given value of an external parameter (Rayleigh number in the case of convection) a periodic solution loses its stability with two complex Floquet multipliers $\exp(\pm i\theta)$ ($\theta \neq 0$, $\pi/2$, $2\pi/3$, π). Equation (264) or (266) describes the nonlinear development of this instability. In the supercritical case one observes the appearance of a second frequency with a finite amplitude. Near the instability point ($\mu = 0$) the system behaves like two independent oscillators.

The circle map (266.b) has a very simple "stroboscopic interpretation". One can use one of the oscillators as a clock to observe the second one. Let us consider for a moment a very simple electromechanical analog: the classical stroboscopic experiment. A disc is rotating at a frequency $\omega_1 = 2\pi/T$. One uses a periodic signal with frequency $\omega_2 = 2\pi/\tau$ to illuminate periodically a dot on this rotating disc. In our intepretation the angular position of the dot corresponds to the absolute phase of the first oscillator. The successive angular positions of the the dot are then given by iteration of the circle map

$$\phi' = \phi + 2\pi\frac{\tau}{T_2} \equiv \phi + \nu. \qquad (270)$$

Once can also use the first oscillator to stroboscope the second and get a mapping analogous to (270) with $\nu = 2\pi(T_2/\tau)$.

4.3. Frequency locking phenomena

In the Rayleigh-Bénard experiment when the Rayleigh number is increased further, one first observes a nonlinear interaction between the two oscillators. This phenomena is clearly seen on the Fourier analysis of a typical signal by the presence of harmonics of the form $n_1 \omega_1 + n_2\omega_2$. Increasing again the Rayleigh number the frequency locking phenomena occurs. We first describe it with the help of the circle map. In the convective experiment the two oscillators have very distinct behaviors. At Rayleigh numbers where the second frequency apperars, the first oscillator is very strong and quite insensitive to the variation of the Rayleigh number. This continues to be true in the range where the frequency locking is observed. For these reasons we represented it by the rotating disk in our mechanical analog. The second oscillator near its threshold is weak and reacts nonlinearly with the first as the Rayleigh number increases. We model this interaction in introducing a feedback effect in the stroboscopic experiment [50]. The successive time flashes are now going to depend on the angular position of the observed dot. The simplest law compatible with the periodic character of the phase $\phi(t)$ is the following

$$\tau(\tau) = \tau + f\sin\phi \tag{271}$$

where τ measures the importance of the feedback effect. The successive angular positions are now obtained by iterating the following circle map

$$\phi' = \phi + \nu + \varepsilon\sin\phi \tag{272}$$

where $\nu = 2\pi(\tau/T)$ and $\varepsilon = 2\pi f/T$. The map known as the Arnold map [51] contains many of the typical phenomena associated with the nonlinear interaction between two oscillators. The connection between (272) and the initial differential equations (237) is the following. For the Rayleigh number values where the convective rolls behave quasiperiodically,the corresponding behavior in the phase space is a winding motion on a two-dimensional torus. The trace of this torus on a typical Poincaré section is a closed curve. With the help of appropriate coordinates, this curve is a circle. The dynamics on this circle is necessarily described by some invertible map. For $\varepsilon < 1$ (small feedback effects)(272) is a good candidate to represent such a map. More generally one can consider maps of the form

$$\phi' = f(\phi) = \phi + \nu + \varepsilon g(\phi) \tag{273}$$

where g is a periodic function of $\phi, g(\phi + 2\pi) = g(\phi)$,which is supposed, for technical reasons [52], to be twice differentiable. ε is supposed to be small enough in order to ensure the invertibility of $f(\phi)$.

The main tool used to study diffeomorphisms of the circle [52] (differentiable and invertible map of the circle) is the rotation number. Its main interest lies in the fact that it is an invariant of conjugation. By this one means that this number does not depend on the system of coordinates chosen on the circle. It is defined as

$$\rho_f = \lim_{k\to\infty} \frac{f^k(\phi_0) - \phi_0}{k} . \tag{274}$$

For a diffeomorphism of the circle ($\phi < 1$ in (272) for example), it does not depend on the initial angle ϕ_0. It describes the mean value of the angular advance. When ρ_f is a *rational* number p/q we have on the circle periodic orbits of period q. These values of the rotation number correspond to the frequency locking phenomena. It is a structurally stable phenomenon in the sense that if $\rho_f = p/q$ for a given value $R_{p/q}$ of the external parameters (Rayleigh number for example), it will be constant in general and equal to p/q for a finite range of parameters around $R_{p/q}$. The existence of a periodic orbit of period q corresponds to the existence of q fixed points for the qth iterate of f

$$f^q(\phi) = \phi. \tag{275}$$

The existence of roots for (275) is a stable phenomena since f is a differentiable function. When ρ_f is *irrational*, the map is conjugate to a pure rotation map with angle ρ_f. This means that there exists an invertible change of coordinate on the circle $\tilde{\phi} = h(\phi)$ such that in the new coordinate system (273) simply reads

$$\tilde{\phi}' = \tilde{\phi} + \rho_f. \tag{276}$$

For the parameter value where this happens the map describes the Poincaré map of two independent oscillators with irrationally related frequencies.

It is convenient to represent the different dynamical behaviors associated with (272) (or (273)) in an ε- ν diagram (Fig. 16). For $\varepsilon = 0$ we have a pure rotation with an angle ν. The quasiperiodic behavior corresponds to irrational values of ν. Hence, when the oscillators are uncoupled this phenomenon is observed on a set of parameters of Lebesgue measure equal to one. When ε increases at each rational value of $\nu = p/q$ a tongue starts in which the rotation number is constant and equal to p/q. These tongues do not overlap as far as (273) is invertible. The quasiperiodic behaviors are located between these tongues. For ε small the values of $\nu(\varepsilon)$ where these quasiperiodic behavior occur have a Cantor set structure with positive Lebesque measure (< 1) [53].

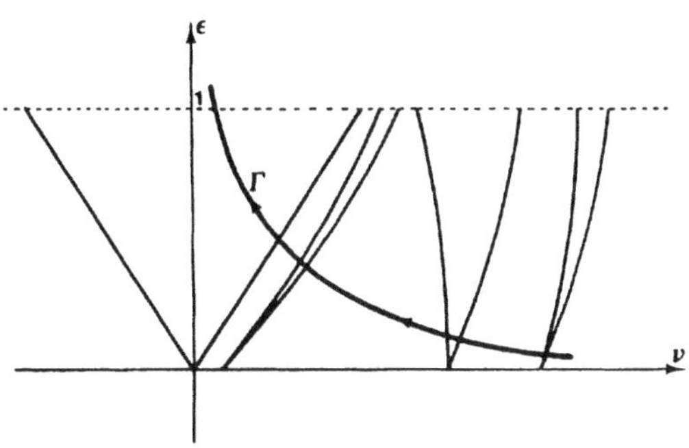

Figure 16

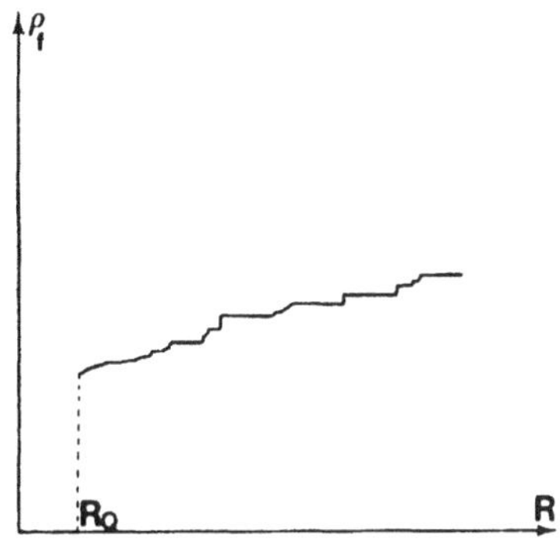

Figure 17

One can model the Rayleigh-Bénard experiment by drawing in this ε - ν diagram a typical path $\Gamma(R)$ parametrized by the Rayleigh number. One expects ε (feedback effect) to increase when the Rayleigh number increases. Figure 17 is a sketch of the graph of the rotation number against the Rayleigh number in following Γ. The experimental interpretation of the Rayleigh number is nothing else but the effective ratio of the frequencies of the two oscillators $2\pi\omega_1/\omega_2$. R_0 represents the value of the Rayleigh number where the second frequency appears. Typical aspects of this curve known as a "Devil staircase" are observed in the Rayleigh-Bénard experiments: a succession of locked states with simple arithmetic properties, and quasiperiodic states in between [54].

5. Chaotic phenomena

5.1. Chaotic behaviors in a typical experiment

We now increase the Rayleigh number further. What we first observe is a period doubling bifurcation occurring on a given locked state with a frequency Ω. On the Fourier spectra one then identifies the presence of a subharmonic at frequency $\Omega/2$ and its harmonics. Its amplitude is much smaller than the amplitude of the fundamental frequency Ω. When one increases again the Rayleigh number the subharmonic grows in amplitude and then a second period doubling bifurcation takes place and is followed by a complete cascade of such events [55, 56]. It is experimentally observed [57] that the successive values of the Rayleigh number follow asymptotically a geometrical law with a rate $\delta \simeq 4.4$. This number will be called bifurcation rate. At the end of the cascade the Fourier analysis of the signal presents peaks at all frequencies $\Omega/2^n$. One also finds that the ratio between the amplitudes of the peaks at the frequency $\Omega/2^n$ and at the frequency $\Omega/2^{n+1}$ is roughly (for n large enough) constant and equal to 5 [57]. When one crosses the accumulation point of the cascade, aperiodic behaviors take place in the convective signal. A weak noise which increases with the Rayleigh number is observed in the Fourier spectrum. It successively washes out the subharmonics peaks at frequencies $\Omega/2^n$ with n decreasing. This phenomena is called the reverse cascade. There exists a definite value of the Rayleigh number where a subharmonic at frequency $\Omega/2^n$ dissappears in the noise [57]. The sequence of Rayleigh numbers where this phenomenon is observed also follows a geometrical law with a rate δ' which is equal to δ. When one increases further the Rayleigh number in the so-called chaotic regime, new phenomena occur. For a finite range of this number, the aperiodic behavior disappears and the convective signal becomes periodic with a frequency Ω/p $(p > 1)$. Several periodic windows are experimentally observed. Their width decreases with p increasing. The largest one is observed for $p = 3$. There is a well-defined way to get in a periodic window and to get out from it. When the Rayleigh number is just above the value where the periodic behavior occurs, one observes the so called intermittency phenomenon. The signal presents pieces of periodic behavior and pieces of chaotic bursts. The duration of the periodic pieces increases when one approaches the periodic window. One gets out from a periodic window by a cascade of period doubling bifurcations in the same way that the fundamental frequency Ω goes to chaos, with the same bifurcation rate δ. The corresponing reverse cascade is also observed.

5.2. Non invertible circle maps

We now have to interpret all these phenomena. Again we are going to do it in its simplest prototypical form. There is no simple way to associate a normal form to them since the cascade of period doublings by itself is not a local phenomena. Renormalization group techniques will appear as a powerful approach to this problem. For the moment we just increase R following the path Γ in the stroboscopic experiment. We first remark that when ε is greater than 1 (too large feedback) (272) is no longer an honest Poincaré map for a differential system like (237) since then it is not invertible in that range of parameters. There exists a simple way [58] to cure this difficulty, for example by replacing (272) by tne following map of the annulus into itself

$$\phi' = \phi + \nu + \varepsilon\sin\phi + \beta\rho, \qquad \rho' = \alpha(\rho + \sin\phi) \tag{277}$$

where $\rho = r - 2/3$ with $1/3 < r < 1$. When $\alpha = 0$ we have an infinite contraction on the circle of radius $r = 2/3$. The dynamics on this circle is then described by (272). Any small α guarantee the invertibility of (277). On a pure heuristic level one then expects that the qualitative dynamics of (272) even when ε becomes greater than one will have something to do with those of (277) which is now a good candidate to be a Poincaré map of equations like (237). Actually this feeling is quite well supported by numerical simultations both done with (272) and (277) [58].

Let us then choose a particular path Γ as sketched on Fig. 18. For $\varepsilon > \varepsilon_0$ with $\varepsilon_0 < 1$, (272) becomes

$$\phi' = \phi + \varepsilon\sin\theta. \tag{278}$$

The scenario we describe now actually occurs in each tongue when one follows a path qualitatively analogous to $\Gamma(R)$. Figure 19 represents a sketch of the graph of $\phi'(\phi)$ for a sequence of increasing values of ε. When ε increases a direct numerical simulation successively shows

- a period doubling bifurcation when $\varepsilon = 2$ which leads to a stable periodic orbit with period two. On a real simulation of the stroboscopic experiment one would see alternatively two dots symmetric with respect to π

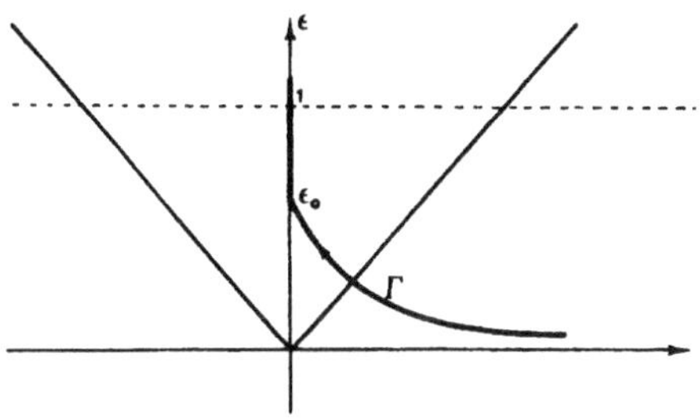

Figure 18

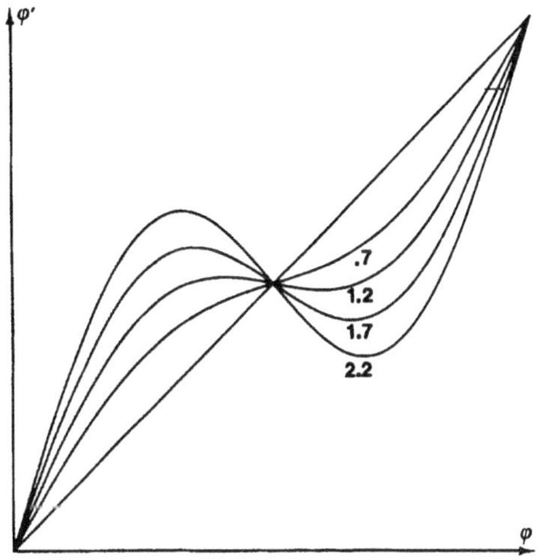

Figure 19

- a symmetry-breaking bifurcation which leads to two distinct, no longer symmetric with respect to π, periodic orbits with period two
- a complete sequence of period doubling bifurcations characterized by a sequence of parameter values which follows the asymptotic law

$$\frac{\varepsilon_{n+1} - \varepsilon_n}{\varepsilon_n - \varepsilon_{n-1}} = (4.669\ldots)^{-1} \tag{279}$$

- the appearance of aperiodic behaviors. In the parameter range where these chaotic behaviors occur, one finds finite windows with periodic behaviors. We are not going to analyze this scenario with the help of (278) since the only important ingredient is its non-invertible character.

5.3. Maps on an interval

The simplest non-invertible map one can consider reads [59]

$$X' = 1 - \mu X^2 \equiv f_\mu(X). \tag{280}$$

Again (280) can be seen as the infinite contraction limit of an invertible map: the Henon map [60]

$$X' = 1 - \mu X^2 - bY, \quad Y' = X. \tag{281}$$

When b goes to 0 (282) becomes (281). One can easily check that when $\mu \in [0,2]$ the interval $I = [X_\mu^+, X_\mu]$, where X_μ^+ is one of the fixed point of f_μ ($f_\mu(X_\mu) = X_\mu$, $X_\mu < 0$), is a trapping region, i.e.

$$f_\mu(I) \subset I. \tag{282}$$

When $\mu = 0$, $I = \mathbb{R}$ and all the initial data converges in one iteration toward the unique fixed point of $f_0(X)$, $X_0^* = 1$

When $\mu = 2$ the dynamics associated with (280) is chaotic. In order to describe the phenomena observed at this parameter value it is convenient to introduce a change of coordinate [61] defined by

$$X = \cos\theta. \tag{283}$$

(280) then reads

$$\theta' = 2\theta , \ \theta < M/2 \ ; \quad \theta' = 2(1-\theta), \quad \theta > M/2. \tag{284}$$

This transformation is known as the Baker transformation. It possesses an invariant measure: the Lebesque measure which is characterized by strong mixing properties. Any small deviation from an initial condition diverges with a factor two at each iteration. The exponent of the exponential splitting of two nearby trajectories known as the Liapunov exponent [7] in then $\ell n2$, more generally defined as the average logarithm of the slope on a given orbit of the map X_0, $X_1 = f(X_0)$, $X_2 = f^2(X_0)\ldots$

$$\lambda = \lim_{n\to\infty} \frac{1}{n} \sum_{i=1}^{n} \ell n f'(X_i). \tag{285}$$

It is constant and equal to $\ell n2$ for all orbits generated by the map (284). This map possesses $N_n = 2^n$ periodic points with period n defined by

$$f^{(n)}(X_i^{(n)}) = X_i^{(n)} . \tag{286}$$

Its topological entropy [7], defined as

$$h_T(f) = \lim_{n\to\infty} \frac{1}{n} \ell n N_n, \tag{287}$$

where N_n represents the number of periodic points with period n, is

$$h_T = \ell n2. \tag{288}$$

This number which measures in some sense the topological complexity of the map is an invariant of conjugation as the rotation number was for the circle maps.

Many phenomena occurred for $\mu \in [0,2]$: saddle mode and period doubling bifurcation, a transition from zero topological entropy to a positive one, known as the transition to chaos. In order to analyze these phenomena, one first introduces the piecewise linear map [62]

$$X' = 1 - \mu|X| \equiv \ell_\mu(X) \tag{289}$$

which is equivalent to f for N = 2. Its main advantage lies in its piecewise character. Chaotic behaviors arise when $\mu \geqslant 1$ since then (289) has the property of sensitivity upon initial conditions. Its Liapunov exponent is simply

$$\lambda = 0 \text{ for } \mu < 1, \quad \lambda = \ell n(1+\nu) \quad \text{for } \mu = 1 + \nu \geqslant 1. \tag{290}$$

We now describe the mechanism for the appearance of periodic points. As a simple example one considers the case of a period three orbit. It is first important to remark that, when a periodic orbit appears, 0 is always one of its periodic points as illustrated in Fig. 20 in the period three case. Using this remark we only need to consider the successive iterates of 0. For the period three case $\ell_\mu^{(3)}(0) = 0$ leads to the equation

$$1 + \mu - \mu^2 = 0 \tag{291}$$

whose solution for $1 < \mu < 2$ is the golden mean $(1+\sqrt{5})/2$. It is convenient to rewrite this equation in the following form

$$\mu = 1 + \frac{1}{\mu} . \tag{292}$$

More generally the equation for the parameter value where an arbitrary periodic orbit with period μ appears reads

$$\mu = 1 + \frac{1}{\mu} \pm \frac{1}{\mu^2} \pm \frac{1}{\mu^3} \ldots \pm \frac{1}{\mu^{\mu-2}} . \tag{293}$$

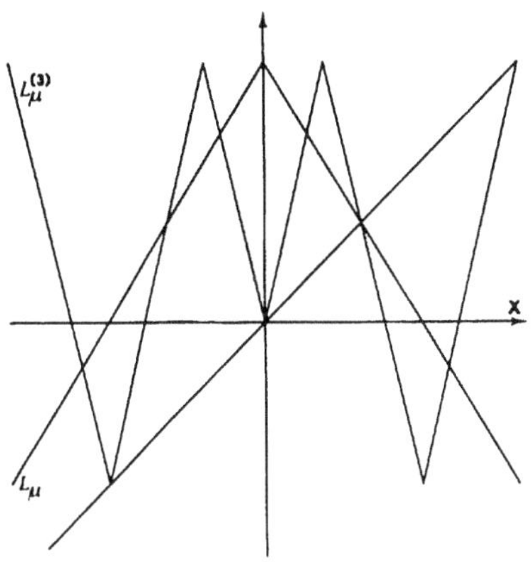

Figure 20

We remark that this equation can be interpreted as an expansion of μ in its own basis (autoexpansion) [62]. One can now organize the whole bifurcation set when $1 < \mu < 2$. Taking a given μ, one computes its autoexpansion

$$\mu = 1 + \frac{1}{\mu} + \ldots \tag{294}$$

-if $1 + \frac{1}{\mu} > \mu$ one writes $\mu = 1 + \frac{1}{\mu} - \frac{1}{\mu^2} + \ldots$ and continues

-if $1 + \frac{1}{\mu} < \mu$ one writes $\mu = 1 + \frac{1}{\mu} + \frac{1}{\mu^2} + \ldots$ and continues

-if $1 + \frac{1}{\mu} = \mu$, μ is equal to the golden mean and a period three orbit appears at this parameter value

When this autoexpansion contains N-2 terms the value of μ corresponds to the appearance of a periodic orbit of period N. The values of μ having a finite autodecomposition are dense on the interval [1,2] but have a Lebesgue measure zero (they are the analogues of the rationals on the interval). Otherwise speaking in general a value of μ will have an infinite autodecomposition.

We now have to discuss what we lost in replacing (280) by (289). It is convenient to introduce the map

$$X'' = 1 - \mu |X|^{1+\varepsilon} \equiv \ell_{\mu,\varepsilon}(X) \tag{294}$$

which realizes a smooth interpolation between f_μ and ℓ_μ. The basic change when one goes from $\varepsilon = 0$ to $\varepsilon \neq 0$ is illustrated in Fig. 21. In the $\varepsilon = 0$ case, the periodic points appear unstable, while in the $\varepsilon \neq 0$ case one has a pair of periodic orbits, one stable, the other unstable. When μ is close enough to the saddle mode bifurcation point, the dynamics associated with the map are locally described by a map in the form of f_μ. This selfsimilarity in the behavior increases the complexity of the bifurcation phenomena by the presence of periodic windows in the chaotic regimes followed by a cascade of period doubling bifurcations.

It is convenient to analyze the period doubling phenomena on the period one orbit. This occurs for $\mu = \mu_1 = 3/4$ when $\left.\frac{\partial f_\mu}{\partial X}\right|_{X=X_\mu^+} = -1$ (Fig. 22). The orbit of period two which appears corresponds to two fixed points for the second iterate $f_\mu^{(2)}$ of f_μ. More striking, two pieces of $f^{(2)}$ look like f. For μ close enough to μ_1 they have two independent trapping domains I_- and I_+ which exchange under f_μ.

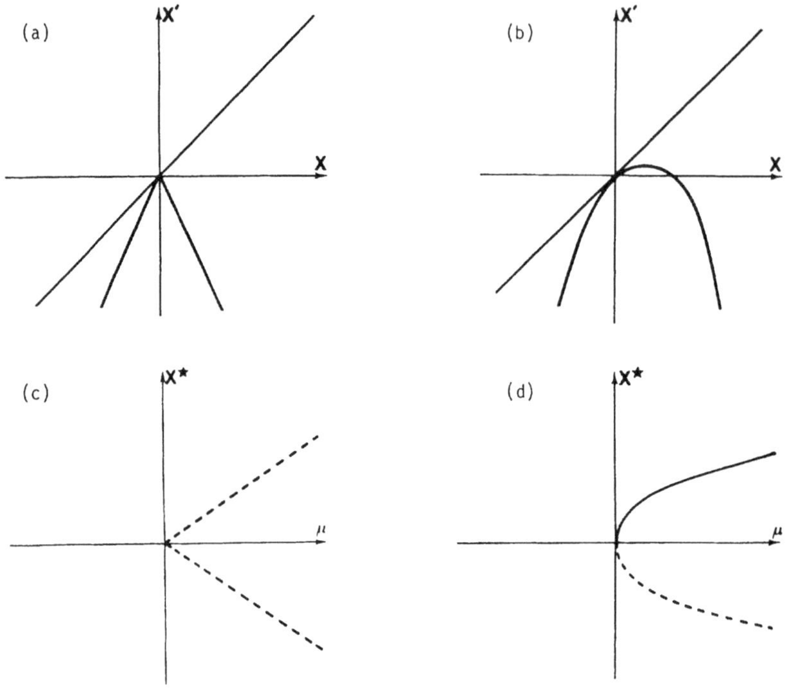

Figure 21

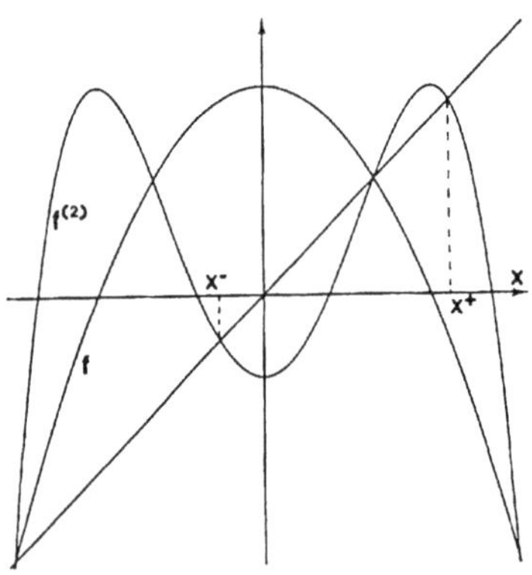

Figure 22

This similarity between the graph of f_μ and two pieces of the graph of $f_\mu^{(2)}$ is clearly the origin of the cascading process. Increasing μ a new period doubling occurs for some $\mu = \mu_2$ leading to a period four orbit. The restriction of $f_\mu^{(4)}$ in I_+ and I_- again presents some analogy with $f_\mu^{(2)}$. We get a complete sequence μ_n of parameter values where the period doubling bifurcations take place. This sequence has the following asymptotic property:

$$\frac{\mu_n - \mu_{n-1}}{\mu_{n+1} - \mu_n} \xrightarrow[n \to \infty]{} 4.669\ldots$$

This number appears numerically to be universal in the sense that it is the same for a whole class of maps. We already mentioned this asymptotic number in the context of the circle map. This kind of universality is reminiscent of critical phenomena where critical indices appear to be the same for whole classes of transitions. Actually there is a deep analogy [63] between the transition to chaos and critical phenomena. For example the asymptotic law for the parameter values can be reinterpreted in terms of a critical law for the diverging period of the orbits

$$T_n \sim (\mu_\infty - \mu_n)^{-\nu} \tag{295}$$

where $\nu = \frac{\log 2}{\log 4.669\ldots}$. This analogy is very useful. It suggests immediately for example that on the other side of the transition ($\mu > \mu_\infty$) some quantities which also have the dimension of a time do behave like

$$\tilde{T}_n \sim (\tilde{\mu}_n - \mu_\infty)^{-\tilde{\nu}} \tag{296}$$

and one expects $\nu = \tilde{\nu}$ (usual scaling law). This quantity is easily identified. It is the inverse of the Liapunov number or the inverse of the topological entropy [64] and the $\tilde{\mu}_n$ values of the parameter are nothing more than the values where we get a merging of the trapping domains of the 2^n iterates of the map, which corresponds to the reverse cascade. Other critical quantities like an order parameter for example can be defined [65]. The role of an external magnetic field in a ferromagnetic transition is played here by an external additive noise [66,67]. The analogy goes further. The same techniques applied in phase transitions can be used here to compute the critical indices and to interpret the universality phenomenon, namely, the renormalization group method. The idea of the construction of a renormalization operation is very simple. One exploits the analogy between the graph of f_μ and the two pieces of the graph $f_\mu^{(2)}$. This leads to two diferent renormalization operators. Using [68] the central part of $f^{(2)}$ the renormalization operation is defined as

$$\mathcal{R}f = \frac{1}{\lambda} ff(\lambda x). \tag{297}$$

Using the right part of $f^{(2)}$ one can define [69] the renormalization as

$$\tilde{\mathcal{R}}f = \frac{1}{\lambda} (ff(\lambda X + \xi) - \xi). \tag{298}$$

In both cases a fixed point is found numerically. (f^* for $\mathcal{R}$, $\tilde{f}^*$ for $\tilde{\mathcal{R}}$). Its linearization gives 4.669... as the only eigenvalue outside of the unit circle. Various approximations have been used to compute the fixed point f^* and the spectrum of the linearization of $\mathcal{R}$ at this point [59]. Rigorous proofs for the existence of this fixed point and the unicity of the eigenvalue outside of the unit circle exist [70, 71]. This establishes using chemical arguments from the renormalization theory, the phenomenon of universality for maps on the interval. Renormalization analysis has also been performed [72] for maps in N dimensions giving the same result, namely a bifurcation rate equal to 4.669... . This result is very important from a physical point of view since it guarantees that invertible maps like (281) have the same universal properties as one-dimensional maps. It explains why this universal bifurcation rate is measured in so many different fields, going from pure electromechanics to the dynamics of the heart, for example.

We have now on hand all the material needed to complete the bifurcation analysis of the map of the interval. Let us come back to the interpolating map (294). It is convenient to introduce a μ - ε diagram in order to represent the various dynamical behaviors of these maps. When $\varepsilon = 0$ the topological entropy is simply $\ln\mu$. For every μ with a finite autodecomposition with p terms one has the appearance of a periodic orbit of finite period p + 1. At each of these values, when ε increases a tongue starts in which the topological entropy is "locked" and equal to $\ln\mu$. The right boundary of the tongue corresponds to the appearance of the periodic orbit of period p, while the left boundary corresponds to the disappearance of the trapping domains I_p for the corresponding iterate f^p of f. In between, for ε constant, one observes all the bifurcation phenomena displayed by f_μ itself when μ varies from 0 to 2. This gives a global idea of the complexity

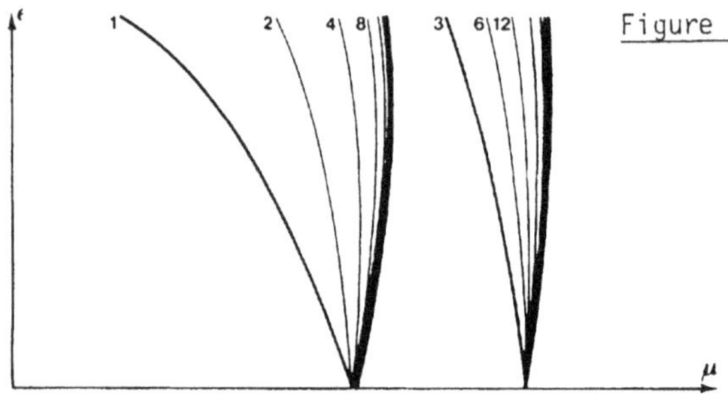

Figure 23

involved in the dynamics of $f_{\mu,\varepsilon}$ when μ varies (Fig. 23). It also shows some analogy between quasiperiodic behavior and chaos. For example frequency locking corresponds to the periodic windows in the chaotic regime.

Unfortunately this analogy breaks down when one considers invertible maps like (281). In that case, new phenomena appear, the most important of which is the so called Newhouse phenomenon [73]. It can roughly be described as a multistability. For some values of the parameter μ an infinite number of periodic orbits are simultaneously asymptotically stable. In general, several competing attractors are likely to exist. In practice this gives, for example, a limitation to the observability of a complete period doubling cascade. For a finite value of the dissipation b in (281) a periodic solution can exist which is, in some sense, more stable than some higher steps of the cascade, making this period doubling phenomena metastable near its accumulation point. This metastability phenomenon is actually observed in experiments and in simple dynamical systems [74]. Let us also remark that the unicity of the attractor is a property that the circle map does not have and the same type of metastability is expected to be observed in such dynamical systems [75]

In the fluid experiment many of the phenomena described below are observed. The period doubling cascade, its metastability, the reverse cascade, the periodic windows. For these values of the Rayleigh number the fluid motion is quite well organized in a roll structure. The aperiodic behavior which appears has nothing to do with real turbulence. On a real signal this aperiodicity is very weak at the onset and seems to increase with the Rayleigh number. A typical Poincaré map looks very close to the iteration of the Henon map. The attractors have the structure of the cross product of a line and a Cantor set. In the reconstructed phase space it has a dimension $2 + \eta$ with $0 < \eta < 1$ (the cross product of a surface and a Cantor set). Since we know that the dimension of the phase space of the fluid is infinite, does the chaotic behavior spread in more and more dimensions when the Rayleigh number is increased? Various methods have been devised to compute the dimension [76] of experimentally [77, 78] reconstructed attractors and the answer seems to be affirmative.

The electromechanical systems provide also nice examples of low dimensional chaotic behavior. Actually they have been known for a long time by engineers to exhibit period doubling cascades and chaos. A typical example of such a system is provided by the parametrically pumped pendulum whose equation reads

$$\ddot{\theta} + \nu\dot{\theta} + \omega^2(1 + f\cos\Omega t)\sin\theta = 0. \qquad (299)$$

It displays chaotic behavior as illustrated in Fig. 24 [74]. We have represented in this figure a projection $\dot{\theta}$ vs θ of the phase portrait associated with (299) ($\omega = 1$, $\Omega = 2$, $\nu = .18$) for f = 1.5, 1.568, 1.575, 1.5782, 1. 5785, 1.5800.

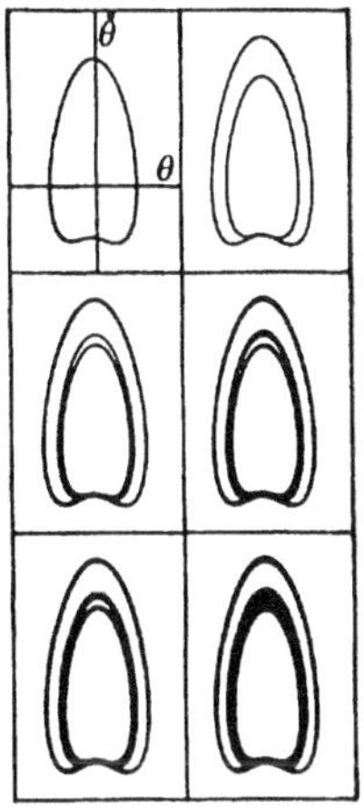

Figure 24

5.4. Triple instability leading to chaos

There is no way to derive a set of differential equations from the Boussinesq equations which would contain all these phenomena. The technical difficulty is that the three main instabilities namely stationary convection, periodic convection, and quasiperiodic convection occur for Rayleigh numbers which are very far from each other. We have seen an example of a system (thermosolutal convection) for which oscillatory convection can occur in the rest state. For more complex systems one can find chaotic behavior as close as one wants from a stable stationary state. These systems are in fact a pretext [79] to derive universal equations which contain all these nonlinear phenomena. Technically they correspond to the singularity ξ^3 already mentioned (225). As an example of such a system let us consider the thermosolutal convection in a rotating frame [80, 81]. In the two-dimensional case (X, Z) the corresponding Boussinesq equations read

$$\begin{aligned}
\partial_t \nabla^2\psi &= \sigma\nabla^2\psi + \sigma R\partial_x\theta + \sigma\tau S\partial_x\phi + \sigma^2 T\partial_z v - J(\psi,\nabla^2\psi) \\
\partial_t \theta &= \nabla^2\theta + \partial_x\psi - J(\psi,\nabla^2\psi) \\
\partial_t \phi &= \tau\nabla^2\phi + \partial_x\psi - J(\psi,\nabla^2\psi) \\
\partial_t v &= \sigma\nabla^2 v - \partial_z\psi - J(\psi,\nabla^2\psi)
\end{aligned} \tag{300}$$

where v represents the y-velocity and T is the Taylor number. Introducing the linear modes of this problem the characteristic equation reads

$$s^4 + A_{n,k}s^3 + B_{n,k}s^2 + C_{n,k}s + D_{n,k} = 0 \tag{301}$$

where B, C, D depend linearly on R, S, T. The triple point is defined by

$$B_{1,k} = C_{1,k} = D_{1,k} = 0. \tag{302}$$

This gives $R = R_c, S = S_c, T = T_c$ and the most unstable wavelength k_c^2. The Jordan form of the restriction of the linear operator associated with (300) on the basis of critical eigenvectors is given by

$$\begin{pmatrix} 0 & 1 & 0 \\ 0 & 0 & 1 \\ 0 & 0 & 0 \end{pmatrix} . \tag{303}$$

This singularity can be found in the chemical problem when three species are present. In that case the characteristic equation reads

$$s^3 + A(k)s^2 + B(k)s + C(k) = 0. \tag{304}$$

The triple point is given by the condition

$$A(k) = B(k) = C(k) = 0. \tag{305}$$

This occurs at $k_c = 0$ (no pattern formation). Near this point the characteristic equation associated with the three marginal modes reads

$$s^3 + \mu s^2 + \nu s + \eta = 0 \tag{306}$$

and the linear amplitude equation reads

$$\dddot{A} + \mu\ddot{A} + \nu\dot{A} + \eta A = 0. \tag{307}$$

The nonlinear development of this triple instability is described by a nonlinear equation which a priori reads

$$\dddot{A} + \mu\ddot{A} + \nu\dot{A} + \eta A = f(A, \dot{A}, \ddot{A}) \tag{308}$$

where f is a nonlinear function of its arguments. Actually a change of variables allows us to suppress two terms at the quadratic order. The resulting normal form is given by

$$\dddot{A} + \mu\ddot{A} + \nu\dot{A} + \eta A = \alpha A^2 + \beta A\dot{A} + \gamma\dot{A}^2 + \delta A\ddot{A} + h.o.t. \,. \tag{309}$$

The method presented in the main body of this paper has been used to compute the various coefficients involved in (309). This equation displays a large number of dynamical behaviors. We are just going to sketch a typical way to get a qualitative information from it [81]. It gives us the opportunity to introduce a very important tool of the theory of dynamical systems, namely the geometrical flow models. We first remark that these equations contain elementary instabilities but also double instabilities (ξ^2 and $\xi\omega$). Let us focus on the case of ξ^2. This occurs when $\nu = \eta = 0$, that is on a line of codimension two in the parameter space μ, ν, η. Near this line (309) can be reduced to its normal form (236). This implies the existence of a surface of codimension one on which we have a global bifurcation for a periodic solution (homoclinic bifurcation already discussed). When μ decreases the reduction from (309) to (236) is no more valid but we expect the homoclinic bifurcation to persist. We have sketched in Fig. 25 the evolution of the homoclinic orbit in such case. The basic hypothesis is the persistence of an homoclinic connection. The stationary solution becomes necessarily a saddle-focus when μ decreases which gives a spiralling homoclinic orbit. Some exact results [82,83] indicate the presence of chaotic behavior in that case.

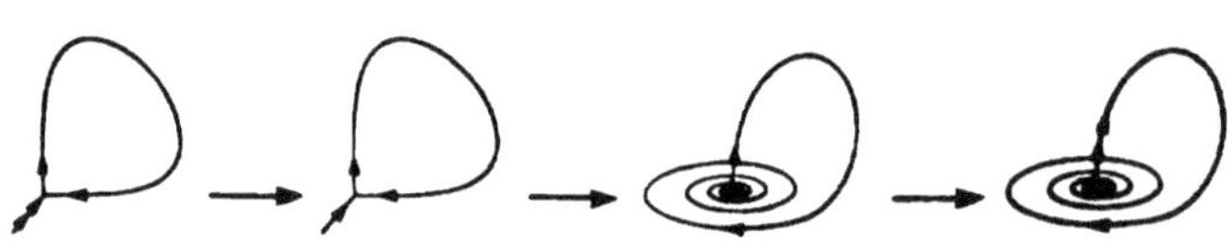

Figure 25

We are now able to construct a geometrical model which contains many of the nonlinear features of (309) near the parameter value where this bifurcation occurs. In some sense it can be seen as a global normal form for the homoclinic bifurcation. One represents [81,84] the flow associated with (309) by a linear flow near the stationary solution and we simply model the reinjection process with a rigid motion. Figure 26 is a sketch of this modeling process. For such a geometrical model one can compute an exact Poincaré map on P_0 which reads [81]

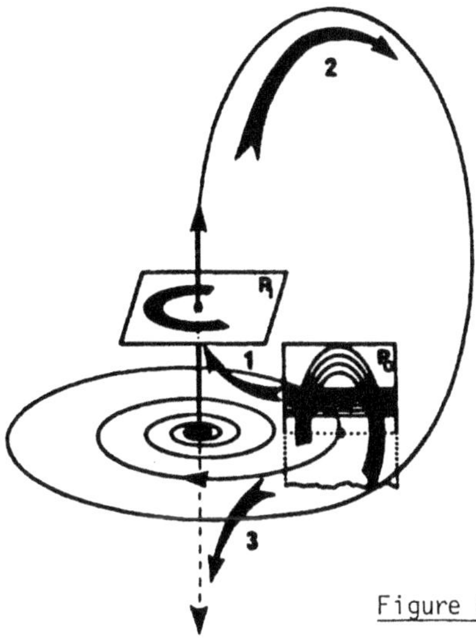

Figure 26

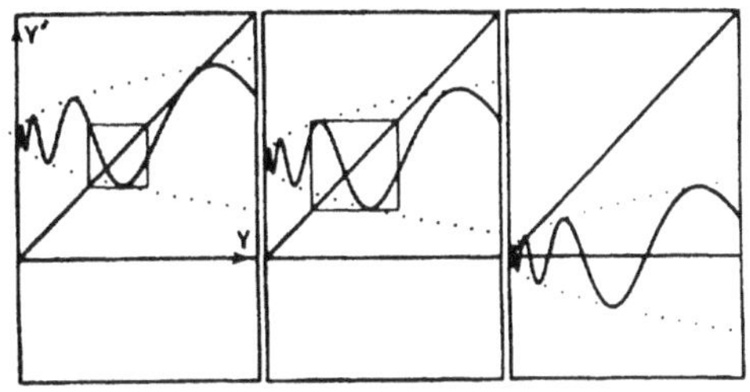

Figure 27

$$X' = X\left(\frac{Z}{h}\right)^{\delta}\cos\left(\xi\ln\frac{Y}{h}+\Phi\right)+\tilde{X}\ , \quad Y' = X\left(\frac{Z}{h}\right)^{\delta}\sin\left(\xi\ln\frac{Y}{h}+\phi\right)+\tilde{Z} \tag{310}$$

where $\delta = r/\lambda$, $\xi = \omega/\lambda$, $-r + i\omega$ and λ being the eigenvalues associated with the stationary solution, h represents the typical size of the linear domain, Φ is a possible rotation in the rigid reinjection map, and $\tilde{X}$, $\tilde{Z}$ is the image of the unstable manifold of the stationary solution. The homoclinic bifurcation occurs when $\tilde{Z} = 0$. This map is very rich and difficult to analize rigorously. In the limit $b\xi \to 0$,
limit $b\xi \to 0$, where $b = (h^{\delta}/\tilde{X})^{1/(\delta-1)}$, the mapping becomes infinitely dissipative and reduces itself to a one dimensional map

$$Z' = C + Z^{\delta}\sin(\xi\ell nZ + \Phi) \tag{311}$$

whose graph has been represented in Fig. 27 in the case $\delta < 1$. This map can be studied in great detail. It displays various nonlinear phenomena like period doubling cascades, intermittencies, etc. It allows us to do several qualitative predictions for the dynamics associated with (309) near the homoclinic bifurcation. Fig. (28) shows a real Poincaré map associated with (309), while the successive iterations of the Poincaré map of the geometrical model correspond to Fig. (29). Many systems [85,88] displaying chaotic behavior present homoclinic bifurcation of the previos type. We have just seen an example of a system for which chaotic behavior is described by a simple normal form. In the case of the convective problem, thanks to the symmetry, quadratic terms have to be replaced by cubic ones. The conclusions are qualitatively the same. Equation (309) exhibits chaos in its prototypical form in the same sense that the complex Landau equation produces harmonic periodic solutions. The whole scenario described earlier is also contained in this equation. The reduction to $\xi\omega$ allows us to get as a secondary

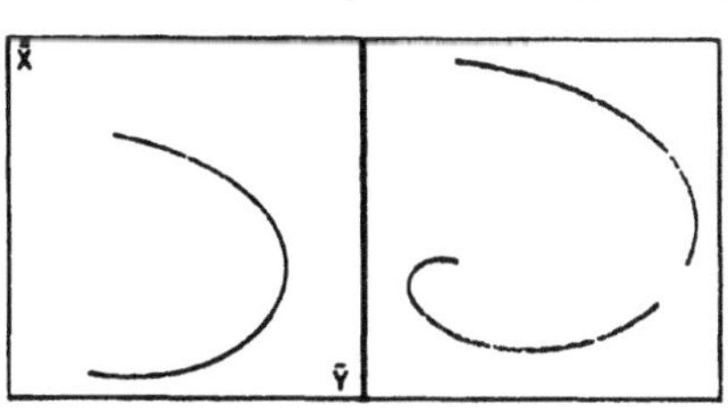

Figure 28

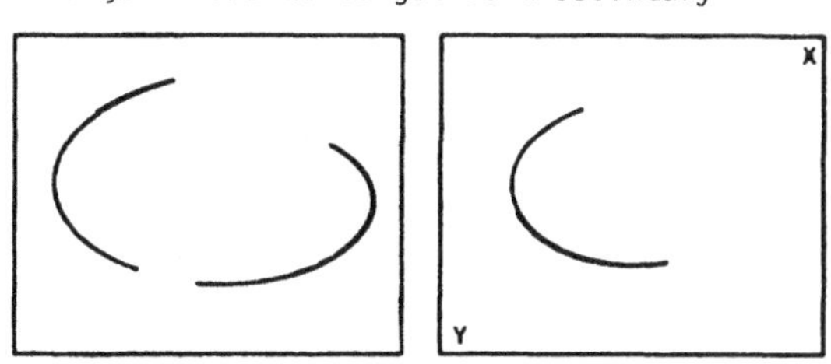

Figure 29

instability a quasiperiodic solution whose ultimate evolution leads to chaos when one approaches the triple point.

5.5. Stable routes to chaos

We have described earlier a scenario for the transition toward chaos which consists of a cascade of period doubling bifurcations. This scenario has been mainly studied with the help of one-dimensional non invertible maps. Other scenarios exist and have been experimentally and numerically identified. In the case of a structurally stable route to chaos, the basic state is of periodic behavior. We are thereby able to classify in general the various routes to chaos by studying non-invertible one-dimensional maps.

A simple example of a transition to chaos is provided by the iteration of the map [89] (Fig. 30)

$$X' = -X^3 + pX + C(p) + \mu = f_{p\mu}, \quad \text{with} \quad C(p) = 2\left(\frac{p-1}{3}\right)^{3/2} . \tag{312}$$

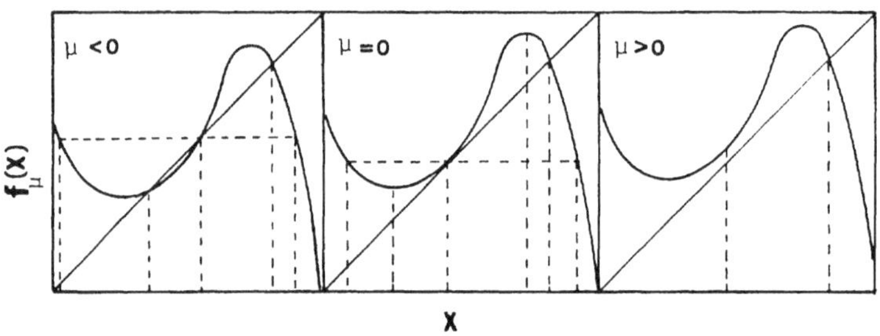

Figure 30

For $\mu = 0$ a tangent bifurcation occurs. When μ is positive one observes chaotic behaviors. For $p < 2.13$ the transition is abrupt and one observes a phenomenon of hysteresis. The same kind of abrupt transition is expected to occur also in the case of subcritical bifurcation (see for example the Lorentz model). More interesting is what happens when $2.13 \leq p < 2.19$. In this case one can show that when the stable fixed point exists ($\mu \lesssim 0$) it attracts almost every initial condition which belongs to the trapping domain of f. The transition to chaos is then without hysterestic effects. On a typical temporal record one observes [90] for $\mu < 0$ a periodic signal which becomes chaotic for $\mu \gtrsim 0$ in exhibiting from time to time some irregular bursts. This is the intermittant route to chaos [91,92]. Figure 31 shows typical intermittent patterns in the successive iterates of (312). A critical index can be attached to the duration of the laminar period in the chaotic regime,

$$\tau \sim \mu^{-1/2}. \tag{313}$$

Since the time is controlled only by the fixed point and its neighborhood, it is enough to do the analysis on the map

$$X' = \mu + X^2 \tag{314}$$

which is the normal form of the saddle mode bifurcation. Since it is a purely local phenomenon the renormalization group analysis [93] of these precesses leads to exact fixed points, with exact eigenvectors. The effect of the noise on this transition has also been investigated [93,94]

Circle maps provide nice examples of scenarios to chaos [58,75]. Except for the period doubling cascade already discussed one can have an intermittant transition for the phase when for $\varepsilon \gtrsim 1$ one crosses the frontier of a tongue. This

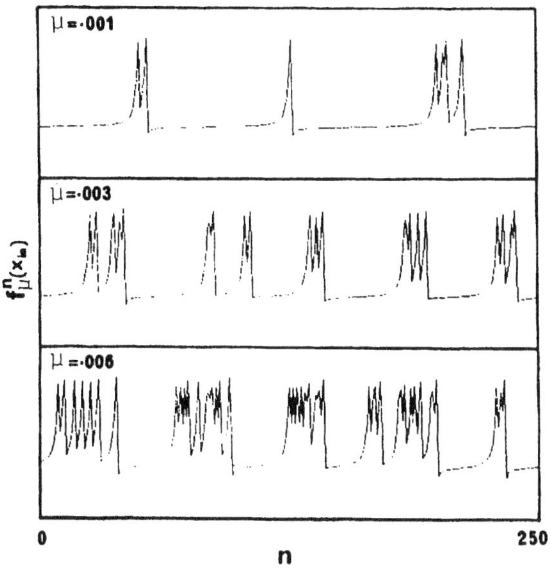

Figure 31

phase intermittency when ε is close to 1 is known as the Newhouse-Palis-Takens scenario [95]. Another possibility is to cross the $\varepsilon = 1$ line by an irrational value of the rotation number [96,97]. One can for example with the help of two parameters follow the line in the ε, ν corresponding to this rotation number. This is the first example of a codimension two scenario. Renormalization group analysis gives in the case of $\rho_f = (1+\sqrt{5})/2$ a fixed point associated with two unstable directions. Hence the critical manifold (stable manifold of the fixed point) has a codimension two and one needs two parameters (scaling fields in the terminology of phase-transition) to cross it in a given family of circle map. We call this scenario an unstable scenario [98,99] in the sense that if it is naturally observed in an experiment, any small changes in the experimental procedure will destroy it. Using two parameters in order to follow a given irrational rotation line is a way to stabilize it.

As a paradigm of an unstable scenario [99] let us consider the iterations of the discontinuous non-invertible map (Fig. 32).

$$f(X) = \begin{cases} 1 - \mu|X|^{1+\varepsilon} & , \ X < 0 \\ -1 + \mu|X|^{1+\varepsilon} & , \ X > 0. \end{cases} \tag{315}$$

These maps arise as the one-dimensional version of the Poincaré map of the geometrical model flow associated with an homoclinic orbit of the Lorenz type [100]. More precisely, one considers third order differential equations which read

$$\partial_t x = \alpha x + f_\mu(x,y,z), \quad \partial_t y = \beta y + g_\mu(x,y,z), \quad \partial_t z = \gamma z + h_\mu(x,y,z) \tag{316}$$

and have the following symmetry property

$$\begin{aligned} f_\mu(x,y,z) &= -f_\mu(-x,-y,z) \\ g_\mu(x,y,z) &= -g_\mu(-x,-y,z) \\ h_\mu(x,y,z) &= h_\mu(-x,-y,z) \ , \end{aligned} \tag{317}$$

and possess 0 as a stationary solution: $f_\mu(0) = g_\mu(0) = h_\mu(0) = 0$. Let us assume that when $\mu = 0$ one has a homoclinic orbit of the type sketched in Fig. 33. Near

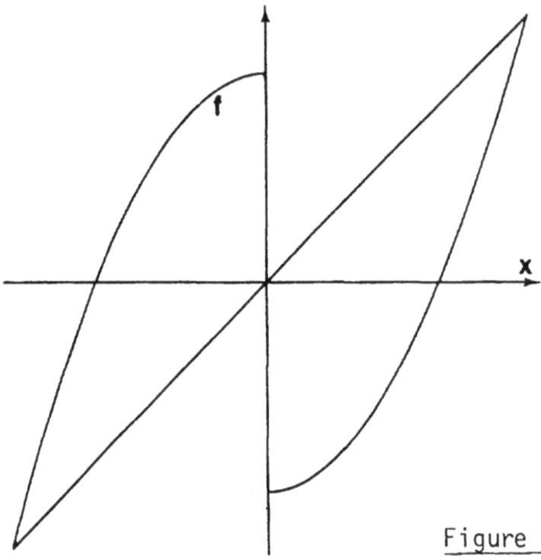

Figure 32

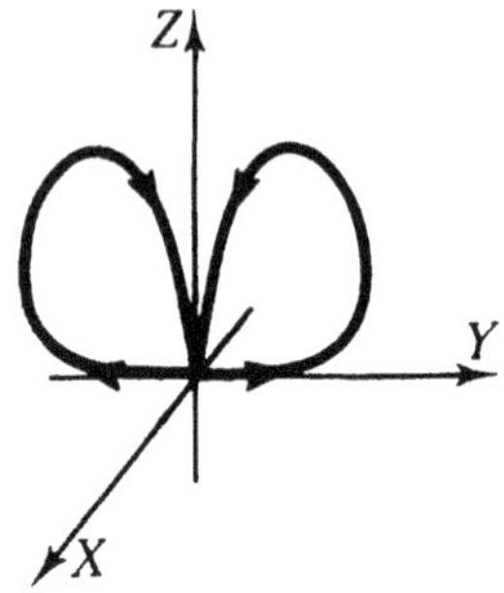

Figure 33

this parameter value and under some technical assumptions, the geometrical flow model obtained by linearizing the flow around the origin and modeling the reinjection by a rigid motion leads to the study of (315). From this map one can predict a cascade of homoclinic bifurcations (glueing bifurcations) leading to chaos, which is actually observed in some differential equations [101] and partial derivative equations [102]. The renormalization group analyzis [99] of this cascade follows the same line as the one for the period doubling cascade. One defines the renormalization operation by

$$\mathcal{R}f = \frac{1}{\eta} ff(\eta x). \tag{318}$$

It is straightforward to verify that the fixed point of the period doubling cascade ϕ^* can be used to compute the fixed point ψ^* of $\mathcal{R}$

$$\psi^*(X) = -\mathrm{sgn}x\phi^*(x). \tag{319}$$

As far as one considers odd perturbations, the spectrum is found to be the same as the one associated with the period doubling cascade. For example, when $\alpha = 2$ the eigenvalue is equal to 4.669 and the bifurcation rate is the same as for the period doubling bifurcation. If one considers even perturbations, which means that the differential equations (316) do not possess any longer the exact symmetry, one finds a second eigenvalue with modulus greater than one. In the space of all possible perturbations the fixed points have now a critical manifold of codimension 2. The practical implication is the instability of this scenario. As soon as the symmetry is broken, the scenario disappears. If one restricts the study to symmetric flows, it appears as a stable one. More exotic scenarios can be imagined but they are also more unstable. As an example one can consider a cascade of period doubling bifurcations or glueing bifurcation for tori. The Table summarizes the known scenarios and their stabilities.

To summarize, we have used Rayleigh-Bénard as a pretext to introduce various nonlinear phenomena. We introduced some techniques to study bifurcations and pattern formations, namely the adiabatic elimination procedure, the normal form theory, the multi space-time scale methods. We have tried to show the interest in studying the interaction between instabilities in order to locate rich nonlinear behaviors close to the instability threshold of a simple steady state. We described phase instabilities and phase turbulence as the fate of spatio-temporal periodic patterns in extended systems. In the case of confined systems, we studied typical phenomenon of the transition to chaos. Renormalization group analysis of the so called scenarios have been sketched in the case of the period doubling cascade. A rapid review of the various scenarios has finally been presented.

Codimension	Scenarios	Stabilizing constraint	Experimental observation
1	Intermittencies	Stable	Yes
	Period doubling	Stable	Yes
2	Quasiperiodicity	Follow a given rotation number	
C > 2	Glueing bifurcations	Symmetry	No
	Intermittency, doubling glueing for Tori	Decoupling of the two frequencies	Yes, no, no

References

1. (Lord) Rayleigh: Phil. mag. 32, 529 (1916)
S. Chandrasekar: Hydrodynamic and hydromagnetic stability (Dover, New York 1981)
2. J. Boussinesq: Theorie analytique de la chaleur, Vol. 2 (Paris, Gauthier Villars 1903)
3. A. Oberdeck: Annalen der Physik und Chemie 7, 271 (1879)
4. F.H. Busse: Rep. Prog. Phys. 41, 1929-1967 (1978)
5. H. Haken: Synergetics, (Springer-Verlag, New York 1968)
6. L. Landau: Dokl. Akad. Nauk SSSR 44, 311 (1944);L. Landau and E.M. Lifshitz: Fluid mechanics,(Pergamon Press, Oxford 1959)
7. J. Guckenheimer and P. Holmes: Nonlinear Oscillations, Dynamical Systems and Bifurcations of Vector Fields, Appl. Math. Sci. 42 (Springer-Verlag, 1983)
8. J. Carr: Application of the center manifold theory, Lecture Notes, Lefshetz Center for Dynamical Systems, Brown Univ.,Providence RI (1975)
9. J.T. Stuart: J. Fluid Mech. 9, 353 (1960)
10. P. Coullet and E.A. Spiegel: SIAM J. Appl. Math. 43, 776 (1983)
11. H. Poincaré: Les Methodes nouvelles de la mecanique celeste: III (Dover, New York 1957)
12. A. Lindstedt: Memoires de l'academie de Saint-Petersbourg (1882)
13. L. Gorkov: JETP 6, 311 (1958)
14. W.V.R. Malkus and G. Verones: J. Fluid. Mech. 4, 225 (1958)
15. G. Iooss and D.D. Joseph: Elementary Stability and Bifurcation Theory (Springer-Verlag,New York, 1980)
16. E. Knoblock and J. Guckenheimer: Phys. Rev. A 27, 1 (1983)
17. A.C. Newell and J.A. Whitehead: J. Fluid Mech. 38, 279 (1969)
18. L.A. Segel: J. Fluid Mech. 38, 203 (1969)
19. Y. Pomeau and P. Manneville: J. Phys. (Paris) Lett. 40, 609 (1979)
20. W. Eckauss: Studies on nonlinear stability theory, Springer Tracts in natural Philosophy, Vol. 6 (Springer, Berlin 1965)
21. A. Pocheau and V. Croquette: J. Physique 45, 35 (1984)
22. Y. Kuramoto: Prog. Theor. Phys. 71, 1182 (1984)
23. L. Kramer and W. Zimmermann: *On the Eckauss instability for spatially Periodic Patterns*, Preprint 84
24. P. Coullet and S. Fauve: *Rayleigh-Bénard Convection: a typical nonlinear system*, to appear in Workshop on Combustion, Flames and Fires held at Les Houches March 1984
25. E. Siggia and A. Zippelius: Phys. Rev. Lett. 47, 835 (1981)
26. Hydrodynamic stability and the transition to turbulence, Ed. by H. Swinney and J.P. Gollub, Topics in Applied Physics 45 (Springer-Verlag 1981)
27. F.H. Busse: J. Fluid Mech. 52, 97 (1972)
28. J.E. Marsden and M. McCracken: The Hopf bifurcation and its applications (Springer-Verlag, Berlin 1976)

30. V.I. Arnold: Geometrical Methods in the Theory of Ordinary Differential Equations (Springer-Verlag, Berlin 1982)
31. V.I. Arnold: Singularity Theory: Selected Papers, London Mathematical Society Lecture Notes, Vol. 53 (Cambridge University Press, Cambridge 1981)
32. R. Thom: Structural Stability and Morphogenesis, W.A. Benjamin (Reading MA 1975)
33. Y. Kuramoto and T. Yamada: Prog. Theor. Phys. 56, 679 (1976)
34. H.T. Moon, P. Huerre and L.G. Redekopp: Physica D 7, 135 (1983)
35. C. Bretherton and E.A. Spiegel: Phys. Lett. 96A, 152 (1983)
36. Y. Kuramoto: Prog. Theor. Phys. Suppl. 64, 346 (1978)
37. Y. Pomeau, A. Pumir and P. Pelce: *Intrinsic stochasticity with many degree of freedom*, Preprint (1984)
38. H.R. Brand, P.C. Hohenberg and V. Steinberg: Phys. Rev. A 27 591 (1983)
39. A. Arneodo, P. Coullet, C. Tresser and E.A. Spiegel: *Asymptotic chaos* (to appear in Physica D)
40. P. Coullet: Phys. Rep. 103, 95 (1984)
41. P. Coullet, C. Elphick and E. Tirapegui : *Normal forms for periodic perturbations of singular vector field*, Preprint (1984)
42. G. Ahlers and R.P. Behringer: Phys. Rev. Lett. 40, 712 (1978)
43. J. Maurer and A. Libchaber: J. Phys. Letters 40, 505 (1979)
44. M. Dubois and P. Berge: Phys. Lett. 72A, 53 (1980)
45. J.P. Gollub and S.V. Benson: J. Fluid Mech. 100, 449 (1980)
46. D. Ruelle and F. Takens: Comm. Math. Phys. 20, 167 (1971)
47. V.I. Arnold: Ordinary differential equations (M.I.T. Press, Cambridge MA 1973)
48. F. Takens: in Dynamical systems and turbulence, Ed. by D.A. Rand and L.S. Young, Lecture notes in Mathematics 898, 366 (Springer-Verlag, Berlin 1980)
49. G. Iooss: Bifurcation of maps and applications, Mathematical studies Vol. 36 (North Holland, Amsterdam 1979)
50. P. Coullet and A. Libchaber: *The Mathematics of time in a convection experiment* (unpublished)
51. V.I. Arnold: A.M.S. Transl. Ser. 2, 46, 213 (1965)
52. A. Denjoy: J. Math. 17, 333 (1932)
53. M.R. Herman: Publ. Math. IHES, 49, 669 (1982)
54. A. Libchaber and J. Maurer: in Nonlinear Phenomena at Phase-transition and Instabilities, Ed. by T. Riste (Plenum, New York) 259
55. J. Maurer and A. Libchaber: J. Phys. Lettres 40, 419 (1979)
56. M. Giglio, S. Musazzi and V. Perini: Phys. Rev. Lett. 47, 243 (1981)
57. A. Libchaber, G. Laroche and S. Fauve: J. Phys. Lettres 43, 211 (1982)
58. P. Coullet, E. Tresser and A. Arneodo: Phys. Lett. 77A, 327 (1980)
59. P. Collet and J.P. Eckmann: Iterated map on the interval as Dynamical Systems, Progress on Physics Vol. I. (Birckhauser-Boston 1980)
60. M. Henon: Comm. Math. Phys. 50, 69 (1976)
61. S.M. Ulam and J. von Neumann: Bull. Amer. Math. Soc. 53, 1120 (1947)
62. B. Derrida, A. Gervois and Y. Pomeau: Ann. Inst. Henri Poincaré, Vol. XXIX, 3 305 (1978)
63. P. Coullet and C. Tresser: in Field Theory, Quantization and Statistical Physics, Ed. by E. Tirapegui (Reidel 1981)
64. C. Tresser and P. Coullet: C.R. Acad. Sc. Paris 287A, 577 (1978)
65. P. Coullet and C. Tresser: J. Phys. Lettres 41, 225 (1980)
66. B.A. Huberman and J. Rudnick: Phys. Rev. Lett. 45, 154 (1980)
67. J. Crutchfield, M. Nauenberg and J. Rudnick: Phys. Rev. Lett. 46, 933 (1981); B. Shraiman, C.E. Wayne and P.C. Martin: Phys. Rev. Lett., 46, 935 (1981)
68. M.J. Feigenbaum: J. Stat. Phys. 19, 25 (1978)
69. P. Coullet and C. Tresser: J. Phys. Coll. (Paris) C5, 39 (1978)
70. P. Collet, J.P. Eckmann and O.E. Lanford: Comm. Math. Phys. 76, 211 (1980)
71. M. Campanino and H. Epstein: Comm. Math. Phys. 79, 261 (1981)
72. P. Collet, J.P. Eckmann and H. Koch: J. Stat. Phys. 25, 1 (1981)
73. S.E. Newhouse: Topology 13, 9 (1974)
74. A. Arneodo, P. Coullet, C. Tresser, J. Maurer, A. Libchaber and D. D'Humieres: Physica D 6, 385 (1983)
75. R.S. Mac Kay and C. Tresser: J. Phys. Lettres 45, 741, (1984)
76. P. Grassberger and I. Procaccia: Physica D 9, 189 (1983)

77. B. Malraison, P. Atten, P. Berge and M. Dubois: J. Phys. Lettres 44, 897 (1983)
78. A. Brandstäter, J. Swift, H.L. Swinney, A. Wolf and J.D. Farmer: Phys. Rev. Lett. 51, 1442 (1983)
79. P. Coullet: in Stochastic phenomena and chaotic behaviour in complex systems, ed. by P. Schuster (Springer-Verlag 1984)
80. A. Arneodo, P. Coullet and E.A Spiegel: Phys. Lett. 92A, 369 (1982)
81. A. Arneodo, P. Coullet, C. Tresser and E.A. Spiegel: *Asymptotic chaos*, to appear in Physica D.
82. L.P. Shilnikov: Sov. Math. Dokl. 6, 163 (1965)
83. L.P. Shilnikov: Math. U.S.S.R. Sbornik 10, 91 (1970)
84. A. Arneodo, P. Coullet and C. Tresser: Commun. Math. Phys. 79, 573 (1981)
85. A. Arneodo, P. Coullet and C. Tresser: J. Stat. Phys. 27, 171 (1982)
86. D.R. Moore, J. Toome, E. Knobloch and N.O. Weiss: Nature 303, 663 (1983)
87. P. Glendinning and C. Sparrow: J. Stat. Phys. 5, 645 (1984)
88. P. Gaspard, G. Nicolis: J. Stat. Phys. 31, 499 (1983);
P. Gaspard, R. Kapral and G. Nicolis: J. Stat. Phys. 5, 697 (1984)
89. C. Tresser, P. Coullet and A. Arneodo: J. Phys. Lettres 41, 243 (1980)
90. P. Berge, M. Dubois, P. Manneville and Y. Pomeau: J. Phys. Lettres 41, 341 (1980)
91. P. Manneville and Y. Pomeau: Phys. Letters 75A, 1 (1979)
92. Y. Pomeau and P. Manneville: Comm. Math. Phys. 74, 189 (1980)
93. J.E. Hirsch, M. Nauenberg and D.J. Scalapino: Phys. Lett. 87A, 391 (1982)
94. J.E. Hirsch, B.A. Huberman and D.J. Scalapino: Phys. Rev. A 25, 519 (1982)
95. S.E. Newhouse, J. Palis and F. Takens: Publ. Math. IHES, 57, 5 (1983)
96. M.J. Feigenbaum, L.P. Kadanoff and S.J. Shenker: Physica 5D, 370 (1982)
97. D. Rand, S. Ostlund, J. Sethna and E. Siggia: Phys. Rev. Letters 49, 182 (1982); S. Ostlund, D. Rand, J. Sethna and E. Siggia: Physica D 8, 303 (1983)
98. P. Coullet: in Chaos and Statistical Methods, Ed. by Kuramoto (Springer-Verlag (1984)
99. P. Collet, P. Coullet and C. Tresser: *Scenarios under constraint*, Preprint (1984)
100. E.N. Lorenz: J. Atmos. Sci. 20, 130 (1963)
101. A. Arneodo, P. Coullet and C. Tresser: Phys. Lett. 81A, 197 (1981)
102. Y. Kuramoto and S. Koga: Phys. Lett. 92A, 1 (1982)

Statistical Mechanics of the sine-Gordon Field: Part I

R.K. Bullough

Department of Mathematics, U.M.I.S.T., P.O. Box 88, Manchester M60 1QD, United Kingdom

1. Introduction

These lectures will be devoted to one particular problem — namely the calculation of the equilibrium statistical mechanics, that is the partition function Z, for the sine-Gordon field.

The sine-Gordon equation (s-G) with velocity $c = 1$ is

$$\phi_{xx} - \phi_{tt} = m^2 \sin\phi \quad ; \tag{1.1}$$

in units with $\hbar = c = 1$, m is a mass, while ϕ_{xx}, etc. means $\partial^2\phi/\partial x^2$, etc. and will do so throughout these lectures.

The s-G is 'Hamiltonian': for equation (1.1) follows as Hamilton's equations of motion from the Hamiltonian

$$H[\phi] \equiv \gamma_0^{-1}\int [\tfrac{1}{2}\gamma_0^2\,\Pi^2 + \tfrac{1}{2}\phi_x^2 + m^2(1-\cos\phi)]\,dx \quad . \tag{1.2}$$

Since these lectures are deliberately pedagogical we shall now show this in detail. To do it we need to know that Π and ϕ are canonically conjugate variables satisfying the 'Poisson bracket' relation $\{\Pi(x,t),\,\phi(x',t)\} = \delta(x-x')$. Note that the Hamiltonian $H[\phi]$ also depends on a dimensionless coupling constant γ_0. Unlike the mass m, γ_0 does not appear in the equations of motion (1.1): its important role is to 'scale' the nonlinearity, as we shall see.

First, we need to understand the form Hamilton's equations must take if $\phi(x,t)$ is a field: $\phi(x,t)$ is defined on some support L (say) i.e. $\phi(x,t)$ is defined for $-\frac{1}{2}L \le x \le \frac{1}{2}L$: then we have to think of $\phi(x,t)$ as a canonical co-ordinate $q(t)$ evolving in time for each label x. There is therefore a continuous infinity of such co-ordinates $\phi(x,t)$ and there will be a corresponding continuous infinity of canonical momenta $\Pi(x,t)$ (say). Thus, there is a double continuous infinity of degrees of freedom associated with the set of pairs $\{(\Pi(x,t),\phi(x,t));-\frac{1}{2}L \le x \le \frac{1}{2}L\}$. Initially, we conveniently think of $L \to \infty$. Boundary conditions are needed, and these we choose to be $\phi \to 0 \pmod{2\pi}$, $\phi_x \to 0$ for $|x| \to \infty$. If ϕ is C^∞ all derivatives ϕ_{xx}, etc. vanish for $|x| \to \infty$.

Lagrange's equations may be a little more familiar than Hamilton's equations. Thus consider the Lagrangian

$$L[\phi] = \int \mathcal{L}[\phi,\phi_x,\phi_t]\,dx \equiv \gamma_0^{-1}\int [\tfrac{1}{2}\phi_t^2 - \tfrac{1}{2}\phi_x^2 - m^2(1-\cos\phi)]\,dx \quad . \tag{1.3}$$

In so far as the integral is the limit of a sum, Lagrange's equations must be of the form

$$-\frac{d}{dt}\left(\frac{\partial\mathcal{L}}{\partial\phi_t}\right) + \frac{\partial\mathcal{L}}{\partial\phi} = 0 \quad . \tag{1.4}$$

However, in that actual sum ϕ_x involves ϕ at neighbouring points x and $x+dx$. Accordingly $\partial\mathcal{L}/\partial\phi$ must be interpreted as the *functional* derivative $\delta L/\delta\phi$. By this we mean the following: if the functional $\mathcal{A}[\phi,\phi_x,\phi_{xx},\cdots]$ is a function of ϕ and its x-derivatives and $A = \int\mathcal{A}[\phi,\phi_x,\cdots]\,dx$ the variation δA induced by an arbitrary but small variation $\delta\phi$ in ϕ is precisely

$$\delta A = \int\left(\frac{\delta A}{\delta\phi}\,\delta\phi\right)dx \quad . \tag{1.5}$$

This means, for example, that if ϕ depends on t as well as x $dA/dt = \int\frac{\delta A}{\delta\phi}\,\phi_t dx$ the limit of the sum $\sum_{i=1}^{N}\frac{\partial A}{\partial\phi_i}\frac{d\phi_i}{dt}$ for 2N degrees of freedom. The rules of ordinary differentiation thus carry over to functional differentiation accordingly. The definition applies to densities $\mathcal{A}$ by noting that $\mathcal{A}[\phi,\phi_x,\cdots] = \int\mathcal{A}[\phi(x'),\phi_{x'}(x')\cdots]\,\delta(x-x')dx'$. Thus if $\mathcal{A} = f(\phi(x))$, $\delta\mathcal{A}/\delta\phi(x') = f'(\phi(x'))\delta(x-x')$ where $f'(\phi) \equiv df/d\phi$.

By successive integrations by parts using $\delta\phi_x = \partial(\delta\phi)/\partial x$, etc., and with $\delta\phi \to 0$ as $|x| \to \infty$ we readily find

$$\delta A = \int\left(\left[\frac{\partial\mathcal{A}}{\partial\phi} - \frac{\partial}{\partial x}\frac{\partial\mathcal{A}}{\partial\phi_x} + \frac{\partial^2}{\partial x^2}\frac{\partial\mathcal{A}}{\partial\phi_{xx}} + \ldots\right]\delta\phi\right)dx \quad . \tag{1.6}$$

Thus, in the case of (1.3) $\delta L/\delta\phi = \gamma_0^{-1}[\phi_{xx} - m^2\sin\phi]$, the γ_0^{-1} factors, and (1.4) is indeed the s-G. Similarly therefore, for Hamilton's equations

$$H[\phi] = \int(\Pi\phi_t)dx - L[\phi] \tag{1.7}$$

with $\Pi \equiv \delta L/\delta\phi_t$ in which however ϕ_t is to undergo independent variations $\delta\phi_t$. Thus $\delta L/\delta\phi_t = \gamma_0^{-1}\phi_t$, $H[\phi]$ takes the form (1.2), and Hamilton's equations of motion must be

$$\Pi_t = -\delta H/\delta\phi \quad , \quad \phi_t = \delta H/\delta\Pi \tag{1.8}$$

which (with (1.2)) yield the s-G.

Finally, we could use Hamilton's principle of stationary action $\delta S[\phi] = 0$. Set

$$S[\phi] = \int L[\phi]\,dt = \int dt\left[\int(\Pi\phi_t)\,dx - H[\phi]\right]. \tag{1.9}$$

Then $\delta S = 0$ means

$$0 = \delta\int dt\left\{\int(\Pi\phi_t)\,dx - H[\phi]\right\} = \int dt\left[\int dx\left\{(\delta\Pi\phi_t + \Pi\delta\phi_t) - \left(\frac{\delta H}{\delta\Pi}\delta\Pi + \frac{\delta H}{\delta\phi}\,\delta\phi\right)\right\}\right] \cdot \tag{1.10a}$$

The independent variations are $\delta\Pi$ and $\delta\phi$ so we use

$$\int dx\int dt\,\Pi\,\delta(\phi_t) = -\int dx\int\delta\phi\,\Pi_t\,dt \tag{1.10b}$$

(the $\delta\phi$ vanish at the beginning and end of the trajectories in t) in order actually to obtain the equations of motion (1.8). By similar steps on $S[\phi] = \int L[\phi]\,dt$, Lagrange's equations (1.4) are $\delta S/\delta\phi = 0$.

So now we ask why it should be important to consider the statistical mechanics of the s-G. Then we must ask how to carry out such a study. There are at least four reasons why the statistical mechanics of the s-G is important: first, there are actual physical examples open to experiment (§3) in which these statistical mechanics or ones related to it actually arise ; second, the s-G is a classical model field theory, it is Lorentz covariant, its quantisation would provide a model quantised covariant field theory of real interest, and mathematical methods for quantisation and statistical mechanics relate closely; third, the quantised s-G in one space and one-time dimension (1+1 dimensions) corresponds, essentially as a 1 : 1 map, to certain solvable lattice models, taken in continuum limit, in the statistical mechanics of 2 Euclidean dimensions — notably the 8-vertex (or octet) model and the 2-dimensional Ising model which is contained within it [1]; fourth, the classical s-G is 'integrable' in the sense used here that it can be completely solved — indeed the Cauchy initial value problem $\phi(x,0) = f_1(x)$, $\phi_t(x,0) = f_2(x)$ on boundary conditions $\phi \to 0 \pmod{2\pi}$, $\phi_x \to 0$ as $|x| \to \infty$) can be solved by the so-called inverse scattering method (§4). It is perhaps because of this that the quantised s-G can be solved completely for its eigenspectrum and eigenvectors [2,3]: in principle it can be solved too for its correlation functions, though this is done so far only for particular choices of γ_0. The properties of the 'integrable' (= solvable) systems are of profound importance, and so are all the relations between them. One such, the Painlevé property [4], will be described by other lecturers at this School.

In order to evaluate Z we must see first how to define it. Because of the continuous infinity of degrees of freedom this will be a 'functional integral' even in classical limit. This is the case we shall mostly consider, so we write this classical partition function in the following way now, and define it carefully later:

$$Z = \int \mathcal{D}\Pi\, \mathcal{D}\phi\, \exp - \beta H[\phi] \tag{1.11}$$

and $\beta^{-1} = k_B T$ (k_B = Boltzmann's constant, T the temperature — often we write $\beta^{-1} = T$ simply). The symbol '$\mathcal{D}$' means functional integration, and what this means is explained in §5.

For the s-G $H[\phi]$ is given by (1.2), but the principal interest of our calculation will be to exploit the integrability of the s-G and replace the canonical pairs (Π,ϕ) by new canonical co-ordinates, the action-angle variables, and evaluate Z with these. This amounts to studying Z, expression (1.11), under a change of variable for functional integration. We shall discover this is not quite simple. However, the transformation is important because it relates the calculational problem directly to the 'quantum inverse method' for quantising and solving integrable systems like the s-G as pioneered by Faddeev [2,3,5] and the Leningrad School. And it also relates to the Bethe ansatz methods for quantising and solving integrable classical field theories [5,6], and to the Bethe ansatz for their quantum statistical mechanics [7,8].

This is our programme. First of all (§2) we give an elementary discussion of action-angle variables for collections of oscillators. Then we look at the experimental applications of the statistical mechanics of the s-G (§3). Then in §4 we look at the s-G as an integrable field, and find its action-angle variables. Finally we turn to the functional integrals.

2. Elementary Aspects of Action-Angle Variables: Action-Angle Variables for the Klein-Gordon Equation

For $\hbar = c = 1$ the linear Klein-Gordon equation (K-G) is

$$\phi_{xx} - \phi_{tt} = m^2\phi \quad . \tag{2.1}$$

This is precisely the linearised form of the s-G, (1.1). One way to reach the K-G from the s-G is to let $\phi \to \gamma_0^{\frac{1}{2}}\phi$ and $\Pi \to \gamma_0^{-\frac{1}{2}}\Pi$ in (2.2) so that

$$H[\gamma_0^{\frac{1}{2}}\phi] = \int \left[\tfrac{1}{2}\Pi^2 + \tfrac{1}{2}\phi_x^2 + m^2\gamma_0^{-1}\left(1 - \cos\left(\gamma_0^{\frac{1}{2}}\phi\right)\right)\right] dx \tag{2.2}$$

with $\{\Pi,\phi\} = \delta(x-x')$: this transformation is a canonical one since e.g. the Poisson bracket $\{\Pi,\phi\}$ is preserved. Now expand the nonlinear term to see that, as $\gamma_0 \to 0$,

$H \to H_{KG}$. Since we need this later we give it as

$$H_{KG}[\phi] = \int \left[\tfrac{1}{2}\Pi^2 + \tfrac{1}{2}\phi_x^2 + \tfrac{1}{2}m^2\phi^2\right] dx \quad . \tag{2.3}$$

Eqn. (2.1) follows from (2.3) through Hamilton's equations (1.7). Thus γ_0 scales the nonlinearity as was foreshadowed in §1. Later we shall see that γ_0 acts like $\hbar$.

The K-G is a collection of oscillators: Fourier transform on $-\infty < x < \infty$ so that $\phi(x,t) \to \bar{\phi}(k,t)$ on $-\infty < k < \infty$: then from (2.1) (and we shall first take the $\bar{\phi}$ as real fields)

$$\bar{\phi}_{tt} + \omega^2(k)\bar{\phi} = 0 \qquad \text{with} \tag{2.4}$$

$$\omega(k) = (m^2 + k^2)^{\frac{1}{2}} \quad . \tag{2.5}$$

The energy of the oscillator (2.4) is $E(k) = \frac{1}{2}\bar{\phi}_t^2 + \frac{1}{2}\omega^2(k)\bar{\phi}^2$ and $E(k)$ is a constant of the motion for each label k. Thus the K-G has a continuous infinity of constants of the motion $\{E(k); -\infty < k < \infty\}$. Define an action $P(k) = E(k)(\omega(k))^{-1}$. The solution of (2.4) is $\bar{\phi} = A(k)\cos(\omega(k)t + \delta(k))$, so $P(k) = \frac{1}{2}\omega A^2(k)$ and $\bar{\phi} = [2\omega^{-1}P]^{\frac{1}{2}}\cos Q(k,t)$, where $Q(k,t) \equiv \omega(k)t + \delta(k)$. If $H \equiv \frac{1}{2}\bar{\Pi}^2 + \frac{1}{2}\omega^2\bar{\phi}^2$, $\bar{\phi}_t = \bar{\Pi}$ and $\bar{\Pi}_t = -\omega^2\bar{\phi}$, so $\bar{\Pi} = -[2\omega P]^{\frac{1}{2}}\sin Q(k,t)$. But this means H has the values $\omega(k)P(k)$ also. Moreover, if $P(k) \to P(k,t)$ and $P(k,t)$, $Q(k,t)$ are thought of as new canonical variables, $H = \omega(k)P(k,t)$ means $Q_t = \partial H/\partial P = \omega(k)$ (whence $Q(k,t) = \omega(k)t + \delta(k)$) and $P_t = -\partial H/\partial Q = 0$, so $P(k,t) = P(k)$ independent of t, a constant of the motion as required.

We check the Poisson bracket: if $A = A(\bar{\Pi},\bar{\phi})$, $B = B(\bar{\Pi},\bar{\phi})$, then $\{A,B\} \equiv (\partial A/\partial\bar{\Pi})(\partial B/\partial\bar{\phi}) - (\partial A/\partial\bar{\phi})(\partial B/\partial\bar{\Pi})$. But it is merely slightly tedious to check that

$$\{A,B\} = \left(\frac{\partial A}{\partial P}\frac{\partial B}{\partial Q} - \frac{\partial A}{\partial Q}\frac{\partial B}{\partial P}\right)\frac{\partial(P,Q)}{\partial(\bar{\Pi},\bar{\phi})} \tag{2.6}$$

always, while for the particular transformation we have made the Jacobian $J \equiv \partial(P,Q)/\partial(\bar{\Pi},\bar{\phi}) = +1$. Evidently $J = +1$ is necessary and sufficient to preserve the Poisson bracket, i.e. make the transformation canonical. The variables $P(k)$, $Q(k)$ are action-angle variables for the oscillator labelled by k. Their phase space is the cylinder $0 \le P(k) < \infty$, $0 \le Q(k) < 2\pi$.

For later work on the s-G it is convenient to introduce the symplectic differential 2-form [5,9,10,11,12] $d\bar{\Pi}_\wedge d\bar{\phi}$: roughly 'd' acts like the derivative operator and wedge '$\wedge$' is skew ie. $d\bar{\Pi}_\wedge d\bar{\phi} = -d\bar{\phi}_\wedge d\bar{\Pi}$. A differential form is what stands under an integral: thus for $\int f dx$, $f dx$ is a 1- form; for $\int G(x,y)\,dxdy$, $Gdxdy$ is a 2-form, etc. Likewise the whole parenthesis in the line integral $\int(Adx + Bdy + Cdz)$ is a differential 1-form; and a p- form is what appears under a p-tuple integral. The surface integral $\iint(Pdydz + Qdzdx + Rdxdy)$, with no $dzdy$, $dxdz$, $dydx$ and particularly no $dxdx$ terms, suggests the skew symmetry ie. use orientated integrals so $dxdy = -dy\,dx$ (written $dx_\wedge dy = -dy_\wedge dx$).

The properties of '$\wedge$' are [10]

(i) $\lambda_\wedge\mu$ is distributive; λ,μ are p,q- forms;

(ii) $\lambda_\wedge(\mu_\wedge\nu) = (\lambda_\wedge\mu)_\wedge\nu$; λ,μ,ν are p,q,r- forms;

(iii) $\mu_\wedge\lambda = (-1)^{pq}\lambda_\wedge\mu$; λ,μ are p, q-forms. Then $dx^j{}_\wedge dx^i = -dx^i{}_\wedge dx^j$ is an elementary 2-form; $dx^k{}_\wedge dx^j{}_\wedge dx^i = -dx^k{}_\wedge dx^i{}_\wedge dx^j$ is an elementary 3-form.

Example: $(Adx + Bdy + Cdz)_\wedge(Edx + Fdy + Gdz)$

$= (BG - CF)dydz + (CE - AG)dzdx + (AF - BE)dxdy =$ "vector product"

(we use $dydz$ for $dy_\wedge dz$) (2.7)

The properties of 'd' are

(i) $d(w+v) = dw + dv$ for p-forms w,v

(ii) $d(\lambda_\wedge\mu) = d\lambda_\wedge\mu + (-1)^{(\deg\lambda)}\lambda_\wedge d\mu$ for p-form λ, q-form μ and $\deg\lambda = p$.

(iii) for each form w, $d(dw) = 0$

(iv) for each function f, $df = \sum_i \frac{\partial f}{\partial x^i} dx^i$.

For co-ordinates x^i, dx^i is the usual dx^i and $d^2x^i = 0$ by (iii). The property (iii) is the content of the Poincaré lemma (an integrability condition). For take any p-form $\omega = a dx^p$ (dx^p is an elementary p-form). Then by (iv)

$$d(d\omega) = d\sum \frac{\partial a}{\partial x^i} dx^i dx^p = \sum_{i\neq j} \frac{\partial^2 a}{\partial x^i \partial x^j} dx^j dx^i dx^p$$

$$= \frac{1}{2}\sum_{i\neq j}\left(\frac{\partial^2 a}{\partial x^i\partial x^j} - \frac{\partial^2 a}{\partial x^j \partial x^i}\right) dx^j dx^i dx^p = 0 \tag{2.8}$$

if, and only if, the integrability conditions $\partial^2 a/\partial x^i\partial x^j = \partial^2 a/\partial x^j\partial x^i$ apply. Check that if $w = Pdx + Qdy + Rdz$, then $dw = \{(\partial R/\partial y) - (\partial Q/\partial z)\}dydz + \ldots$, the 'curl'. Similarly, the divergence follows from d acting on the 2-form $Adydz + Bdzdx + Cdxdy$ [10].

This digression allows us to note the following: $\varpi = d\bar{\Pi}(k)_\wedge d\bar{\phi}(k)$ is a differential 2-form: evidently $d\varpi = 0$. Evidently too

$$d\bar{\Pi} = -\omega^{\frac{1}{2}}(2P)^{-\frac{1}{2}} dP\sin Q - \omega^{\frac{1}{2}}(2P)^{\frac{1}{2}}\cos Q\, dQ$$

$$d\bar{\phi} = \omega^{-\frac{1}{2}}(2P)^{-\frac{1}{2}} dP\cos Q - \omega^{-\frac{1}{2}}(2P)^{\frac{1}{2}}\sin Q\, dQ \tag{2.9}$$

so $d\bar{\Pi}_\wedge d\bar{\phi} = dP_\wedge dQ$, ie. the symplectic form ϖ is invariant (and closed $d\varpi = 0$). Rather than check the Poisson brackets, it is easier to show that ϖ is invariant and closed for a canonical transformation.

It is now obvious that for N oscillators labelled by $k_1, k_2, \cdots, k_N$, $H = \sum_{i=1}^{N} \omega(k_i)P(k_i)$ in action-angle variables. The Jacobian $J = \prod_{i=1}^{N} J(k_i) = +1$. The phase space is the N-product of cylinders and the actual motion lies on an N-dimensional torus.

It is convenient to think of all this being done first of all on the K-G with periodic boundary conditions of period $L < \infty$, and then let $L \to \infty$. For then $N = \infty$ for $L < \infty$ and $k_n = 2\pi n L^{-1}$, n integers, $-\infty < n < \infty$. It is convenient now to label $P(k_n) \to P_n$ and $\omega(k_n) \to \omega_n$: for then we can set $(L/2\pi)P_n \to P(k)$ as $n,L \to \infty$ so that $k_n \to k$ dense in $-\infty < k < \infty$. Then $P(k)$ is a continuous function of k, and $\sum_{n=-\infty}^{\infty} \omega_n P_n \to$

$$\int_{-\infty}^{\infty} P(k)\,\omega(k)\,dk \equiv H_{KG}[p] \tag{2.10}$$

in which $\omega(k)$ is given by (2.5).

It is plain that (2.10) is H_{KG} now expressed in terms of action-angle variables. Since $\{P_m,Q_m\} = \delta_{mn}$, $\{P(k), Q(k')\} = \delta(k-k')$. The phase space is a continuously infinite product of cylinders $0 \le P(k) < \infty$, $0 \le Q(k) < 2\pi$, and the motion is on a continuously infinite dimensional torus. Plainly Hamilton's equations are

$$Q_t(k,t) = \delta H_{KG}/\delta P(k,t) = \omega(k);\ P_t(k,t) = -\delta H_{KG}/\delta Q(k,t) = 0 \tag{2.11}$$

and the $P(k)$ are constants of the motion. Evidently the Jacobian $\partial(\bar{\Pi},\bar{\phi})/\partial(P,Q) = 1$ (being the continuously infinite product of unity). It is plain too that $\mathfrak{D} \equiv \int(d\bar{\Pi}(k)_\wedge d\bar{\phi}(k))\,dk = \int(dP(k)_\wedge dQ(k))\,dk$. And it is easy to check that $\mathfrak{D}$ is also equal to $\int(d\Pi(x)_\wedge d\phi(x))\,dx$ providing $\bar{\Pi}(k) = (2\pi)^{-1}\int\Pi(x)e^{-ikx}dx$ when $\bar{\phi}(k) = \int\phi(x)e^{ikx}dx$. Then $H_{KG}[\phi]$ transforms to

$$\frac{1}{2\pi}\int\Big[\tfrac{1}{2}(4\pi^2\bar{\Pi}(k)\bar{\Pi}(-k)) + \tfrac{1}{2}\omega^2(k)\,(\bar{\phi}(k)\bar{\phi}(-k))\Big]\,dk \tag{2.12}$$

and the equations of motion (2.4) again follow. This analysis shows that the Fourier transform is canonical with this definition of $\bar{\Pi}(k)$ and (2.12) expresses H_{KG} in the new, Fourier transformed, canonical variables: $\{\bar{\Pi}(k),\bar{\phi}(k')\} = \delta(k-k')$. Thus, the route we have followed from $(\Pi,\phi) \to (P,Q)$ is also canonical since it is a sequence of two canonical transformations. By implication, the Jacobian of this transformation is unity. Note that $\bar{\Pi}$, $\bar{\phi}$ in (2.12) are now complex fields.

It is also plain that $\{P(k),P(k')\} = 0$, so one continuous infinity of constants of the motion $\{P(k);\ -\infty < k < \infty\}$ are 'in involution' (their Poisson brackets vanish). It is also plain that the system can be integrated, for plainly $Q(k,t) = \omega(k)t + \delta(k)$ ($\delta(k)$ fixed by initial conditions). Then $\bar{\phi}(k,t)$ can be constructed and $\phi(x,t)$ found by Fourier inversion.

Thus the K-G, eqn. (2.1), with Hamiltonian (2.3) comes into the class of 'completely integrable' Hamiltonian systems. Technically a Hamiltonian system with 2N degrees of freedom is completely integrable if it has N first integrals $\{I_n\}$ which are constants of the motion in involution [12]: in this case, by a method of generating functions involving quadratures only, action-angle variables (P_n,Q_n) can be found [12] and H can be expressed as

$$H[p;N] \equiv H(\{P_n;\ n = 1,2,\cdots,N\}).$$

$$\text{Then}\quad P_{n,t} = -\partial H/\partial Q_n = 0$$

$$\text{and}\quad Q_{n,t} = \frac{\partial H}{\partial P_n} = \omega_n(\{P_m\}) = \text{constant} \tag{2.13}$$

and this system can be integrated. The K-G is unusual in that the number of degrees of freedom is continuously infinite. It is, however, a particularly simple case — it is *separable*, ie. $H = \sum_n \omega_n P_n$ or $H = \int\omega(k)P(k)\,dk$, a linear sum of action variables alone each multiplied by a frequency $\omega(k)$ independent of the $P(k)$. We shall see (§4) that the s-G has a Hamiltonian expressible in a similar separable form. Consequently, this would seem to simplify the calculation of Z: a separable Hamiltonian acts like an ideal (non-interacting) gas. However, in practice, we shall find that because of the structure of the functional integral itself, this simplicity is illusory: the effective H is not separable for the s-G and the 'gas' does interact. But our confrontation of this problem in these lectures is still some way away yet!

It will also be helpful to compare all of this discussion with the corresponding case of a *lattice* with spacing a and periodic boundary conditions, period L. Thus $N = La^{-1}$ and $H = \sum_{-\frac{1}{2}N}^{+\frac{1}{2}N}\tilde{\omega}_n P_n$. The equation of motion is

$$\phi_{\nu,tt} = a^{-2}(\phi_{\nu+1} + \phi_{\nu-1} - 2\phi_\nu) - m^2\phi_\nu\ ;\quad 1 \le \nu \le N\ ; \tag{2.14}$$

and normal co-ordinates are $\phi_\nu = e^{-i\omega_n t}e^{i\nu k_n a}$ providing

$$\tilde{\omega}_n^2 = m^2 + \left(\frac{\sin\frac{1}{2}k_n a}{\frac{1}{2}a}\right)^2\ ;\quad k_n = \frac{2\pi n}{L}\ ,\quad -\tfrac{1}{2}N \le n \le \tfrac{1}{2}N\ . \tag{2.15}$$

For $a \to 0$ this reduces to the periodic case. At fixed $a > 0$, the dispersion relation (2.15) persists as $n, L \to \infty$ so that $k_n \to k$. The Hamiltonian becomes

$$H_{KG}[p;a] \equiv \int_{-\pi/a}^{+\pi/a} \tilde{\omega}(k)\, P(k)\, dk \tag{2.16}$$

with Poisson brackets for $P(k)$, $Q(k)$ as before.

We complete this §2 with a remark on the 'conservation laws' for the K-G. For later reference we work in light-cone co-ordinates: set

$$\tfrac{1}{2}(x+t) = \xi \ , \ \tfrac{1}{2}(x-t) = \eta \quad . \tag{2.17}$$

Then (2.1) is $\phi_{\xi\eta} = m^2\phi$. Think of η as 'time': then

$$\begin{aligned} &\frac{\partial}{\partial\eta}\tfrac{1}{2}\phi_\xi^2 - m^2\frac{\partial}{\partial\xi}\tfrac{1}{2}\phi^2 = 0 \\ &\frac{\partial}{\partial\eta}\tfrac{1}{2}\phi_{\xi\xi}^2 - m^2\frac{\partial}{\partial\xi}\tfrac{1}{2}\phi_\xi^2 = 0 \\ &\quad -------- \\ &\frac{\partial}{\partial\eta}\tfrac{1}{2}\phi_{n\xi}^2 - m^2\frac{\partial}{\partial\xi}\tfrac{1}{2}\phi_{n-1\xi}^2 = 0 \\ &\quad -------- \end{aligned} \tag{2.18}$$

where $\phi_{n\xi} \equiv \phi_{\xi\cdots\xi}$ n times. Thus, for periodic boundary conditions $\phi_{m\xi}(\xi,\eta) = \phi_{m\xi}(\xi+L,\eta)$

$$\frac{d}{d\eta}\int_0^L \tfrac{1}{2}(\phi_{n\xi})^2\, d\xi = 0, \quad n = 1, 2, \cdots \tag{2.19}$$

and the K-G in light-cone co-ordinates has a countably infinite set of 'polynomial' conserved densities $T^{(2n)} \equiv \frac{1}{2}\phi_{n\xi}^2$. The corresponding conserved densities for the s-G in light-cone co-ordinates are [13,14,15]

$$\begin{aligned} T^{(2)} &= \tfrac{1}{2}\phi_\xi^2 \ , \quad T^{(4)} = -\left(\tfrac{1}{2}\phi_{\xi\xi}^2 - \tfrac{1}{8}\phi_\xi^4\right) \\ T^{(6)} &= \tfrac{1}{2}\phi_{\xi\xi\xi}^2 - \tfrac{5}{4}\phi_{\xi\xi}^2\phi_\xi^2 + \tfrac{1}{16}\phi_\xi^6 \quad , \\ T^{(8)} &= \quad ---- \quad ---- \end{aligned} \tag{2.20}$$

Note that the $T^{(2n)}$ do not include the Hamiltonians of the K-G or s-G. In light cone co-ordinates $H_{KG} = \int(\frac{1}{2}m^2\phi^2)\,d\xi$, $H_{S-G} = m^2\int(1-\cos\phi)\,d\xi$. Then, for canonical variables (ϕ_ξ,ϕ) ([9,11,14,17] and see §4) one finds $\phi_{\xi\eta} = \delta H_{KG}/\delta\phi = m^2\phi$, $\phi_{\xi\eta} = \delta H_{S-G}/\delta\phi = m^2\sin\phi$. The densities $\frac{1}{2}m^2\phi^2$ and $m^2(1-\cos\phi)$ prove to be the first of a countably infinite sequence of *non-local* conserved densities in each case [14, 15]: only the first members are local.

All of these results ultimately follow from the 'complete integrability' of the K-G and s-G respectively. Note that the polynomial conserved densities $T^{(2n)}$ though (countably) infinite in number omit the Hamiltonians in each case, so they cannot be complete. What constitutes completeness for the conserved densities in ϕ, ϕ_ξ, $\cdots$ is not known (I think) [15,17]. But the conserved quantities $P(k)$, the action variables (which are not denumerable), seem to be a complete set in involution because the integration can be carried out. This was plain for the K-G and a similar situation surrounds the s-G as we show in §4.

Next we look at a physical application of the s-G eqn. (1) in laboratory co-ordinates.

3. The Ferromagnetic $CsNiF_3$; A Physical Example of the sine-Gordon Equation

Caesium nickel fluoride, $CsNiF_3$, has a hexagonal crystal structure in which Ni^{++} ions form chains along the c-axis. The ions have spin $S = 1$ and the system is a ferromagnetic at low enough temperatures T.[1] The exchange coupling along $\underline{c}$ is stronger than along $\underline{a},\underline{b}$. The magnetic correlation length is linear in T - a prediction for a one-dimensional chain. This and other evidence suggests that Ni^{++} forms strictly one-dimensional $S = 1$ spin chains along $\underline{c}$. There is a phase-transition at $T \sim 3^oK$ apparently due to interchain coupling.[1] But above this, the interchain coupling is weak. A phenomenological Hamiltonian has been proposed, namely [16]

$$H = - J \sum_n \vec{S}_n \cdot \vec{S}_{n+1} + A \sum_n (S_n{}^z)^2 - g\mu_B B^x \sum_n S_n{}^x \tag{3.1}$$

with $J(> 0) = 23.6^oK$ and $A(> 0) = 5^oK$. The linear lattice n runs along z, parallel to $\underline{c}$, and a *transverse* magnetic field B^x has been imposed.

The Hamiltonian

$$H = - J \sum_n \vec{S}_n \cdot \vec{S}_{n+1} \tag{3.2}$$

is the one-dimensional Heisenberg ferromagnet for $J > 0$ and is antiferromagnetic for $J < 0$. The lattice for $J > 0$ is an integrable lattice, whose equations of motion in classical and continuum limits reduce to [18,19,20] (and see below)

$$\vec{S}_t = Ja^2 \vec{S} \times \vec{S}_{zz} \tag{3.3}$$

in which a is the lattice spacing. This system is also integrable, and transforms for $z \to za^{-1}$ to the non-linear Schrödinger equation (NLS)

$$iu_t = u_{zz} + \chi u|u|^2 , \quad \chi > 0 \tag{3.4}$$

$$u = \kappa(z,t) \exp i \int^z \tau(z',t)dz' \tag{3.5}$$

in which $\kappa(z,t) \propto (\vec{S}_z)^2$, an energy density; $\tau(z,t) \propto \vec{S}_z \cdot \vec{S}_t$, a momentum density [18,19]. Any magnetic field B can be transformed away whether it is longitudinal $(B = B^z)$ or transverse $(B = B^x)$ and (3.4) is regained. The NLS is of interest to these lectures in its own right, because the problem of the impenetrable Bose gas studied by YANG and YANG [21] and others [22] is the quantum statistical mechanics of (3.4) in the case $\chi < 0$ (with $\chi \to \infty$).

Reverting to $CsNiF_3$ and the Hamiltonian (3.1) because $A > 0$ and B^x *is* transverse (3.1) becomes the Hamiltonian $H[\phi]$ of the s-G, eqn. (1.2), in classical and continuum limits. The antiferromagnetic TMMC $((CH_3)_4N\ Mn\ Cl_3)$ which has $J < 0$ also reduces in effect to a case of the s-G in this limit [23,24]. For $CsNiF_3$ the Heisenberg lattice (3.2) with $J > 0$ is modified by the single site anisotropy term $A\sum_n (S_n{}^z)^2$: $A = 5^oK$ classically and $A = 9^oK$ quantally: in the latter case $\vec{S}$ satisfies angular momentum commutation relations

$$[S_m{}^\alpha, S_n{}^\beta] = i\varepsilon^{\alpha\beta\gamma} S_n{}^\gamma \delta_{mn} \quad . \tag{3.6}$$

For no magnetic field H is minimised with $S_n{}^z = 0$ but the spins are then free in the $\underline{a},\underline{b}$ (= XY) plane. The magnetic field B^x now introduces a "mass" and in classical terms we can expect the system to approximate to the s-G. More precisely, for $T < (AJ)^{\frac{1}{2}}$ (see below), since $A > 0$, S^z is small and we have "XY" behaviour i.e. motion only in the X-Y plane. If $A/JS(S+1)$ is small enough i.e. S is big enough we expect classical spin behaviour. Thus at the nth site

$$\vec{S}_n = S(\sin\theta_n \cos\phi_n, \sin\theta_n \sin\phi_n, \cos\theta_n) \tag{3.7}$$

with all $\theta_n \approx \frac{1}{2}\pi$. Then H reduces to

$$H \sim - J S^2 \sum_n \cos(\phi_{n+1} - \phi_n) + A \sum_n (\tfrac{1}{2}\pi - \theta_n)^2 - g\mu_B B^x S \sum_n \cos\phi_n \quad . \tag{3.8}$$

---

[1] *$CsNiF_3$ is a ferromagnetic in that J in (3.1) is positive. However, it is actually paramagnetic for $T > 3^oK$ and it becomes antiferromagnetic for $T < 3^oK$ because of interchain coupling.*

The excitations at temperature T are studied by neutron scattering of momentum transfer q (say) [25]. In the long wavelength "limit" $qa \ll (2A/J)^{\frac{1}{2}}$ we shall see that it is appropriate to set

$$H = J\,Sa^2 \int a^{-1}dz\,\{\tfrac{1}{2}\phi_z^2 + (AS^2/JS^2a^2)\,[\Theta(z)]^2 + m^2(1-\cos\phi)\} \tag{3.9}$$

in which

$$m^2 \equiv (g\mu_B\,B^x/JSa^2) \quad ; \tag{3.10}$$

m is a reciprocal length. We have set $\Theta \equiv \frac{1}{2}\pi - \theta$. We do not yet have the canonical variables, but can infer these from the equations of motion.

The classical Poisson brackets corresponding to (3.6) are

$$\{S_m^\alpha, S_n^\beta\} = -\,\varepsilon^{\alpha\beta\gamma}\,S_m^\gamma\,\delta_{mn} \quad . \tag{3.11}$$

With these or Heisenberg's quantum equations of motion with commutators (3.6) we can reach

$$\dot{S}_n^{\,x} = -\,A[S_n^{\,y}S_n^{\,z} + S_n^{\,z}S_n^{\,y}] + J[S_n^{\,z}S_{n+1}^{\,x} + S_n^{\,z}S_{n-1}^{\,x} - S_n^{\,y}S_{n+1}^{\,z} - S_n^{\,y}S_{n-1}^{\,z}], \text{etc.} \tag{3.12}$$

Then in classical and continuum "limits" these yield

$$\dot{\theta} = -\,J\,Sa^2[2(\cos\theta)\theta'\phi' + \sin\theta\phi''] - g\mu_B B^x\sin\phi$$

$$\sin\theta\,\dot{\phi} = J\,Sa^2[\theta'' - \sin\theta\,\cos\theta\cdot\phi'^2] + 2AS\sin\theta\,\cos\theta + g\mu_B B^x\cos\theta\,\cos\phi \quad . \tag{3.13}$$

Conveniently dot is $\partial/\partial t$, dash is $\partial/\partial z$. In the experiments [25] $B^x = 5$kG and $g\mu_B B^x \sim 0.5^\circ$K. Since classically, $2AS \sim 18^\circ$K, $g\mu_B B^x \ll 2AS$ and then (3.13) is

$$\dot{\phi} = 2AS\cos\theta$$

$$\dot{\theta} = -\,J\,Sa^2[2(\cos\theta)\theta'\phi' + \sin\theta\cdot\phi''] - g\mu_B B^x\sin\phi \quad . \tag{3.14}$$

This actually requires θ', θ'', and ϕ'^2 all small enough. But (as we shall see) $\phi_z^2 \approx m^2$, so $g\mu_B B^x \ll AS < JS$ implies ϕ_z small enough. And if, in addition, $\Theta \equiv \frac{1}{2}\pi - \theta$ is small with θ', θ'' small

$$\dot{\phi} = 2AS\Theta$$

$$\dot{\Theta} = J\,Sa^2\phi'' - g\mu_B B^x\sin\phi \tag{3.15}$$

which is the s-G with m^2 given by (3.10) and a velocity c given by $c^2 = 2AJ\,Sa^2$. Thermal agitation controls Θ so that $A\langle\Theta^2\rangle \propto k_BT$. With A in $^\circ$K the characteristic scaling proves to be $(2A/J)^{\frac{1}{2}}$ so that $\langle\Theta^2\rangle$ small requires $T \ll (2AJ)^{\frac{1}{2}}$. Alternatively, the energy scale is $(2AJ)^{\frac{1}{2}}$ (see below) so the same condition obtains. Since $(2AJ)^{\frac{1}{2}} \approx 15^\circ$K and the experiments were performed up to $T \approx 10^\circ$K [25] this condition is not necessarily well satisfied. In this case we have to consider significant out-of-plane motions i.e. significant motion in Θ. We consider this possibility in more detail below.

So far (3.15) is certainly the s-G. Moreover with $S\Theta a^{-1} \equiv \Pi$, and $\{\Pi,\phi\} = \delta(z-z')$ Hamilton's equations on H, eqn. (3.9), yield (3.15) precisely, so justifying the scaling adopted there. However, by restoring $\hbar$ to the problem and finally setting $\hbar = c = 1$ *with* $a = 1$ one reaches $H[\phi]$, eqn. (1.2), with $\gamma_0 = \sqrt{2A/J}$: m is unchanged, but $mc^2 = m$ $(c = 1)$ is also an energy. Since $c = \sqrt{2AJ} = 1$ for $S = a = 1$ the energy is scaled by $\sqrt{2AJ}$: this is the scale referred to above, and this is the scale to be put on H in (3.9) (so that $H[\phi]$ of (1.2) then follows). Notice that $\Pi \equiv \Theta$ precisely now. I know of no *a priori* reason why Θ, namely $\frac{1}{2}\pi - \theta$, plays the role of a canonical momentum — still it does!

The analysis shows the classical s-G may be a good model of $CsNiF_3$ for $3^oK < T < 10^oK$. But since Θ may be too large for $T \approx 10^oK$ we need to consider this. And we should look too at possible quantal corrections. For these latter, a measure is the coupling constant $\gamma_0 = \sqrt{2A/J}$. This has value ~ 0.65 for $CsNiF_3$ while $\gamma_0 \gtrsim 1$ is the quantal range. To see this, we need to consider the classical s-G in more detail first. We return to the large Θ problem after a brief investigation of quantal effects.

On infinite support, $-\infty < z < \infty$, and boundary conditions $\phi \to 0 \pmod{2\pi}$, $\phi_z \to 0$ as $|z| \to \infty$, the s-G (1.2) has *four* and only four distinct sorts of bounded C^∞ solutions (for the existence of the spectral transform (§4) one assumes e.g. $\int_{-\infty}^{\infty}(1+|z|)\phi(z)dz < \infty$): these are the 'kink' and 'antikink' solutions

$$\phi = 4\tan^{-1}\exp[\pm m(z - Vt)(1 - V^2)^{-\frac{1}{2}}] \quad ; \tag{3.16}$$

the breather solution

$$\phi = 4\tan^{-1}[\tan\theta\sin\{m\cos\theta(t - Vz)(1 - V^2)^{-\frac{1}{2}}\}\operatorname{sech}\{m\sin\theta(z - Vt)(1 - V^2)^{-\frac{1}{2}}\}] \tag{3.17}$$

in which θ is a parameter $(0 < \theta < \frac{1}{2}\pi)$ and nothing to do with θ above; and the "phonon" or "magnon" solutions, which are vanishingly small amplitude harmonic modes

$$\phi = \varepsilon e^{-i(\omega t - kz)} + 0(\varepsilon^2) \tag{3.18}$$

satisfying the dispersion relation (2.5). From this view the solutions (3.18) to $0(\varepsilon)$ are solutions of the linearised equation namely the K-G, eqn. (2.1).

It is instructive to consider the breather solution as a bound kink-antikink pair. In the rest frame $V = 0$, the solutions (3.16) have rest energy $M \equiv 8m\gamma_0^{-1}$ (put (3.16) into $H[\phi]$). The breather (3.17) has rest mass $2M\sin\theta$ however. For $\theta = \frac{1}{2}\pi$ we can expect the breather becomes a kink plus antikink: indeed, in the rest frame the breather (3.17) has internal frequency $m\cos\theta \to 0$ as $\theta \to \frac{1}{2}\pi$. On the other hand for $\theta \to 0$

$$\phi \to 4\theta\sin(\omega t - kz) \tag{3.19}$$

with dispersion relation (2.5), so the breathers become the harmonic modes (3.18). This fact will cause us a great deal of trouble later!

The classical breather rest energies $2M\sin\theta$ lie in $0 < \theta < \frac{1}{2}\pi$: we shall find a *canonical* co-ordinate Θ (nothing to do with Θ above) which acts as an internal momentum. More precisely, $4\gamma_0^{-1}\Theta$ is that momentum; and there is a canonical Φ, $0 \le \Phi < 8\pi$; $\{4\gamma_0^{-1}\Theta_m, \Phi_n\} = \delta_{mn}$. Naive Bohr quantisation then tells us

$$4\gamma_0^{-1}\int_0^{8\pi}\Theta\, d\Phi = n2\pi \quad ; \tag{3.20}$$

the 2π arises since $\hbar = 1$ so $h = 2\pi$. But Θ is a constant of the motion (θ does not evolve in time in (3.17) either). So $\Theta = n\gamma_0/16$ and the Bohr quantised mass spectrum is

$$16m\gamma_0^{-1}\sin(n\gamma_0/16) \quad ; \quad n = 1,2,\cdots, N \tag{3.21}$$

where N is the integral part of $8\pi\gamma_0^{-1}$. This was the result found by DHN [26] by semi-classical quantisation of the s-G except that $\gamma_0 \to \gamma_0'' = \gamma_0/[1 - \gamma_0/8\pi]$ and we comment on this below. The lowest breather level is $m + 0(\gamma_0^2)$ and for small n the eigenenergies are $mn + 0(\gamma_0^2)$ corresponding to n quantised particles of the field. However, the total number of levels is $N \le 8\pi\gamma_0^{-1}$ and with $\gamma_0 = 0.65$ $N \sim 38$. Thus if 38 levels of the quantised theory are indistinguishable from the breather band continuum, the quantum aspects of the s-G are irrelevant.

Still, it is worth asking what happens to the classical 'phonons' (3.18): in the quantised theory these go into the structure of the vacuum and their only consequence is vacuum fluctuations. It is these that make $\gamma_0 \to \gamma_0''$ found by DHN [26]. Interestingly enough, the statistical mechanics we develop needs to generalize these vacuum fluctuations to include the thermal ones. We see how this happens in §§7, 10-12.

It is not necessarily clear that our demonstration, that the quantum s-G approximates to the classical s-G for $\gamma_0 \approx 0.65$ means that $CsNiF_3$ does not show significant quantal corrections. However, an analysis of the strictly quantum problem for $B^x = 0$, for which case (3.1) is quantum-integrable in continuum limit [27] shows that for $A = 9^\circ K$, $B = 23.6^\circ K$ the system is indeed effectively classical. I do not know if the *lattice* (3.1) with $B^x \equiv 0$ is quantum integrable, but we can expect that it is (compare [3]). For in classical and continuum limits, with $B^x \equiv 0$, (3.1) reduces to the integrable Landau-Lifshitz equation [3,28]. This equation covers the large out-of-plane angle Θ as we now show — but only in the case $B^x \equiv 0$.

The Hamiltonian of the Landau-Lifshitz equation is

$$H = \tfrac{1}{2}\int_{-\infty}^{\infty} [a^2 J \vec{S}_z^{\,2} - \vec{S}\cdot\mathcal{J}\cdot\vec{S} + J_1]\,dz \tag{3.22}$$

in which $\mathcal{J} = \mathrm{diag}(J_1, J_2, J_3)$. The Poisson brackets are

$$\{S^\alpha, S^\beta\} = -\varepsilon^{\alpha\beta\gamma} S^\gamma \delta(x - x') \tag{3.23}$$

and the equations of motion prove to be

$$\vec{S}_t = Ja^2 \vec{S} \times \vec{S}_{zz} + \vec{S} \times (\mathcal{J}\cdot\vec{S}) \tag{3.24}$$

which reduces to (3.3) in the isotropic case $\mathcal{J}$ is the unit tensor. The spins $\vec{S}$ are classical spins given by (3.7) and (3.24) means

$$\dot{\theta} = -JSa^2[2\cos\theta\cdot\theta'\phi' + \sin\theta\phi''] + (J_1 - J_2)S\sin\theta\sin\phi\cos\phi \tag{3.25a}$$

$$\sin\theta\,\dot{\phi} = JSa^2[\theta'' - \sin\theta\cos\theta\phi'^2] + (J_1 - J_2)S\sin\theta\cos\theta\cos^2\phi + (J_2 - J_3)S\sin\theta\cos\theta \quad . \tag{3.25b}$$

When $J_1 = J_2$, and $J_2 - J_3 = 2A$, equations (3.25) coincide exactly with (3.13) providing $B^x \equiv 0$.

Evidently, when $J_1 = J_2$ and $A = \tfrac{1}{2}(J_2 - J_3) = 0$, equations (3.25) are the Heisenberg ferromagnet in classical and continuum limit, namely (3.3). By setting $S_\pm = S^x \pm iS^y$ and $b = g\mu_B B_z$ for a *longitudinal* field B^z that field can be transformed away by $S_\pm \to S_\pm e^{\mp\frac{1}{2}ibt}$ even when $A \neq 0$. Unfortunately, the field B^x can be transformed away in similar fashion only if $A = 0$ that is for the Heisenberg case. Thus, the Landau-Lifshitz equation is not applicable to $CsNiF_3$ in a transverse field B^x.

If $\Theta = \tfrac{1}{2}\pi - \theta$ is small (an XY-model), eqns. (3.25) are

$$\dot{\Theta} = JSa^2\phi'' + \tfrac{1}{2}(J_1 - J_2)S\sin 2\phi$$
$$\dot{\phi} = (J_2 - J_3)S\Theta - JSa^2\Theta\phi'^2 \quad . \tag{3.26}$$

If $p = J\Theta$, $z \to JS^{\frac{1}{2}}az$, and $S(J_1 - J_2)J \to m^2$, $S(J_2 - J_3)J^{-1} \to 1$, $S(J_1 - J_3)J^{-1} \to 1$ eqn. (3.26) becomes as $J \to \infty$

$$P_t = \phi'' + \tfrac{1}{2}m^2\sin 2\phi \quad , \quad \phi_t = p \tag{3.27}$$

and this is the s-G in which the mass is generated from the anisotropy $(J_1 - J_2)$. This is a very different s-G from that derived from (3.1) where $J_1 = J_2$ and the mass is derived from the transverse magnetic field while the anisotropy $J_2 - J_3 = 2A$

keeps the canonical momentum $\Pi = S\Theta a^{-1}$ in the problem (see (3.9)) while the coupling constant $\gamma_0 \propto A^{\frac{1}{2}}$.

By putting $J_1 - J_3 = J_1 - J_2 = \gamma R$ and $\psi = R^{\frac{1}{2}} e^{i\gamma R\tau} S_+$, $t = R^{\frac{1}{2}}\tau$, one can also obtain the NLS [28]. But again, this is not the transform of the isotropic problem which is the Heisenberg chain (3.3) mapping to the NLS (3.4)($R \to \infty$ for the NLS [28]).

The result (3.27) associates with the fact that it is a classical and continuum limit of the quantum spin-$\frac{1}{2}$ XYZ model for which

$$H = -\sum_n (J^x S_n^x S_{n+1}^x + J^y S_n^y S_{n+1}^y + J^z S_n^z S_{n+1}^z) \quad . \tag{3.28}$$

The S_n^α satisfy the spin commutation relations (3.6). In continuum (but quantum) limit this maps to the quantised massive Thirring model

$$H = \int dz \{-i(\psi_1^+ \partial_z \psi_1 - \psi_2^+ \partial_z \psi_2) + m_0(\psi_1^+ \psi_2 + \psi_2^+ \psi_1) + 2g(\psi_1^+ \psi_2^+ \psi_2 \psi_1)\} \tag{3.29}$$

in which $m_0 \propto J^x - J^y$ and $g \propto J^z$. The ψ satisfy

$$\{\psi^+\psi + \psi\psi^+\} \equiv [\psi^+, \psi]_+ = \delta(z - z') \tag{3.30}$$

and this fermi system is *equivalent* to (that is up to expectation values and correlation functions [29,30]) the quantised s-G. This result is particularly relevant to our work in so far as TAKAHASHI and SUZUKI [31] used the Bethe ansatz method (see [6]) to derive the quantum statistical mechanics of the spin-$\frac{1}{2}$ XYZ model. And FOWLER [7] and IMADA [8] have used the mapping to the s-G to derive the quantum statistical mechanics of the s-G. We return to this in §13. We note in passing the following: the spin-$\frac{1}{2}$ XY model is (3.28) with $J^z \equiv 0$. This is equivalent in continuum limit to a free massive fermion system ($g \equiv 0$) and corresponds to the quantised s-G with $\gamma_0 = 4\pi - \varepsilon$. In this case $\gamma_0'' = 8\pi - \varepsilon$, and this is exactly the condition for no quantised breather states ie. N = integral part of $1 - \varepsilon = 0$. We also note that the 8-vertex model of 2-dimensional statistical mechanics and the one-dimensional spin-$\frac{1}{2}$ XYZ model are equivalent [1]: in particular, the lowest eigenvalue of the spin-$\frac{1}{2}$ XYZ model determines the 8-vertex model's free energy: the case $J^z = 0$ (corresponding to $\gamma_0 = 4\pi - \varepsilon$ above in the continuum limit) corresponds to the square lattice Ising model.

These remarks have taken us some way from $CsNiF_3$: our conclusion so far is that in classical and continuum limits this is the classical integrable sine-Gordon system for $3^\circ K < T \lesssim 10^\circ K$, that quantum effects are (probably) unimportant, that out of XY-plane motions may be important for $T \gtrsim 10^\circ K$, and that here, the system is Landau-Lifshitz and integrable if and only if $B^x \equiv 0$. We then indicated connections of the Landau-Lifshitz equation with the quantum spin-$\frac{1}{2}$ XYZ model: the L - L' is the classical and continuum limit of this model.

The experimental position obtained by neutron scattering from $CsNiF_3$ for $3^\circ K < T \lesssim 10^\circ K$ is as follows [25,32]: we need to calculate dynamical form factors

$$S_{\parallel}(z,t) = \langle(1-\cos\phi(z,t))(1-\cos\phi(0,0))\rangle, \; S_{\perp}(z,t) = \langle\sin\phi(z,t)\sin\phi(0,0)\rangle \tag{3.31}$$

where $\langle \cdots \rangle$ denotes thermal average. The Fourier transforms $S_{\parallel}(q,\omega)$; $S_{\perp}(q,\omega)$ are observed. The problem of calculating correlation functions like $S_{\parallel,\perp}(z,t)$ is a difficult one. However, we know that if $CsNiF_3$ is s-G, its thermal excitations will be the kink, antikink, breather and "phonon" excitations of the s-G. MIKESKA [16] assumed a dilute gas of kinks and antikinks alone and derived (up to factors and with $c = 1$)

$$S_{\parallel}(q,\omega) = 64(m^2 q)^{-1} (\pi m\beta)^{\frac{1}{2}} n_k \left\{\exp\left[-\frac{4m\beta\omega^2}{q^2}\right]\right\} \times \{(\pi q/2m)(\sinh[\pi q/2m])^{-1}\}^2 \tag{3.32}$$

where

$$n_k(T) = (2m\beta)\frac{M}{2\pi}\sqrt{\frac{2\pi}{M\beta}}\, e^{-M\beta} = (2m\beta)\, 2\, (m^{\frac{1}{2}}\pi^{-\frac{1}{2}}\gamma_0^{-\frac{1}{2}}\beta^{-\frac{1}{2}})\, e^{-8m\gamma_0^{-1}\beta} \tag{3.33}$$

is an equilibrium kink density, the number of kinks per unit length. It is also the antikink density.

Mikeska's result (3.32) means that at finite $q \lesssim 4m/\pi$ there is a 'central' Gaussian peak in ω with width $\Delta\omega = q/(4\beta m)^{\frac{1}{2}}$; $\Delta\omega$ is linear in q and $\propto T^{\frac{1}{2}}$, and there is no adjustable parameter. Moreover, the integrated intensity $I(q) = \int_0^\infty S_{||}(q,\omega)d\omega \propto T^{-\frac{1}{2}} \exp - \beta M$ at fixed q, B^x; and, at fixed $q, T, I(q) \propto B^{-\frac{1}{4}} \exp - (\text{const.}\, B^{\frac{1}{2}})$ (where $B \equiv B^x$). Thus, in particular at fixed q, B^x, $\ln(I(q)T^{\frac{1}{2}}) = MT^{-1}$ if M is measured in $^{\circ}$K. The absolute scale of $I(q)$ is not determined. But the data fitted to the topmost temperature point gave a good straight line with $M \approx 27^{\circ}$K instead of $M \approx 34^{\circ}$K for the given $B^x = 5$ kG. However, consistency with the q-dependence (giving $M \sim 69^{\circ}$K [32]) was not obtained. Moreover, the analysis ignores the breathers: since n_k has Boltzmann factor $e^{-M\beta}$ and massless breathers for which $2M\sin\theta \to 0$ exist, we expect breather contributions to dominate for $T \ll 34^{\circ}$K ie. the experimental range $T \lesssim 10^{\circ}$K. In fact, my colleague J. Timonen and myself found a contribution $\sim T^2$ as $T \to 0$ [24]. But these T dependences are controlled by the kink, antikink, and breather densities n_k, $n_{\bar{k}}$, and n_b; and these densities can be derived from the partion function Z. The calculation of Z is the subject of §7 and following sections. We shall see that the effective n_b is

$$n_b = m\left[\tfrac{1}{4}(M\beta)^{-1} + \tfrac{1}{8}(M\beta)^{-2} + \tfrac{3}{16}(M\beta)^{-3} + \cdots\right] \tag{3.34}$$

and this is actually *linear* in T — so breathers certainly dominate at low T.

Even so, it has not been possible to find consistent agreement with the neutron-scattering observations. Coupled spin wave theory succeeds quite well in some respects [33]. And others have argued that out-of-plane motion must be seriously taken into account [34]. Thus, to date, the neutron scattering from $CsNiF_3$ has not been shown as unequivocal evidence for the (indirect) observation of sine-Gordon soliton excitations. The statistical mechanics, quantum and classical, is nevertheless plainly of sufficient interest to warrant its serious investigation. The rest of these lectures is concerned with this problem. As projected in §1, the proposal is to exploit the complete integrability of the s-G and compute the partition function Z this way. Thus, we need to demonstrate complete integrability for the s-G. And to do this we need some results on the classical 'inverse scattering method'.

4. Complete Integrability of the Classical sine-Gordon Equation

It is convenient to follow habit and use co-ordinates x,t (not z,t). The inverse method is derived for 'infinite support' $-\infty < x < \infty$. For the s-G the dependent variable is conveniently $\phi_x \equiv u$ (say). For then, in light-cone co-ordinates x,t the s-G is the *nonlinear evolution equation* $u_t = m^2 \sin(\int_{-\infty}^{x} u\, dx')$: the boundary conditions are $u \to 0$ for $|x| \to \infty$ (or $\phi = 0 (\text{mod } 2\pi)$, $\phi_x, \phi_{xx}, \ldots, \to 0$ as $|x| \to \infty$). In nonlinear evolution form the s-G is solvable via a 2x2 matrix spectral transform (called variously the Dirac, Zakharov-Shabat or AKNS spectral problem [14, 35]). This takes the form

$$\hat{L}v = \zeta v \; ; \quad \hat{L} = +i\begin{pmatrix} \partial/\partial x & -q \\ r & -\partial/\partial x \end{pmatrix} \tag{4.1}$$

with $r = -q = -q^* = \frac{1}{2}\phi_x$ (i.e. r, q, $u = \phi_x$ all real): v is the 2-spinor $\begin{pmatrix} v_1 \\ v_2 \end{pmatrix}$, ζ is a complex valued "eigenvalue" or spectral parameter. With $\hat{L}$ one associates

$$v_t = \hat{A}v \; ; \quad \hat{A} = \frac{im^2}{4\zeta}\begin{bmatrix} \cos\phi & \sin\phi \\ \sin\phi & -\cos\phi \end{bmatrix} . \tag{4.2}$$

It is easily checked that the integrability condition $\hat{v}_{tx} = v_{xt}$ on the pair of linear relations (4.1) and (4.2) for *any* 2x2 matrix $\hat{A}$ means

$$\hat{L}_t - \zeta_t = [\hat{A},\hat{L}] \quad ; \tag{4.3}$$

from (4.2) for $\hat{A}$ it follows that

$$\phi_{xt} = m^2 \sin\phi \tag{4.4}$$

together with $\zeta_t = 0$. In this sense the 'flow' (4.4) is 'isospectral', and $\hat{L}$ and $\hat{A}$ constitute a 'Lax pair' of linear operators for the s-G. Experts in the theory of differential forms may like to write (4.1) and (4.2) as $dv = \Omega v$ in which Ω is a matrix of 1-forms. Then (4.3) is the 'vanishing curvature' condition $d\Omega + \Omega_\wedge\Omega = 0$ [5,9,14,15]. Since the spectral problem (4.1) determines all the constants of the motion (see below) and all the integrable flows are constructed from them [9, 14,15], $\hat{L}$, and $\hat{L}$ alone, is sufficient to determine that pair. We therefore focus on the spectral problem (4.1).

For the two independent fields q,r in general we can solve (4.1) for the given initial data $q(x,0)$, $r(x,0)$. It is convenient to introduce the idea of scattering data S associated with this initial data. The mapping $q, r \to S$ is 1:1 and the inverse mapping regains q,r from S. The key to the inverse method is to find S from the data at $t = 0$, then derive $q(x,t)$, $r(x,t)$ from S at time t.

Consider then for arbitrary t the two 'Jost' function solutions of (4.1), $\psi(x,t)$ and $\bar{\psi}(x,t)$. Set $\xi = \mathrm{Re}\zeta$, $\eta = \mathrm{Im}\zeta$. Since $q, r \to 0$ as $x \to +\infty$ (the boundary condition)

$$\psi \sim \begin{bmatrix}0\\1\end{bmatrix} \exp(+i\xi x) \ , \quad \bar{\psi} \sim \begin{bmatrix}1\\0\end{bmatrix} \exp(-i\xi x) \tag{4.5}$$

for $x \to \infty$. Correspondingly, since $q, r \to 0$ as $x \to -\infty$, there is a pair of Jost functions $\bar{\phi},\phi$ asymptotic to the same forms as $x \to -\infty$. Since there can only be one linearly independent pair

$$\phi = a\,\bar{\psi} + b\psi \quad , \quad \bar{\phi} = \bar{b}\,\bar{\psi} + \bar{a}\psi \tag{4.6}$$

where $a,b,\bar{a},\bar{b}$ depend only on ξ (and perhaps t). Two solutions of (4.1) with row vectors $\tilde{u} = (u_2,u_1)$, $\tilde{v} = (v_2,v_1)$ (the transposed form is chosen this way to define the natural inner product $\langle u,v\rangle \equiv \int(u_2v_1 + u_1v_2)dx$ [9,13,14,36]) have Wronskian $W(u,v) = u_1v_2 - u_2v_1$ independent of x. It is easy to show that $a = W(\phi,\psi)$, $b = -W(\phi,\bar{\psi})$, $\bar{a} = W(\bar{\phi},\bar{\psi})$, $\bar{b} = W(\bar{\phi},\psi)$ and

$$a\,\bar{a} - b\,\bar{b} = 1 \quad . \tag{4.7}$$

Note that $|a|$ and $|b|$ are unbounded: since, with $w = b\,a^{-1}$, $w\,\bar{w} + (a\,\bar{a})^{-1} = 1$, w and a^{-1} are natural reflexion and transmission coefficients. Since ϕ,ψ can be continued in the upper half ζ-plane and $\bar{\phi},\bar{\psi}$ in the lower half plane $a(\zeta)(\bar{a}(\zeta))$ is analytic in the upper (lower) half plane: $b(\zeta)(\bar{b}(\zeta))$ is analytic in the upper (lower) half plane only for finite support $L < \infty$. The zeros of $a(\zeta)$ in the upper half plane determine bound state eigenvalues: here both ϕ and ψ are exponentially damped as $|x| \to \infty$. Similar remarks apply to the roots of $a(\zeta) = 0$ in the lower half plane. For bound states $\hat{L}$ cannot be Hermitean (i.e. $r \neq q^*$): otherwise the eigenvalues are solely real.

With $a(\zeta_j) = 0$ it is convenient to write b_j for the value of b at $\zeta = \zeta_j$ and to set $b_j = ik_j\,\dot{a}(\zeta_j)$; likewise $\bar{b}_j = -i\ell_j\,\dot{\bar{a}}(\bar{\zeta}_j)(\dot{f}(\zeta_j) \equiv (df/d\zeta)_{\zeta=\zeta_j})$. The scattering data S are then [11,36]

$$S = \{\zeta_j,\bar{\zeta}_s,k_j,\ell_s;a(\xi),\bar{a}(\xi),w(\xi),\bar{w}(\xi) ;$$

$$\xi \in R;\ j = 1, 2, \ldots, M;\ \ s = 1, 2, \ldots, N\} \quad . \tag{4.8}$$

This is one possible choice with the key property that it is complete for inversion by a linear integral equation (the Marchenko equation [9,13,14,36]) from which q and r are found. In this sense, the spectral transform (4.1) is invertible and since the time evolution (4.2) determines the time evolution of S, $q(x,t)$, $r(x,t)$ at $t > 0$ can be found from $q(x,0)$, $r(x,0)$ so solving the initial value problem.

From (4.1) if $q = r^*$, $\hat{L}$ is Hermitean, $M = N = 0$ and all eigenvalues lie on the real axis $-\infty < \xi < \infty$. If $r = -q$, $M = N$, $\bar{a}(\zeta) = a(-\zeta)$, $-\bar{b}(\zeta) = b(-\zeta)$, $\bar{\zeta}_j = -\zeta_j$, $\bar{b}(\bar{\zeta}_j) = -b(\zeta_j)$; if $r = -q^*$, $\bar{a}(\zeta) = a^*(\zeta^*)$, $\bar{b}(\zeta) = b^*(\zeta^*)$, $\bar{\zeta}_k = \zeta_k^*$, $\bar{b}(\bar{\zeta}_k) = \{b(\zeta_k)\}^*$ (* denotes complex conjugate). In the case of the s-G, $q = -r = -r^* = -\frac{1}{2}\phi_x$ since ϕ is real. Thus $\zeta_j = i\,\eta_j$, $\bar{\zeta}_j = -i\,\eta_j$ (η_j real > 0); or they arise as pairs (ζ_j, $-\zeta_j^*$) with $(\bar{\zeta}_j, -\bar{\zeta}_j^*) = (\zeta_j^*, -\zeta_j)$. The η_j determine the 'speeds' of the kink (or antikink) solutions in light cone-co-ordinates $\phi = 4\tan^{-1}\exp\{-(k_j x + k_j^{-1}t)\}$ with $k_j = 2i\zeta_j = -2\eta_j$; the pairs $(\zeta_j, -\zeta_j^*)$ determine the 'speeds' of breather solutions through $|\zeta_j|$ and their internal oscillation frequencies through $\arg\zeta_j$.

The quantity $\Omega(\zeta) = -m^2/4i\zeta$ for the s-G (4.4) is related to its linearised dispersion relation $\omega(k) = m^2k^{-1}$ by $\Omega(\zeta) = {}^1/2i\ \omega(-2\zeta)$ [9,11,36]. This dispersion relation $\Omega(\zeta)$ more generally determines the infinite class of integrable equations to which the s-G (4.4) also belongs [9,11]. For $x \to \infty$ we have from (4.5) that

$$(\hat{A} - \Omega(\zeta)\hat{I})\phi \sim \begin{bmatrix} 0 & 0 \\ 0 & -2\Omega(\zeta) \end{bmatrix} \begin{bmatrix} a(\xi)e^{-i\xi x} \\ b(\xi)e^{i\xi x} \end{bmatrix} = \begin{bmatrix} a_t(\xi)e^{-i\xi x} \\ b_t(\xi)e^{i\xi x} \end{bmatrix} \quad . \tag{4.9}$$

We can add $-\Omega(\zeta)\hat{I}$ ($\hat{I}$ is the unit matrix operator) to $\hat{A}$ in (4.2) since this does not change (4.3) and this choice is now forced on us by the need to satisfy (4.5) for the Jost functions. By this choice $a(t) = a(0)$, i.e. $a(\zeta)$ is a constant of the motion and $b(\xi,t) = b(\xi,0)e^{-2\Omega(\xi)t}$. For the bound states $a(\zeta_j) = 0$ and $b(\zeta_j, t) = b(\zeta_j,0)e^{-2\Omega(\zeta_j)t}$. Thus, for the s-G $b_j(t) = b_0e^{-t/2\eta_j}$ with $\eta_j > 0$. The triviality of the integration in time relies on the boundary conditions q, $r \to 0$ as $|x| \to \infty$ and this invokes infinite support $L \to \infty$.

From §2, $T^{(2)} = \frac{1}{2}\phi_x^2$ and $m^2(1 - \cos\phi)$ are conserved densities for the motion (4.4). We set

$$H = \frac{-m^2}{4\gamma_0}\int_{-\infty}^{\infty}\left\{\cos\left(-2\int_{-\infty}^{x} q\,dx'\right) + \cos(-2\gamma_0 p) - 2\right\}dx \tag{4.10}$$

$$P = \frac{1}{2\gamma_0}\int_{-\infty}^{\infty}\{q^2 + \gamma_0{}^2\,p_x{}^2\}\,dx \tag{4.11}$$

where

$$q = -\tfrac{1}{2}\phi_x, \quad \text{and} \quad p = \gamma_0{}^{-1}\int_{-\infty}^{x} q\,dx' = -\tfrac{1}{2}\gamma_0{}^{-1}\phi \quad . \tag{4.12}$$

The Poisson bracket is $\{A,B\} = \int_{-\infty}^{\infty}\{(\delta A/\delta p)(\delta B/\delta q) - (\delta A/\delta q)(\delta B/\delta p)\}dx$ and $\{p,q\} = \delta(x - x')$. We can now check that (4.10) means that

$$\begin{aligned} -q_t(= \tfrac{1}{2}\,\phi_{xt}) &= -\,\delta H/\delta p = \tfrac{1}{2}\,m^2\sin u \\ -p_t(= \tfrac{1}{2}\gamma_0{}^{-1}\,\phi_t) &= \delta H/\delta q = (m^2/2\gamma_0)\int_{-\infty}^{x}\sin\phi\,dx' \end{aligned} \tag{4.13}$$

and these equations are the s-G in both cases. If H is written in terms of q alone, the bracket is [9,14] $\{A,B\} = \gamma_0\int\{(\delta A/\delta q)\,\partial(\delta B/\delta q)/\partial x\}\,dx$. Then if $A = \int q(x',t)\,\delta(x - x')dx' = q(x,t)$, $-q_t = \{H,q\} = \partial(\delta H/\delta q)/\partial x$. In this case $\frac{1}{2}\phi_{xt} = \frac{1}{2}m^2 \sin\phi$, and the skew symplectic operator is $\partial/\partial x$.

I show elsewhere [11] that, under the spectral transform (4.1) in the case $r = -q = -q^*$, $p = \gamma_0{}^{-1}\int_{-\infty}^{x} q\,dx'$ that $\mathfrak{W} = \int_{-\infty}^{\infty}(dp(x)\wedge dq(x))dx$ on the symplectic manifold with co-ordinates $\{p(x), q(x)\}$ transforms to

$$\varpi = \sum_{j=1}^{N} dp_{j\wedge}dq_j + \int_{-\infty}^{\infty} d\xi \, (dP(\xi)_\wedge \, dQ(\xi)) \tag{4.14}$$

on a manifold with co-ordinates $\{p_j, q_j; P(\xi), Q(\xi); \; j = 1, \ldots, N; \quad -\infty < \xi < \infty\}$. These complex canonical co-ordinates are defined by

$$p_j = \gamma_0^{-1} \ln \zeta_j, \quad q_j = 2 \ln b_j \; ; \quad 1 \le j \le N \quad ;$$

$$P(\xi) = (2\pi\xi\gamma_0)^{-1} \ln \{a(\xi)\bar{a}(\xi)\} \, , \quad Q(\xi) = \arg b(\xi); \quad -\infty < \xi < \infty \; ; \tag{4.15}$$

(and $\bar{a}(\xi) = a^*(\xi)$).

An important object in our theory will be $a(\zeta)$ in the form

$$a(\zeta) = \prod_{j=1}^{N} \left(\frac{\zeta - \zeta_j}{\zeta - \zeta_j^*}\right) \exp\left\{ - \frac{\zeta}{\pi i} \int_{-\infty}^{\infty} \frac{\ln|a(\xi)|}{\zeta^2 - \xi^2} \, d\xi \right\} \quad . \tag{4.16}$$

We derive it below. Since $a(\zeta)$ is a constant of the motion, $\ln a(\zeta)$ is a Hamiltonian. By expanding $\ln a(\zeta)$ about $\zeta = \infty$ by using $\hat{L}$ we find the polynomial conserved densities $T^{(2n)}$ [11,14,15] and P, (4.11). By expanding about $\zeta = 0$ (a choice dictated by the singularity in $\Omega(\zeta)$) we find H, (4.10), and a sequence of nonlocal densities [11,14,15]. By expressing $a(\zeta)$ in terms of scattering data S through (4.16) and expanding about $\zeta = 0$ and $\zeta = \infty$ we express H and P in terms of S in the forms

$$H = - \tfrac{1}{2} m^2 \gamma_0^{-1} i \left\{ 2i \sum_{j=1}^{M} \eta_j^{-1} - 2 \sum_{\ell=1}^{L} (\zeta_\ell^{-1} - \zeta_\ell^{*-1}) + i\gamma_0 \int_{-\infty}^{\infty} d\xi (\xi^{-1} P(\xi)) \right\} \tag{4.17}$$

$$P = 2i\gamma_0^{-1} \left\{ - 2i \sum_{j=1}^{M} \eta_j - 2 \sum_{\ell=1}^{L} (\zeta_\ell - \zeta_\ell^*) - i\gamma_0 \int d\xi (\xi P(\xi)) \right\} \quad . \tag{4.18}$$

The integral in H, (4.17) is

$$\tfrac{1}{2} m^2 \int_{-\infty}^{\infty} d\xi (\xi^{-1} P(\xi)) \tag{4.19}$$

independent of γ_0. This is $H_{KG} = \int_0^\infty \omega(k) P(k) dk$ with $\omega(k) = m^2 k^{-1}$ for light-cone co-ordinates. We have used $\xi^{-1}P(\xi) = (-\xi)^{-1} P(-\xi)$ and so in effect lose half of the canonical co-ordinates: however they, and their motions, are constrained through $P(-\xi) = -P(\xi)$ corresponding to the dependence of ϕ and ϕ_x and hence of p and q through (4.12). To be more precise given (4.12) we should take (4.19) over $0 \le \xi < \infty$ with $\omega(\xi) = \frac{1}{2} m^2 \xi^{-1}$ and define $P(-\xi) = -P(\xi)$: then $\{P(\xi), Q(\xi')\} = \delta(\xi - \xi')$, $-\infty < \xi < \infty$ with $Q_t (= (\arg b)_t) = \frac{1}{2} m^2 \xi^{-1}$ for $\xi > 0$ as required.

The distinct sums over $j = 1, \ldots, M$ and $\ell = 1, \ldots, L$ in (4.17) and (4.18) arise by grouping the ζ_j into $i\eta_j$ ($\eta_j > 0$) and the pairs $(\zeta_\ell, \zeta_\ell^*)$. It is plain that ϖ, (4.14), is invariant if we use $p_j = \gamma_0^{-1}\zeta_j$, $q_j = 2\zeta_j^{-1} \ln b_j$. Thus, for the s-G in light cone co-ordinates, we can use

$$4\gamma_0^{-1} \eta_j, \; (2\eta_j)^{-1} \ln |b_j|; \quad 8\gamma_0^{-1}|\zeta_\ell|, \quad (2|\zeta_\ell|^{-1}) \ln |b_\ell|;$$

$$\gamma_0^{-1}\Theta_\ell, \quad 4 \arg b_\ell \quad . \tag{4.20}$$

We now have the tricky problem of moving from light-cone to laboratory (Lorentz covariant) co-ordinates corresponding to the s-G (1.1). We observe that P and H can be written in the forms

$$P = \sum_{j=1}^{M} p_j + \sum_{\ell=1}^{L} \hat{p}_\ell + \int_{-\infty}^{\infty} d\xi \; p(\xi) \; P(\xi) \tag{4.21}$$

$$H = \sum_{j=1}^{M} h_j + \sum_{\ell=1}^{L} \hat{h}_\ell + \int_{-\infty}^{\infty} d\xi\, h(\xi)\, P(\xi) \quad . \tag{4.22}$$

We now define covariant quantities P', H' by $P' = H-P$, $H' = H+P$; that is we set

$$h(\xi) = \tfrac{1}{2}\{h'(\xi) + p'(\xi)\}$$

$$p(\xi) = \tfrac{1}{2}\{h'(\xi) - p'(\xi)\}$$

and similarly for (h_j, p_j) and $(\hat{h}_\ell, \hat{p}_\ell)$. We define a new mass $m' = 2m$, and a new coupling constant $\gamma_0' \equiv 4\gamma_0$. Then

$$h_j' = [M^2 + p_j'^2]^{\frac{1}{2}} \ ; \qquad \hat{h}_\ell' = [4M^2 \sin^2\Theta_\ell + \hat{p}_\ell'^2]^{\frac{1}{2}} \ ;$$

$$h'(\xi) = [m'^2 + p'(\xi)^2]^{\frac{1}{2}} \tag{4.23}$$

with $M \equiv 8m'/\gamma_0'$.

We relabel $p'(\xi) \equiv k$: since $p'(\xi) = h(\xi) - p(\xi) = [m'^2/8\xi - 2\xi]$, the mapping $\xi \to k$ is 2:1 since both $\{\xi;\ 0 < \xi < \infty\}$ and $\{\xi;\ -\infty < \xi < \infty\} \to \{k;\ \infty > k > -\infty\}$. We meet this by using $\xi > 0$ in the integrals in (4.22) and (4.21). This discards one half of the canonical variables $(P(\xi), Q(\xi))$ as for the K-G previously. Since $b(\xi) = b^*(\xi)$ for $r = r^* = -q$, $\arg b(-\xi) = Q(-\xi) = -Q(\xi)$. Thus, the motions for $\xi < 0$ are determined from those for $\xi > 0$ and the constraints $P(\xi) = -P(\xi)$, $Q(-\xi) = -Q(\xi)$. The mapping $\xi \to k$ takes $d\xi\, P(\xi) \to dk((d\xi/dk)\, P(\xi))$. We define $P(k) = (-(d\xi/dk)P(\xi))$. The symplectic form $\int_0^\infty (dP(\xi)_\wedge\, dQ(\xi))d\xi$ goes to $\int_{-\infty}^{\infty} (dP(k)_\wedge dQ(k))dk$ with $Q(k)$ defined by $Q(k) = Q(\xi)$. We drop the primed notation finally and write what was P' and H' as

$$P = \sum_{i=1}^{N_k} p_i + \sum_{j=1}^{N_{\bar{k}}} \bar{p}_j + \sum_{\ell=1}^{N_b} \hat{p}_\ell + \int_{-\infty}^{\infty} k\, P(k)dk \tag{4.24}$$

$$H = \sum_{i=1}^{N_k} [M^2 + p_i^2]^{\frac{1}{2}} + \sum_{j=1}^{N_{\bar{k}}} [M^2 + \bar{p}_j^2]^{\frac{1}{2}}$$
$$+ \sum_{\ell=1}^{N_p} [4M^2 \sin^2 \Theta_\ell + \hat{p}_\ell^2]^{\frac{1}{2}} + \int_{-\infty}^{\infty} dk(m^2 + k^2)^{\frac{1}{2}}\, P(k) \ \equiv \ H[p]. \tag{4.25}$$

We are distinguishing between 'kinks' momenta p_i and 'antikinks' momenta p_j.

The mapping $\xi \to k$ with the prime dropped was

$$[m^2/8\xi - 2\xi] = k \quad . \tag{4.26}$$

From this $\xi = \frac{1}{4}(-k \pm \omega(k))$, and the k-plane is cut at $k = \pm im$ where $\omega(k) = 0$. Thus, we cover $-\infty < \xi < \infty$ by using both sheets of the Riemann surface for k. For present purposes the single covering in (4.24) and (4.25) is enough, but we need more when we ask how $a(\xi)$ maps. For the canonical co-ordinates we use p'_j, etc: since $p_j = 4\gamma_0^{-1}\eta_j$ becomes $p'_j = \gamma_0^{-1}(m^2\eta_j^{-1} - 4\eta_j)$, $dp'_j = -16\gamma_0'^{-1}\eta_j^{-2}(m'^2/16 + \eta_j^2)\, d\eta_j$. Hence $dp_j \wedge dq_j = dp'_j \wedge dq'_j$ if $q'_j = -8\eta_j(m'^2 + 16\eta_j^2)^{-1} \ln|b_j|$. This way we reach new canonical co-ordinates with the primed notation dropped, which are

$$\{p_i, q_i;\quad \bar{p}_j, \bar{q}_j;\quad \hat{p}_\ell, \hat{q}_\ell;\quad 4\gamma_0^{-1}\Theta_\ell, \Phi_\ell;\quad P(k), Q(k);$$
$$1 \le i \le N_k,\quad 1 \le j \le N_{\bar{k}},\quad 1 \le \ell \le N_b;\quad -\infty < k < \infty\} \quad . \tag{4.27a}$$

They satisfy

$$\{p_i, q_j\} = \delta_{ij}, \{\bar{p}_j, \bar{q}_i\} = \delta_{ji}, \{\hat{p}_\ell, \hat{q}_m\} = \delta_{\ell m}$$

$$\{4\gamma_0^{-1}\Theta_\ell , \Phi_m\} = \delta_{\ell m}, \quad \{P(k), Q(k')\} = \delta(k - k') \quad . \tag{4.27b}$$

The phase spaces are

$$-\infty < p_i, q_i; \quad \bar{p}_j, \bar{q}_j; \quad \hat{p}_\ell, \hat{q}_\ell < \infty; \quad 0 < 4\gamma_0^{-1}\Theta_\ell < 2\pi\gamma_0^{-1};$$

$$0 < \Phi_\ell < 8\pi; \quad 0 \le P(k) < \infty; \quad 0 \le Q(k) < 2\pi \quad . \tag{4.27c}$$

These are the action-angle variables and their phase spaces for the s-G equation (1.1) and Hamiltonian (1.2) in Lorentz covariant form. They apparently express H for the covariant s-G in the form (4.25), that is as a separable Hamiltonian (or ideal gas) for N_k kinks, $N_{\bar{k}}$ antikinks, N_b breathers, and the "phonons" (or magnons) with dispersion relation (2.5). Plainly, the co-ordinates (4.27a) are collective co-ordinates and the Hamiltonian (1.2) is diagonalised by them. Incidentally, we have shown that the s-G is a completely integrable Hamiltonian system with a continuous infinity of action variables p_i, $\bar{p}_j$, $\hat{p}_\ell$, $4\gamma_0^{-1}\Theta_\ell$, $P(k)$, which are constants of the motion in involution: $\{p_i, P(k)\} = \{P(k), P(k')\} = 0$. In the co-ordinates (4.27a) Hamilton's equations can be integrated, and there is the inverse map from the proper k-sheet to $P(\xi)$, $Q(\xi)$ on $0 < \xi < \infty$ and from these to the same on $-\infty < \xi < \infty$ - and so back to (1.1) with (1.2) by the inverse spectral transform. TAHTADZJIAN and FADDEEV [37] first found (4.25) by the spectral transform on the covariant s-G (1.1). My colleague R.K. DODD and myself [38] found (4.17) and (4.18) and thence (4.25) [9,11,38]. The relation between the two forms is analysed by TAHTADŽJIAN and FADDEEV [17].

It remains to show that the support for the phase-co-ordinates q_i, $\bar{q}_j$, $\hat{q}_\ell$ coincides with L for $L \to \infty$. Evidently, under the mapping (4.26) corresponding to $\phi \sim \begin{bmatrix}0\\1\end{bmatrix} b_j(t)\, e^{-\eta_j x}$ in light-cone co-ordinates will be

$$\phi \sim \begin{bmatrix}0\\1\end{bmatrix} b_j(o) \exp\{-[m'^2 - 16\eta_j^{\,2}]t/8\eta_j\} \exp\{-[m'^2 + 16\eta_j^{\,2}]x/8\eta_j\} \tag{4.28}$$

in covariant laboratory co-ordinates. Then, if x is displaced to $x + X$ in the laboratory frame, $b_j(o)$ gains the factor $\exp\{-[m'^2 + 16\eta_j^{\,2}]X/8\eta_j\}$ and $q_j = -8\eta_j[m'^2 + 16\eta_j^{\,2}]^{-1}\ell n|b_j|$ gains a shift of precisely X, i.e. $q_j \to q_j + X$. Thus if $-\frac{1}{2}L < x < \frac{1}{2}L$ for $L \to \infty$, $-\frac{1}{2} < q_j < \frac{1}{2}L$ for $L \to \infty$. So the supports for q_i, $\bar{q}_j$, $\hat{q}_\ell$ coincide with the support L for x as $L \to \infty$. A spectral operator L for the covariant s-G (1.1) is $-i\partial/\partial x + \frac{1}{2}\zeta\sigma_3 - (1/8\zeta)\sigma_3\cos\phi - (1/8\zeta)\sigma_1\sin\phi + \frac{1}{4}(\phi_x + \phi_t)\sigma_2$, where the σ_i are Pauli matrices. The spectral problem has precisely the Jost function behaviour of (4.28) and agrees with TAHTADZJIAN and FADDEEV [17] after a reinterpretation of their eigenvalues (cf. also DE VEGA [39]).

In the last part of this §4 we derive (4.16) and comment on it. We simply use Cauchy's theorem:- Consider [36]

$$f(\zeta) = a(\zeta) \prod_{i=1}^{N} \frac{\zeta-\zeta_i^*}{\zeta-\zeta_i}, \qquad \bar{f}(\zeta) = \bar{a}(\zeta) \prod_{i=1}^{N} \frac{\zeta+\zeta_i^*}{\zeta+\zeta_i};$$

$$\bar{a}(\zeta) = a(-\zeta) \; ; \qquad \bar{f}(\zeta) = f(-\zeta) \; ; \qquad a(\zeta_i) = \bar{a}(-\zeta_i) = 0 \quad . \tag{4.29}$$

Then $f(\zeta)(\bar{f}(\zeta))$ is analytic in the upper (lower) half plane. We take contours Γ, $\bar{\Gamma}$ which are big semi-circles traversing from right to left just above (Γ) and left to right just below $(\bar{\Gamma})$ the real ζ axis, ξ. Then, for ζ in the upper half plane

$$\ln f(\zeta) = \frac{1}{2\pi i}\int_{\Gamma} \frac{\ln f(\zeta')}{\zeta - \zeta'} d\zeta' \; ; \qquad 0 = \frac{1}{2\pi i}\int_{\bar\Gamma} \frac{\ln \bar f(\zeta')}{\zeta - \zeta'} d\zeta' \quad . \tag{4.30}$$

For $\zeta = \xi$, the end points of $\Gamma(\bar\Gamma)$ are retained as above (below) the real axis but Γ is deformed below $\zeta = \xi$ and $\bar\Gamma$ likewise. Although f is not analytic below $\zeta = \xi$ it has a finite jump there. So we can reach

$$\ln f(\xi) - \tfrac{1}{2}\ln f(\xi)\,\bar f(\xi) = \frac{-P}{2\pi i}\int_{-\infty}^{\infty} \frac{d\xi' \,\ln f\bar f}{(\xi + \xi')} \tag{4.31a}$$

or

$$\ln f(\xi) = \frac{-1}{2\pi i}\int_{-\infty}^{\infty} \frac{d\xi' \,\ln f\bar f}{\xi + \xi' + i\varepsilon} \tag{4.31b}$$

after using the symmetry on $f\bar f$. Indeed, because of the symmetry of the zeros ζ_i, $-\zeta_i$ of a and $\bar a$ (for the s-G) when $\zeta = \xi =$ real, $f\bar f = |a|^2$. Hence (4.31a) means that

$$a(\xi) = |a(\xi)| \prod_{j=1}^{N} \frac{(\xi - \zeta_j)}{(\xi - \zeta_j^*)} \exp\left[\frac{-P}{2\pi i}\left\{\xi \int_{-\infty}^{\infty} \frac{d\xi' \,\ln|a|^2}{(\xi^2 - \xi'^2)}\right\}\right] \tag{4.32}$$

in which we again use the symmetry $|a(\xi)| = |a(-\xi)|$. The same analysis for ζ in the upper half-plane does not indent the contours or the real axis and the result is (4.16).

From the definition of $P(\xi)$ in (4.15) the integral in (4.32) is $i\gamma_0 P[\xi\int_{-\infty}^{\infty}\xi' \, d\xi' \, P(\xi')(\xi^2 - \xi'^2)^{-1}$. We use the mapping (4.26) to show that

$$\frac{m^2\xi\xi'}{m^2(\xi^2-\xi'^2)} \rightarrow \frac{-\,m^2}{2(k\omega(k')-\omega(k)k')} \tag{4.33}$$

with $\omega(k)$ given by (2.5). This way and after replacing γ_0 by $\frac{1}{4}\gamma_0$ we find a contribution to the <u>phase</u> of $a(k)$

$$-\tfrac{1}{8}\gamma_0 m^2 P\left\{\int_{-\infty}^{\infty} \frac{dk' \, P(k')}{k\omega(k')-\omega(k)k'}\right\} \quad . \tag{4.34}$$

It is well known that 'solitons' phase shift each other in pairs when they collide: a two-soliton solution of the s-G (1.1), for example, is the 2-kink solution [40]

$$\phi(x,t) = 4\tan^{-1}[\sinh\{\tfrac{1}{2}(\Theta_1 + \Theta_2)\}(a_{12})^{-\frac{1}{2}}\,\mathrm{sech}\{\tfrac{1}{2}(\Theta_1 + \Theta_2)\}]$$

$$\sim 4\tan^{-1}\exp(\Theta_1 + \delta_1^{\pm}) + 4\tan^{-1}\exp(\Theta_2 + \delta_2^{\pm}) \quad \text{as} \quad t \to \pm\infty \quad . \tag{4.35}$$

Here $\Theta_i = m(x - V_i t)(1 - V_i^2)^{-\frac{1}{2}}$ and $\delta_1^{\pm} = -\delta_2^{\pm} = \pm\frac{1}{2}\ln a_{12}$; $a_{12} \equiv [(a_1 - a_2)/(a_1 + a_2)]^2$ and $a_i^2 = (1 - V_i)/(1 + V_i)$. The phase shifts are therefore

$$\delta_1^{+} - \delta_1^{-} = -(\delta_2^{+} - \delta_2^{-}) = \ln a_{12} \tag{4.36}$$

and phase shift is 'conserved' i.e. sums to zero. Since $a_i = 2\eta_i$, $a_{12} = \{(\eta_1 - \eta_2)(\eta_1 + \eta_2)^{-1}\}^2$ and we shall be able to find this in $a(\zeta)$ for $\zeta_i = i\eta_i$ (§9). So far we have the important result (4.34) which we can view as a *phonon-phonon* phase shift, i.e. as the shift of mode k by all modes $k' \neq k$. Note that it depends on the nonlinearity through γ_0 and will not arise in the K-G.

We also note that under the same mapping

$$\arg\left(\frac{\xi - i\,\eta_i}{\xi + i\,\eta_i}\right) \rightarrow 2\tan^{-1}\{m^2[k\omega(p) - p\omega(k)]^{-1}\} \equiv -\Delta_k(k, p_i) \tag{4.37}$$

where $p = mV_i\gamma_i$, $\omega(p) = (m^2 + p^2)^{\frac{1}{2}}$, $\gamma_i = (1 - V_i^2)^{-\frac{1}{2}}$ and $MV_i\gamma_i = p_i$, the momentum of the kink (or antikink) labelled by i. The result (4.37) is the phase shift of a phonon labelled by k due to a kink of velocity V_j as calculated by CURRIE [41]. A factor 2 in (4.37) is inserted through the mapping. In a similar way the pairs

$$\arg\left(\frac{\xi - \zeta_\ell}{\xi - \zeta_\ell^*} \cdot \frac{\xi + \zeta_\ell^*}{\xi + \zeta_\ell}\right) \rightarrow 4\tan^{-1}\{m^2\sin\Theta[k\omega(p) - p\omega(k)]^{-1}\} \equiv -\Delta_b(k, p_\ell, \Theta) \quad (4.38)$$

where $\Theta = \arg\zeta_\ell$ and p now refers to the translational momentum of the breather with translational velocity V_ℓ as computed by CURRIE [41]. When $V_j = V_\ell = 0$, (4.37) and (4.38) reduce to $2\tan^{-1}(mk^{-1})$ and $4\tan^{-1}(m\sin\Theta\, k^{-1})$, the phase shifts computed by Rubinstein [42] for weak harmonic modes and by DHN [26] as the phonon limit of breather-breather phase shifts respectively.

We therefore have an interesting result: the quantity $a(k)$ has a phase

$$-\arg a(k) = \sum_{i=1}^{N_k} \Delta_k(k, p_i) + \sum_{j=1}^{N_{\bar{k}}} \Delta_{\bar{k}}(k, \bar{p}_j) + \sum_{\ell=1}^{N_b} \Delta_b(k, \hat{p}_\ell, \Theta_\ell)$$
$$+ \tfrac{1}{8}\gamma_0 m^2 P\int \frac{dk' P(k')}{k\omega(k') - k'\omega(k)} \quad . \quad (4.39)$$

This is precisely the total phase shift of the mode labelled by k due to the N_k kinks of momentum p_i, the $N_{\bar{k}}$ antikinks, momentum $\bar{p}_j$, the N_b breathers momenta $\hat{p}_\ell$, $4\gamma_0^{-1}\Theta_\ell$, and all the other phonons labelled by $k' \neq k$. Note that $0 \leq P(k') < \infty$, so these phonons can be of arbitrarily large amplitude: they are not the vanishingly weak modes (3.18) therefore, and the transformation to action-angle variables has decoupled the nonlinearity in this way.

We shall find that the result (4.39) has an important role to play in the statistical mechanics. It will therefore be helpful to draw a connection with the quantum inverse method [5].

We note first that the phonon-phonon shift $\frac{1}{8}\gamma_0 m^2[k\omega(k')-\omega(k)k']^{-1}$ parametrized in rapidities through $k = \sinh\lambda$, $k' = \sinh\lambda'$ is $\frac{1}{8}\gamma_0 m^2[\sinh(\lambda-\lambda')]^{-1}$ and this is the classical limit $(\gamma_0 \to 0)$ of the corresponding S-matrix element of the massive Thirring model [43]. We can sense how this comes about through the quantum inverse method by computing the classical "r-matrix" [5]. We can use $\ln|a(k)| + i\arg(k)$ to find classical commutation relations for all the scattering data of the covariant problem. Our results evidently show that for the covariant problem (1.1) and (1.2) there is a set of scattering data $\{a(k), a^*(k), b(k), b^*(k)\}$ and a 'transition matrix'

$$T = \begin{bmatrix} a(k) & b(k) \\ b^*(k) & a^*(k) \end{bmatrix} \quad (4.40)$$

with $\det T = 1$. We have found that $P(k) = (\pi\omega(k)\gamma_0)^{-1}\ln|a(k)|$, $Q(k) = \arg b(k)$ and

$$\{a(k), a(k')\} = \{a(k), a^*(k')\} = \{b(k), b(k')\} = 0 \quad . \quad (4.41a)$$

Then through (4.39) we find

$$\{a(k), b(k')\} = \tfrac{1}{8}m^2\gamma_0 a(k)\, b(k')\, [k\omega(k') - k'\omega(k)]^{-1}$$
$$+ \pi i\gamma_0\omega(k)\, a(k)\, b(k)\, \delta(k - k') \quad (4.41b)$$

$$\{a(k), b^*(k')\} = -\tfrac{1}{8}m^2\gamma_0 a(k)\, b^*(k')[k\omega(k') - k'\omega(k)]^{-1} \quad (4.41c)$$

$$\{b(k), b^*(k')\} = \tfrac{1}{4}\pi i\gamma_0\omega(k)|a(k)|^2\, \delta(k - k') \quad . \quad (4.41d)$$

In rapidity variables λ, λ' these change to

$$\{a(\lambda), a(\lambda')\} = \{a(\lambda), a^*(\lambda')\} = \{b(\lambda), b(\lambda')\} = 0 \qquad (4.42a)$$

and

$$\{a(\lambda), b(\lambda')\} = \tfrac{1}{8}\gamma_0 a(\lambda)\, b(\lambda')/\sinh(\lambda - \lambda' + i\varepsilon) \qquad (4.42b)$$

$$\{a(\lambda), b^*(\lambda')\} = -\tfrac{1}{8}\gamma_0 a(\lambda)\, b^*(\lambda')/\sinh(\lambda - \lambda' + i\varepsilon) \qquad (4.42c)$$

$$\{b(\lambda), b^*(\lambda')\} = \tfrac{1}{4}\pi i\gamma_0 |a(\lambda)|^2\, \delta(\lambda - \lambda') \qquad (4.42d)$$

while $P(\lambda) = (\tfrac{1}{8}\pi\gamma_0)^{-1} \ln|a(\lambda)|$, $Q(\lambda) = \arg b(\lambda)$ and $\{P(\lambda), Q(\lambda')\} = \delta(\lambda - \lambda')$. These results agree entirely with those quoted for the classical r-matrix on infinite support $L \to \infty$ [5,28,44]. They correspond, however [5,44] to

$$\hat{L} = -\frac{\partial}{\partial x} + i[\frac{\gamma_0}{4}\phi_x\sigma_3 + m\sinh\lambda\cos\tfrac{1}{2}\gamma_0\phi\sigma_1 + m\cosh\lambda\sin\tfrac{1}{2}\gamma_0\phi\sigma_2], \qquad (4.43)$$

in which the σ_i are Pauli matrices, and an equation of motion $\phi_{xx}-\phi_{tt} = m^2\gamma_0^{-\frac{1}{2}} \sin\gamma_0^{\frac{1}{2}}\phi$. With this in mind, we can establish formal connection with the quantum inverse method using the different parametrisation we adopted for $\hat{L}$ in (4.1) and $\hat{A}$ in (4.2): this parametrisation leads to the $\hat{L}$ quoted below (4.28) for the s-G in covariant form, and it differs from (4.43) most significantly in the choice of the matrix basis.

Following FADDEEV [5] we define 4x4 matrices $T'(\lambda) = T \otimes I$, $T''(\lambda) = I \otimes T$ in which $\otimes$ is Kronecker product (so that eg. $T''(\lambda) = \mathrm{diag}(T,T)$). By $\{T(\lambda) \overset{\otimes}{,} T(\lambda')\}$ we mean the 4x4 matrix of Poisson brackets read from T as Kronecker products. Then we can write (4.42) as

$$\{T(\lambda) \overset{\otimes}{,} T(\lambda')\} = r_+(\lambda-\lambda')T'(\lambda)T''(\lambda')-T'(\lambda)T''(\lambda'')r_-(\lambda-\lambda') \qquad . \qquad (4.44)$$

The r-matrices $r_\pm$ are

$$r_\pm(\lambda-\lambda') = \frac{1}{16}\gamma_0 \left[P\frac{I \otimes I - \sigma_3 \otimes \sigma_3}{\sinh(\lambda-\lambda')} \pm i\pi\delta(\lambda-\lambda')(\sigma_+ \otimes \sigma_- - \sigma_- \otimes \sigma_+)\right] \quad . \qquad (4.45)$$

The brackets (4.44) are for infinite support $L \to \infty$. Thus, for finite support under periodic boundary conditions period L we might infer [5,44] that there is also a classical r-matrix

$$r(\lambda-\lambda') = \frac{1}{16}\gamma_0 \{[\sigma_1 \otimes \sigma_1 + \sigma_2 \otimes \sigma_2]\coth(\lambda-\lambda') + [I \otimes I - \sigma_3 \otimes \sigma_3] (\sinh(\lambda-\lambda'))^{-1}\} \quad . \qquad (4.46)$$

We shall need to return to the problem of periodic boundary conditions later (§9). However, the thesis adopted there (which, however, is still to be firmly demonstrated) is that, although we *need* to adopt periodic boundary conditions with period $L \to \infty$, we can nevertheless actually use the action-angle variables for infinite support $L \to \infty$. We shall see that in this respect the situation is still being explored and the issue is therefore still open. Reference to [5], [28] and [44] shows nevertheless that after taking account of the changed basis (4.44) to (4.46) agree completely with the classical Poisson brackets associated with the quantum inverse method. Quantisation is achieved by finding the proper quantum forms of (4.44) and (4.46) [5,28,44]. Reference [28] in particular establishes the connection of the classical Landau-Lifshitz equation and its Poisson brackets with the quantum but integrable spin-½ XYZ model and the work of Baxter [1].

We turn now to the statistical mechanics of the sine-Gordon system. We first look more generally at the apparatus needed for the study of both quantum and classical statistical mechanics.

5. Functional Integrals for the Partition Function Z

We define '$\mathcal{D}$' and what we mean by a functional integral like Z eqn. (1.11). Essentially, and following FEYNMAN, it is a limit of a sequence of products of ordinary integrals. We introduce the quantum propagator as a functional integral first of all.

Consider the usual Dirac kets $|q\rangle$, $|p\rangle$: $\hat{q}|q\rangle = q|q\rangle$, $\hat{p}|p\rangle = p|p\rangle$: moreover, $\langle q'|q\rangle = \delta(q' - q)$, $\langle p'|p\rangle = \delta(p' - p)$, $\langle q|\hat{p}|p\rangle = p\langle q|p\rangle = -i\partial/\partial q\ \langle q|p\rangle$ so that $\langle q|p\rangle = (2\pi)^{-\frac{1}{2}}e^{ipq}$ (the normalisation follows from $\langle q'|q\rangle = \int\langle q'|p\rangle\langle p|q\rangle dp$). Now consider $\langle q_2,t + \Delta t|q_1,t\rangle \equiv \langle q_2|e^{-i\Delta tH}|q_1\rangle$ for which we get $\delta(q_2 - q_1)$ as $\Delta t \to 0$. We choose here

$$H = \hat{p}^2(2m)^{-1} + V(\hat{q}) \tag{5.1}$$

so there are two degrees of freedom. To neglect of commutators involving higher powers of Δt, $e^{-i\Delta tH} \sim e^{-i\Delta t(\hat{p}^2/2m)}\, e^{-i\Delta tV(\hat{q})}$. Then

$$\begin{aligned}\langle q_2|e^{-i\Delta tH}|q_1\rangle &\sim \int\frac{dp}{2\pi}\, e^{-i\Delta t(p^2/2m)}\, e^{ip(q_2-q_1)}\, e^{-i\Delta tV(q)}\\ &\approx \left(\frac{m\, e^{-i\pi/2}}{2\pi\ \Delta t}\right)^{\frac{1}{2}} \exp i\Delta t[\tfrac{1}{2}mq_t^{\,2} - V(q)]\end{aligned} \tag{5.2}$$

(since $(q_2-q_1)/\Delta t \to q_t$ as $\Delta t \to 0$).

Now use

$$\langle q_f,\, t_f|q_i,\, t_i\rangle \equiv \langle q_f|e^{-i(t_f-t_i)H}|q_i\rangle = \lim_{N\to\infty}\int\prod_{p=1}^{N-1} dq_p \prod_{p=0}^{N-1}\langle q_{p+1},t_{p+1}|q_p,t_p\rangle$$

with

$$t_p = t_i + p\, N^{-1}(t_f - t_i),\quad t_o \equiv t_i,\ t_N \equiv t_f \quad . \tag{5.3}$$

Then

$$\begin{aligned}\langle q_f,\, t_f|q_i,\, t_i\rangle = \lim_{N\to\infty}\int\prod_{p=1}^{N-1} dq_p \left|\frac{Nme^{-i\pi/2}}{2\pi(t_f-t_i)}\right|^{\frac{1}{2}N}\\ \exp\left\{i\int_{t_i}^{t_f}(\tfrac{1}{2}mq_t^{\,2} - V(q))dt\right\} \quad .\end{aligned} \tag{5.4}$$

We *define* the right side as a functional integral

$$\langle q_f,\, t_f|q_i,\, t_i\rangle = \int\mathcal{D}q \exp i\, S[q] \tag{5.5}$$

where the action $S[q] = \int_{t_i}^{t_f}L(q,q_t)dt$ is the *classical* action. Note that the measure $\mathcal{D}q$ contains within it the normalisation involving $N^{\frac{1}{2}N}$: one of the problems we face (but solve! - see especially §8) is finding the correct measures for the functional integral Z. Note that integrals in q_p are evaluated for $1 \le p \le N - 1$: the end points $q(t_0) = q(t_i)$ and $q(t_N) = q(t_f)$ are fixed. Between these fixed end points $q(t_p)$ takes on all possible values, so in that sense all possible *paths* between these end points are followed. In practice, smoothness conditions (Lipschitz conditions) restrict the acceptable paths.

The object (5.5) is a Green's function: for certainly $|\psi(t_i)\rangle = \int |q_t\rangle \psi(q_i,t_i)dq_i$. Then $\psi(q_f,t_f) = \int \langle q_f,t_f)|q_i,t_i\rangle\, \psi(q_i,t_i)dq_i$. Indeed if $\langle q_f,t_f|q_i,t_i\rangle = K$ for $t_f - t_i \equiv T > 0$, and is zero for $T < 0$, then

$$i\, \partial K/\partial T - HK \quad = \quad \delta(q_f - q_i)\, \delta(t) \qquad . \tag{5.6}$$

It is plain that trajectories leading to stationary phase $\delta S[q] = 0$ are the important trajectories in (5.5). This way, the quantum effects arise as fluctuations about the classical trajectories. Steepest descents tend to provide more manageable mathematics corresponding to the shift to imaginary time $\tau = -i\alpha t$ (α = real). Eqn. (5.6) is then a diffusion equation. In this case with $\alpha = 2$ and $m = 1$ EDWARDS and GULYAEV [45] evaluated (5.5) for 2 space co-ordinates (4 degrees of freedom) for a free particle $V(\vec{q}) = 0$. They evaluated it with both cartesian, (x,y), and polar, (r, θ), co-ordinates. Though the cartesian case readily produced the known result for K corresponding to (5.6), they found they needed to work to higher terms in Δt (i.e. $O(\Delta t^2)$ could not be neglected against $O(\Delta t)$) in the polar case. Thus the Jacobian $\partial(r,\theta)/\partial(x,y)$ was insufficient for the transformation of the integrals making up the functional integral. This example must be a salutary warning when we come to decide on the proper choice of measure for a functional integral like (1.11) under a canonical change of variables: the Jacobian may not just be unity [39]. Nevertheless, in the case [45] mentioned, the problem was subsequently solved [46] by using an extended form of functional integral which we now derive.

For a compact argument we follow ITZYKSON and ZUBER [47] as we have done already. We suppose $\langle p|H|q\rangle = h(p,q)(2\pi)^{-\frac{1}{2}}\, e^{-ipq}$. Hamiltonian (5.1) certainly leads to this form. Then $\langle p|e^{-i\Delta tH}|q\rangle = e^{-i\Delta t h(p,q)}(2\pi)^{-\frac{1}{2}}\, e^{-ipq}$ and

$$\langle q_f, t_f|q_i, t_i\rangle \quad = \quad \lim_{N\to\infty} \int \prod_{j=1}^{M} dp_j \prod_{k=1}^{M-1} dq_k\, \langle q_M|p_M\rangle\, \langle p_M|e^{-iTM^{-1}H}|q_{M-1}\rangle$$

$$\ldots\, \langle q_1|p_1\rangle\, \langle p_1|e^{-iTM^{-1}H}|q_0\rangle \qquad . \tag{5.7}$$

The limit is

$$\lim_{N\to\infty} \int \prod_{j=1}^{M} dp_j \prod_{k=1}^{M-1} dq_k\, (2\pi)^{-M} \exp i\, S[q] \tag{5.8a}$$

where

$$S[q] \quad = \quad p_M(q_M - q_{M-1}) + \ldots + p_1(q_1 - q_0) - TM^{-1}\, h(p_M,q_{M-1}) \ldots - TM^{-1}\, h(p_1,q_0)$$

$$\to \int_{t_i}^{t_f} (p\, q_t - h(p,q))dt \qquad . \tag{5.8b}$$

Note that, if H is quadratic in $\hat{p}$ (as in (5.1)), (5.7) with (5.8) reduces to (5.4) by integrating out $p_i - (q_i - q_{i-1})\, MT^{-1} \to (p - q_t)$. The route to (5.8b) may seem formal, but the normalisation in (5.7) is very natural (recall $h = 2\pi$), and we expect it provides a yardstick for finding the proper measure: (5.8) already surmount the particular Jacobian problem mentioned. Note the extra integral over p_M in (5.7): this leaves $q(t_i)$ and $q(t_f)$ fixed as end points as required.

We must now extend the definition (5.7) to fields $(\Pi(x,t), \phi(x,t))$: if we write $dp_j = dp(t_j)$ in (5.7) we need for 4 degrees of freedom $dp_1(t_j)\, dp_2(t_j)$: thus for fields $\Pi(x_s,t_j)$ we get a measure $\prod_{j=1}^{M}\prod_{s=1}^{N} d\Pi(x_s,t_j)$ with $x_s = -\frac{1}{2}Na + sa$ and $Na = L$. We then need the limit $M,N \to \infty$ $(a \to 0)$. Each pair $\Pi(x,t)$, $\phi(x,t)$ takes all acceptable time trajectories joining $\phi(x,t_i)$ and $\phi(x,t_f)$. The corresponding functional integral will be written

$$G(\phi, \phi_0; T) \quad = \quad \int \mathcal{D}\Pi \mathcal{D}\phi \exp i\, S[\phi] \tag{5.9a}$$

where

$$S[\phi] = \int_0^T dt[\int dx\Pi(x, t)\ \phi_t(x, t) - H[\phi]] \quad . \tag{5.9b}$$

We set $t_i = 0$, $t_f = T$ and $\phi \equiv \phi(x,T)$, $\phi_0 \equiv \phi(x,0)$; $H[\phi]$ is an integral (like (1.2)) over the Hamiltonian density $\mathcal{H}[\phi]$. Thus G is a functional depending on $\{\phi(x,T), \phi(x,0); -\frac{1}{2}L < x < \frac{1}{2}L\}$. Note that, at fixed $t = t_j$, $(\phi(x,t_j), \Pi(x,t_j))$ take all acceptable x-trajectories between $\phi(-\frac{1}{2}L, t_j)$ and $\phi(+\frac{1}{2}L, t_j)$.

We shall be interested in the trace of G:

$$\mathrm{Tr}\, G(\phi, \phi_0; T) = \int G(\phi_0, \phi_0; T)\mathcal{D}\phi_0$$

$$\equiv \lim_{N\to\infty} \int \prod_{s=1}^{N} d\phi(x_s, 0) \prod_{s=1}^{N} G(\phi_0, \phi_0; T) \tag{5.10}$$

which concerns $\phi(x,t)$ periodic in time with period T. Since

$$\mathrm{Tr}\, e^{-iHT} = \int G(\phi_0, \phi_0; T)\mathcal{D}\phi_0 \tag{5.11}$$

the Fourier transform $G(E) = i\,\mathrm{Tr}\int_0^\infty dT \exp[i(E-H)]T$ can be evaluated from (5.10) and its poles constitute the quantum eigenspectrum. Note that Gaussian integration on $\frac{1}{2}(\Pi - \phi_t)^2$ converts (5.9) to

$$G(\phi, \phi_0; T) = \int\mathcal{D}\phi' \exp i\, S[\phi] \tag{5.12}$$

and the trace according to (5.11) yields the eigenspectrum. This was the method of DHN [26] for the eigenspectrum of the covariant s-G. Note that the measure $\mathcal{D}\phi'$ in (5.12) is not the measure $\mathcal{D}\phi$ in (5.9a), and $S[\phi]$ is in Lagrangian form.

By similar arguments to those used above the correlation functions, time-ordered through an operator $\hat{T}$, are obtained through the relation

$$\langle\phi, t_f|\hat{T}\, 0_1(t_1)\ 0_2(t_2)\ \ldots\ |\phi_0, t_i\rangle$$

$$= \int\mathcal{D}\Pi\mathcal{D}\phi 0(\Pi_1, \phi_1)\ 0(\Pi_2, \phi_2)\ \ldots \exp i\, S[\phi] \tag{5.13}$$

when the operators $0_i(t_i)$ have the representation $\langle\Pi(x, t_i|0_i(t_i)|\phi(x,t_i)\rangle = 0_i(\Pi(x, t_i), \phi(x,t_i))\ \langle\Pi(x,t_i)|\ \phi(x,t_i)\rangle$ (compare by (5.6)). The relation extends to operator densities $0_j(x_i,t_i)$ eg. by ordering on t for each ordered x and performing the functional integrals on time first of all.

We shall be more interested here in the Wick rotation $t \to -it = \tau$ with period $\beta\hbar \equiv T$ ($\hbar$ is explicitly restored because we want to take the classical limit $\hbar \to 0$. Note that β^{-1} is now the temperature not T):

$$\bar{G}(\phi, \phi_0; T) = \int\mathcal{D}\Pi\mathcal{D}\phi \exp\{-\hbar^{-1}\int_0^{\beta\hbar} d\tau[H[\phi] - i\int\Pi\phi_\tau\, dx]\} \quad . \tag{5.14}$$

The quantum statistical partition function must be $Z = \mathrm{Tr}\, \bar{G}(\phi,\phi_0;T)$. Its classical limit sets $T = \beta\hbar \to 0$, and in this case $\int_0^T dt\Pi\phi_\tau - \Pi d\phi$. Then $\phi(x,\tau)$ will not evolve since $d\phi = \phi(x,T) - \phi(x,0) = 0\ \forall\ x$. Thus the classical limit is (1.11) namely

$$Z = \int\mathcal{D}\Pi\mathcal{D}\phi \exp - \beta H[\phi] \tag{5.15}$$

involving functional integration only over labels x. The trace is taken up by the observation that (5.10) is itself a functional integral, and it is this which corresponds to the $\mathcal{D}\phi$ in (5.15): the $\mathcal{D}\Pi$ is not concerned with end points since (5.7) is not. There are boundary conditions on ϕ: they prove to be crucial, for we have to choose between periodic boundary conditions on period L or, for the s-G (1.1), $\phi \equiv 0 \pmod{2\pi}$ with $\phi_x \to 0$ as $|x| \to \frac{1}{2}L$ with $L \to \infty$. Since we have only considered the latter case so far, we think in these terms until we need to do otherwise!

In principle, we are now able to use G to find quantum eigen spectra, S-matrix, and correlation functions at zero temperature for a (normally ordered) quantised s-G by using the classical $H[\phi]$, eqn. (1.2). From the quantum partition function $Z = \mathrm{Tr}\ \tilde{G}(\phi,\phi_0;T)$ we can find the quantum free energy and correlation functions at finite temperatures $\beta^{-1} > 0$. We can therefore compare in principle with the quantum inverse method - connection with which was begun at the end of §4 - with the Bethe ansatz, and, for the s-G, DHN's functional integral calculation of the eigenspectrum. For $\beta^{-1} > 0$ we can compare with the Bethe ansatz [7,8,31] for the s-G. However, we have still to learn how to evaluate the functional integrals.

The rest of these lectures are concerned with evaluating the classical partition function (5.15) for the s-G eqn. (1.1). To this end we need to know the result which (5.15) gives for the K-G eqn., so we evaluate this next.

6. The Partition Function for the Linear Klein-Gordon Equation

There are at least 4 ways of proceeding in this case: canonical transform to Fourier space, canonical transform to action-angle variables, steepest descents, and transfer integral method (TIM). The TIM is an important method which we can apply to the s-G itself, so we consider this as a whole in §7.

(i) We evaluate (5.15) with $H_{KG}[\phi]$, (2.3) for the K-G Fourier transformed to (2.12) namely

$$(2\pi)^{-1} \int \{\tfrac{1}{2}\, 4\pi^2\, \bar{\Pi}(k)\, \bar{\Pi}(-k) + \tfrac{1}{2}\omega^2(k)\, \bar{\phi}(k)\, \bar{\phi}(-k)\}dk \qquad . \tag{6.1}$$

The $4\pi^2$ is introduced so that $\{\bar{\Pi}(k), \bar{\phi}(k')\} = \delta(k-k')$ (see §2). This Hamiltonian, conveniently re-expressed in terms of real fields, discretizes to

$$(1/4\pi)(2\pi\, L^{-1}) \sum_{j=-n}^{+n} \left\{\omega^2(k_j)\, \bar{\phi}_j^{\,2} + 4\pi^2\, L^2 (2\pi)^{-2}\, \bar{\Pi}_j^{\,2}\right\} \tag{6.2}$$

with $k_j = 2j\pi L^{-1}$. The new canonical momenta $\bar{\Pi}_j$ are introduced so that $\{\bar{\Pi}_j, \bar{\phi}_k\} = \delta_{jk}$. The measure on each integral is $h = 2\pi$ for each j. The result for the $(2n+1)$ dimensional integral from (4.2) (which is simply a product of integrals) can be put in the form

$$(\pi\beta)^{-2n}\, a^{2n}\, e^{2n}\, (Lm)^{-1} \prod_{j=1}^{n} (1 + m^2L^2/4\pi^2 j^2)^{-1} \qquad . \tag{6.3}$$

We are concerned here to check on invariance of the functional integrals under canonical transformation so note that the discretization parameter is $2\pi L^{-1} \to 0$ as $L \to \infty$. However, we introduce a second parameter a by $na = \frac{1}{2}L$. The product $\prod_{j=1}^{n}(1 + m^2L^2/4\pi^2 j^2)^{-1} \sim (\frac{1}{2}mL/\sinh\frac{1}{2}mL)$ for *small enough* a, so that the contribution to the free energy per unit length is

$$-\beta^{-1}L^{-1}\ln Z \;=\; a^{-1}\beta^{-1}(\ln(\beta\pi a^{-1}) + \tfrac{1}{2}ma - 1) + O(a) \qquad . \tag{6.4}$$

(ii) We evaluate

$$Z \;=\; \int \mathcal{D}P(k)\, \mathcal{D}Q(k)\, \exp - \beta H_{KG}[p] \tag{6.5a}$$

where $H_{KG}[p] \equiv \int_{-\infty}^{\infty}\omega(k)\, P(k)dk$, (2.10), in action-angle variables. From the remarks there, and ignoring specific consideration of the lattice for the moment, $H_{KG}[p]$

discretizes to $\sum_{j}^{n} \omega(k_j)P_j$ with $k_j = 2\pi jL^{-1}$, $P_j = 2\pi L^{-1}P(k_j)$, and $n = \frac{1}{2}La^{-1}$, so that $Z = \prod_{j=-n}^{n} (\beta\omega(k_j))^{-1}$ which becomes

$$Z = \exp\left\{ -\frac{L}{2\pi} \int_{-\pi/a}^{+\pi/a} ln(\beta\omega(k))dk \right\} \tag{6.5b}$$

and the free energy per unit length is again (6.4). There are now two important lessons here: one is the emergence of the extensive variable L in the exponent in (6.5) as the $\sum_{-n}^{n} ln(\beta\omega(k_j)) \to L(2\pi)^{-1} \int_{-\pi/a}^{+\pi/a} ln(\beta\omega(k))dk$ at fixed a; the other is that the answer (6.4) does not admit $lim\ a \to 0$.

This means we should really look at the lattice with support $L = 2n\,a$ and spacing a. From (2.15) we need

$$-\frac{L}{2\pi} \int_{-\pi/a}^{+\pi/a} dk\ ln[\beta\{m^2 + (\frac{\sin \frac{1}{2}ka}{\frac{1}{2}a})^2\}^{\frac{1}{2}}]$$

$$= -L\left[a^{-1}\ ln(\beta a^{-1}) - \frac{m}{2\pi} \int_{0}^{\pi/ma} d\xi\ ln[m^2a^2\{1 + (\frac{\sin \frac{1}{2}am\xi}{\frac{1}{2}am})^2\}]\right]$$

$$= -L(a^{-1}\ ln(\beta a^{-1}) + \tfrac{1}{2}m) + O(a) \tag{6.6}$$

which gives a free energy per unit length differing from (6.4) by $a^{-1}\beta^{-1}(ln\pi - 1)$. We interpret this result below.

(iii) Method of steepest descents. We use $H_{KG}[\phi]$, (2.3), without canonical transformation. The Π integration is Gaussian and can be done immediately: we actually do it in §7. From (5.7) the normalisations will be $h^{-1} = (2\pi)^{-1}$ on each integral so one finds from the Π integration a free energy

$$-\tfrac{1}{2}\beta^{-1}a^{-1}\ ln(a/2\pi\beta\gamma_0) \tag{6.7}$$

per unit length. However, for the K-G we have already set $\gamma_0 \equiv 1$ (in §7 we find (6.7) for the s-G with $\gamma_0 \neq 1$: however, the *total* contribution then proves to be independent of γ_0).

We are left to evaluate

$$\int \mathcal{D}\phi\ \exp - \beta \int \{\tfrac{1}{2}\ \phi_x^2 + \tfrac{1}{2}\ m^2\phi^2\}\ dx \equiv H[\phi] \tag{6.8}$$

We first consider briefly the general case in which the K-G potential $\frac{1}{2}m^2\phi^2$ is replaced by $V(\phi)$. This potential will have, perhaps degenerate, minima ϕ_{min} (the s-G is an example of degenerate minima ϕ_{min}). The boundary condition is taken to be $\phi \to \phi_{min}$, $x \to -\infty$. For steepest descents dominant contributions to (6.8) arise from neighbourhoods of those ϕ which minimise $H[\phi]$. These satisfy

$$0 = \delta H[\phi]/\delta\phi = -\phi_{xx} + dV/dX \tag{6.9}$$

and this equation determines static, that is time-independent, solutions of the non-linear K-G $\phi_{xx} - \phi_{tt} = dV/d\phi$. We functionally expand $H[\phi]$ about such solutions ϕ_{class} (say). Since $\delta H/\delta\phi = 0$

$$H[\phi] = H[\phi_{class}] + \tfrac{1}{2}(\delta\phi,\ (\delta^2H/\delta\phi^2)\delta\phi) + \ldots \tag{6.10}$$

for real variations $\delta\phi = \phi - \phi_{class}$. The second functional derivative is formally the operator

$$\frac{\delta^2 H}{\delta\phi^2} = \frac{-d^2}{dx^2} + \left.\frac{d^2V}{dx^2}\right|_{\phi=\phi_{class}} \tag{6.11}$$

And this operator has eigenvalues ω_n^2 and real eigenfunctions $\eta_n(x)$ (for appropriate boundary conditions). If we write $\delta\phi = \sum_{n=-\infty}^{\infty} c_n\eta_n(x)$

$$H[\phi] = H[\phi_{class}] + \tfrac{1}{2}\, a \sum_{n=-\infty}^{\infty} c_n^2\omega_n^2 + O(a^3) \tag{6.12}$$

(we have taken a normalisation $a^{-1}\int_{-\infty}^{\infty}\eta_m\eta_n dx = \delta_{mn}$). We suppose the measure is such that $\mathcal{D}\phi \to \prod_{n=-\infty}^{\infty} dc_n$ with $-\infty < c_n < \infty$. Then (6.8) becomes

$$e^{-\beta H[\phi_{class}]} \prod_{n=-\infty}^{\infty} \sqrt{\frac{2\pi}{a\beta\omega_n^2}} \tag{6.13}$$

providing no $\omega_n = 0$. In the case of the s-G there are ω_n which are zero. In the case of the K-G, $\phi_{class} = 0$, $H[\phi_{class}] = 0$ and $d^2V/d\phi^2|_{\phi=\phi_{class}} = m^2$, so (6.11) has eigenvalues $\omega_n^2 = m^2 + k_n^2$ with $k_n = 2n\pi L^{-1}$ for periodic boundary conditions and $\omega_n \neq 0$ for any n. We now replace

$$\prod_{n=-\infty}^{\infty} \sqrt{\frac{2\pi}{a\beta\omega_n^2}} \to \exp\left\{-\frac{L}{2\pi}\int_{-\pi/a}^{+\pi/a} dk\; ln[\{\frac{\beta a}{2\pi}(k^2+m^2)\}^{\frac{1}{2}}]\right. \tag{6.14}$$

to see that the free energy per unit length is given again by (6.4) once the contribution (6.7) from the Π integration (with $\gamma_0 = 1$) is included. Since $\delta^3H/\delta\phi^3 = 0$ for the K-G there are no further corrections from the expansion (6.10).

In the case of the lattice, we expect to need to find the eigenvalues and eigenfunctions of

$$-a^{-2}(\eta_{\nu+1} + \eta_{\nu-1} - 2\eta_\nu) + m^2\eta_\nu = \omega_\nu^2\eta_\nu\ ; \qquad -n \le \nu \le n\ ; \tag{6.15}$$

under periodic boundary conditions the eigenvalues are given by (2.15) and (6.14) changes to $\exp\{-L(2a)^{-1} ln(\beta/2\pi a) - \frac{1}{2}mL\}$. When (6.7) (with $\gamma_0 = 1$) is added, the total contribution to the free energy per unit length proves to be (6.6), which is (6.4) corrected by $\beta^{-1}a^{-1}(ln\pi - 1)$.

The conclusion from this analysis is that for this linear problem the boundary conditions play no very significant role, but the lattice discretization does so. The difficulty is this: the functional integral requires us to take the lim $a \to 0$. But this limit cannot be taken either in the form (6.6) or in the form (6.4) for the calculated free energy. Since the dispersion relation (2.15) becomes $\omega_n^2 = m^2 + k_n^2$ for $a \to 0$ it is plain that if the $lims$ $a \to 0$ of (6.6) and (6.4) existed they would coincide. On the other hand, the limits do not exist: the problem is intrinsic and is simply the familiar ultra violet divergence problem. We shall meet it for the s-G, where it also emerges, by scaling Z against the K-G that is by a renormalisation. In this sense, the problem is swept away. An alternative procedure is always to use the lattice: this is in effect what the TIM does (§7).

The second important point we have noticed is that the extensive variable L emerges in the passage from a sum to an integral in exponentiating the form of Z. We need to exploit this fact later.

We now evaluate the partition function Z, (5.15), for the s-G by the TIM.

7. Transfer Integral Method for the Classical Partition Function of the sine-Gordon Equation

Temporarily, we work on support $0 \le x \le L$ divided into discrete steps of length a: $x \to ja$ with $0 \le j \le N$, and $Na = L$. We work with $H[\phi]$ given by (1.2) and find

$$\int \mathscr{D}\Pi \exp\{-\beta\gamma_0^{-1} \int \tfrac{1}{2}\Pi^2(x)\gamma_0^{\,2}dx$$

$$\rightarrow \int \Big(\prod_{j=1}^{N} (2\pi)^{-1}\, d\Pi_j \exp - \beta a\gamma_0^{-1} \sum_{j=1}^{N} \tfrac{1}{2}\Pi_j^{\,2}\gamma_0^{\,2}a^{-2} = (a/2\pi\beta\gamma_0)^{\frac{1}{2}N}. \qquad (7.1)$$

This leads to the free energy contribution (6.7). Notice that, after discretization with spacing a, it is necessary to define new momenta $\Pi_j \equiv a\Pi(ja)$ in order to preserve the equations of motion (these are (7.3) below).

The remaining part of the functional integral is discretized to

$$\int \Big(\prod_{j=1}^{N} d\phi_j\Big) \exp\{-\beta\gamma_0^{-1}a \sum_{j=1}^{N} \tfrac{1}{2}a^{-2}(\phi_j - \phi_{j+1})^2 + m^2(1 - \cos\phi_j)\} \quad . \qquad (7.2)$$

Using the discretized form of (7.1) with the new momenta, one finds through the Hamiltonian now appearing in (7.2) that

$$\phi_{j,tt} - a^{-2}(\phi_{j+1} + \phi_{j-1} - 2\phi_j) = - m^2 \sin\phi_j \qquad (7.3)$$

and this becomes the s-G as $a \rightarrow 0$. On the other hand if $a \neq 0$ (7.1) - (7.3) concern a lattice. However the lattice s-G (7.3) is not an integrable lattice and no action-angle variables are available for it. It is an undetermined and interesting question what damage (if any) is done to a functional integral for an integrable field when the integrability is broken by discretization. It seems to be possible to follow another route to the s-G by working with integrable lattices for which action-angle variables are becoming available [5,48,49] but this is still to be explored. The Toda lattice is integrable and action-angle variables are available for it [50].

The transfer integral method (TIM) has a long history, but we refer the reader to SCALAPINO, SEARS and FERRELL [51] and KRUMHANSL and SCHRIEFFER [52] in the present context. One can also draw comparison with transfer matrix methods [1]. The TIM evaluates N-dimensional integrals like (7.2) in the following way: periodic boundary conditions are placed on ϕ so that $\phi_{N+1} = \phi_1$. This is done by introducing $\delta(\phi_{N+1} - \phi_1)$ and using $\int d\phi_{N+1}\, \delta(\phi_{N+1} - \phi_1) = 1$: the range of any ϕ_j is $-\infty < \phi_j < \infty$. Introduce a complete set of functions ψ_m so that

$$\delta(\phi_{N+1} - \phi_1) = \sum_{\ell} \psi_\ell^{\,*}(\phi_{N+1})\, \psi_\ell(\phi_1). \qquad (7.4)$$

The N-dimensional integral (7.2) becomes

$$\sum_{\ell} \int d\,\phi_{N+1}\, \psi_\ell^{\,*}\, (\phi_{N+1}) \int d\,\phi_N \exp\{- \beta f(\phi_{N+1}, \phi_N)\} \,\cdot$$

$$\cdot \int d\,\phi_{N-1} \exp\{- \beta f(\phi_N, \phi_{N-1})\}\cdot \;\ldots\; \cdot \int d\,\phi_2 \exp\{- \beta f(\phi_3, \phi_2)\} \,\cdot$$

$$\cdot \int d\,\phi_1 \exp\{- \beta f(\phi_2, \phi_1)\;\; \psi_\ell(\phi_1) \qquad (7.5a)$$

in which

$$f(\phi_{j+1}, \phi_j) \equiv \gamma_0^{-1}\, a[\tfrac{1}{2}a^{-2}(\phi_j - \phi_{j+1})^2 + m^2(1 - \cos\phi_j)] \quad . \qquad (7.5b)$$

The ϕ_m are then chosen as eigenfunctions of the eigen condition for the 'transfer integral operator'

$$\int d\,\phi_j \exp\{- \beta f(\phi_{j+1}, \phi_j)\;\; \psi_\ell(\phi_j) = e^{-\beta a\gamma_0^{-1}m^2\varepsilon_\ell}\psi_\ell(\phi_{j+1}) \quad . \qquad (7.6)$$

From this the contribution to the partition function is

$$\sum_{\ell} e^{-\beta\gamma_0^{-1}m^2L\varepsilon_\ell} \sim e^{-\beta\gamma_0^{-1}m^2L\varepsilon_0} \quad \text{as} \quad L \to \infty. \tag{7.7}$$

Hence the contribution to the free energy per unit length is $\gamma_0^{-1}m^2\varepsilon_0$ and it remains to calculate the smallest eigenvalue ε_0 of (7.6).

This eigenvalue problem can be replaced, to terms $O(a)$, by

$$[-\frac{1}{2m^*}\frac{d^2}{d\phi^2} + (1 - \cos\phi) + V_0]\,\psi_\ell(\phi) = \varepsilon_\ell\psi_\ell(\phi) \tag{7.8}$$

in which $m^* \equiv \beta^2m^2\gamma_0^{-2}$ and

$$V_0 \equiv \beta^{-1}\gamma_0 a^{-1}m^{-2}\, ln(\beta\gamma_0^{-1}/2\pi a)^{\frac{1}{2}} \quad . \tag{7.9}$$

It is assumed that the range for ϕ is $-\infty < \phi < \infty$ and the usual Schrödinger eigenvalue conditions are imposed on $\psi_\ell(\phi)$. Evidently V_0 shifts the zero of ε_ℓ, which is bounded below, and it can be removed and added to (6.7): V_0 and (6.7) together provide precisely $\beta^{-1}a^{-1}\, ln(\beta a^{-1})$ to the free energy: this is independent of γ_0 and corresponds to a K-G contribution: the $\frac{1}{2}m$ is found below - so we are indeed dealing with the lattice.

We have now to find the lowest eigenvalue of the Mathieu equation (7.8) with V_0 removed. For (7.8) to have periodic solutions of period 2π an asymptotic series for ε_0 exists [53], and the contribution to the free energy, found by perturbation theory [53], proves to be

$$m\beta^{-1}[\tfrac{1}{2} - \tfrac{1}{4}(M\beta)^{-1} - \tfrac{1}{8}(M\beta)^{-2} - \tfrac{3}{16}(M\beta)^{-3} \ldots] \quad . \tag{7.10}$$

The leading term is the $\frac{1}{2}m$ contribution, and is the harmonic approximation to (7.8). With considerably more difficulty, one finds that there is a further contribution

$$-\, 8m\beta^{-1}(\pi\gamma_0)^{-\frac{1}{2}}(m\beta)^{\frac{1}{2}}e^{-\beta M}(1 - \tfrac{7}{8}(M\beta)^{-1} + \ldots) \quad . \tag{7.11}$$

Naive WKB is not good enough to get this result: we can write (7.8) in the form

$$y'' + (a + 16q\cos 2x)y = 0 \quad ; \tag{7.12}$$

$a \equiv 8\beta^2\gamma_0^{-1}(\varepsilon_0 - m^2)$; $q \equiv \frac{1}{2}m^* \gg 1$ for $\beta^{-1} \to 0$ and a is large there also. The usual WKB solutions are found from $y''(x) + h^2p(x)y$ with $h \gg 1$ by putting $y = \exp\{\int^x\phi(x')dx\}$, so that $\phi' + \phi^2 + h^2p = 0$. Then if $p(x) = O(1)$ in h, $\phi(x) = h\,\phi_0(x) + \phi_1(x) + \ldots$ yields WKB solutions $\propto p^{-\frac{1}{4}}\exp\{\pm ih\int^x(p(x'))^{\frac{1}{2}}dx'\}$. These solutions are matched as linear combinations across the 'turning points' (the zeros of $p(x)$) and consistency leads to a tunneling contribution to the eigenvalue (7.11) with a spurious factor $(e/\pi)^{\frac{1}{2}}$. The error is simply that $\varepsilon_0 = O(\beta^{-1}) = O(h^{-1})$ and $p(x)$ is $O(h^{-1})$ not $O(1)$.

Coleman [54] makes the same point: in (7.8) we are concerned with barrier penetration which splits the harmonic oscillator energy $\frac{1}{2}m^*$. There is always the problem of correcting for turning points in WKB: the well is deep (m^* is large), two turning points are a few wavelengths apart and linear connection across the turning points is insufficient. Coleman [54] actually calculates for the double well (not the periodic potential (7.8)): he carefully matches WKB solutions outside the wells to the solution to the harmonic approximation to the wells inside them. The reader is referred to an entertaining calculation [54]. However, to get (7.10) we must work outside harmonic approximation and since we shall become interested in the number $-\frac{7}{8}$ in (7.11) (§12) we need to do rather more. We have therefore extended the methods of matched asymptotic expansions pioneered by GOLDSTEIN [55] to get not only this coefficient but terms $O(e^{-2\beta M})$ also. They are of interest in connection with certain fermion properties of the theory ('classical fermions', §14). This is not the place to give the matched asymptotic expansion analysis but reference to Goldstein [55] will give the idea.

We therefore sum up the result of this application of the TIM: the contribution to the free energy per unit length is

$$F = -\beta^{-1}m8(m\beta/\pi\gamma_0)^{\frac{1}{2}}\,e^{-\beta M}[1 - \tfrac{7}{8}(M\beta)^{-1} + \ldots]$$

$$-\beta^{-1}m[\tfrac{1}{4}(M\beta)^{-1} + \tfrac{1}{8}(M\beta)^{-2} + \tfrac{3}{16}(M\beta)^{-3} + \ldots]$$

$$+\beta^{-1}a^{-1}(ln\beta a^{-1} + \tfrac{1}{2}ma) \quad . \tag{7.13}$$

To this must be added the term in $e^{-2\beta M}$ which is

$$16(2\pi)^{-1}\beta^{-1}m(M\beta)\{\gamma + ln(4M\beta)(1 - \tfrac{5}{4}(M\beta)^{-1} - \ldots) - (\tfrac{3}{8}(M\beta)^{-1} + \ldots)\}\,e^{-2\beta M} \tag{7.14}$$

in which γ is Euler's constant; and there are of course terms in $e^{-3\beta M}$ etc.

It is intuitive that the term in a is the contribution of the 'phonons' that is the free K-G - albeit for a lattice: the limit $a \to o$ cannot be taken. The first terms with Boltzmann factor $e^{-\beta M}$ evidently represent the kink and antikink contributions; so the series must be the breather contribution. For thermal excitations $F = -\beta n$, where n is a particle number density. Thus n_b is given by (3.34) and (compare (3.33))

$$n_k = n_{\bar{k}} = 4m(m\beta/\pi\gamma_0)^{\frac{1}{2}}\,e^{-\beta M}[1 - \tfrac{7}{8}(M\beta)^{-1} + \ldots] \quad . \tag{7.15}$$

These two results reinforce the view that breathers dominate at sufficiently low temperatures in the neutron scattering cross-section.

The result (7.15) is of sufficient importance to our later work that we now rederive it by steepest descents [54,56]. For a direct comparison with the work of steepest descents in §6 we use the Hamiltonian (2.2), $H[\gamma_0^{\frac{1}{2}}\phi]$, with $\{\phi_t,\phi\} = \delta(x-x')$ and set $\gamma_0 = \lambda m^{-2}$. A solution of (6.9) is $\phi = \phi_{class} = 4m\lambda^{-\frac{1}{2}}\tan^{-1}(\exp{+mx})$. This is the static form of the kink solution (3.16) for which $H[\phi_{class}] \equiv M = 8m^3\lambda^{-1}$. The Π integration is done as before, and we are left with the remaining contribution. The eigenvalue problem corresponding to (6.11) is

$$\left\{-\frac{d^2}{dx^2} + m^2(1 - 2\,\text{sech}^2 mx)\right\}\eta_n(x) = \omega_n^{\,2}\,\eta_n(x) \quad . \tag{7.16}$$

There is now a non-degenerate eigenvalue $\omega_0^{\,2} = 0$ with eigenfunction $\eta_0(x) = 2m^2\lambda^{-\frac{1}{2}}$ sech $mx = \partial\phi_{class}/\partial x$. It corresponds to the translational invariance of $H[\phi_{class}] = M$: a static kink can be placed at any point x_0 $(-\infty < x_0 < \infty)$ when $L \to \infty$. Accordingly, one uses variables $(x_0;\, b_n,\, n = \pm1, \pm2, \ldots)$ instead of the c_n used in §6, and expands

$$\varphi(x) = \phi_{class}(x - x_0) + \left\{\sum_{n=1}^{\infty} b_n\,\tilde{\eta}_n(x - x_0) + b_{-n}\,\tilde{\eta}_{-n}(x - x_0)\right\} \quad . \tag{7.17}$$

Dimensional arguments suggest that the element dx_0 is associated with a Jacobian $\propto m^{\frac{1}{2}}$. The proper choice, which is important even to a factor 2 in what follows, can be inferred from Coleman [54]: $\prod_n dc_n \to (a^{-1}M)^{\frac{1}{2}}\,dx_0 \prod_{n,n\neq0} db_n$, the factor a^{-1} coming from our particular scaling. The contribution to the functional integral is then

$$e^{-\beta M}(Ma^{-1})^{\frac{1}{2}}\int_{-\frac{1}{2}L}^{+\frac{1}{2}L} dx_0 \prod_{n=1}^{\infty}\left\{\int db_n \exp - \tfrac{1}{2}a\beta\tilde{\omega}_n^{\,2}b_n^{\,2}\right\}\cdot\left\{\int db_{-n}\exp - \tfrac{1}{2}a\beta\tilde{\omega}_{-n}^{\,2}\,b_{-n}^{\,2}\right\}$$

$$= Le^{-\beta M}(Ma^{-1})^{\frac{1}{2}}\prod_{n=1}^{\infty}\left\{(2\pi/a\beta\tilde{\omega}_n^{\,2})^{\frac{1}{2}}\,(2\pi/a\beta\tilde{\omega}_{-n}^{\,2})^{\frac{1}{2}}\right\} \quad . \tag{7.18}$$

We have used the normalisation $a^{-1}\int_{-\infty}^{\infty}\eta_n\eta_m dx = \delta_{mn}$ as before.

It is important to our understanding to scale the result (7.18) against the free Klein-Gordon equation. The ratio is

$$Le^{-\beta M} m(\beta M/2\pi)^{\frac{1}{2}} \prod_{n=1}^{\infty} (\tilde{\omega}_n \tilde{\omega}_{-n})^{-1} (\omega_n \omega_{-n})$$

$$= Le^{-\beta M} m(\beta M/2\pi)^{\frac{1}{2}} \exp - \frac{L}{2\pi} \int_{-\pi/a}^{\pi/a} dk \; ln[\tilde{\omega}(k)/\omega(k)]$$

$$\equiv 2Lm\{\beta m/\pi\gamma_0\}^{\frac{1}{2}} e^{-\beta M+\sigma} \equiv Z_k \tag{7.19}$$

with

$$\sigma \equiv -\frac{L}{2\pi} \int_{-\pi/a}^{\pi/a} dk \; ln[\tilde{\omega}(k)/\omega(k)] \quad . \tag{7.20}$$

We show below σ does not depend on L.

In the language of functional integration we have just evaluated the ratio of two determinants $\det\{-d^2/dx^2 + V''(x)\}/\det\{-d^2/dx^2 + m^2\}$ [54]; $V''(x) = m^2(1 - 2\text{sech}^2 mx)$. The translational mode with eigenvalue $\omega_0 = 0$ evidently plays a crucial role: it introduces the factor $2m(\beta m/\pi\gamma_0)^{\frac{1}{2}}$. This factor is made up of $M^{\frac{1}{2}}$ from $dc_0 \to (a^{-1}M)^{\frac{1}{2}} dx_0$ and then a factor $(a\beta/2\pi)^{\frac{1}{2}}\omega_0 = (a\beta/2\pi)^{\frac{1}{2}} m$ since the $n = 0$ is *missing* from (7.18). Otherwise it is assumed that both $\omega_n = (m^2 + k_n^2)^{\frac{1}{2}}$ and $\tilde{\omega}_n$ become dense in k-space when $L \to \infty$. These modes, and the frequencies ω_n, $\tilde{\omega}_n$ determined by them, also play an important role because they *dress* the mass M with the β-dependent correction $\beta^{-1}\sigma$. We wish to show next that Z_k in (7.19) makes an exponential contribution to the final partition function.

Consider two well-separated kinks for which $\phi_{2,\text{class}}(x) = \phi_{\text{class}}(x - x_1) + \phi_{\text{class}}(x - x_2)$ with $|x_1 - x_2| \gg m^{-1}$. Then $H[\phi_{2,\text{class}}] = 2M$ and the eigenvalue problem for the operator (6.11) has two well-separated but identical single kink potential wells. The bound state eigenvalues are unchanged, but become 2-fold degenerate i.e. $\omega_0{}^2 = 0$ is 2-fold degenerate. This corresponds to independent translation of the two kinks, and this is not actually possible for the s-G since at most one kink can be brought to rest, while the second phase-shifts as it passes through the first at rest. Exactly what this means to the mathematics is analysed in §9: the condition of distinct momenta leads to fermion properties and thence across to the massive Thirring model (§3 and (3.29)) and this is looked at in a later section (§13). For present purposes the argument will be good for $|x_1 - x_2| \gg m^{-1}$. There are now two location co-ordinates x_1 and x_2, and it will be assumed $\int dx_1$ and $\int dx_2$ can be executed independently - thus ignoring $|x_1 - x_2| \gg m^{-1}$. There is an error here perhaps $O((mL)^{-1})$. Evidently we are making a dilute gas assumption - of instantons [54].

We show below that $\sigma = 2\times$ (σ for a single kink). Then the contribution to Z is $(2!)^{-1}Z_k{}^2$: the factor $(2!)^{-1}$ allows for invariance under interchange of x_1 and x_2. Since any number of well-separated kinks solves (6.11) (minimizes H) the contribution to Z is finally $\sum_{N_k=0}^{\infty} (N_k!)^{-1} (Z_k)^{N_k} = \exp Z_k$ and the contribution to the free energy is $-\beta^{-1}Z_k$.

We have now to show that the σ's add for individual kinks and antikinks and calculate them. We require the eigenvalues $\tilde{\omega}_n{}^2$ of (7.16) which are the eigenvalues of K-G type modes in the presence of one kink. Solutions $\eta_n(x) \sim e^{i\tilde{k}_n x}$ for $x \to -\infty$ $\sim a(\tilde{k}_n)e^{-i\tilde{k}_n x} + b(\tilde{k}_n)\, e^{-i\tilde{k}_n x}$ for $x \to +\infty$ so that $\tilde{\omega}_n{}^2 = m^2 + \tilde{k}_n{}^2$. However, we do not have the $\tilde{k}_n$. We solve this problem in §9. Here we simply take the results $\omega_n = (m^2 + k_n{}^2)^{\frac{1}{2}}$, $k_n L = 2\pi n$ for free K-G; and

$$\tilde{\omega}_n = (m^2 + k_n^2)^{\frac{1}{2}}, \quad \tilde{k}_n L = 2\pi n + \Delta(k_n) \tag{7.21}$$

in the presence of the kink. We show later *why* we should now follow the prescription of DHN [26] and take $\Delta(k_n)$ as the phase-shift induced by a single kink (at rest) in the wave number of the K-G mode labelled by k_n. This phase-shift is given by (4.37) namely $\Delta_k(k, p_i)$ with $p_i = 0$. We know from the example (4.35) that these phase-shifts add. We show that σ is linear in $\Delta(k_n)$, and so, since the $\Delta(k_n)$ add, the σ add as was assumed. Thus we need consider only the 1-kink phase-shifts.

After substituting for ω_n and $\tilde{\omega}_n$ and expanding $\tilde{\omega}_n$ about ω_n we find

$$- \sigma = L(2\pi)^{-1}\int_{-\pi/a}^{+\pi/a} ln[\tilde{\omega}(k)/\omega(k)]dk = (2\pi)^{-1}[\Delta(k) ln(1 + k^2m^{-2})^{\frac{1}{2}}]_{-\pi/a}^{+\pi/a}$$

$$- (2\pi)^{-1}\int_{-\pi/a}^{+\pi/a} dk(d\Delta/dk)\, ln(1 + k^2m^{-2})^{\frac{1}{2}} \tag{7.22}$$

which is indeed independent of L. From (4.37), $\Delta(k) = \Delta_k(k,0) = -2\tan^{-1}(mk^{-1})$; there is no contribution from the square bracket in (7.22) and the integral is precisely $ln2$ (for $a \to 0$). This introduces a factor 2 to Z_k and a further factor 2 follows by including the antikinks. Thus the contribution to the free energy per unit length is

$$- \beta^{-1}L^{-1}Z_k = - \beta^{-1}8m(\beta m/\pi\gamma_0)^{\frac{1}{2}}e^{-\beta M} \tag{7.23}$$

in agreement with (7.13). It is a simple matter to continue the expansion (6.10) to $\delta^4H/\delta\phi^4$: this depends only on $d^4V/d\phi^4$. There is a contribution to the exponents of the Gaussian integrals in (7.18): the exponentials in $b_n{}^2b_m{}^2$ can be factored and expanded to leading order to pick up terms $O((M\beta)^{-1})$: the coefficient is $-\frac{7}{8}$ in agreement with (7.13) and [57].

We can make a valuable comparison with the contribution of a free kink-antikink gas: for a single kink the partition function will be

$$Z_k = (2\pi)^{-1}\int_{-\infty}^{\infty} \exp\{-\beta[M^2 + p^2]^{\frac{1}{2}}\}dp \int_{-\frac{1}{2}L}^{+\frac{1}{2}L} dq$$

$$= \pi^{-1}K_1(M\beta)LM \sim 2mL(\pi\gamma_0)^{-\frac{1}{2}}(m\beta)^{-\frac{1}{2}}\, e^{-\beta M}(1 + \tfrac{3}{8}(M\beta)^{-1} + \dots) \quad . \tag{7.24}$$

The $K_1(M\beta)$ is a Bessel function. The free energy per unit length will therefore be

$$- 4m(\pi\gamma_0)^{-\frac{1}{2}}\, (m\beta)^{-\frac{1}{2}}\, e^{-\beta M}(1 + \tfrac{3}{8}(M\beta)^{-1} + \dots) \quad . \tag{7.25}$$

To first order there is a correction in (7.23) to the mass appearing in (7.25) of $M \to M - \beta^{-1}\, ln(2m\beta)$ - a result found by Maki [23] from a quantum theory and by Fowler [7] by the quantum Bethe ansatz. We see precisely how it arises here: the $ln2$ comes from the dressing σ of the kink by the K-G modes of eigenfrequencies $\tilde{\omega}_n$ in the presence of that kink: the $ln(m\beta)$ really arises at the first line of (7.19) where, from (7.18), the free K-G has the $n = 0$, $k_0 = 0$ mode making a contribution; the coupled case does not. Evidently the $k_0 = 0$, $\omega_0 = m$ mode simply drops down to $\omega_0 = 0$, namely to the translational mode of the kink. The phase space for the K-G mode, which is a cylinder for $\omega_0 \neq 0$ (§2) opens to a sheet traversed by the kink in an infinite time ($\omega_0 = 0$). These facts will substantially influence our thinking in what follows. We shall actually see factors $(m\beta)$ associated with each degree of freedom of momentum moving into the kink, antikink, and breather prefactors in this fashion: the actual factor $(m\beta)^{\frac{1}{2}}$ arising in this steepest descents calculation is a feature of that particular analysis.

I do not know how to do a really satisfactory steepest descents analysis for the breather: in the frame of translational rest, the breather solution (3.17) oscillates at frequency $m\cos\theta$ and the solution does not satisfy (6.9). The reader is referred to DHN [26] for their very comparable approach to the zero temperature quantum problem.

Everything we have said so far is available in some form in the published literature. Apart from §9 everything which follows is new work. All of it is joint work with my two colleagues J. Timonen of the University of Jyväskylä in Finland and my assistant D.J. Pilling. It is only proper to acknowledge this fact and the very real debt I owe to them. Needless to say, all of the lectures so far have benefitted from these two people also. Thus I would like to terminate the lectures at this point, call them 'Part I' and follow this with a 'Part II' in which the joint authorship of all the work can be explicitly displayed. The work of the Part II is ongoing work and in some places is tantalisingly incomplete. If it is speculative in places I make no apology: the speculations may help students at this School to find the answers.

All the references to both parts I and II will be found at the end of the Part II.

Statistical Mechanics of the sine-Gorden Field: Part II

R.K. Bullough and D.J. Pilling

Department of Mathematics, U.M.I.S.T., Manchester M60 1QD, United Kingdom

J. Timonen

Department of Physics, University of Jyväskylä, 40100 Jyväskylä, Finland

We shall simply follow on from the §7 above with the next §8.

8. Evaluation of the Partition Function Z for the sine-Gordon Field in Terms of Action-Angle Variables: Construction of the Measure

From the work of the Part I we are now in a position to address ourselves to the main problem posed in these lectures - the evaluation of Z, (1.11), for the s-G field after canonical transformation to the action-angle variables (4.27).

A naive argument is the following: we have to evaluate

$$Z = \int \mathcal{D}\mu \exp - \beta H[p] \tag{8.1}$$

for H[p] in (4.25). Presumably the first shot at the measure would be

$$\mathcal{D}\mu = \prod_{i=1}^{N_k} (2\pi)^{-1} dp_i dq_i \prod_{j=1}^{N_{\bar{k}}} (2\pi)^{-1} dp_j dq_j \prod_{\ell=1}^{N_b} (2\pi)^{-2} 4\gamma_0^{-1} d\Theta_\ell d\Phi_\ell \mathcal{D}P(k)\mathcal{D}Q(k) \tag{8.2}$$

since the Jacobian is unity. The Hamiltonian is separable and Z_{KG} factors out. Indeed $Z = Z_k Z_{\bar{k}} Z_b Z_{KG}$: $Z_k = Z_{\bar{k}}$ and both are given by (7.24): Z_b proves [24] to be

$$Z_b = \pi^{-1}\{I_0(M\beta)\, K_1(M\beta) - I_1(M\beta)\, K_0(M\beta)\}\; 8\pi\; LM : \tag{8.3}$$

we use the spaces $0 < \Phi < 8\pi$ for Φ and $-\frac{1}{2}L < q_i, \bar{q}_j, \hat{q}_\ell < \frac{1}{2}L$ proved in §4 for the translational co-ordinates. We introduce chemical potentials $\mu_k, \mu_{\bar{k}}, \mu_b$:

$$Z_k \rightarrow \sum_{N_k=0}^{\infty} (N_k!)^{-1} e^{\beta\mu_k N_k} (Z_k)^{N_k} = \exp\{e^{\beta\mu_k} Z_k\} \quad , \tag{8.4}$$

etc. This way we reach a free energy per unit length

$$\begin{aligned} F = & -\beta^{-1} m 4 (m\beta\pi\gamma_0)^{-\frac{1}{2}} e^{-\beta M}[1 + \tfrac{3}{8}(M\beta)^{-1} + \ldots] - \beta^{-1} m[32\gamma_0^{-2}(M\beta)^{-2} + \ldots] \\ & + a^{-1}\beta^{-1}\{\ln(\beta\pi a^{-1}) + \tfrac{1}{2}ma - 1\} \quad . \end{aligned} \tag{8.5}$$

We immediately recognize a number of things about this solution: the free K-G contribution, the term in a, is not the lattice contribution delivered by the TIM; the dressing of the kink and antikink contribution as established by steepest descents does not appear; and the breather series has leading term $O((M\beta)^{-2})$ and is entirely in even powers of $(M\beta)^{-1}$. It will be important later to know that this series is manifestly due to breather contributions - for it is directly derived from the breather contributions in (4.25). The origin of the series in the TIM result (7.13) is less clear; but it is certainly both even and odd in $(M\beta)^{-1}$, has leading term $O((M\beta)^{-1})$, and has a different structure for its coefficients.

Evidently the measure (8.2) is not correct and we must construct it more carefully: in this general sense we have Jacobian trouble (refer §5). The point seems to be the following:- the §5 shows that a functional integral is defined over N x-points for $N \rightarrow \infty$. The TIM explicitly uses a lattice of points spacing a, with $Na = L$

and periodic boundary conditions. This way it achieves a well-defined thermodynamic limit as $L \to \infty$. We have to impose these facts on the functional integral (8.1).

We now have a considerable problem, for we are concerned with a finite number N of degrees of freedom, a lattice with spacing a, periodic boundary conditions, and the condition that the Hamiltonian H reduces to $H[\phi]$ on $-\frac{1}{2}L \le x \le \frac{1}{2}L$ for $a \to 0$. Although there are promising results for integrable systems with some of these properties [5,49,50] no useable action-angle variables are available for the problem posed. We therefore have to find them, bearing in mind we can finally exploit the condition $L \to \infty$. We remember from §6 that we pick up an extensive free energy for the K-G as a sum of mode contributions goes over to an integral; and we remember from §7 that we can pick up terms $O(L^{-1})$ this way so as to give the L-independent dressings σ. We also know we cannot set $a \to 0$.

We first construct a measure $\mathcal{D}\mu$ as

$$\mathcal{D}\mu\mathcal{D}'P\mathcal{D}'Q = (N_k!N_{\bar{k}}!N_b!)^{-1}\prod_{i=1}^{N_k}(2\pi)^{-1}dp_i dq_i \prod_{j=1}^{N_{\bar{k}}}(2\pi)^{-1}d\bar{p}_i d\bar{q}_i \prod_{\ell=1}^{N_b}(2\pi)^{-2}d\hat{p}_\ell d\hat{q}_\ell 4\gamma_0{}^{-1}d\Theta_\ell d\Phi_\ell \mathcal{D}'P\mathcal{D}'Q \tag{8.7}$$

in which $\mathcal{D}'P\mathcal{D}'Q \to \mathcal{D}P\mathcal{D}Q$ for functional integration as $N \to \infty$. We use the discretized form

$$H_{kG}[p,N_{ph}] = \sum_{-\frac{1}{2}N_{ph}}^{+\frac{1}{2}N_{ph}} \tilde{\omega}_n P_n ; \qquad \tilde{\omega}_n = (m^2 + \tilde{k}_n{}^2)^{\frac{1}{2}} \tag{8.8}$$

for H_{kG} with $N_{ph}+1$ modes $\le N$. The number N is given by

$$N = N_k + N_{\bar{k}} + 2N_b + N_{ph} + 1. \tag{8.9}$$

We do not specify yet how to determine the $\tilde{k}_n$. For each N, N_{ph} the functional integral $\mathcal{D}'P\mathcal{D}'Q$ is discretized with spacing a. We define the Hamiltonian $H[N_k,N_{\bar{k}},N_b]$ by

$$H[N_k,N_{\bar{k}}, N_b] = H[p,N] - H_{KG}[p,N_{ph}] \quad . \tag{8.10}$$

By this we mean what remains in $H[p]$, (4.25), once the K-G contribution is removed. We normalise the problem about the *free* K-G so that

$$Z/Z_{kG} = \lim_{N\to\infty} \sum_{N_k,N_{\bar{k}},N_b} \int \mathcal{D}\mu \exp\{-\beta H[N_k,N_{\bar{k}},N_b] + \sigma[N_k,N_{\bar{k}},N_b]\} \tag{8.11}$$

Note that the measure (8.7) employed in this fashion automatically exponentiates the kink, antikink and breather contributions (and no chemical potentials are needed). It seems that this is the proper *mathematical* formulation of the measure. We have no deductive proof of this fact, and its validity will be demonstrated by comparison with the TIM results. DE VEGA [39] (in effect) first wrote this measure down.

The problem now is how to calculate the dressing $\sigma = \sigma[N_k,N_{\bar{k}},N_b]$. Evidently this must concern the phase-shifts due to N_k kinks, $N_{\bar{k}}$ antikinks, and N_b breathers and will depend on the set of momenta $\{p_i,\bar{p}_j,\hat{p}_\ell,4\gamma_0{}^{-1}\Theta_\ell\}$ in consequence. But the suggestive result (4.39) means that we are concerned with phonon-phonon phase-shifts too.

In order to employ the result (4.39) we have to consider the periodic problem. Evidently we have re-discovered all of the problems negotiated in the quantum inverse method [3,5] and some extra ones.

9. Floquet Theory for Periodic Problems

Fortunately Professor FLASCHKA has treated the problem for the Toda lattice at some length in his lectures [50]. It is therefore convenient to work with him, and to present this short summary of Floquet theory for a *scalar second order* differential operator $\hat{L}$: the extension to the 2×2 matrix operator $\hat{L}$, (4.1), is straightforward and almost immediate. We are therefore concerned here with Hill's operator $\hat{L} = -d^2/dx^2 + q(x)$; $q(x + 2P) = q(x)$ so it is $2P$ - periodic ($2P = L$). The Mathieu operator (7.8) is a special case of this operator. The eigenvalue problem is $\hat{L}\psi = \lambda\psi$. As the lectures [50] show, the particular questions we usefully ask in Floquet theory are (and see [58]) are the solutions ψ bounded or unbounded for $-\infty < x < \infty$ but $L < \infty$; are they periodic of period $2P$ or $4P$, and so on. The two functions of λ, $\rho(\lambda)$, the Floquet multiplier, and $\Delta(\lambda)$ the discriminant, contain this information.

We define a first set of fundamental solutions $\psi_1(x,\lambda)$, $\psi_2(x,\lambda)$ of $\hat{L}\psi = \lambda\psi$ by the normalisations

$$\psi_1(-P,\lambda) = 1, \quad \psi_1'(-P,\lambda) = 0; \quad \psi_2(-P,\lambda) = 0, \quad \psi_2'(-P,\lambda) = 1 \quad . \tag{9.1}$$

Any solution of $\hat{L}\psi = \lambda\psi$ can be written as a linear combination of these: $\psi = \alpha_1\psi_1 + \alpha_2\psi_2$. If $\psi_1(-P) = \alpha_1$, $\psi'(-P) = \alpha_2$, then

$$\begin{bmatrix}\psi(P)\\ \psi'(P)\end{bmatrix} = \begin{bmatrix}\psi_1(P,\lambda) & \psi_2(P,\lambda)\\ \psi_1'(P,\lambda) & \psi_2'(P,\lambda)\end{bmatrix} \begin{bmatrix}\alpha_1\\ \alpha_2\end{bmatrix} \quad . \tag{9.2}$$

The matrix (called U) is a transition matrix (cf. (4.40)), or monodromy matrix taking $\begin{bmatrix}\alpha_1\\ \alpha_2\end{bmatrix} = \begin{bmatrix}\psi(-P)\\ \psi'(-P)\end{bmatrix}$ to $\begin{bmatrix}\psi(P)\\ \psi'(P)\end{bmatrix}$. Since $\det U$ is the Wronskian, $\det U = 1$. If ψ is periodic $\psi(P) = \psi(-P)$ and U is the unit matrix. More generally, we are interested in the Floquet solutions for which [58]

$$\psi(P) = \rho\psi(-P), \quad \psi'(P) = \rho\psi'(-P). \tag{9.3}$$

The initial data $\begin{bmatrix}\alpha_1\\ \alpha_2\end{bmatrix}$ satisfy $U\begin{bmatrix}\alpha_1\\ \alpha_2\end{bmatrix} = \rho\begin{bmatrix}\alpha_1\\ \alpha_2\end{bmatrix}$. Thus ρ is determined from $\det(U - \rho) = 0$ and

$$\rho^2 - \Delta\rho + 1 = 0 \; ; \qquad \Delta = \mathrm{Tr}\, U \quad . \tag{9.4}$$

Both ρ and Δ are independent of the particular basis. For $q = 0$, $\Delta(\lambda) = 2\cos(2P\lambda^{\frac{1}{2}})$; in general $\Delta(\lambda)$ will be an analytic function of $\lambda^{\frac{1}{2}}$ and asymptotic to $2\cos(2P\lambda^{\frac{1}{2}})$ as $|\lambda| \to \infty$. Then the roots $\Delta = \pm 2$ correspond to periodic solutions period $2P$ ($\Delta = +2$, $\rho = +1$) and antiperiodic solutions period 4ρ ($\Delta = -2$, $\rho = -1$). If $|\Delta(\lambda)| < 2$ both roots of (9.4) are complex of modulus unity and solutions are bounded through (9.3). When $|\Delta(\lambda)| > 2$ solutions are real, one being greater than unity, the other less than unity (if ρ solves (9.4), ρ^{-1} solves (9.4)): thus the only solution of $\hat{L}\psi = \lambda\psi$ is exponentially large at $-\infty$, $+\infty$, or both. The regions of λ where $|\Delta(\lambda)| > 2$ are the 'gaps' [50,58]: they are not admissible in the spectrum of $\hat{L}$ which is to have bounded eigenfunctions. Gaps can shrink to a point corresponding to a double root of $\Delta = +2$. Operators with only finitely many non-degenerate gaps have been studied in depth [59]: the countable case has been studied also [60]. Note that the eigenvalues for the 'box boundary conditions' $\psi(-P) = \psi(P) = 0$ (the auxiliary spectrum [50]) lie in the gaps. They solve $\psi_2(P,\lambda) = 0$ for $\lambda_1, \lambda_2, \ldots$: these λ_i lie in the gaps, for from $\det U = 1$, $\psi_1(P,\lambda) \times \psi_2'(P,\lambda) = 1$ so $\Delta(\lambda_j) = \psi_1(P,\lambda_j) + (\psi_1(P,\lambda_j))^{-1}$ and $|\Delta(\lambda_j)| \geq 2$. Moreover $\psi_2(x,\lambda_j)$ is a Floquet solution: for $\psi = \psi_2(x + 2P;\lambda)$ solves $\hat{L}\psi = \lambda\psi$ since q is periodic: $\psi(-P,\lambda) = 0$ means ψ is proportional to $\psi_2(x,\lambda)$: then because $\psi'(-P) = \psi_2'(P,\lambda_j)$ the proportionality constant is $\psi_2'(P,\lambda_j)$. Thus $\psi_2(x + 2P,\lambda_j) = \psi_2'(P,\lambda_j)\,\psi_2(x,\lambda_j)$ and $\psi_2(x,\lambda_j)$ is a Floquet solution with multiplier $\rho = \psi_2'(P,\lambda_j)$.

It is obvious that the basis functions for the spectral operator (4.1) are now

$v_1 = \begin{bmatrix}0\\1\end{bmatrix}$, $v_2 = \begin{bmatrix}1\\0\end{bmatrix}$. The analysis now goes through essentially unchanged. The condition for a periodic solution of period 2P (4P) is $\Delta = \pm 2$ [61].

For the lattice it is convenient to use $2N + 1$ points temporarily, not N. For the lattice Hill's operator one can use [50,58]

$$\psi_1(-N) = 1, \quad \psi_1(-N+1) = 0; \qquad \psi_2(-N) = 0, \quad \psi_2(-N+1) = 1 \tag{9.5}$$

for the basis set: these become $v_1(-N) = \begin{bmatrix}0\\1\end{bmatrix}$, $v_2(-N) = \begin{bmatrix}1\\0\end{bmatrix}$ in the 2×2 case. There is a natural correspondence with the continuum case: one uses [58] $\psi_2'(P) \to \psi_2(N+2)$ for the lattice Hill's operator and Floquet solutions satisfy $\psi(n + 2N + 1) = \rho\psi(n)$: then ρ satisfies (9.4) with Δ given by $\Delta = \psi_1(N+1,\lambda) + \psi_2(N+2,\lambda)$: the 2×2 case is similar [61].

We now change the basis: for the lattice Hill's case let $2\lambda = z + z^{-1}$ [50] and choose the basis by boundary conditions

$$\psi(n, z) = z^{-n}, \qquad \bar{\psi}(n, z) = z^{n} \qquad (n = -N) \quad . \tag{9.6}$$

Then there exist coefficients $a_L(z)$, $b_L(z)$ so that near $n = +N$

$$\begin{bmatrix}\psi(n, z)\\ \bar{\psi}(n, z)\end{bmatrix} = \begin{bmatrix}a_L(z) & b_L(z)\\ b_L(z^{-1}) & a_L(z^{-1})\end{bmatrix} \begin{bmatrix}z^{-n}\\ z^{+n}\end{bmatrix} \quad ; \tag{9.7}$$

$a_L(z)$, $b_L(z)$ will be analytic for all z. It is plain that

$$\Delta = a_L(z^{-1})\, z^{2N+1} + a_L(z)\, z^{-(2N+1)} \ . \tag{9.8}$$

One also finds ψ_1, ψ_2 (normalised at $-N$) in terms of ψ and $\bar{\psi}$. From these one finds

$$\psi_2(N+2, z) = (z - z^{-1})^{-1}\{z^{N}[a_L(z^{-1})\, z^{N+2} + b_L(z^{-1})\, z^{-(N+2)}]$$
$$- z^{-N}[a_L(z)\, z^{-(N+2)} + b_L(z)\, z^{N+2}]\} \quad . \tag{9.9}$$

We now set $z = e^{i\zeta a}$: recall $2Na = L$ (a is the lattice spacing). So

$$\Delta = a_L(-\zeta)\, e^{i(L+a)\zeta} + a_L(\zeta)\, e^{-i(L+a)\zeta} \quad . \tag{9.10}$$

On the line $-\pi/a < k < \pi/a$ (k real) in the ζ-plane z lies on the unit circle: for $a \to 0$ this line is the real ζ-axis. The upper (lower) half ζ-planes lie outside (inside) the unit circle in z. For eigenvalues λ-1, the region $(\lambda - 1) < 0$ lies on $-1 < z < 1$: these contain the bound states which as we shall see lie in the gaps in λ. When the spacing $a \to 0$

$$\Delta = a_L(-k)\, e^{iLk} + a_L(k)\, e^{-iLk} \tag{9.11}$$

while $\lambda - 1 \sim -\frac{1}{2}\zeta^2 a^2$: thus bound states lie on $\zeta = \pm i\eta$ (η real, $\eta > 0$). The lattice spectral operator corresponding to (4.1) in the 2×2 case is [62]

$$\begin{bmatrix}v_{1,n+1}\\ v_{2,n+1}\end{bmatrix} = \begin{bmatrix}z & q_n\\ r_n & z^{-1}\end{bmatrix} \begin{bmatrix}v_{1,n}\\ v_{2,n}\end{bmatrix} \quad . \tag{9.12}$$

By extending to this 2×2 case, the definition (9.6) shows that the transition matrix

$$\begin{bmatrix}a_L(k) & b_L(k)\\ b_L(k) & a_L(-k)\end{bmatrix} \tag{9.13}$$

goes over to (4.40) for $L \to \infty$. Although $a_L(\zeta)$, $b_L(\zeta)$ are analytic everywhere, $a(\zeta)(a(-\zeta))$ are analytic in the upper (lower) half plane; $b(\zeta)(b(-\zeta))$ have no simple analyticity. Both $a_L(\zeta)$ and $b_L(\zeta)$ have expression as Cauchy integrals: in particular, $a_L(k)$ is expressible in a form like (4.16). However the zeros ζ_i of $a_L(\zeta)$ are not action variables until $L \to \infty$: nor is a properly scaled $ln|a_L(k)|$. For $L < \infty$ we could choose $b_L(k)$, $b_L^*(k)$ as variables in line with the inverse method [5]. They are not canonical for $L < \infty$ (or for $L \to \infty$ cf. (4.41d)) but for $L \to \infty$ can be replaced by the action-angle variables $P(k)$, $Q(k)$. A choice of variables for $L < \infty$ has not been followed through in the present context of functional integration and we report here on how far one can get by using the $P(k)$ and $Q(k)$.

From (9.10) with $a_L(-k) = a_L^*(k)$ the condition for a periodic mode labelled by $\tilde{k}$ is

$$1 = |a_L(\tilde{k})| \cos\{L\tilde{k} - \arg a_L(\tilde{k})\} \quad . \tag{9.14}$$

This condition is already given by FLASCHKA in his lectures [50]: $\arg a_L(\tilde{k})$ is the 'spectral shift'. From (9.14) it follows that

$$L\tilde{k}_n - \arg a_L(\tilde{k}_n) = 2\pi n + \cos^{-1}\{|a_L(\tilde{k}_n)|^{-1}\} \quad . \tag{9.15}$$

We must expect that $ln|a_L(\tilde{k}_n)| \to \frac{1}{2}\gamma_0\omega(\tilde{k}_n)\, P_n \to \gamma_0\omega(\tilde{k}_n)\, P(\tilde{k}_n)\,\pi L^{-1}$ as $L \to \infty$. Consequently $\cos^{-1}\{|a_L(\tilde{k}_n)|^{-1}\} \to \cos^{-1}\{\exp(-\gamma_0\omega(\tilde{k}_n)P(\tilde{k}_n)\pi L^{-1})\} \to 2\pi n$ for $L \to \infty$. This does not change the $2\pi n$ in (9.15). Thus the $\tilde{k}_n$ are apparently determined by $\tilde{k}_n = 2\pi n L^{-1} + L^{-1} \arg a_L(\tilde{k}n)$. However, a non-trivial change of sign is still needed corresponding to the mapping (4.26) in moving from light-cone to covariant co-ordinates — and this means finally that the $\tilde{k}_n$ are determined by

$$\tilde{k}_n = \frac{2\pi n}{L} - \frac{1}{L}\arg a_L(\tilde{k}_n) \quad . \tag{9.16}$$

For $\arg a_L(\tilde{k}_n)$ we take (4.39). This way we reach

$$\begin{aligned}\tilde{k}_n = {} & \frac{2\pi n}{L} + \frac{1}{L}\sum_{i=1}^{N_k} \Delta_k(\tilde{k}_n, p_i) + \frac{1}{L}\sum_{j=1}^{N_{\bar{k}}} \Delta_{\bar{k}}(\tilde{k}_n, \bar{p}_j) \\ & + \frac{1}{L}\sum_{\ell=1}^{N_b} \Delta_b(\tilde{k}_n, \hat{p}_\ell, \Theta_\ell) + \frac{1}{L}\sum_{\substack{m=-\frac{1}{2}N_{ph} \\ m\neq n}}^{m=+\frac{1}{2}N_{ph}} G(\tilde{k}_n, \tilde{k}_m) P_m\end{aligned} \tag{9.17}$$

where

$$G(k,k') \equiv \frac{\gamma_0 m^2}{8} \frac{1}{k\omega(k') - k'\omega(k)} \quad . \tag{9.18}$$

Since $\tilde{k}_n$ is expressed in terms of the canonical variables for $L \to \infty$ it will be correct to $O(L^{-2})$. However, the important places are those where combinations $L^{-1}\sum_{m=-\frac{1}{2}N_{ph}}^{m=+\frac{1}{2}N_{ph}}$ arise. Consequently, we believe it makes sense to solve (9.16) by iteration: we shall see that this procedure will bring correct results in principle at e.g. the $-\frac{7}{8}(M\beta)^{-1}$ term that is at order $(M\beta)^{-1}$ in the free energy (7.13) of the kink, for example. Whether it will be correct at order $(M\beta)^{-2}$ after a further iteration is being investigated now. Note that the integral equations (9.16) for $\tilde{k}_n$ represents the Bethe ansatz integral equation for modes k_n in the quantum theory: the new feature here is that in (9.18) the P_m appear and these of course are subsequently eliminated by functional integration.

There are corresponding results for the allowed periodic modes associated with kinks, antikinks, or breathers of the $L \to \infty$ problem. For kinks interacting with other kinks and phonons but no breathers, we find an eigenvalue $2i\eta$ given by

$$2\eta L = \sum_{i=1}^{N_k} ln\left(\frac{\eta+\eta_i}{\eta-\eta_i}\right)^2 - \frac{m^2\gamma_0}{8}\sum_n \frac{P_n}{\eta\,\omega(\eta)\tilde{\omega}_n - \eta\tilde{k}_n} \quad . \tag{9.19}$$

The extension to include effects of antikinks and breathers is plain. Since η must be close to η_j for $L \to \infty$ we suppose $\eta = \eta_j\, e^{\alpha_j - \eta_j L}$. Then by choosing the α_j

$$2\eta = 2\eta_j + \frac{1}{L}\sum_{i\neq j}^{N_k} \ln\left(\frac{\eta_j+\eta_i}{\eta_j-\eta_i}\right)^2 - \frac{m^2\gamma_0}{L}\sum_n \frac{P_n}{\omega(\eta)\tilde{\omega}_n-\eta\tilde{k}_n} \tag{9.20}$$

to exponential order. In this we can recognize the kink-kink phase-shifts $\Delta_{k,k}(p_i, p_j) \equiv -\ln\{(2\eta_j + 2\eta_i)/(2\eta_j - 2\eta_i)\}^2$ given by (4.36) for pairs of kinks of momentum p_i, p_j respectively. This means that the η_j are shifted at order $O(L^{-1})$ by these phase-shifts.

On the other hand, there is apparently a canonical momentum close to η_i: we know this for $L \to \infty$ from (4.20) for example, where it is an action variable. But we can say more than this. We follow [58]: the spectral problem (4.1) is not that used in [58] so that we shall finally make an *ad hoc* adjustment to reach final agreement with (4.20). Following [58], canonical pairs are $\mu_i = \frac{1}{2}\gamma_0^{-1}(z_i + z_i^{-1})$ and $f_i = -2\alpha\ln|\psi_2(N + 2, z_i)|$ where z_i is a root of $\psi_2(N + 2, z_i) = 0$ and $\alpha = \alpha(z_i)$ will be adjusted. From this one proves $d\mu_i \wedge df_i = \gamma_0^{-1} d\zeta_i \wedge 2\alpha\ln|\psi_2|$ with $\psi_2 \equiv \psi_2(N + 2, \zeta_i)$ corresponding to the transformation $z = e^{i\zeta a}$. The root ζ_i is from (9.9) therefore a root of

$$a(-\zeta)e^{i\zeta(L+a)} + b(-\zeta)e^{ia} = a(\zeta)e^{-i(L+a)\zeta} + b(\zeta)e^{-ia} \tag{9.21}$$

and for ζ in the upper half plane we must have $a(\zeta_i) \to 0$ as $L \to \infty$. As a further check, if $a(\zeta_i) = 0$, $\psi_2(N + 2,\zeta_i) \to -2|b_i| \cos(\zeta_i a + \arg b_i)$ and canonical pairs are therefore $\gamma_0^{-1}\zeta_i$ and $-2\alpha(\zeta_i) \ln|b_i|$ up to a trivial canonical transformation. The choice $\alpha(\zeta_i) = -\zeta_i^{-1}$ then allows us to use $4\gamma_0^{-1}\eta_i$ and $(2\eta_i)^{-1} \ln|b_i|$ as canonical pairs in agreement with (4.20). The *ad hoc* adjustment α is consistent with the effect of the symmetry condition $r = -q$ imposed on the action-angle variables for the spectral operator (4.1) [11]. Note again that the bound states $\zeta_i = i\eta_i$ (η_i real) in the ζ-plane (with $a \to 0$) lie on $-1 < z < +1$ in the z-plane (which explains a factor i [11]). The conclusion from this analysis for the kinks (and by implication for the antikinks and breathers too) is that we can retain the $L \to \infty$ canonical variables p_i, q_i etc. of (4.27) at least to $O(L^{-1})$ and perhaps to $O(e^{-L})$. Certainly (9.17) suggests 'interactions' through phase-shifts occur at this order, but we have not yet been able to explore their role in this respect. The remaining part of these lectures is concerned with the consequences of (9.16) and (9.17), the integral equation for $\tilde{k}_n$. With $k_n = 2\pi n L^{-1}$ and one iteration

$$\begin{aligned}\tilde{k}_n - k_n = {} & \frac{1}{L}\sum_{i=1}^{N_k} \Delta_k(k_n,p_i) + \frac{1}{L}\sum_{j=1}^{N_{\bar{k}}} \Delta_{\bar{k}}(k_n,p_j) \\ & + \frac{1}{L}\sum_{\ell=1}^{N_b} \Delta_b(k_n,\hat{p}_\ell,\Theta_\ell) + \frac{1}{L}\sum_{\substack{m=-\frac{1}{2}N_{ph}\\ m\neq n}}^{m=+\frac{1}{2}N_{ph}} G(k_n,k_m)P_m\end{aligned} \tag{9.22}$$

correct to $O(L^{-2})$: $G(k,k')$ is given by (9.18). This is the place from which we shall work.

10. Partition Function for the sinh-Gordon Equation

This is the 'easy' case: the sinh-Gordon equation is obtained from (1.1) for $\phi \to -i\phi$: consequently its Hamiltonian is found from (1.2) by $\phi \to i\phi$ and $\gamma_0 \to -\gamma_0$. The corresponding spectral problem (4.1) has $q = r^*$ so there are no bound states and no solitons: in this respect it corresponds to the NLS (3.4) in the repulsive case $\gamma < 0$ [5]. It follows that in action-angle variables

$$H[p,N] = \sum_{-\frac{1}{2}N}^{+\frac{1}{2}N} \tilde{\omega}_n(\tilde{k}_n)P_n \quad : \tag{10.1}$$

$\tilde{\omega}_n(\tilde{k}_n)$ is the lattice dispersion relation (2.5) evaluated for wave number $\tilde{k}_n$. At first sight since for $a \to 0$ and $L \to \infty$ $H[p] = \int_{-\infty}^{\infty} \omega(k)\, P(k)dk$ it seems we have just the free K-G! However, this ignores the spectral shifts: (10.1) shows that to $O(L^{-1})$ the k_n are not the free K-G modes but the shifted ones.

By TIM one finds [53]

$$F = +\beta^{-1}m[\tfrac{1}{4}(M\beta)^{-1} - \tfrac{1}{8}(M\beta)^{-2} + \tfrac{3}{16}(M\beta)^{-3} + \ldots] + \beta^{-1}a^{-1}(\ln\beta a^{-1} + \tfrac{1}{2}ma). \quad (10.2)$$

This is actually a surprising result, since if we look at (7.13) for the s-G we see that it can be obtained by setting $\gamma_0 \to -\gamma_0$ and dropping only the kink and anti-kink contributions: we must retain the 'breather' series. But we show that the corresponding 'breather' series for the sinh-G is constructed from the shifted K-G modes, and it is not a breather series at all!

To $O(L^{-1})$ we observe that

$$H[p,N] = \sum_{-\frac{1}{2}N}^{+\frac{1}{2}N} \omega_n P_n - \frac{1}{L}\sum_{-\frac{1}{2}N}^{+\frac{1}{2}N} \sum_{m\neq n} \frac{\partial\tilde{\omega}_n}{\partial k_n} G(k_n,k_m)P_nP_m \quad (10.3)$$

with $G(k_n,k_m)$ given by (9.18) and the sign $(\gamma_0 \to -\gamma_0)$ taken into account: $\tilde{\omega}_n$ is given by (2.15) evaluated at k_n. Evidently Hamiltonian (10.3) is not separable and of course (10.1) is not separable.

Since

$$Z = \left\{\int_0^{2\pi}\int_0^{\infty}\right\} \prod_n (2\pi)^{-1}\, dQ_n dP_n \exp - \beta H[p,N] \quad (10.4)$$

and to $O(L^{-1})$

$$Z = \left\{\int_0^{\infty}\right\} \prod_n dP_n \left[1 + \beta L^{-1} \sum_{m\neq n} \frac{\partial\omega_n}{\partial k_n} G(n,m)P_nP_n\right] \exp(-\beta \sum_n \omega_n P_n) \quad (10.5)$$

we find $Z = \exp(-\beta LF)$ where

$$\beta LF \equiv \sum_{n=-\frac{1}{2}N}^{n=+\frac{1}{2}N} \ln(\beta\omega_n) - (\beta L)^{-1} \sum_{m\neq n} \omega_m^{-1}\, G(n,m)\frac{\partial}{\partial k_n}\ln\omega_n \quad . \quad (10.6)$$

Thus for $N, L \to \infty$

$$F = (2\pi\beta)^{-1}\int_{-\pi/a}^{+\pi/a} dk\, \ln(\beta\omega(k)) - (2\pi\beta)^{-2}\int dk\omega^{-1}(k)\int dqG(k,q)\frac{\partial}{\partial q}\ln\omega(q) \quad (10.7)$$

$$= \beta^{-1}a^{-1}(\ln(\beta a^{-1}) + \tfrac{1}{2}ma) + \beta^{-1}m[\tfrac{1}{4}(M\beta)^{-1} + \ldots] \quad (10.8)$$

in agreement with (10.2).

To reach this form we use the lattice dispersion relation only for the evaluation of the first term: we could use it elsewhere but, to $O(a^2)$, $\omega_n = (m^2 + k_n^2)^{\frac{1}{2}}$ can just as well be used. For the further terms of (10.7) we need further terms in the iteration of (9.17) and further terms in the Taylor series development of $\omega(k_n)$: evidently to every term $L^{-1}\sum_n$ there will correspond a term $(M\beta)^{-1}$ in the development of (10.7). The evidence is that this is enough so that iteration of (9.17) will yield *all* terms at all orders: this is still being explored as was noted in §9.

We must also recognize that $G(k_n,k_m)$ given by (9.18) adopts the result (4.39): we have not mentioned before that this result is valid (with the factor $\frac{1}{8}\gamma_0$) only if one sheet of the two-sheeted k-plane is used: if both sheets are used, the factor is $\frac{1}{4}\gamma_0$ and the coefficient $\frac{1}{4}$ in the $\frac{1}{4}(M\beta)^{-1}$ in (10.8) is replaced by $\frac{1}{2}(M\beta)^{-1}$. In the case of the s-G the breather eigenvalues lie in pairs on both sheets of the k-plane. Thus it seems that for the s-G both sheets *must* be used, and the factor in (4.39) is $\frac{1}{4}\gamma_0$. The fact that sinh-Gordon needs one sheet and the factor $\gamma_0/8$ has still to be understood (there could be a trivial error of a factor 2 but we have not found it yet!).

As noted, the relationship between sinh-G and s-G otherwise parallels that between the repulsive and attractive cases of the NLS (3.15) ($\chi < 0$ and $\chi > 0$). The repulsive NLS includes the impenetrable Bose gas whose quantum statistical mechanics was pioneered by YANG and YANG [21]. For such systems we expect the classical limit to take the form

$$F = (2\pi\beta)^{-1} \int ln(\beta\bar{\omega}(k))dk, \tag{10.9a}$$

$$\bar{\omega}(k) = \omega(k) - (\pi\beta)^{-1} \int dqG(k,q) \frac{\partial}{\partial q} ln\, \bar{\omega}(q) \quad . \tag{10.9b}$$

By iteration we find

$$F = (2\pi\beta)^{-1} \int_{-\pi/a}^{+\pi/a} ln(\beta\omega(k))dk - 2(2\pi\beta)^{-2} \int dk\omega^{-1}(k)\int dqG(k,q) \frac{\partial}{\partial q}(ln\omega(q)). \tag{10.10}$$

This disagrees with (10.7) by the factor 2 in the second term. However, by subsequent iterations we find

$$F = \beta^{-1}m[\tfrac{1}{2}(M\beta)^{-1} - \tfrac{1}{8}(M\beta)^{-2} + \tfrac{3}{16}(M\beta)^{-3} + \ldots] + \beta^{-1}a^{-1}(ln(\beta a^{-1}) + \tfrac{1}{2}ma). \tag{10.11}$$

Thus the TIM result seems to mean that, for the sinh-G, the correct classical limit is not (10.9a) but is $F - m(4M\beta^2)^{-1}$. This curious result means of course that the guess (10.9a) for the classical limit of the Bethe ansatz was incomplete: it has no bearing on the agreement established between our result (10.8) and the TIM result (10.2) (as far as this has currently been calculated which is to the $\frac{1}{4}(M\beta)^{-1}$ term exhibited).

Work is currently in hand in establishing connections with the Bethe ansatz formulations (10.9).

11. Sine-Gordon Statistical Mechanics Without Phonon-Phonon Phase-Shifts

According to the arguments of §§9 and 10, to significant order in L^{-1} the Hamiltonian is

$$H[p,N] = \sum_{i=1}^{N_k} (M^2 + p_i^2)^{\frac{1}{2}} + \sum_{j=1}^{N_{\bar{k}}} (M^2 + \bar{p}_j^2)^{\frac{1}{2}} + \sum_{\ell=1}^{N_b} (4M^2 \sin^2\Theta_\ell + \hat{p}_\ell^2)^{\frac{1}{2}} + \sum_{n=-\frac{1}{2}N_{ph}}^{n=+\frac{1}{2}N_{ph}} \omega_n(1 + \frac{k_n\Delta_n}{L\omega_n^2})P_n \quad ; \tag{11.1}$$

from (4.39) and (9.22)

$$\Delta_n = \sum_{i=1}^{N_k} \Delta_k(k_n,p_i) + \sum_{j=1}^{N_{\bar{k}}} \Delta_{\bar{k}}(k_n,\bar{p}_j) + \sum_{\ell=1}^{N_b} \Delta_b(k_n,\hat{p}_\ell,\Theta_\ell) \tag{11.2}$$

with the phonon-phonon shifts omitted. The Hamiltonian (11.1) is not separable since Δ_n depends on all of the action variables $\{p_i, \bar{p}_j, \hat{p}_\ell, 4\gamma_0^{-1}\Theta_\ell\}$.

The functional integral on the P_n yields

$$\prod_{-\frac{1}{2}N_{ph}}^{+\frac{1}{2}N_{ph}} [\beta\omega_n(1 + \frac{k_n\Delta_n}{L\omega_n^2})]^{-1} \quad . \tag{11.3}$$

Scaled against the free K-G with N modes k_n one finds

$$(\beta m)^{N_k+N_{\bar{k}}+2N_b}\left\{ \prod_{-\frac{1}{2}N_{ph}}^{+\frac{1}{2}N_{ph}}(1 + \frac{k_n\Delta_n}{L\omega_n^2})^{-1}\right\} \left\{\prod_{n=+\frac{1}{2}N_{ph}}^{+\frac{1}{2}N} m^{-2}(\omega_n\omega_{-n})\right\} \quad . \tag{11.4}$$

We suppose $\lim_{N,N_{ph}\to\infty} \prod_{n=\frac{1}{2}N_{ph}}^{\frac{1}{2}N} m^{-2}(\omega_n\omega_{-n}) = 1$. The idea is that as $N, N_{ph} \to \infty$ there are always modes to match the two sets of modes in pairs: however $N - N_{ph}$ modes drop out from the $k = 0$ region - corresponding to zero frequency modes for the kinks and antikinks (one each) and breathers (two each). The phase spaces open up from cylinders to sheets for translation and to the compact sheet $0 < \Theta < 2\pi\gamma_0^{-\frac{1}{2}}$; $0 < \Phi < 8\pi$ for the breathers. This is the origin of the factor $(\beta m)^{N_k+N_{\bar{k}}+2N_b}$ already foreshadowed in §7.

With the last factor in (11.4) equal to unity, the remaining part exponentiates to

$$\exp\left\{ \sum_{-\frac{1}{2}N_{ph}}^{+\frac{1}{2}N_{ph}} \ln\Big(1 + \frac{k_n\Delta_n}{L\omega_n^2}\Big)^{-1} + (N_k + N_{\bar{k}} + 2N_b)\ \ln\ \beta m \right\}$$

$$= \exp\left\{ - \sum_{-\frac{1}{2}N_{ph}}^{+\frac{1}{2}N_{ph}} \frac{k_n\Delta_n}{L\omega_n^2} + (N_k + N_{\bar{k}} + 2N_b)\ \ln\ \beta m \right\} \tag{11.5}$$

to $O(L^{-1})$ in the exponent. This becomes

$$\exp\left\{ - \frac{1}{2\pi}\int_{-\pi/a}^{+\pi/a} d(\ln\omega(k))\Delta(k) + (N_k + N_{\bar{k}} + 2N_b)\ \ln\ \beta m \right\} \quad ; \tag{11.6}$$

$\Delta(k)$ is Δ_n (11.2) with k replacing k_n. By parts on the integral gives

$$\exp\left\{ + \frac{1}{2\pi}\int_{-\pi/a}^{+\pi/a} [\ln(\omega(k))\ \frac{d\Delta}{dk}\ ]\ dk + (N_k + N_{\bar{k}} + 2N_b)\ \ln\ \beta m \right\}$$

$$\equiv \exp\{\sigma[N_k, N_{\bar{k}}, N_b]\} \tag{11.7}$$

and σ is that quantity in (8.11). Note how the step to the integral in the first line of (11.7) takes up the L^{-1} in (11.5): we can expect from the previous work (§7) that σ does not depend on L.

The by parts result (11.7) uses the fact that $\Delta(\pm\pi/a) \to 0$ as $O(a)$ as $a \to 0$. This result (11.7) together with the measure (8.7) with (8.11) is the first rigorous justification of phenomenological theories [63,64,65] introduced to dress the kinks, antikinks and breathers *ad hoc*. It also justifies for the first time the otherwise arbitrary procedures adopted by DHN [26] and used earlier in the §7. We find we need to check the by parts step in the course of the details following.

These details are that, first of all, on infinite support $L \to \infty$, $\Delta(k)$ is additive in terms of single kink, single antikink, and single breather, phase-shifts. Thus the integral contributing to $\sigma[N_k,N_{\bar{k}},N_b]$ is additive and given by

$$\sum_{i=1}^{N_k} \sigma_k(p_i) + \sum_{j=1}^{N_{\bar{k}}} \sigma_{\bar{k}}(\bar{p}_j) + \sum_{\ell=1}^{N_b} \sigma_b(\hat{p}_\ell,\Theta_\ell) \tag{11.8}$$

where

$$\sigma_k(p) = \sigma_{\bar{k}}(p) = \frac{1}{2\pi}\int_{-\pi/a}^{+\pi/a} [\ln(\omega(k))\ \frac{d}{dk}\Delta_k(k,p)]dk \tag{11.9a}$$

$$\sigma_b(p,\Theta) = \frac{1}{2\pi}\int_{-\pi/a}^{+\pi/a} [\ln(\omega(k))\ \frac{d}{dk}\Delta_b(k,p,\Theta)]dk \tag{11.9b}$$

The phase-shifts $\Delta_k(k,p) = \Delta_{\bar{k}}(k,p)$ and $\Delta_b(k,p,\Theta)$ are given by (4.37) and (4.38) respectively.

It is now obvious that the measure (8.7) employed in the form (8.11) exponentiates the single but dressed kink, antikink, and breather contributions which are

$$\tilde{Z}_k = \tilde{Z}_{\bar{k}} = \frac{L}{2\pi}\int_{-\infty}^{\infty} dp[-\beta(M^2+p^2)^{\frac{1}{2}} + \sigma_k(p) + ln\,\beta m] \tag{11.10a}$$

$$\tilde{Z}_b = \frac{L}{4\pi^2}\int_0^{8\pi} d\phi \int_{-\infty}^{\infty} dp \int_0^{\pi/2} 4\gamma_0^{-1} d\Theta \exp[-\beta(4M^2\sin^2\Theta + p^2)^{\frac{1}{2}} + \sigma_b(p,\Theta) + 2ln\beta m] \tag{11.10b}$$

respectively. We consider $\tilde{Z}_k$ with $\tilde{Z}_{\bar{k}}$ first.

The integral for $\sigma_k(p)$ can be evaluated: the result is expressed in terms of a single velocity V through $p = MV\gamma$, $\gamma \equiv (1-V^2)^{-\frac{1}{2}}$, and is

$$\sigma_k(p) = ln[\gamma^{-1}(1+\gamma)] \quad . \tag{11.11}$$

One then uses

$$\frac{1}{2\pi}\int_{-\infty}^{\infty} dp\, \gamma^n\, e^{-\beta(M^2+p^2)^{\frac{1}{2}}} \sim \pi^{-\frac{1}{2}}M(2\beta M)^{-\frac{1}{2}}e^{-\beta M}\left\{1 + (\frac{n+1}{2} - \tfrac{1}{8})(M\beta)^{-1} + \ldots\right\} \tag{11.12}$$

(which can also be given in closed form in terms of Bessel functions). Then the total contribution to the free energy per unit length made by the kinks and antikinks together is

$$-\,2\beta^{-1}L^{-1}\,ln\,\tilde{Z}_k = -\,\beta^{-1}m8(\beta^{\frac{1}{2}}m^{\frac{1}{2}}\pi^{-\frac{1}{2}}\gamma_0^{-\frac{1}{2}})e^{-\beta M}(1 + \tfrac{1}{8}(M\beta)^{-1} + \cdots) \quad . \tag{11.13}$$

The result (11.13) agrees with the first term of (7.13) *except* in the coefficient at order $(M\beta)^{-1}$. This error is due to phonon-phonon phase-shifts (§12)!

Elsewhere [66] we reported the coefficient $+\frac{5}{8}$ (not $+\frac{1}{8}$) in the absence of phonon-phonon phase-shifts. The coefficient $+\frac{5}{8}$ is also obtained in the phenomenological theory [65]; but there it was assumed that $\sigma_k(p)$ was to be evaluated as a simple P-integral through the zero of $k\omega(p) - p\omega(k)$ at $k = p$. In this case, however, one must take account of the jump of the branch of the $\tan^{-1}$ function for $\Delta_k(k,p)$ (4.37), and in moving the d/dk by integration by parts as at (11.7) above there is a correction from the integrated part. In [66] we were guided by the result of taking a classical limit of the Bethe ansatz analysis [67]: there we found such a simple P-integral, but it is already evident that that classical limit analysis cannot be complete since it did not exhibit any aspect of the phonon-phonon shifts (§12) - as was noted at the time [67]. These tricky manipulative points are necessarily still under review; but (11.11) seems correct and without phonon-phonon shifts so therefore does the coefficient $\frac{1}{8}$ in (11.13). An appendix to these lectures (Appendix III) develops these remarks much further.

The analysis of the breather contributions is substantially more difficult than the kink-antikink contributions, so we shall not attempt to give any details here: moreover, we shall follow the Bethe ansatz analysis [67] and thus take the P-integral interpretation just referred to now extended to the breathers: the appendix (Appendix III) elaborates further on this choice. In this interpretation

$$\sigma_b(p,\Theta) = 2\,ln\,(\gamma + \sin\Theta) \tag{11.14}$$

and disagrees with (11.11) for a kink-antikink pair when $\Theta = \frac{1}{2}\pi$. In this form, the problem is now that the Θ-integration in (11.10b) diverges at its lower limit. It seems obvious what the trouble is: we are mixing small mass breathers and phonon contributions (*vide* the remark in §3 near (3.19)).

By reference to the Bethe ansatz [67] and its classical limit, we have found a cut-off procedure. The Θ-integral is cut-off in terms of a rapidity variable $x(\cosh\frac{1}{2}x$

$= \gamma$) at the temperature dependent value

$$\Theta_b(x) = \kappa\beta^{-1}(2M\cosh(\tfrac{1}{2}x))^{-1} \ ; \quad \kappa \approx 4(\pi\, ln\,(\tfrac{1}{2}e))^{-1} \sim 4.1 \quad . \tag{11.15}$$

In this form

$$F_{breather} = -m\pi^{-1}\beta^{-1}(M\beta)^2 \int_{-\Lambda}^{\Lambda} dx\, \cosh(\tfrac{1}{2}x) \int_{\Theta_b(x)}^{\pi/2} d\Theta\, \sin\Theta\, (\cdots)$$

$$(\cdots) \equiv (\cosh\tfrac{1}{2}x + \sin\Theta)^2 \exp\{-2\beta M \cosh(\tfrac{1}{2}x)\sin\Theta\} \tag{11.16}$$

and the cut-off on rapidity x is $\Lambda = 2\sinh^{-1}(\pi/am)$. The integral is extraordinarily sensitive to the cut-off $\Theta_b(x)$ and the number κ on which it depends: for the value $\kappa = 4.1$ quoted we find the contribution $F_{breather}$ is $-m\beta^{-1}(0.22)(M\beta)^{-1}$ which compares with the $\frac{1}{4}$ in the series in the second line of (7.13). However, a more searching analysis of the argument [67] for the cut-off yields $-m\beta^{-1}(0.25\cdots)(M\beta)^{-1}$ as well as a contribution in a from the cut-off Λ which reduces the discrepancy $(ln\pi - 1)$ between the free energy for the free K-G, (6.4), and the TIM result, (7.13), to almost nothing.

This analysis raises almost as many questions as the problems it solves, since from §10 we must expect that phonon-phonon interactions also generate a breather series. We readdress but do not solve this question in the penultimate section, §14. Appendix III reviews the choice of the P-integral interpretation for (11.14) as noted.

12. Sine-Gordon Statistical Mechanics with Phonon-Phonon Phase-Shifts Included

Corresponding to (11.1) the Hamiltonian will be

$$\begin{aligned} H[p,N] = {} & \sum_{i=1}^{N_k} (M^2 + p_i^{\,2})^{\frac{1}{2}} + \sum_{j=1}^{N_{\bar{k}}} (M^2 + \bar{p}_j^{\,2})^{\frac{1}{2}} \\ & + \sum_{\ell=1}^{N_b} (4M^2\sin^2\Theta_\ell + \hat{p}_\ell^{\,2})^{\frac{1}{2}} + \sum_{n=-\frac{1}{2}N_{pn}}^{+\frac{1}{2}N_{ph}} \omega_n \left(1 + \frac{k_n\Delta_n}{L\omega_n^{\,2}} \right) P_n \\ & - \frac{1}{L^2} \sum_{\substack{m,n = -\frac{1}{2}N_{ph} \\ m\neq n}}^{+\frac{1}{2}N_{ph}} k_n\omega_n^{-1}\,\Delta_m \frac{\partial G}{\partial k_n}(k_n,k_m) P_n P_m \\ & - \frac{1}{L} \sum_{\substack{m,n = -\frac{1}{2}Np_h \\ m\neq n}}^{+\frac{1}{2}N_{ph}} k_n\omega_n^{-1}\, G(k_n,k_m) P_n P_m \end{aligned} \tag{12.1}$$

The quantities Δ_n and Δ_m are still given by (11.2). The new terms are the last one, which is expressible alternatively in the form in (10.3) (with sign changed) and is the phonon-phonon shift, and the next to the last one which is a soliton (kink, antikink and breather)-phonon shift mediated by a phonon-phonon shift. These terms are generated by the shift of both $\tilde{k}_n$ to k_n *and* k_m to k_m in (9.17). Evidently this involves a second iteration of (9.17) so we are now working at $O(L^{-2})$. We are not working consistently at this order since Δ_m in the next to the last term in (12.1) should be extended by a phonon term; but this term cannot dress the solitons. There are possibly other 'loose' terms $O(L^{-2})$ also; but none of these associate with sums $\sum_{n=-\frac{1}{2}N_{ph}}^{+\frac{1}{2}N_{ph}}$.

By repeating the line of argument in §11 we have been able to show in particular that before integrating by parts

$$-\sigma_k(p) = \frac{1}{2\pi} \int_{-\infty}^{\infty} \frac{\bar{\omega}'(k)}{\bar{\omega}(k)} \Delta_k(k,p)dk \tag{12.2}$$

where

$$\frac{\bar{\omega}'(k)}{\bar{\omega}(k)} \equiv \frac{k}{m^2+k^2}\left(1 + \frac{\frac{1}{4}\gamma_0 m\beta^{-1}}{(m^2+k^2)}\right) \quad . \tag{12.3}$$

From this we find by integrating by parts that there are two shifts

$$-\frac{1}{4\pi}\int_{-\infty}^{\infty} ln(m^2+k^2)\frac{d}{dk}\Delta_k(k,p)\,dk \quad , \tag{12.4}$$

which reduces to (11.11), and a second one proportional to γ_0 which is

$$\frac{\gamma_0 m}{8\pi\beta}\int_{-\infty}^{\infty}\frac{1}{(m^2+k^2)}\;\frac{d}{dk}\Delta_k(k,p)\,dk \tag{12.5}$$

and reduces to $(\gamma_0/4\beta m)\,(\gamma^{-2} - (\gamma+1)^{-1})$. By using (11.12) one then finds, after a little manipulation, that the kink with antikink free energy is

$$-\beta m 8(\beta^{\frac{1}{2}}m^{\frac{1}{2}}\pi^{-\frac{1}{2}}\gamma_0^{-\frac{1}{2}})e^{-\beta M}(1 - \tfrac{7}{8}(M\beta)^{-1}\ \ldots) \;. \tag{12.6}$$

The contribution of the phonon-mediated term to the dressing of the kink is precisely $-1(M\beta)^{-1}$ correcting the $+\frac{1}{8}$ in (11.13) to $-\frac{7}{8}$ as required.

This analysis certainly demonstrates that phonon-mediated dressing of the kinks and antikinks is important at $O((M\beta)^{-1})$ as expected. It appears to confirm the results of both the TIM and steepest descents obtained in §7. However, the crucial result is (12.3), and here we reach the number $\frac{1}{4}\gamma_0$ by using both sheets of the k-plane (introducing one factor 2) and a second factor of 2 still unaccounted for! This analysis is therefore still under scrutiny, and the coeffficient at $O((M\beta)^{-2})$ (which proves to be $59/128(M\beta)^{-2}$ from the TIM analysis extended as described in §7) is also being checked.

A tentative conclusion is, nevertheless, that we can successfully use the idea that terms $O(L^{-n})$ can be handled at each order $n \geq 2$: they are significant only in the combination $L^{-1}\sum_{-\frac{1}{2}N_{ph}}^{+\frac{1}{2}N_{ph}}$, then induce $O((M\beta)^{-n})$ in the free energy F as phonon-phonon contribution (§10), or as $O((M\beta)^{-(n-1)})$ in phonon mediated soliton contributions as demonstrated in this §12.

The corresponding situation surrounding the breathers is still too complicated and incomplete for us to report it here. We note that the last term exhibited in (12.1) is the phonon-phonon term at $O(L^{-1})$ and will (by analogy with sinh-Gordon, §10) generate the first term $-\beta^{-1}m\frac{1}{4}(M\beta)^{-1}$ of the breather series. But as noted at the end of §11 we must already have a dressed breather series from the breather term in (12.1). This paradox is not yet resolved (§14).

13. Classical Fermions

We have repeatedly compared the present analysis with the Bethe ansatz [6,7,8]. The Bethe ansatz statistical mechanics for the s-G exploits the equivalence between the quantum spin-$\frac{1}{2}$ XYZ model in continuum limit (§3) the quantum massive Thirring model (§3, (3.29)) which is a fermi system, and the quantum sine-Gordon equation [29,30] which is a bose system. The quantum statistical mechanics will exhibit fermi statistics for the following reason: the Bethe ansatz works in k-space and introduces harmonic modes under periodic boundary conditions labelled by k_n [6]: the allowed k_n are determined by an integral equation analoguous to (9.17): the k_n are (believed to be) distinct [5,6]. It follows that eigenstates are labelled by momenta $k_1, k_2, \ldots, k_N$ which are distinct and the quantum systems obey fermi statistics.

Since we work in the classical limit (5.15), there may be no vestige of fermi statistics remaining. However, the canonical transformation via the spectral transform (4.1), ultimately from variables $(\Pi(x,t), \phi(x,t))$ to the action-angle variables

(4.27), contains a crucial assumption: the eigenvalues ζ are nondegenerate, and the zeros ζ_i of $a(\zeta)$, (4.16), corresponding to the bound states are simple zeros. Evidently this condition has to be imposed on the phase spaces for the action-angle variables.

First we change the measure: instead of (8.7)

$$\mathcal{L}_\mu = \prod_{i=1}^{N_k} \frac{d\bar{p}_i d\bar{q}_i}{2\pi} \prod_{j=1}^{N_{\bar{k}}} \frac{d\bar{p}_j d\bar{q}_j}{2\pi} \prod_{\ell=1}^{N_b} \frac{d\hat{p}_\ell d\hat{q}_\ell 4\gamma_0^{-1} d\Theta_\ell d\Phi_\ell}{4\pi^2} \tag{13.1}$$

(the phonons are already non-degenerate). Fermi statistics then arises through the following analysis (which is essentially the usual one [68]). Consider

$$A \equiv \sum_{N=0}^{\infty} \prod_{i=1}^{N} \int dp_i \alpha(p_i) \quad . \tag{13.2}$$

The functions $\alpha(p_i)$ will be $\exp\{-\beta(M^2+p_i^2)^{\frac{1}{2}} + \sigma(p_i)\}$ for the kinks. But since $\zeta_i = i\eta_i$ $(\eta_i > 0)$ determines momenta p_i for kinks and momenta $\bar{p}_i$ for antikinks we shall need to impose the further condition that p_i and $\bar{p}_i$ cannot coincide. We do this after first evaluating (13.2) as it stands. Note that no such difficulty arises with the breathers: their associated eigenvalues arise in the pairs $(\zeta_j, -\zeta_j^*)$ and never coincide with the imaginary axis $\zeta = i\eta$ of the ζ-plane.

We interpret the integrals in (13.2) as formal sums so that that expression becomes

$$A \equiv \sum_{N=0}^{\infty} \prod_{i=1}^{N} \sum_{p_i} \alpha(p_i) \tag{13.3}$$

(any scaling on the $\sum_{p_i}$ is compensated by return to an integral below). The product in (13.3) is

$$\sum_{p_1} \sum_{p_2} \cdots \sum_{p_N} \alpha(p_1) \cdots \alpha(p_N) \tag{13.4}$$

and the condition $p_1 \neq p_2 \neq p_3 \neq \ldots \neq p_N$ implies

$$A \equiv \sum_{N=0}^{\infty} \prod_{i=1}^{N} \sum_{p_i} \alpha(p_i) = \prod_p (1 + \alpha(p))$$

$$(= 1 + \sum_p \alpha(p) + \sum_{p_1} \sum_{p_2, p_2 \neq p_1} \alpha(p_1)\alpha(p_2) + \ldots) \quad . \tag{13.5}$$

Then $A = \exp(\ln A)$ and

$$\ln \prod_p (1 + \alpha(p)) = \sum_p \ln(1 + \alpha(p)) \to \int dp \ln(1 + \alpha(p)) \quad . \tag{13.6}$$

If we impose the further condition that the kinks and antikinks together have no momenta in common, we find a total contribution $\int dp \ln(1 + 2\alpha(p))$ from these two sorts of excitation.

From this analysis

$$F = F_{KG} - \beta^{-1}(2\pi)^{-1} \int_{-\infty}^{\infty} dp \ln\{1 + 2\exp[-\beta(M^2 + p^2)^{\frac{1}{2}} + \sigma_k(p) + \ln\beta m]\}$$

$$- \beta^{-1} 8(\gamma_0\pi)^{-1} \int_{-\pi/a}^{+\pi/a} dp \int_{\Theta_b(p)}^{\pi/2} d\Theta \ln\{1 + \exp[-\beta(4M^2\sin^2\Theta + p^2)^{\frac{1}{2}}$$

$$+ \sigma_b(p,\Theta) + 2\ln\beta m]\} \tag{13.7}$$

where F_{KG} is given by (6.4). We have derived precisely this form from the classical limit of the Bethe ansatz for the quantum statistical mechanics of the s-G as reported by IMADA et al. [8]. However the expressions for $\sigma_k(p)$ and $\sigma_b(p,\Theta)$ coincide with (11.9a) and (11.9b) and additional dressings mediated by the phonons as described in §12 were neither analysed nor found. This aspect of the classical limit of the Bethe ansatz is therefore still to be studied: it seems particularly difficult in that there are no quantum phonons in that analysis - the classical phonons go into the vacuum at $T = 0$, and it is therefore necessary to extract F_{KG} from the small mass breather contributions [67] and this is the origin of the cut-off $\Theta_b(p)$.

We can note two things about the result (13.7): first the $\ln(1 + \alpha(p))$ forms are certainly fermi, but for $\beta M >> 1$ the two logarithms expand to order $e^{-\beta M}$ and yield precisely the classical expressions we have been using: indeed, if phonon-phonon terms are ignored, the free energy (13.7) developed to $O(e^{-\beta M})$ coincides identically with that of §11: if phonon-phonon terms are included it will coincide with that of §12.

Second, these fermions owe nothing to quantum theory: they follow from a strictly classical analysis and the constraints on the phase space. So these fermions are indeed classical fermions!

Moreover by expanding the logarithms further we shall evidently find terms $O(e^{-2\beta M})$ and $O(e^{-3\beta M})$, etc. in the final contributions to the free energy, and in this respect the theory begins to coincide still more closely with the results of the TIM, §7.

Unfortunately, to neglect of phonon-phonon shifts, it is possible to evaluate the first integral in (13.7) at every order in $e^{-\beta M}$: the leading term (in $e^{-\beta M}$) coincides with (11.13), but the term in $e^{-2\beta M}$ is

$$\tfrac{1}{2}\gamma_0(\pi)^{-\frac{1}{2}}\, m\beta^{-1}(M\beta)^{3/2}\left\{1 - \tfrac{1}{16}(M\beta)^{-1} + \cdots\right\} e^{-2\beta M} \quad . \tag{13.8}$$

and this does not coincide with (7.14).

The reason for this discrepancy is not understood yet. Once again phonon-phonon shifts may change the picture; otherwise, the following points may be relevant:- it is not necessarily the case that the zeros of $a(\zeta)$ are distinct: indeed, it is obvious that $a(\zeta)$ may have multiple zeros. In this case the solutions are not solitons (i.e. neither kinks, antikinks nor breathers). The actual solutions have been studied in terms of the Bäcklund transformation for the s-G [69]. The problem is that the spectral transform as a canonical transform is analysed so far only under the condition that $a(\zeta)$ has simple zeros [11]: thus, the whole problem has to be reworked (a Ph.D. thesis for somebody!). It may be that multiple zeros contribute zero measure to the free energy - but against this we do know from (13.7) that an assumption of simple zeros already has a significant effect.

14. Unsolved Problems

The subject matter of these lectures has been the evaluation of the functional integral (1.11) in terms of action-angle variables. Evidently there are still five unsolved problems surrounding that calculation. First of all, should we use phonon-phonon shifts given by the propagator (9.19), coefficient $\frac{1}{8}\gamma_0$, or should that coefficient at least be doubled? the rationale is, do we use one sheet or both sheets of the two-sheeted k-plane? Certainly the sinh-Gordon (§10) appeals to $\frac{1}{8}\gamma_0$ and one k-sheet: the s-G (§12) seems to need both k-sheets *and* a factor 2!

Secondly, we appear to duplicate the breather series in F for the s-G: the 'breather series' for sinh-Gordon (§10) can only arise (and it does) from the spectral shift $\arg(k_n)$, (9.16), evaluated for phonon-phonon interactions alone; but for the s-G the dressed breathers provide a breather series (§11) and additional phonon-phonon terms (§12) surely provide another one.

Thirdly, although the classical fermion property (§13) must be a feature of the theory, detailed results beyond $O(e^{-\beta M})$ for kinks and antikinks do not coincide with the TIM at these orders.

Fourthly, we have not yet established in detail that we can use action-angle variables beyond $O(L^{-1})$ in corrections to the Hamiltonian H. Here, however, prospects are good: the evidence (§12) is that as long as $O(L^{-n})$ combines with n sums $\sum_{-\frac{1}{2}N_{ph}}^{+\frac{1}{2}N_p}$ appropriately the contribution is important and is $O((M\beta)^{-n})$ or $O((M\beta)^{-(n-1)})$ in the free energy F: otherwise $O(L^{-n})$ is simply negligible as $L \to \infty$ for any $n > 0$.

Fifthly, detailed scrutiny of the paper by ARTHURS [46] referred to in §5 will show that 'Jacobian trouble' may still be possible, even though double integration $\int \mathcal{D}\Pi \mathcal{D}\phi$ (§5) is otherwise consistently used.

All of these open problems should be resolved fairly soon.

15. Conclusion

We have come a long way in order to negotiate one relatively simple functional integral! I hope we have learnt quite a lot about the sine-Gordon equation and its solitons in the process; and I hope we have learnt quite a lot about functional integrals themselves. The astute student must nevertheless already be saying "Well, look - why go to all this trouble? The TIM you develop in §7 may involve you in matched asymptotic expansions; but these are trivial compared with the complications you unearth by canonically transforming to action-angle variables. Moreover the TIM obviously works for non-integrable systems like the 'ϕ-four' equation $\phi_{xx} - \phi_{tt} = \phi - \phi^3$ [52] - so why not use it and forget the spectral transform!"

Of course, my answer to this is really given already in the introduction in §1: as a process for the quantisation and solution of the classically integrable field theories, the functional integral method must be capable of expression in terms of the scattering data if it is to stand any natural comparison with the quantum inverse method [2,3,5,28,44] and the method of Bethe ansatz [5,6,7,8,44]: the two latter methods are already brought into line [3,5,44]. This situation becomes even more pressing for the quantum statistical mechanics, where less work has been done in terms of Bethe ansatz and none, (I think), in terms of the quantum inverse method: in some ways the functional integral seems better adapted for the quantum statistical mechanics than the two other methods. Moreover, in the case of the s-G in classical limit we have a 'test bed' available, because we can check term by term against the TIM - always assuming the term-by-term validity of the asymptotic expansion of the Mathieu equation needed for that method. The work of the Part II shows we have had to twist and turn rather a lot to follow out this programme, and the summary, §14, of open problems remaining shows there is still some way to go.

Nevertheless, a lot of problems have been solved: the §8, tempered by the fermion analysis of §13, shows us how to construct the fundamental measure; the spectral shift §4, (4.39), and the §9, (9.16) - (9.18) shows, as we must have expected, that a Bethe ansatz type integral equation underpins and constrains the whole theory; and in comparison with the TIM result (7.13) for the free energy the actual results reported in §§10, 11 and 12 show a remarkable advance on the first naive calculations developed at (8.5) in §8.

For, whatever the discrepant details, the analysis confirms the significance of a proper 'thermodynamic limit': periodic boundary conditions and finite support period L, with $L\to\infty$ subsequently, are a necessity of the theory; and by studying integrable systems under periodic boundary conditions (§9) these finally find a natural place in the functional integral expressed in terms of action variables. So indeed does discretization and the use of lattices. Consequently, it is no surprise that the quantum inverse method was obliged to accommodate the same conditions [2,3,5,28,44] - though one feature of our analysis is that the limit of zero lattice spacing $a \to 0$ is not possible without renormalisation against the free lattice K-G.

In doing all of this we establish, albeit without rigorous proof, a number of essentially mathematical results about changes of variable in functional integration. Again, the TIM 'test bed' provides the check and guide for proper measures for such transformations.

It is obvious, of course, that the same methods as have been used for the classical partition function (1.11) apply with little change to the quantisation of the s-G at $T = 0$ through the functional integral (5.9). And these methods will apply to other integrable systems like the NLS (3.4), the Landau-Lifshitz equation (3.24), the Heisenberg ferromagnet (3.3), the integrable lattices [3], and especially the Toda lattice [50,58]. But on all of these we must report elsewhere.

The reader should find that despite the tough nature of the calculations and their somewhat incomplete character, the material of these lectures fits in well with Professor FLASCHKA's presentation [50]. He should read the material [50] first and these lectures second. He may indeed then begin to speculate how the completely integrable Hamiltonian systems begin to break down as such under non-Hamiltonian perturbations; how the infinite dimensional tori go; and how, in that respect, they relate to the deterministic 'chaos' described by other lecturers at this very stimulating School ELAF '84.

Evidently there is a great deal to be done!

Appendices

Appendix I: The Method of Bethe Ansatz

The method has a long history [6]: it was given as an ansatz by BETHE [70] for one-dimensional spin systems: recently, it has been realised that it solves the quantum integrable systems and coincides (in essence) with the quantum inverse method [5].

As a quick illustration, we treat the problems of N bosons on a line with δ-function interactions. In first quantisation the Schrödinger equation is

$$\left(-\sum_{i=1}^{N} \partial^2/\partial x_i^2 + 2c \sum_{i \neq j} \delta(x_i - x_j)\right)\psi = E\psi \qquad \text{(AI.1)}$$

and $c > 0$ is a real coupling constant. The equation is *solved* by the ansatz

$$\psi = \sum_{P} a(P)\, e^{i \sum_{j=1}^{N} k_{P_j} x_j} \quad ; \qquad x_1 < x_2 \ldots < x_N \; ; \qquad \text{(AI.2)}$$

$$a(P) = \Pi\, e^{i\theta_{ij}} \quad ; \qquad \text{(AI.3)}$$

and $e^{i\theta_{ij}}$ is the phase-shift for transposition $i \leftrightarrow j$ making up the permutation P. This is easily calculated for one pair by working in relative co-ordinates $x_i - x_j = r_{ij}$; one finds

$$e^{i\theta_{ij}} = -\frac{c - i(k_i - k_j)}{c + i(k_i - k_j)} \quad . \qquad \text{(AI.4)}$$

The bose symmetry defines (AI.2) for $x_2 < x_1 < x_3 < \cdots < x_N$, etc.

On the other hand the energy E and momenta P are

$$E = \sum_{i=1}^{N} k_i^2 \quad , \qquad P = \sum_{i=1}^{N} k_i \qquad \text{(AI.5)}$$

and the *set* $\{k_i;\ i = 1, \ldots, N\}$ is a constant of the motion. Comparison with (4.24), (4.25) and (4.26) in the main text is immediate.

Under periodic boundary conditions the condition for the allowed k_j is

$$k_j L + \sum_{\ell \neq j} \theta_{j\ell} = 2\pi n_j \ ; \qquad n_j = \text{integer} \quad . \tag{AI.6}$$

This is an example of the Bethe ansatz integral equation referred to in the text (§9) and the comparison with (9.17) is plain. Indeed, if the action variables P_n in (9.17) can be integrated out (as they are during the functional integration) this integral equation bears some exact equivalence (still not explicitly achieved) to the corresponding Bethe ansatz integral equation for the s-G quoted below. The other new feature in (9.17) over the Bethe ansatz form is, of course, the contribution of the solitons as such to the spectral shift: the Bethe ansatz has to construct its own solitons, if they arise, out of 'strings' [6].

From (AI.4)

$$\theta_{ij} = -i\, \ln\left\{-\left|\frac{c-i(k_i-k_j)}{c+i(k_i-k_j)}\right|\right\} = -i\, \ln\left|1 + \frac{2ic}{k_i-k_j-ic}\right| \simeq \frac{2c}{k_i-k_j-ic} \tag{AI.7}$$

as $c \to 0$. Reference to [5] shows the connection between this result 'in classical limit' $c \to 0$ and the classical r-matrix for the nonlinear Schrödinger equation (3.4) with $c = -\chi$. This is no accident since the problem (AI.1) for N bosons on a line with δ-function interactions in first quantisation becomes the problem of the quantised NLS

$$i\,\frac{\partial\psi}{\partial t} = \psi_{xx} - c\,\psi^{\dagger}\psi\psi \tag{AI.8}$$

with $[\psi, \psi^{\dagger}] = \delta(x - x')$ and $c = -\chi$ in second quantisation: note the normal order. The quantised NLS with coupling constant $c > 0$ is studied and solved at length by the quantum inverse method in [3] and [5]; and the relation to the Bethe ansatz established there.

For reference, the corresponding Bethe ansatz integral equation for the s-G is [7,44]

$$Lk_j = Lm \sinh\lambda_j = -2\pi n_j - \sum_{\ell \neq j} \phi(\lambda_j - \lambda_\ell)$$

$$\phi(\lambda) = -i\, \ln\left|\frac{-\sinh \frac{1}{2}(\lambda - 2i\mu)}{\sinh \frac{1}{2}(\lambda + 2i\mu)}\right| \tag{AI.9}$$

In this $\mu = \frac{1}{2}(\pi + g)$ and $2g$ is the coupling constant for the massive Thirring model ((3.29) of the text). The classical limit is obtained for $\mu \to 0$. As $\mu \to 0$, $\mu \to \frac{1}{8}\gamma_0$ as $\gamma_0 \to 0$, and $\theta(\lambda) \to \pi - 2\mu\cosh\frac{1}{2}\lambda/\sinh\frac{1}{2}\lambda = \pi - \gamma_0/4\ \coth\frac{1}{2}\lambda = \pi - \gamma_0/4$ $(1 + \cosh\lambda/\sinh\lambda)$. A connection with the r-matrix (4.46) is apparent.

YANG and YANG [21] show that at finite temperature $T > 0$ under periodic boundary conditions the free energy per unit length for the Bose gas (AI.1) is given by

$$F = -\frac{1}{2\pi\beta}\int_{-\infty}^{\infty} \ln(1 + e^{-\beta\varepsilon(k)})\,dk \tag{AI.10}$$

where $\varepsilon(k)$ is a solution of the integral equation

$$\varepsilon(k) = k^2 - \frac{c}{\pi\beta}\int_{-\infty}^{\infty} \frac{dq}{(k-q)^2+c^2}\, \ln(1 + e^{-\beta\varepsilon(k)}) \quad . \tag{AI.11}$$

By defining

$$1 + e^{-\beta\varepsilon(k)} = \frac{1}{1 - e^{-\beta\omega(k)}}$$

one can show [71] (with some missing rigour) that in natural classical limit

$$\omega = k^2 - \frac{1}{\pi\beta}\int \frac{c}{(k-q)}\frac{d}{dq}(ln\beta\omega(q))dq \qquad (AI.12)$$

with

$$F = \frac{1}{2\pi\beta}\int_{-\infty}^{\infty} ln\beta\omega(k)dk \qquad (AI.13)$$

The two results (AI.12) with (AI.13) motivate the analysis from (10.9a,b) (§10) which leads to the TIM result for F for the sinh-Gordon equation. The result (10.7) (as far as it is taken) appears to justify (10.9a,b) by our methods of functional integration to the correction to (10.9a) made.

Appendix II: Integrability and the Quantum Inverse Method

In his recent set of lecture notes [5] L.D. FADDEEV shows how classical r-matrices like (4.46) are fundamental to the classical integrable systems on finite support L and periodic boundary conditions, and how corresponding r-matrices $r_{\pm}$ like (4.45) are fundamental to the $L \to \infty$ limit. Thus, in the notation of §4 for finite support L and with lattice spacing $a \to 0$ one has the Poisson brackets

$$\{T_L(\lambda) \overset{\otimes}{,} T_L(\lambda')\} = [r(\lambda - \lambda'), \; T_L(\lambda) \otimes T_L(\lambda')] \qquad (AII.1)$$

where $r(\lambda - \lambda')$ is the r-matrix and $T_L(\lambda)$ is a monodromy matrix related to (9.13), here expressed in rapidities through $k = \sinh\lambda$ and $a_L(-k) \to a_L^*(\lambda), b_L(-k) \to b_L^*(\lambda)$, by

$$\begin{bmatrix} a_L(\lambda) & b_L(\lambda) \\ b_L^*(\lambda) & a_L^*(\lambda) \end{bmatrix} = E(-\tfrac{1}{2}L, \lambda)\, T_L(\lambda)\, E(\tfrac{1}{2}L, \lambda) \qquad (AII.2)$$

with $E(x, \lambda) \equiv e^{-ik\sigma_3 x}$. For the s-G $r(\lambda - \lambda')$ is given by (4.46) and the corresponding Poisson brackets by (AII.1). Thus for $L \to \infty$ one regains (4.44) with $r_{\pm}$ given by (4.45). And, as shown in the text, these are the brackets (4.42).

These r-matrices are replaced by matrices R in the quantum inverse method: just as the r-matrices determine the Poisson brackets for the scattering data as in (AII.1) so the R-matrices determine commutators for the scattering data read as operators. An R-matrix figures in the work of BAXTER [1,72] on the 8-vertex and spin-$\frac{1}{2}$ XYZ models where he used it to demonstrate commutativity of the trace of the spin transfer matrices he was considering. The 'complete integrability' (solvability) of the 8-vertex model was used to demonstrate the complete integrability of the spin-$\frac{1}{2}$ XYZ model (3.28) (§3): the R-matrix necessarily figures in the theory of this quantum model and its classical limit r determines the Poisson brackets of the classical Landau-Lifshitz model (3.22).

I illustrate the situation by following SKLYANIN [28] (his §4), who establishes the connection between (3.28) and (3.22). It is convenient to use spin commutation relations $[S_m^{\alpha}, S_n^{\beta}] = i\, 2\varepsilon^{\alpha\beta\gamma} S_n^{\gamma}\delta_{mn}$ rather than (3.6): thus the quantum spin-$\frac{1}{2}$ operators are interpreted as Pauli operators. In order to keep the Pauli operators in usual form, we switch indices $1 \leftrightarrow 3$ compared with (3.22) and use x instead of z: thus we shall use J_{α}^{q} $(\alpha = 1,2,3)$ for the coupling constants J^{β} $(\beta = z, y, x)$ in (3.28) and J_{α}^{c} $(\alpha = 1,2,3)$ for their classical limits J_{β} $(\beta = 3,2,1)$ in (3.22): the lattice has spacing a.

As in §3 we set $\hbar = a$ but consider the classical and continuum limit $\hbar = a \to 0$. We let $J_{\alpha}^{q} \to 1$ in such a way that

$$J_{\alpha}^{q} \sim 1 + \tfrac{1}{2}J_{\alpha}^{c}a^2 \quad . \qquad (AII.3)$$

Then, if $\vec{S}_{n+1} = \vec{S}(x) + a\vec{S}_x(x) + \frac{1}{2}a^2\vec{S}_{xx}(x)$ is put into (3.28) multiplied by $\hbar$ (= a), then for $a \to 0$, after a renormalisation using the boundary conditions $\vec{S}(x) \to (0,$

$0, \pm I)$ for $|x| \to \infty$ we obtain the classical Hamiltonian (3.22) for the L-L' equation with $Ja^2 = 1$. So this establishes *that* connection.

BAXTER [72] established the complete integrability of the spin-½ XYZ model: JOHNSON, KRINSKY and McCOY [72] found the spectrum. The connection between BAXTER's method and the quantum inverse method is established in [2,3,5,44] as well as in [28]. In [28] SKLYANIN starts from a matrix operator $L_n(u)$ (called $L_i(\lambda)$ in [5]):

$$L_n(u) = \sum_{\alpha=0}^{3} W_\alpha^q(u)\, S_n^\alpha \sigma_\alpha \tag{AII.4}$$

where the σ_α are Pauli matrices, $\sigma_0 = I$, and

$$\begin{aligned} W_0^q(u) &= 1, \qquad W_1^q(u) = \mathrm{sn}(\xi, k)/\mathrm{sn}(u+\xi, k) \\ W_2^q(u) &= \mathrm{dn}(u+\xi, k)\,\mathrm{sn}(\xi, k)/\mathrm{dn}(\xi, k)\,\mathrm{sn}(u+\xi, k) \\ W_3^q(u) &= \mathrm{cn}(u+\xi, k)\,\mathrm{sn}(\xi, k)/\mathrm{cn}(\xi, k)\,\mathrm{sn}(u+\xi, k) \quad . \end{aligned} \tag{AII.5}$$

The functions sn, cn, dn are Jacobian elliptic functions with modulus k and arguments u or $u + \xi$: u (or ξ) acts as spectral parameter. The matrix $L_n(u)$ appears in [72] with a slightly different parametrisation [28].

In the notation of (4.44) we define $\sigma_\alpha' = \sigma_\alpha \otimes \sigma_0$ and $\sigma_\alpha'' = \sigma_0 \otimes \sigma_\alpha$ and define L_n', L_n'' by (AII.4) with σ_α replaced by σ_α' and σ_α'' respectively. Then [1,72]

$$L_n'(u)\, L_n''(v)\, R(v-u) = R(v-u)\, L_n''(v)\, L_n'(u) \tag{AII.6a}$$

in which

$$R(u) = \sum_{\alpha=0}^{3} W_\alpha^q(u)\, \sigma_\alpha \otimes \sigma_\alpha \quad . \tag{AII.6b}$$

If [1,28] $J_3^q : J_2^q : J_1^q = \mathrm{cn}2\xi : \mathrm{dn}2\xi : 1$ it follows from (AII.6) that there is an operator Lax pair corresponding to the Heisenberg equation of motion from (3.28)

$$\frac{d\vec{S}_n}{dt} = i[H, \vec{S}_n] = \vec{S}_n \times \mathcal{J}\cdot(\vec{S}_{n-1} + \vec{S}_{n+1}) \quad . \tag{AII.8}$$

For, because of the symmetry between the matrices σ_α and the operators S_n^α, (AII.6a) can be written in terms of 2×2 matrices in the form

$$L_{n-1}(u)\, L_n(v)\, R_{n-1,n}(v-u) = R_{n-1,n}(v-u)\, L_n(v)\, L_{n-1}(u) \tag{AII.9}$$

in which $R_{n-1,n}(u)$ is the scalar operator

$$R_{n-1,n}(u) = \sum_{\alpha=0}^{3} W_\alpha^q(u)\, S_{n-1}^\alpha\, S_n^\alpha \;; \qquad S_n^0 = I \quad . \tag{AII.10}$$

Differentiate (AII.9) with respect to v and set $v = u$:

$$\begin{aligned} &L_{n-1}(u)\, L_n'(u)\; 2P_{n-1,n} + L_{n-1}(u)\, L_n(u)\, R_{n-1,n}'(0) \\ &= 2P_{n-1,n}\, L_n'(u)\, L_{n-1}(u) + R_{n-1,n}'(0)\, L_n(u)\, L_{n-1}(u) \end{aligned} \tag{AII.11}$$

in which $L_n'(u)$ means derivative and $P_{n-1,n} \equiv \frac{1}{2}(I + \vec{S}_{n-1} \cdot \vec{S}_n) = \frac{1}{2}R_{n-1,n}(0)$ is the operator interchanging $(n-1)$th and nth spins. Multiply (AII.10) on the left by $\frac{1}{2}P_{n-1,n}$. Then, if $H_{n-1,n} \equiv \frac{1}{2}(P_{n-1,n}R_{n-1,n}'(0))$

$$[H_{n-1,n}, L_n(u)\, L_{n-1}(u)] = L_n(u)\, L_{n-1}'(u) - L_n'(u)\, L_{n-1}(u) \quad . \tag{AII.12}$$

The result (AII.12) is not changed by writing

$$H_{n-1,n} \equiv \tfrac{1}{2}(P_{n-1,n}\, R_{n-1}'(0) - \tfrac{1}{2}\mathrm{tr}P_{n-1,n}\, R_{n-1,n}(0)) \tag{AII.13}$$

and this allows it to be written in the form $H_{n-1,n} = -\sum_{\alpha=1}^{3} J_\alpha{}^q S_{n-1}{}^q S_n{}^q$: one can check that the $J_\alpha{}^q$ correspond to these quantities in (AII.8) so that $H_{n-1,n}$ is the Hamiltonian density corresponding to (3.28).

Now define the matrix M_n by either

$$M_n L_{n-1} = i[H_{n-1,n}, L_{n-1}] - iL'_{n-1} \qquad \text{(AII.14a)}$$

$$L_n M_n = i[H_{n-1,n}, L_n] - iL_n' \qquad \text{(AII.14b)}$$

The two definitions are equivalent because of (AII.12) as can be checked by calculating $L_n M_n L_{n-1}$. Using (AII.12) and (AII.14) one obtains

$$\frac{d}{dt} L_n = i[H, L_n] = i[H_{n-1,n} + H_{n,n+1}, L_n] = M_{n+1} L_n - L_n M_n \quad . \qquad \text{(AII.15)}$$

This is the compatibility condition for the 2-spinors v_n, v_{n+1}

$$v_{n+1} = L_n v_n, \quad \frac{dv_n}{dt} = M_n v_n \qquad \text{(AII.16)}$$

so L_n, M_n constitute a quantum lattice Lax pair for the Heisenberg equation of motion from (3.28) which is (AII.8). Thus (AII.6) implies the existence of this Lax pair as stated.

Under the classical limit (AII.3) the quantum Lax pair reduces [28] to the Lax pair for the L - L' equation (3.24). The classical Lax pair is an 'integrability condition' (solvability condition) in the sense that corresponding to (AII.16) for the lattice is $v_x = Lv$, $v_t = Mv$ which has the compatibility condition $L_t - M_x + [L,M] = 0$. This is the evolution equation (4.3) if $u \to \zeta$ and $\hat{L} = \partial/\partial x + \zeta - L$, $M \equiv \hat{A}$. The existence of the $\hat{L}$, $\hat{A}$ representation for this evolution equation implies that it is a completely integrable Hamiltonian system as §4 shows for the s-G. Technically one needs the existence of an inverse mechanism, but this is available for $\hat{L}$ (4.1) in the Marchenko equation and more generally through the Riemann-Hilbert problem [5,74]. Thus the quantum 'integrability condition' (AII.6) contains classical integrability in this sense as its classical limit.

The condition (AII.6) contains both quantum and classical integrability also, in that it determines the quantum commutators for the scattering data read as operators and the classical Poisson brackets for the classical scattering data. Define

$$T_{n_1}^{n_2} = L_{n_2} L_{n_2-1} \cdots L_{n_1+1} L_{n_1} , \qquad n_1 < n_2 \quad . \qquad \text{(AII.17)}$$

Then (AII.6) means

$$T_{n_1}^{n_2'}(u)\, T_{n_1}^{n_2''}(v)\, R(v-u) = R(v-u)\, T_{n_1}^{n_2''}(v)\, T_{n_1}^{n_2'}(u) \qquad \text{(AII.18)}$$

corresponding to (AII.1). Indeed, for $\hbar = a \to 0$, $T_{n_1}^{n_2}(u)$ becomes $T_{x_1}^{x_2}(u)$ and for $x_2 = \frac{1}{2}L$, $x_1 = -\frac{1}{2}L$, $T_{x_1}^{x_2}(u)$ becomes $T_L(\mu)$ introduced at (AII.1) - in the present case however for the Landau-Lifshitz equation. For this case $E(x,u)$ from (AII.2) is given by $E(x,u) = e^{-iw_3(u)\sigma_3 x}$ with $w_3(u) = \rho(cn(u,k)/sn(u,k))$ and $\rho = \frac{1}{2}(J_3 - J_1)$, $k = (J_2 - J_1)^{\frac{1}{2}}/(J_3 - J_1)^{\frac{1}{2}}$ $(J_3 > J_2 > J_1)$ [28] and for $L \to \infty$ one obtains results corresponding to (4.44) with (4.45). Indeed the classical r-matrix will arise through

$$R(u) = 1 - ar(u) + O(a^2); \quad r(u) = \sum_{\alpha=1}^{3} w_\alpha{}^c(u)\sigma_\alpha \otimes \sigma_\alpha \qquad \text{(AII.19)}$$

and [28] $w_\alpha{}^c(u) = \rho\, \partial/\partial\xi\, w_\alpha{}^q(u,\xi)|_{\xi=0}$. The condition (AII.3) obtains since this is guaranteed by setting $\xi = -i\rho a$, and this choice is actually made to obtain the classical Lax pair for the L - L' equation as $a \to 0$ [28]. Since

$$[T_{n_1}^{n_2'}, \; T_{n_1}^{n_2''}] \;=\; -i\hbar\{T_{n_1}^{n_2'}, \; T_{n_1}^{n_2''}\} + O(\hbar^2), \qquad \text{(AII.20)}$$

then for $\hbar = a \to 0$ and $x_1 < x_2$

$$\{T_{x_1}^{x_2'}, \; T_{x_1}^{x_2''}\} \;=\; \tfrac{1}{2}[r(u-v), \; T_{x_1}^{x_2'}(u)\, T_{x_1}^{x_2''}(v)] \qquad \text{(AII.21)}$$

which is equivalent to (AII.1) for the s-G (the factor of ½ arises because we took Pauli operator commutation relations, but want to reach Poisson brackets (3.23)). Then, for $x_1 \to -\infty$, $x_2 \to +\infty$ we regain a form like (4.44), and this way we find [28] the classical Poisson brackets for the L - L' equation

$$\begin{aligned}
\{a(u), a(v)\} &= \{a(u), a^*(v)\} = \{b(u), b(v)\} = 0, \\
\{a(u), b(v)\} &= w_3(u - v + i\varepsilon)\, a(u)\, b(v) \\
\{a(u), b^*(v)\} &= -\, w_3(u - v + i\varepsilon)\, a(u)\, b^*(v) \\
\{b(u), b^*(v)\} &= 2\pi i\rho\, |a(u)|^2\, \delta(u - v);
\end{aligned} \qquad \text{(AII.22)}$$

action-angle variables are defined by $P(u) = (\pi\rho)^{-1} \ln|a(u)|$, $Q(u) = \arg b(u)$. In the limit considered at (3.26) and (3.27) (with $1 \leftrightarrow 3$) the modulus k of the elliptic functions $\to 1$ and $w_3 = \rho(\mathrm{cn}(u,k)/\mathrm{sn}(u,k)) \to \rho(\sinh(u))^{-1}$ so these Poisson brackets coincide with (4.42), with ρ replacing $\frac{1}{8}\gamma_0$ ($\rho \to$ a positive constant in the limit considered at (3.26)). Thus, as stated, (AII.6) plays the role of the Lax pair in this respect also.

The R-matrix satisfies [3,5] the 'Yang-Baxter' (or 'factorisation equation' or 'star-triangle relation' [1])

$$\begin{aligned}
&{}_{\alpha\alpha'}R_{\gamma\gamma'}(u-v)\; {}_{\gamma\alpha}R_{\beta\gamma''}(u)\; {}_{\gamma'\gamma''}R_{\beta'\beta''}(v) \\
&= {}_{\alpha'\alpha''}R_{\gamma'\gamma''}(v)\; {}_{\alpha\gamma''}R_{\gamma\beta''}(u)\; {}_{\gamma\gamma'}R_{\beta\beta'}(u-v)
\end{aligned} \qquad \text{(AII.23)}$$

in an obvious 4-index notation with summation over repeated indices. It seems to be the fundamental integrability condition for the integrable systems. We do not use (AII.6) or (AII.23) as such in the functional integral method: we pay for this by the need to iterate the integral equation (9.17) and the Hamiltonians. On the other hand, in the calculation of Z at $T > 0$ the best that has been obtained is a system of nonlinear coupled integral equations (like (AI.11) or (AI.12) for the NLS) as in [31] for the spin-½ XYZ model and as in [8] for the s-G. In classical limit, these integral equations correspond closely to the functional integral analysis carried through in these lectures [71].

Appendix III: P-Integral Interpretation of Phase-Shifts- the 'Fermion' and 'Boson' Views

The work on the breather contributions to the free energy of the s-G reported in §11, which depends on the choice of the dressing $\sigma_b(p,\Theta)$ of (11.14), needs explanation in the light of remarks on $\sigma_k(p,\Theta)$ and its other than P-integral interpretation made on the previous page. There is an unsolved problem here (adding to the five problems listed in §14).

The essential of the matter is that there are two possible choices for the phase-shifts, just as there are two possible choices for phase-shifts in the Bethe ansatz. In Ref. [6] pp.257-8 a 'fermion' interpretation is imposed on the phase-shifts of the Bethe ansatz for the NLS. From (AI.7) $\theta_{ij} = -2\tan^{-1}(ck_{ij}^{-1})$ with $k_{ij} \equiv k_i - k_j$: the 'fermion' interpretation chooses $\theta_{ij} \to 0$ as $k_{ij} \to \infty$ and has no discontinuity along the real axis so that $\theta_{ij} \to -2\pi$ as $k_{ij} \to -\infty$. Thus $\theta_{ij} \to$ step function as $c \to 0$ and cannot represent the true phase-shift. In contrast with this, there is the 'boson' interpretation invoked below, which replaces θ_{ij} by $\theta_{ij} - 2\pi\theta(k)$ where $\theta(k)$ is the step-function: then $\theta_{ij} \to -2\pi$, $k_{ij} \to \pm\infty$, so $\theta_{ij} \to 0$ (mod 2π), $k_{ij} \to \pm\infty$, but has a jump of 2π at $k_i = k_j$.

It is shown in [6] that, in the 'fermion' interpretation, (AI.6) leads to integral equations which, for $\beta^{-1} > 0$, become the pair of relations (AI.10), (AI.11) expressed in fermion form. The steps (AI.11) to (AI.12) amount to replacing the fermion interpretation with the boson one wholly equivalent to it: in these steps, the jump from the step-function becomes included in the definition of the effective phase-shift [71]. The boson interpretation then has a natural classical limit which is (AI.12) with (AI.13), motivating the analysis from (10.9a,b), (10.10) in the text. Remember, all of this applies to the Bethe ansatz for the repulsive case of the NLS, and thence, in our interpretation of §10, to the sinh-G: in neither case are there classical solitons.

It is intuitive that the same alternatives are available in a quantum theory which includes soliton phase-shifts; but in classical limit an underlying *boson*-phonon shift seems appropriate. Thus for $\Delta_k(k,p_i)$, eqn. (4.37), we take $\Delta_k \to 0$, $k \to \pm\infty$ with jump -2π at $k = p$. The jump induces $-2\pi\delta(k-p)$ in $d\Delta_k/dk$ in (11.9a) the contribution from which is included in (11.11) and leads to the number $+\frac{1}{8}$ in (11.13): consistent application of *this* boson description then apparently requires the corrections due to phonon-phonon shifts (§12) and these change the $+\frac{1}{8}$ to $-\frac{7}{8}$ (up to factors of 2) as described. However, the argument is not consistent: the derivation of (4.37) assumes $a(\zeta)$ analytic in the upper half plane — so ((mod 2π)) $\arg a(k)$ must be analytic in $-2\pi < \arg(k) < 0$ for $-\infty < k < \infty$; but now the integral (11.16) is not defined for $a \to 0$, and there are corrections depending on the cut-off π/a after integration by parts to (11.7). Analyticity of $a(\zeta)$ thus seems to force a 'fermion' description on the theory, and by imposing the 'boson' description, we apparently make a renormalisation of the classical theory.

The problem becomes slightly different in the quantum theories themselves: $ln\,\omega(k)$ does not diverge as $ln|k|$ (since $ln\,\beta\omega(k) \to ln(1-e^{-\beta\omega(k)})$ [67,71]; but in this case zero point energies now force a renormalisation. We should stress that the classical dressing analysis first made through eqn. (7.22) in §7 also uses the 'boson' description: the $\tan^{-1}(mk^{-1})$ vanishes for $k \to \pm\infty$ and has a jump of 2π at $k = 0$. It is a consistent application of this point of view in the associated quantum theory [26] which produces the renormalisation of mass and of the coupling constant $\gamma_0 \to \gamma_0'' = \gamma_0/(1-\gamma_0/8\pi)$ found [26] from the zero point energies of the phonons: it is interesting to see that the dressing term $ln\,\beta m$ in (11.10a), for example, now appears as $ln(1-e^{-\beta m})$ and vanishes as $\beta^{-1} \to \infty$ (temperature $T \to 0$).

In the classical limit of the Bethe ansatz analysis [66,67], we naturally took the fermion viewpoint of that quantum theory, did not see (so far anyway) the phonon-phonon shifts, and reached (in a classical limit)

$$\sigma_k(p) = ln\,(1+\gamma) \tag{AIII.1}$$

instead of (11.11) by rejecting the δ-function contribution in $d\Delta_k/dk$ (the P-integral interpretation referred to below (11.13): this produces $+\frac{5}{8}(M\beta)^{-1}$ instead of the $\frac{1}{8}(M\beta)^{-1}$ in (11.13). It is within *this* context that the subsequent analysis of the breather contribution based on (11.14) is made. Note that (11.14) is now $2\sigma_k(p)$ given by (AIII.1) for $\Theta = \pi/2$, so the fermion viewpoint is being consistently applied in classical limit. It contains the problem that the breather contributions to F now diverge. But, in the classical limit of the Bethe ansatz analysis [67] referred to, a cut-off $\Theta_b(x)$, (11.15), emerges in such a way as to define the photon contribution F_{KG} given by (6.4); and in the functional integral analysis based on it the results described below (11.16) are found.

Evidently there is a boson analysis which replaces (11.14) by $2\,ln\,(1+\gamma^{-1}\sin\Theta)$ in agreement with (11.11) for $\Theta \simeq \pi/2$; and this induces a correction of (11.16) by $\gamma^{-2} = \mathrm{sech}\,\frac{1}{2}x)^2$ under the integral sign (actually making the integral *convergent*). In addition, the boson analysis would seem to need to include (i) an effect of phonon-mediated breather phase-shifts certainly contained in (12.1) but not otherwise treated in these lectures; and, (ii), the s-G series equivalent to the sinh-G series, (10.8), derived from the phonon-phonon shifts also implicit in (12.1).

It is an open question whether the 'fermion' and 'boson' analysis can be brought into line in classical limit. This must be possible at the quantum level — that is, before taking any classical limit.

Note that the 'classical fermions' of §13 arise as inevitable consequences of the constraints on the phase spaces as described there: connection with the present alternative phase-shift analyses is not established.

Appendix IV

In these lectures, the functional integral method has not been used to treat any quantum theory as such: plainly, for any detailed comparison with Bethe ansatz or quantum inverse method we need such an analysis. We therefore sketch here the functional integral method for the quantum K-G: the method extends to the quantum s-G but details are not yet complete.

From (5.14) it is plain that the new feature is the $\int_0^{\beta\hbar} i(\int \Pi(x)\phi_\tau(x)dx)d\tau$ in the Wick rotated action $S[\phi]$: thus for the K-G in action-angle variables (with $\hbar = 1$)

$$S[p] = i\int_0^\beta \left(\int_{-\infty}^{\infty} P(k)Q_\tau(k)dk\right) d\tau - \beta H_{KG}[p] \quad ;$$

$$H_{KG}[p] = \int_{-\infty}^{\infty} \omega(k)P(k)\,dk \quad . \tag{AIV.1}$$

For each label k we impose the WKB quantisation

$$\int_0^\beta P(k)Q_\tau(k)d\tau = \oint P(k)dQ(k) = 2\pi(n(k)+\tfrac{1}{2}) \tag{AIV.2}$$

so that $P(k) = n(k)+\frac{1}{2}$ and $n(k)$ an integer ≥ 0. The period β and the period for one circuit of the torus supposedly coincide. The quantisation seems semi-classical and therefore approximate: it will be exact for the K-G, a bunch of oscillators (§2): however, it seems to be exact for the s-G too, and by implication for all the integrable systems.

The functional integral is

$$Z = \int \mathcal{D}P\,\mathcal{D}Q\,\exp S[\phi] = \prod_k \sum_{n(k)} \exp -\beta(n(k)+\tfrac{1}{2})\omega(k) \quad . \tag{AIV.3}$$

Steps are these: the $\int\mathcal{D}Q$ provides 2π for each label k but the measure is scaled by $h^{-1} = (2\pi)^{-1}$ for each k (§5); the $\int\mathcal{D}P$ is replaced by $\prod_k \int dP(k)$ as usual, but, since $P(k)$ has values $n(k)+\frac{1}{2}$ only, integration for each label k is replaced by sums over these values; the contribution of (AIV.2) to (AIV.1) is $\exp 2\pi i(n(k)+\frac{1}{2})$ and there is a phase $e^{i\pi} = -1$ for each label k (which will be discarded).

We therefore reach

$$F = -L^{-1}\beta^{-1}\,ln\,Z = \frac{1}{2\pi}\int dk\left[\tfrac{1}{2}\omega(k) + \beta^{-1}\,ln\,(1-e^{-\beta\omega(\tilde{k})})\right] \quad . \tag{AIV.4}$$

The finite temperature contribution is convergent but the zero-point contribution is $\frac{1}{4\pi}\int_{-\pi/a}^{+\pi/a} dk(m^2+k^2)^{\frac{1}{2}}$: this illustrates the remark on the ultra-violet divergence (§6 below (6.16)). In classical limit $F = (2\pi\beta)^{-1}\int_{-\pi/a}^{+\pi/a} ln\,(\beta\omega(k))dk$ which is F_{KG} (given by eqn. (6.4)). It is also plain that in the quantum statistics of the s-G

$$\sigma[N_k,N_{\bar{k}},N_b] = -\beta\left\{\frac{L}{2\pi}\int dk\left[\tfrac{1}{2}(\omega(\tilde{k})-\omega(k)) + \beta^{-1}\,ln\,\frac{(1-e^{-\beta\omega(\tilde{k})})}{(1-e^{-\beta\omega(k)})}\right]\right\}$$

$$\rightarrow -\frac{L}{2\pi}\int dk\,[\,ln\,\beta\omega(\tilde{k}) - ln\,\beta\omega(k)\,] \tag{AIV.5}$$

in classical limit: σ in this limit is introduced at (8.11) in the text: in both the forms (AIV.5) it includes the terms $ln\ \beta m$ so that the counting of modes $\tilde{k}$ against modes k must be carried out as in the classical analysis (§§11 and 12): thus modes $k = 0$ which are replaced by solitons induce terms $ln(1 - e^{-\beta m}) \to ln(\beta m)$ in classical limit but $\to 0$ for $\beta^{-1} \to 0$ — illustrating a point made in Appendix III.

Otherwise, the new feature is the zero-point contribution. In the steepest descent analysis (§7) one calculates for each kink (since the integrand is odd in k)

$$-\frac{\beta}{2\pi}\int_0^{\pi/a} \frac{\partial\omega(k)}{\partial k}\Delta_k(k,0)\,dk = -\frac{\beta}{2\pi}\left[\omega(k)\Delta_k(k,0)\right]_0^{\pi/a} + \frac{\beta}{2\pi}\int_0^{\pi/a}\omega(k)\frac{\partial\Delta}{dk}(k,0)dk \quad \text{(AIV.6)}$$

The square bracket provides $\beta m\pi^{-1} - \frac{1}{2}m\beta$ for $a \to 0$: the $\frac{1}{2}m\beta$ is just sufficient to eliminate the $k = 0$ zero-point contribution from the free K-G which does not arise for the coupled modes $\tilde{k}$: the $\beta m\pi^{-1}$ takes $e^{-\beta M} \to e^{-\beta(M - m\pi^{-1})}$ and $M - m\pi^{-1} = M(1 - \frac{\gamma_0}{8\pi})$ so $\gamma_0 \to \gamma_0'' = \gamma_0(1 - \gamma_0/8\pi)^{-1}$, the result in [26] first referred to in §3 below (3.21). The remaining integral in (AIV.6) diverges as $ln\ \pi/a$ and renormalises the mass m. These results assume mode shifts by soliton-phonon phase-shifts alone: the work of §12 suggests phonon mediated soliton-phonon phase-shifts should also be included in the steepest descents analysis, but this is still to do: however, they do not arise in the mapping from spin - ½ XYZ to quantum s-G [30]. All the details of zero-point shifts in the more general context of §12 where shifts $\Delta_k(k,p)$ with $p \neq 0$ must be used have still to be worked out too.

References

1. R.J. Baxter: Exactly Solved Models in Statistical Mechanics (Academic Press 1982).
2. L.D. Faddeev, L.A. Takhtadzhyan: Usepekhi Mat. Nauk. 34, 13 (1979).
3. P.P. Kulish, E.K. Sklyanin: "Quantum Spectral Transform Method. Recent Developments", in Proc. of the Tvärminne Symposium, Finland, 1981. J. Hietarinta and C. Montonen eds. (Springer Verlag, Heidelberg 1982).
4. J. Weiss, M. Tabor, G. Carnevale: J. Math. Phys. 24, 522 (1983); J. Weiss: J. Math. Phys. 24, 1405 (1983). (In the event the Painlevé property was not mentioned at the School).
5. L.D. Faddeev: in Proc. Ecole d'Eté de Physique Theorique, Les Houches 1982 R. Stora and J.B. Zuber eds. (North Holland, Amsterdam 1983).
6. H.B. Thacker: Rev. od. Phys. 53, 253 (1981).
7. M. Fowler, X. Zotos: Phys. Rev. B24, 2634 (1981).
8. M. Imada, K. Hida, M. Ishikawa: J. Phys. C16, 35 (1983).
9. R.K. Bullough: in Bifurcation Phenomena in Mathematical Physics (D. Reidel Publ. Co., Dordrecht, Holland 1980).
10. H. Flanders: Differential Forms (Academic Press, New York 1963).
11. R.K. Dodd, R.K. Bullough: Physica Scripta 20, 514 (1979).
12. V.I. Arnold: Mathematical Methods of Classical Mechanics (Springer-Verlag, Heidelberg 1978) (for example describes Liouville's theorem).
13. R.K. Dodd, R.K. Bullough: Proc. Roy. Soc. London A352, 481 (1977).
14. R.K. Bullough, P.J. Caudrey: in Solitons R.K. Bullough and P.J. Caudrey eds. (Springer-Verlag, Heidelberg 1980) and the other chapters there.
15. R. Sasaki, R.K. Bullough: Proc. Roy. Soc. London A376, 401 (1981).
16. H.J. Mikeska: J. Phys. C11, L29 (1978).
17. L.A. Takhtadzhyan, L.D. Faddeev: Proc. Steklov Inst. of Math Issue 3, 277 (1979).
18. M. Lakshmanan: Phys. Lett. 61A, 53 (1977).
19. L.A. Takhtadzhyan: Phys. Lett. 64A, 235 (1977).
20. M. Lakshmanan, R.K. Bullough: Phys. Lett. 80A, 287 (1980).
21. C.N. Yang, C.P. Yang: J. Math. Phys. 10, 1115 (1969).
22. M. Jimbo, T. Miwa, Y. Mori, M. Sato: Physica 1D, 80 (1980).
23. K. Maki: J. Low Temp. Phys. 41, 327 (1980).

24. J. Timonen, R.K. Bullough: Phys. Lett. 82A, 183 (1981).
25. J.K. Kjems, M. Steiner: Phys. Rev. Lett. 41, 1137 (1978).
26. R.F. Dashen, B. Hasslacher, A. Neveu: Phys. Rev. D11, 3424 (1975).
27. J. Timonen, A. Luther: "Continuum Limit Correlation Functions for the Spin One Anisotropic Heisenberg Chain", J. Phys. C. (to appear).
28. E. Sklyanin: LOMI preprint E-3-1979 (1979).
29. S. Coleman: Phys. Rev. D11, 2088 (1975).
30. A.H. Luther: in Solitons ref. [14].
31. M. Takahashi, M. Suzuki: Prog. Theor. Phys. 48, 2187 (1972).
32. K. Kakurai, M. Steiner, J.K. Kjems: "Q Dependence of the Soliton Response in $CsNiF_3$ at T=10K and H=5kG" p.141 in Proc. NATO Adv. Stud. Inst. Geilo, Norway (1979) (Plenum 1980).
33. G. Reiter: Phys. Rev. Lett. 46, 202, 518 (1981).
34. J.M. Loveluck, T. Schneider, E. Stoll, H.R. Jauslin: Phys. Rev. Lett. 45, 1505 (1980); J. Phys. C: Solid State Phys. 15, 1721 (1982).
35. V.E. Zakharov, A.B. Shabat: Sov. Phys. JETP 34, 62 (1972).
36. M.J. Ablowitz, D.J. Kaup, A.C. Newell, H. Segur: Stud. Appl. Maths. LIII, 249 (1974).
37. L.A. Takhtadzhyan, L.D. Faddeev: Teor. Mat. Fiz. 21, 160 (1974).
38. R.K. Dodd, R.K. Bullough: in Synergetics A WOrkshop, H. Haken ed. (Springer-Verlag, Heidelberg 1977).
39. H. De Vega: Phys. Rev. D22, 2400 (1980).
40. P.J. Caudrey, J.C. Eilbeck, J.D. Gibbon: Nuovo Cimento 25B, 497 (1975).
41. J.F. Currie: Phys. Rev. A16, 1692 (1977).
42. J. Rubinstein: J. Math. Phys. 11, 258 (1970).
43. V.E. Korepin: Teor. Mat. Fiz. 41, 169 (1979).
44. E.K. Sklyanin, L.A. Takhtadzhyan, L.D. Faddeev: Teor. Mat. Fiz. 40, 194 (1979).
45. S.F. Edwards, Y.V. Gulaev: Proc. Roy. Soc. London A279, 229 (1964).
46. A.M. Arthurs: Proc. Roy. Soc. A313, 445 (1969).
47. C. Itzykson, J.B. Zuber: Quantum Field Theory (McGraw Hill, New York 1980).
48. A.G. Izergin, V.E. Korepin: Lett. Math. Phys. 5, 199 (1981).
49. A.G. Izergin, V.E. Korepin: in Problems in Quantum Field Theory and Statistical Physics, 3 P.P. Kulish and V.N. Popov eds. LOMI Vol 120 "Nauka", Leningrad. (Also the article by V.O. Tarasov there).
50. H. Flaschka: Lectures at ELAF '84.
51. D.J. Scalapino, M. Sears, R.S. Ferrell: Phys. Rev. B6, 3409 (1972).
52. J.A. Krumhansl, J.R. Schrieffer: Phys. Rev. B11, 3535 (1975).
53. M. Stone, J. Reeve: Phys. Rev. D18, 4746 (1978).
54. S. Coleman: in The Whys of Sub-Nuclear Physics, A. Zichichi ed. (Plenum, New York, 1977).
55. S. Goldstein: Proc. Roy. Soc. Edinburgh 49, 210 (1929).
56. R. Rajaraman, Raj Lakshmi: Phys. Rev. B25, 1866 (1981).
57. M. Lowe, M. Stone: Nucl. Phys. B136, 177 (1978).
58. H. Flaschka, D.W. McLaughlin: Prog. Theor. Phys. 55, 438 (1976).
59. H.P. McKean, Jr., P. van Moerbecke: Inventions Math. 30, 217 (1975) and references, B.A. Dubrovin, S.P. Novikov: Zh. Eksp. Teor. Fiz. 67, 2131 (1974) and other references in [58].
60. H.P. McKean, E. Trubowitz: Comm. Pure. Appl. Math. 29, 143 (1976); Bull. AMS 84, 1042 (1978).
61. M.G. Forest, D.W. McLaughlin: J. Math. Phys. 23, 1248 (1982).
62. M.J. Ablowitz, H. Segur: Solitons and the Inverse Scattering Transform (SIAM 1981).
63. J.F. Currie, J.A. Krumhansl, A.R. Bishop, S.E. Trullinger: Phys. Rev. B22, 477 (1980).
64. A.R. Bishop, J.A. Krumhansl, S.E. Trullinger: Physica 1D, 1 (1980).
65. A.R. Bishop: in Physics in One Dimension J. Bernasconi and T. Schneider (Springer-Verlag, Heidelberg 1983).
66. R.K. Bullough, D.J. Pilling, J. Timonen in Nonlinear Waves L. Debnath ed. (CUP, New York 1983).
67. R.K. Bullough, D.J. Pilling, J. Timonen: in Proc. 2nd International Workshop in Nonlinear and Turbulent Processes in Physics, Kiev, October 1983. R.Z. Sagdeev, A.V. Gaporov, A.S. Davydov, V.E. Zakharov eds. (Gordon and Breach, New York 1984) in press.

68. K. Huang: Statistical Physics (J. Wiley and Sons, 1963).
69. C. Pöppe: Ph.D. Thesis Univ. of Heidelberg, 1980.
70. H.A. Bethe: Z. Physik 71, 205 (1931).
71. R.K. Bullough, D.J. Pilling, J. Timonen: "Statistical mechanics of the quantum and classical sine-Gordon fields" in Proc. III Intl. Symposium on Selected Topics in Statistical Mechanics. Dubna, August 1984, N.N. Bogolubov Jr. ed. (JINR, Dubna 1984). To appear; "Quantum and classical statistical mechanics of the non-linear Schrödinger, sinh-Gordon and sine-Gordon equations" in Proc. 7th Kyoto Summer Institute: Dynamical Problems in Soliton Systems. S. Takeno ed. Springer Series in Synergetics (Springer-Verlag, Heidelberg, 1984). To appear.
72. R.J. Baxter: Ann. Phys. 70, 193, 323 (1972).
73. J.D. Johnson, S. Krinsky, B.M. McCoy: Phys. Rev. A8, 2526 (1973).
74. P.J. Caudrey: Physica 6D, 126 (1982).

Probabilistic Cellular Automata

B.A. Huberman
Xerox Palo Alto Research Center, Palo Alto, CA 94304, USA

1. Introduction

In spite of the recent progress made in understanding the non-linear dynamics of systems with few degrees of freedom, little is known about the time evolution of more complex structures. Systems with many degrees of freedom, which are pervasive in both the natural and artificial worlds, give rise to properties such as self-organization and computation which, in many cases, seem quite immune to external perturbations. Typical examples are found in biological organisms and modern computing structures, where the intractability of their dynamical behavior is partly linked to our limited knowledge about the global properties of solutions of non-linear partial differential equations.

These issues are becoming a topic of current concern with the appearance of large scale integrated circuits capable of powerful information processing, but whose complex dynamic behavior is largely unknown. Examples of complex structures can be found in compact logic circuits[1-2], novel computing architectures[3], and neural networks[4]. In all these instances, one deals with large collections of threshold elements whose states (active or passive) at a given time can in turn excite or inhibit other elements of the network at a later time. As the number and density of these elements increases, the likelihood of failure at the component level forces us to consider the overall behavior of the system, even when its parts can at times fail. Thus, as either the connectivity or the coupling strength between nonlinear elements increases, a sustained, unintended activity of the network can in principle be excited and which can at times supersede the programmed behavior. If such collective modes do indeed exist, it is of interest to study their nature (e.g., periodic or erratic) and their stability as a function of parameters, such as threshold behavior and connectivity. These questions are relevant to seemingly unrelated problems, such as the behavior of neural nets, the halting problem in computing machines based on probabilistic algorithms, and the dynamics of large geological faults.

In what follows, I will summarize results that we have obtained by applying non-equilibrium statistical mechanics techniques to the study of the global dynamics of cellular automata made up of threshold elements[5]. Such networks provide simple models for the study of systems ranging from neural nets and probabilistic computers, to VLSI structures and problems of crystal growth. I will show how, by writing down a master equation for the system, one can obtain analytic expressions for the time dependence of its activity as a function of parameter values. For wide ranges of non-linearity and connectivity, non-linear automata with both excitatory and inhibitory couplings can display collective behavior which can be either multiple periodic or deterministic chaotic. The latter type consists of an intense and sustained erratic activity which results from the relaxation of initial configurations of a system with many degrees of freedom to a strange attractor. Besides these diffuse reverberating modes, one can also obtain simple asymptotic dynamics which are relevant to the study of many other systems, as we shall see when dealing with the stick-slip behavior of faults.

2. Automata dynamics

We start by describing the non-linear elements which make up the network. These are synchronous threshold cells which, when active, send signal pulses to their connected neighbors. The signal pulses (e.g., voltage, digital packets or elastic stresses) can be either positive (excitatory) or negative (inhibitory), and we will assume that each of the identical elements has the same number of inputs (p excitatory connections plus r inhibitory ones) as outputs. Each threshold device, analogous to the formal neurons of logic networks[6], will fire with probability p if after a summing period τ the sum total of its inputs exceeds a threshold value V_0. If to each neuron we assign a variable s_i such that it is $+1$ if it fires, and $s_i = -1$ if it does not fire, the threshold behavior is described by specifying a probability function which we choose to be given by

$$p(s'_i) = \left\{ \exp[-\epsilon s'_i(V_i - V_0)] + 1 \right\}^{-1} \tag{1}$$

where s'_i denotes the value of s'_i at $t+\tau$, ϵ determines the width of the threshold region and V_i is the input potential, which can be written as

$$V_i = \sum_j V_{ij} \frac{(s_j + 1)}{2} \tag{2}$$

with V_{ij} denoting the signal (e. g. voltage) input to the ith neuron coming from the jth neuron, and the sum j is over the $r+p$ incoming connections.

A network of N neurons is then described at each instant of time t by an N vector $\alpha = (s_1 \ldots s_N)$ specifying which neurons are active ($s_i = +1$) and which ones are not ($s_i = -1$). If $P(\alpha,t)$ denotes the probability that at time t the network is in state α, its value at $t+\tau$ is determined by the master equation

$$P(\alpha,t+\tau) - P(\alpha,t) = -\sum_{\{\beta\}} \omega(\alpha\to\beta)P(\alpha,t) + \sum_{\{\beta\}} \omega(\beta\to\alpha)P(\beta,t) \tag{3}$$

where $\omega(\alpha\to\beta)$ denotes the transition probability in time τ from a state α to a state β. For the particular case of single neuron processes, it is determined by Eq. (1). Processes including simultaneous transitions of many neurons can be simply calculated from Eqs (1) and (3), but since they do not lead to different dynamics, we will concentrate on single element updates. Introducing new variables,

$$\tilde{J}_{ij} \equiv \epsilon V_{ij}/2, \quad \tilde{H} = \epsilon/2 \; [\; \sum_j V_{ij} - 2V_0 \;] \quad \text{and } E_i \equiv \sum_{j=1} \tilde{J}_{ij} s_j + \tilde{H}, \;\; \omega(\alpha\to\beta)$$

Eq. (1) can be written as

$$\omega_L(s_L) = \frac{\exp[-E_L s_L]}{\sum_{s'_L} \exp[E_L s_L']} \tag{4}$$

The quantity of interest is the activity of the network, which determines the number of elements active at a given time, and which we will define as

$$A = \frac{\langle s_k \rangle + 1}{2} \tag{5}$$

with the expectation value of the kth spin at time t given by $\langle s_k \rangle \equiv \sum_{\{s\}} s_k P(s_1 \ldots, s_N, t)$, where { } implies a sum over the 2^N possible values of the set $\{s_1 \ldots, s_N\}$, and $\langle s_k \rangle$ ranges from −1 to +1. Using Eq. (3) the expectation value can be written in terms of the following equation of motion

$$\langle s_k \rangle_{t+\tau} - \langle s_k \rangle_t = -2 \langle s_k \omega_k (s_k) \rangle_t \tag{6}$$

with $\omega_k(s_k)$ given by Eq. (4). It now remains to calculate the right hand side of this equation. In the limit of small excess voltages ($V_i - V_0$ small), or broad threshold regimes (small ϵ), it is possible to generate a series expansion in powers of E_k for Eq. (4). To third order in E_k, we obtain

$$\omega_k = \frac{\left[1 - E_k s_k + E^3_k s_k/3\right]}{2} \tag{7}$$

which can in turn be used to evaluate the sum in Eq. (6). In the spirit of a mean field theory, we shall assume that $\langle s_i s_j \rangle = \langle s_i \rangle \langle s_j \rangle$ if $i \neq j, \langle s_j \rangle = \langle s \rangle$,

and write $\sum_j \tilde{J}_{kj} = \tilde{J}$, $\sum_j \tilde{J}_{kj}{}^2 = \tilde{J}'^2$, and $\sum \tilde{J}_{kj}{}^3 = \tilde{J}''^3$ with

$$\tilde{J} \equiv p\tilde{J}_1 - r\tilde{J}_2 \tag{8a}$$

$$\tilde{J}'^2 \equiv p\tilde{J}_1{}^2 + r\tilde{J}_2{}^2 \tag{8b}$$

$$\tilde{J}''^3 \equiv p\tilde{J}_1{}^3 - r\tilde{J}_2{}^3 \tag{8c}$$

where J_1 and J_2 are the strengths of the excitatory and inhibitory potentials, respectively. A lengthy but straightforward calculation then gives the following equation of motion

$$\begin{aligned}\langle s \rangle_{t+\tau} = {} & \tilde{H}(1 - \tilde{J}'^2 - \tilde{H}^3/3) + (\tilde{J} - \tilde{J}\tilde{J}'^2 + 2\tilde{J}''^3/3 - \tilde{H}^2\tilde{J})\langle s \rangle_t \\ & - \tilde{H}(\tilde{J}^2 - \tilde{J}'^2)\langle s \rangle_t^2 - (\tilde{J}^3/3 - \tilde{J}\tilde{J}'^2 + 2\tilde{J}''^3/3)\langle s \rangle_t^3\end{aligned} \tag{9}$$

Equation (9), together with Eq. (5), describes the dynamical behavior of the network to third order in $\epsilon(V_i - V_0)$. Rather than giving a detailed analysis of its solutions, I will now describe several important regimes in the global behavior of the activity. Consider first a network with only purely excitatory (i.e., ferromagnetic-like) or inhibitory (i.e., antiferromagnetic-like) interactions. In such a limit, either $\tilde{J}_1$ or $\tilde{J}_2$ are zero, and $\tilde{J}^2 \gg \tilde{J}'^2$. It then follows that Eq. (9) becomes $\langle s \rangle_{t+\tau} = \tilde{H} \pm |\tilde{J}| \langle s \rangle_t + |\tilde{J}|^3 \langle s \rangle_t^3/3$. The solutions of this equation are such that for J<1 all initial values of the spin configuration relax to the fixed point $\langle \bar{s} \rangle \simeq \tilde{H}$ and therefore $A \simeq (1 + \tilde{H})/2$. This static limit is similar to the dynamics of Ising-like models using a continuous approximation to the master equation Eq. (3)[7]. If $\tilde{H}$ is vanishingly small (i.e.; the average input equals the threshold),we conclude that the long time activity of a network with no competing interactions consists of a simple time-independent fixed point, in which almost half of the neurons are firing at a given time.

The second and most interesting case is the one in which the network has the same amount of randomly distributed excitatory and inhibitory connections (the neuron glass limit). In this limit $\tilde{J}$ is negligible compared to $\tilde{J}\tilde{J}'^2$ or $\tilde{J}''^3$, and Eq. (9) becomes (in the limit of $\tilde{H} \to 0$)

$$\langle s \rangle_{t+\tau} = \tilde{R}(\langle s \rangle_t - \langle s \rangle_t^3) \tag{10}$$

where $R = (2\tilde{J}''^3/3 - \tilde{J}\tilde{J}'^2)$, and which for realistic values of the parameters displays non-trivial dynamics. Consider a network with an initial activity A . For values of $\tilde{R}$ in the range $-2<R<2$ its long time behavior corresponds to a stable, time-independent, fixed point whose magnitude increases monotonically with R. At $R = 2$ a pitchfork bifurcation takes place, whereby initial configurations now relax to a periodic attractor with period 2τ. Therefore, the activity will not only be large and finite, but consist of collective modes which oscillate periodically in time. As R is increased even further, one encounters a full cascade of period doubling bifurcations, at each of which the period of the asymptotic oscillations lengthens by a factor of two. Thus, a computation of a property of the network (such as its energy) at fixed R would reveal an apparent finite set of configurations whose number depends on the period of the activity. For values beyond $R = 2.3$ a new regime is encountered. This regime is characterized by an intense and erratic activity with a noisy power spectrum[8], in which initial configurations of the network relax towards a strange attractor. The figure below shows an example of the time evolution of the activity of the network for $R = 2.34$, as generated by Eqs (5) and (10) after an initial transient of 1000 time steps. As can be seen, the collective behavior of the system is chaotic, with a peak to peak intensity which can be determined by the first and second iterates of Eq. (10) evaluated at the maximum of its right hand side.

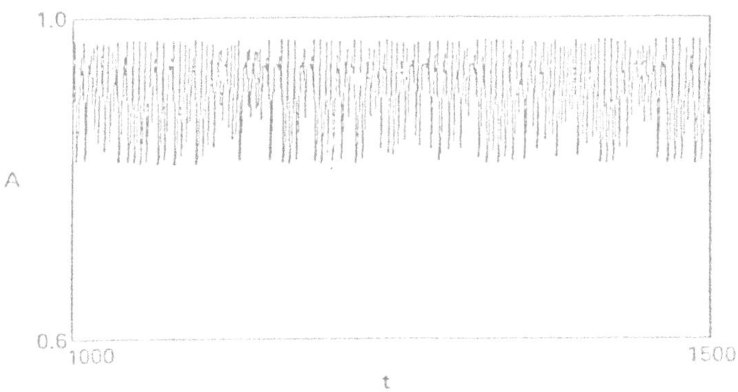

Fig. 1

Furthermore, this deterministic erratic behavior is characterized by a broad band component of its power spectrum, which grows like $[R - (\bar{R})_c]^\sigma$ with $\sigma \simeq 1.525$[8]. As the value of R is further increased, the system may also exhibit laminar regimes,which in some cases will be preceded by intermittent behavior[9], followed by an even more chaotic activity at larger values of $\tilde{R}$.

The effects just described provide a plausible explanation for the diffuse reverberations which have been observed in simulations of neural networks, and the reports of sustained oscillations in similar systems[10]. Also, since the appearance of such collective modes in our theory seems to be a direct consequence of both finite time steps and the non-linearity of the threshold function, we expect them to be pervasive in many other systems with similar characteristics, such as in Monte Carlo simulations of many-body systems with competing interactions[11].

3. The Stick-Slip Behavior of Faults

We now turn to a different set of problems which illustrate the power of the methods developed in the previous section.There is evidence that a large number of active faults, the northern and southern locked sections of the San Andreas providing a likely example[12], appear to behave in a stick-slip rather than a stable sliding manner. This has led to a number of studies of frictional effects on the onset of earthquakes, such as those of Stuart and Mavko[13], Weertman[14], and Newman and Knopoff[15]. Recently, Smalley et al.[16] have modelled a fault system as an array of asperities with a statistical distribution of strengths. By assuming that the stress on each failed asperity can be transferred to one or more neighboring asperities within the same "cell", they were able to use renormalization group tranformations to evaluate the critical stress beyond which an earthquake sets in. A problem with their model is that besides being static (i.e. no time evolution for the stress-transfer) it contains no interactions between asperities in different but neighboring cells.

We now remedy these problems by studying a model of a fault by considering it as a dynamical array of asperities. We use the fact that when the stress on a given asperity exceeds a critical value, it is transferred in a finite amount of time to its neighboring ones, regardless of their belonging to any arbitrary cells. We also consider the time delay in stress propagation and introduce conditional probabilities which allow for a dynamic description of the fault. The resulting equations governing the time evolution of any initial stress configuration are similar to those describing the dynamics of probabilistic cellular automata which I outlined in my previous section. In the limit of a broad distribution of stresses, we obtain exact solutions of those equations and show how the initial failure of an asperity leads to the propagation of an enhanced probability pulse, which in turn makes the fault slip or stick in the course of time.

Consider a fault as a one-dimensional array of N asperities with a given distribution of strengths acting on them. An asperity may be locked or unlocked according to whether the applied stress is smaller or larger than the failure strength of that asperity. Within this model,the ith asperity can have two states (locked and unlocked), which can be represented by a binary variable, $v_i = 0,1$ (0 for the locked state, 1 for the unlocked state ; $i = 1, 2, 3, \ldots, N$). The state of the fault is then characterized by the N-tuple $\alpha \equiv (v_1, v_2, \ldots, v_N)$. We next notice that when asperities fail, i.e. change their states from "0" to "1", the stress in the failed region is transferred to adjacent regions as cracks are introduced into an elastic solid. This allows us to express the total stress applied to the ith asperity as

$$\sigma_i = \sum_j \sigma_{ij} v_j + \sigma_i^o, \tag{11}$$

where σ_i^o is the stress on the ith asperity itself regardless of the failure of the neighboring asperities, $i \neq j$, and σ_{ij} is the stress that is transferred from the jth asperity to the ith asperity when the jth asperity fails. If σ^f denotes the failure strength, the ith asperity will fail if $\sigma_i - \sigma^f > 0$, a condition which, for a one-dimensional array, can be written as

$$\sigma_i - \sigma^f = \sigma^o + \sigma_1 v_{i-1} + \sigma_2 v_{i+1} - \sigma^f > 0, \tag{12}$$

where it has been assumed that $\sigma_i^o = \sigma^o$, $\sigma_{i,i-1} = \sigma_1$, and $\sigma_{i,i+1} = \sigma_2$ do not depend on i. We further assume that the statistical distribution of failure strengths σ^f can be effectively taken into account by regarding σ^f as a constant, and introducing probabilities in determining whether or not an asperity will fail.

Since a finite sound velocity implies the existence of finite time delay between the failure of an asperity and the corresponding transfer of the stress to its neighbors, the state of each asperity at time t is determined by probabilities that depend on the states of the asperity itself and its neighboring ones at time $t-\tau$. Here τ is the basic time step and its magnitude is of the order of the distance between neighboring asperities divided by the sound velocity. Denoting the conditional probability for failure or locking at time t based on the state of the fault at time $t-\tau$ by $p(v',t|\, v_i\, v_{i-1}\, v_{i+1}\, , t-\tau)$, we can write its components explicitly. For a locked asperity at time t, $(v_i=0)$ they are

$$p(1|000) = x \tag{13}$$

$$p(1|010) = y_1$$

$$p(1|001) = y_2$$

$$p(1|011) = z$$

with the normalization condition $p(0|0,v_{i-1},v_{i+1}) = 1 - p(1|0v_{i-1}v_{i+1})$. Similarly, we consider the conditional probabilities that a failed asperity be locked again after a time τ. They can be written as

$$p(0|111) = x' \tag{14}$$

$$p(0|110) = y_1'$$

$$p(0|101) = y_2'$$

$$p(0|100) = z'$$

together with the condition $p(1|1\; v_{i-1}\; v_{i+1}) = 1 - p(0|1\; v_{i-1}\; v_{i+1})$. The values of the quantities x,y_1,y_2,z and their primed counterparts are determined by the statistical distribution of failure strengths. This, in turn, allows the right hand sides of Eqs. (13) and (14) to be expressed in terms of σ^0, σ_1, σ_2 and σ^f. Since it follows from the definition of v_i that $v_i^2 = v_i$, the most general form for the conditional probability can be written as a function of the state of the ith asperity in the form

$$p(v_i' \mid v_i\; v_{i-1}\; v_{i+1}) = a - bv_i' + (2v_i' - 1)\; (c_0\; v_i + c_1 v_{i-1} + c_2 v_{i+1} + dv_{i-1}\; v_{i+1} + ev_{i-1}\; v_i + e_2 v_i v_{i+1} - fv_{i-1} v_i v_{i+1}) \tag{15}$$

where

$$a \equiv 1 - x,\ b \equiv 1 - 2x,\ c_0 \equiv 1 - (x + z'),\ c_1 \equiv y_1 - x,\ c_2 \equiv y_2 - x,\ d \equiv (z + x) - (y_1 + y_2)$$

$$e_1 \equiv (z' + x) - (y_1' + y_1),\ e_2 \equiv (z' + x) - (y_2' + y_2),\ \text{and}\ f \equiv (z + x) - (y_1 + y_2) + (z' + x') - (y'_1 + y'_2)$$

We can now write an equation describing the dynamics of an array of asperities within our model. The discrete master equation for $P(\alpha,t)$, the probability that at time t the fault is in a state $\alpha \equiv (v_1,v_2\; .\; .\; .v_N)$ is given by

$$P(\alpha,t+\tau) - P(\alpha,t) = - \sum_{\beta} \Big[\, p(\alpha\rightarrow\beta)\, P(\alpha,t) - p(\beta\rightarrow\alpha)\, P(\beta,t) \Big] \tag{16}$$

where the transition probability $p(\alpha\rightarrow\beta)$ is given by

$$p(\alpha\rightarrow\beta) \equiv p(\beta,\, t+\tau \mid \alpha,t) = \prod p(v'_i \mid v_i\, v_{i-1}\, v_{i+1}) \tag{17}$$

$\beta \equiv (v'_1, v'_2 \ldots v'_N)$, and $p(v'_i \mid v_i\, v_{i-1}\, v_{i+1})$ is determined by Eq. (15).

The quantity of interest is the probability that the kth asperity will be found in the unlocked state (i. e. $v_k = 1$) at time t . It is given by $\langle v_k \rangle_t \equiv \sum v_k\, p(v_1 \ldots v_N, t)$, and it can be calculated by multiplying Eq. (16) by v_k and summing over α. We thus obtain

$$\langle v_k \rangle_{t+\tau} = \langle v_k \rangle_t + \langle (1-2v_k)\, p(1-v_k \mid v_k\, v_{k-1}\, v_{k+1}) \rangle_t \tag{18}$$

where we have used Eq. (17) together with the normalization condition $\sum p(v'_i | v_i\, v_{i-1}\, v_{i+1}) = 1$.

Substituting the expression for $p(1-v_k \mid v_k\, v_{k-1}\, v_{k+1})$ given by Eq. (15), we finally obtain

$$\langle v_k \rangle_{t+\tau} = x + c_0 \langle v_k \rangle_t + c_1 \langle v_{k-1} \rangle_t + c_2 \langle v_{k+1} \rangle_t + d \langle v_{k-1} v_{k+1} \rangle_t + e_1 \langle v_{k-1}\, v_k \rangle_t + e_2 \langle v_k v_{k+1} \rangle_t - f \langle v_{k-1}\, v_k v_{k+1} \rangle_t \tag{19}$$

which gives the behavior of the kth asperity at time t if the states of both that asperity and its nearest neighbors are specified at time $t-\tau$.

In general, Eq. (19) does not admit of an exact solution, due to the existence of the last four correlation terms. For the case of a broad distribution of failure strengths, however, we can solve the equation in exact fashion. In this limit $d = e_1 = e_2 = f = 0$, and a solution can then be found for an infinite chain, or a finite one with periodic boundary conditions. Introducing the generating function

$$F(\lambda, t) = \sum_{L=-\infty}^{\infty} \lambda^L \langle v_L \rangle_t \tag{20}$$

one obtains a solution for $\langle v_L \rangle$ as a function of time for $t = n\tau$ in the form

$$\langle v_k \rangle_{t=n\tau} = \sum_{p=0}^{n} \sum_{q=0}^{p} \binom{n}{p} \binom{p}{q} c_0^{\,n-p}\, c_1^{\,p-q}\, c_2^{\,q} \langle v_{k-p+2q} \rangle_0 + \frac{1-(c_0+c_1+c_2)^n}{1-(c_0+c_1+c_2)} x \tag{21}$$

Equation (21) gives the explicit time evolution of the probability that the kth asperity will fail given a particular initial condition, as specificied by $\langle v_{k-p+2q} \rangle_0$.

For the case when the initial information about all of the asperities is the same $\langle v_i \rangle_0 = \langle v \rangle_0$, it becomes

$$\langle v_k \rangle_{t=n\tau} = (c_0 + c_1 + c_2)^n \langle v \rangle_0 + \frac{1-(c_0+c_1+c_2)^n}{1 - (c_0+c_1+c_2)} x \tag{22}$$

which does not display relaxational behavior if $c_0 + c_1 + c_2 = 1$. A particularly simple solution appears in the case when,initially,only one asperity is unlocked, i.e., $\langle v_i \rangle_0 = \delta_{il}$. Eq. (21) can then be written as

$$\langle v_k \rangle_{t=n\tau} = \sum_{q=0}^{n} \binom{n}{k-l+2q} \binom{k-l+2q}{q} c_0^{\,n-k+l-2q}\, c_1^{\,k-l+q}\, c_2^{\,q} + \frac{1-(c_0+c_1+c_2)^n}{1-(c_0+c_1+c_2)} x, \tag{23}$$

where the binomial coefficients with negative arguments are defined to be zero. This particular solution yields a simple picture of stress propagation after the initial asperity fails. Once the lth asperity fails, the stress acting on it propagates, and reaches the kth asperity after a time $|k - l|\tau$. At that particular instant, the probability for failure of the kth asperity increases, leading to a more probable scenario for its own failure. As time increases, the stress propagates beyond the kth asperity, while its own failure probability settles down to its asymptotic value. In the meantime, a probability pulse moves through the fault, increasing the likelihood that it will slip as a whole. We should add that, for initial conditions such that many asperities fail, the time evolution of the stress propagation pulse will lead to a more certain slip behavior of the fault.

I have just presented a simple dynamical model for stick-slip behavior of faults which can be solved exactly for one-dimensional faults in the limit where the distribution of stresses acting on a given asperity is fairly broad. As to the approximations that were introduced, the assumption of a wide stress distribution is quite reasonable in view of the wide variety of scales acting on any fault. On the other hand, an exact solution was only produced in the one-dimensional case, which is quite restrictive. Nevertheless, it can be used to gain insight into the approximate treatments which can be generated for more realistic models of a fault. Also, since the model has a formal similarity to probabilistic cellular automata of the type considered above, it leads to the interesting possibility of more complicated behavior, such as oscillations and chaotic wandering of the stress failure probability pulse. Finally, the computer simulation of these arrays of asperities could be implemented by using parallel architectures with simple local rules.

These lectures have shown how one can deal with the dynamics of probabilistic cellular automata by using non-equilibrium statitistical mechanics. The results obtained can be applied to a number of problems which I have not had time to mention, such as systems with multiple states per cell, and the analysis of Monte Carlo algorithms[11]. At a different level, they also show how the introduction of delays into many-body problems leads to digital time formulations with novel dynamical solutions.

The work I have described was done in collaboration with M. Y. Choi, whom I thank for his continued discussions and assistance. It was also supported in part by a grant from the Office of Naval Research, N00014-82-0699

4. References

1. For a description of such systems see, for example, C. Mead and L. Conway *Introduction to VLSI Systems*, Addison-Wesley (1980), Chapters 5-7.
2. The dynamics of deterministic Boolean networks has been considered in detail by, F. Fogelman-Soulie, E. Goles-Chacc and G. Weisbuch, Bull. Math. Biol. 44, 715 (1982).
3. B. A. Huberman and T. Hogg, Phys. Rev. Lett. 52, 1048 (1984).
4. A thorough review of neural networks is given by R. J. McGregor and E. R. Lewis, *Neural Modelling*, Plenum Press (1977).
5. M. Y. Choi and B. A. Huberman, Phys. Rev. A28, 1204 (1983).
6. W. S. McCullough and W. Pitts, Bull. Math. Bio. 5, 115 (1943).
7. R. J. Glauber, J. Math. Phys. 4, 294 (1963); and M. Suzuki and R. Kubo, J. Phys. Soc. Japan 24, 51 (1968).
8. B. A. Huberman and A. Zisook, Phys. Rev. Lett. 46, 626 (1981).
9. J. Hirsch, B. A. Huberman and D. J. Scalapino, Phys. Rev., B25, 519 (1982).

10. D. R. Smith and C. H. Davidson, J. Assoc. Comput. Mach. 9, 268 (1962) and P. A. Anninos, Kybernetic 11, 11 (1972).
11. M. Y. Choi and B. A. Huberman, Phys. Rev. B28, 2547 91983) and B29, 2796 (1984).
12. K. E. Sieh, J. Geophys. Res. 83, 3907 (1978).
13. W. D. Stuart and G. M. Mavko, J. Geophys. Res. 84, 2153 (1979).
14. J. Weertman, J. Geophys. Res. 84, 2146 (1979).
15. W. I. Newman and L. Knopoff, Geophys. Res. Lett. 10, 305 (1983).
16. R. F. Smalley, D. L. Turcotte, and S. A. Solla, Preprint (1984).

Part II **Quantum Optics**

Chaos, Generalized Multistability and Low Frequency Spectra in Quantum Optics

F.T. Arecchi
Department of Physics, University of Firenze, and
Istituto Nazionale di Ottica, I-50125 Firenze, Italy

A crucial problem in non-equilibrium statistical mechanics is that regarding the insurgence of ordered structures starting from a chaotic (maximum entropy) condition, in a system strongly perturbed at its boundary as a quantum optical system. For still higher perturbations, the ordered structures become more and more complex, until reaching deterministic chaos. Some experimental situations for lasers are analyzed. We stress the coexistence of several basins of attraction (generalized multistability) and their coupling by external noise. This coupling induces a low frequency branch in the power spectrum. Comparison is made between the spectra of noise-induced jumps over independent attractors and that of deterministic diffusion within subregions of the same attractor. At the borderline between the two classes of phenomena a scaling law holds, relating the control parameter and the external noise in their effect on the mean escape time from a given stability region.

1 Introduction: order and chaos in quantum optics

Quantum optics deals with lasers and laser-like phenomena. At an elementary level, they can be understood in terms of perturbation theory at lowest orders, that is, in terms of competition between stimulated and spontaneous emission processes.

If one tries to build a nonperturbative picture, one is struck by the complexity of the problem. There are two conceptual escape ways: on one hand to increase the size of the system to infinity while keeping the density finite and the temperature uniform, and look for an asymptotic (thermodynamic) solution; on the other hand to drastically simplify the boundary conditions making it possible to excite only one or a few radiation modes. This was the original idea of Schawlow and Townes, when extending the Maser principle to optical frequencies by use of a Fabry-Perot cavity. This is also done in other classes of nonlinear field problems such as hydrodynamical instabilities, where one works with small "aspect ratios" , that is, with cells with comparable sizes in all three dimensions, in order to excite few Fourier components of velocity and thus deal with a finite number of coupled equations. In quantum optics this procedure leads to a set of quantum equations still unsoluble. A further approximation is the so-called "semi-classical" one, leading to the Maxwell-Bloch equations. Starting in the early sixties, the introduction of photon statistics methods has made possible the study of fluctuations and coherence in lasers: how and why 10^{20} atoms or molecules, rather than radiating e.m. field in a chaotic fashion, decide to "cooperate" to a single coherent field mode; then, for still higher excitation, how and why they organize in a complex pattern (many modes), each highly coherent per-se, but with little correlation with one another. The lack of large scale correlations is shown to correspond to the onset of deterministic chaos.

It is worthwhile to stress the role of nonlinearities as well as that of the phase space topology (number of relevant dynamical variables) for the onset of order and chaos. A detailed experimental analysis of the onset of order in a pumped

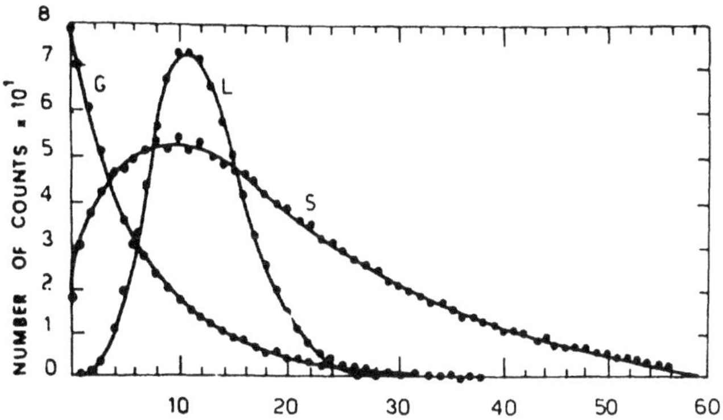

Figure 1 Photocount distributions of three radiations fields. L, Laser Field; G, Gaussian field; S, linear superposition with L and G onto the same space mode

system was given in my 1965-67 investigations on the passage from incoherent to coherent light in a laser [1]. Fig. 1 shows the photon statistics (P.S.) for a radiation field below and above the threshold point (that is, the point where the gain provided by the stimulated emission processes of excited atoms compensates for the losses due to the escape of radiation from the laser volume). The G curve is fitted by a Bose-Einstein distribution describing the fluctuations of the photon number in a black-body around the average value $\langle n \rangle$, given by Planck's formula; the L curve is fitted by a Poisson distribution. The two distributions correspond to fields with the same color, direction and intensity, so no classical optics measurement discriminates between them, yet P.S. does. The reason why is given in Fig. 2. For a uniform field, since photons are Bose particles with zero mass, and hence delocalized, the associated photon detection processes at different points in space-time have no correlations. Therefore, in a volume filled with a coherent, or ordered, field (that is, a field with a δ -like statistics, as in Fig. 2a), the associated probability, p(n), of detecting photons at a given point over a time T is a Poisson distribution.

If the field now has a zero-average Gaussian distribution, as for a thermal equilibrium or maximum entropy situation, weighting each probability element with the detection statistics one obtains the Bose-Einstein distributions (Fig. 2b). Between these two limiting cases of full order and maximum chaos, one can trace a continuous manifold of intermediate cases (Fig. 2c). This smooth behavior is analogous to second order phase transitions, as when a thermodynamic system undergoes a continuous change of state around a critical temperature. The explanation implies the essential role of nonlinearities. The elementary description of a laser in terms of Einstein stimulated emission processes compensating for losses is not sufficient. Indeed, this would just provide a linear polarization $P = \chi E$, and a quadratic free energy

$$F(E) = -P\cdot E = -\chi E^2. \tag{1.1}$$

In a thermodynamic system, open with respect to the field variable E, this has a statistical distribution given by

$$W(E) = Ne^{-F(E)/kT}, \tag{1.2}$$

(N = normalization constant).

Similarly a nonequilibrium system with a dynamical variable E, driven by a nonlinear force f(E) and by stochastic noise with a short correlation time and correlation amplitude D, has a stationary distribution [2,3].

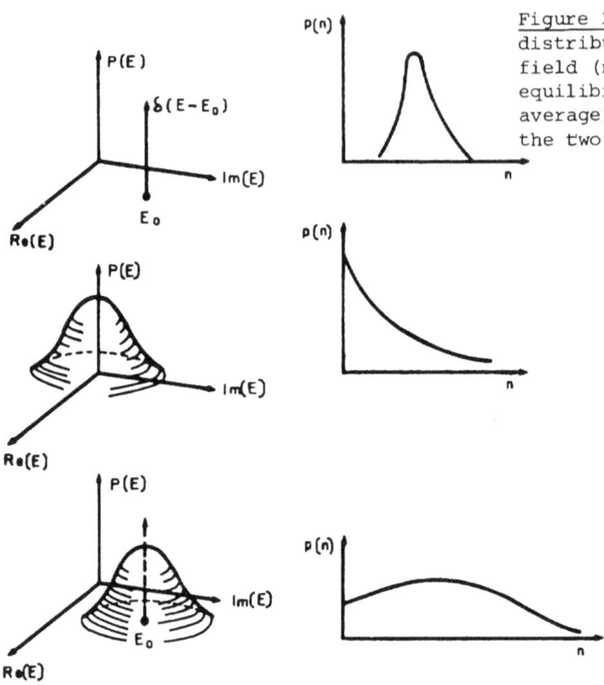

Figure 2 Field and photon statistical distributions for an ideal coherent field (no fluctuations), for a thermal equilibrium field (Gaussian with zero average), and for the superposition of the two (shifted Gaussian)

$$W(E) = Ne^{-\int f(E)dE/D} . \qquad (1.3)$$

In the absorbing case (force of the field proportional to the polarization $P = -\alpha E$), $\int f(E)dE = -1/2\,\alpha E^2$. By (1.3) the field has a Gaussian stationary distribution, as should be expected from thermodynamics. In the linear emitting case the distribution is undefined ($P = \alpha E$; $e^{\alpha E^2}$ is not normalizable).

But an atom is still exposed to photons after emission. The lowest correction is cubic in the field (Fig. 3)

$$P = \alpha E - \beta E^3$$

and it is sufficient to describe the passage from Gaussian chaos to a narrow distribution around nonzero fields

$$\pm E_o = \pm\sqrt{\alpha/\beta} . \qquad (1.4)$$

The spontaneous symmetry breaking does not assign the phase of E_o. We have two equivalent states, 180° apart. To lift the degeneracy we must apply an external field, γ, which assigns a reference phase. The two states are no longer equivalent (optical bistability) (Fig. 3d).

As shown by the locus of stable points, at the threshold the thermodynamic branch (Fig. 3a) becomes unstable and the "coherent" branches appear (Fig. 3c). In mathematical terms this is a bifurcation. Still higher order bifurcations could appear, making the ordered branch unstable, and leading to new "orders".

Before discussing this multiple sequence of bifurcations it is important to decide how many degrees of freedom we have to deal with. In physics we deal in general with nonlinear equations for a field $\vec{q}(x,t)$

$$\frac{\partial \vec{q}}{\partial t} = f\left(\vec{q}, \frac{\partial}{\partial \vec{x}}\vec{q}\right) . \qquad (1.5)$$

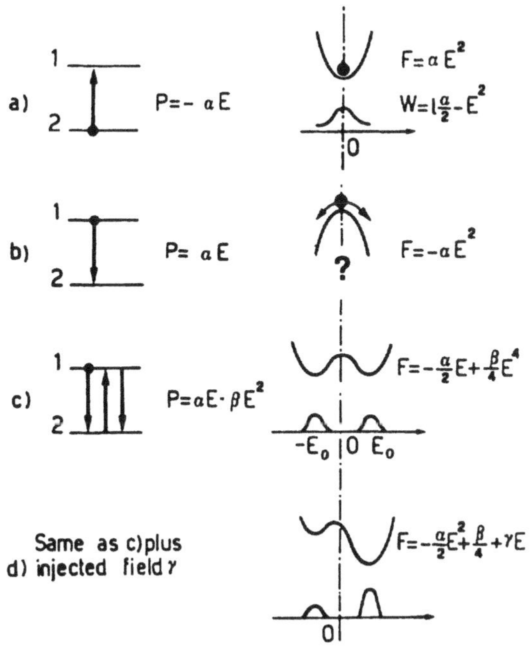

Figure 3 a) Absorbing atom, linear polarization; parabolic pseudo-potential; Gaussian probability. b) Emitting atom, linear polarization; parabolic pseudo-undefined potential; undefined probability. c) Emitting atom, cubic polarization; quadratic pseudo-potential; probability peaks around values ±E. d) As c) but with a symmetry-breaking odd term in potential, arising from an external field

Such are Navier-Stokes and Fourier equations for a velocity field coupled to a temperature field in a convective fluid instability. For a rectangular cell of small aspect ratio (ratio of two lateral sizes with respect to the fluid height) and for a temperature difference ΔT between lower and upper plate near the onset of the instability, (1.5) reduce to three coupled equations for a velocity mode and two temperature modes [4]. In suitable units, these three equations are [5]

$$\begin{aligned}\dot{x} &= -\sigma(x-y)\\ \dot{y} &= -y+rx-xz\\ \dot{z} &= -(8/3)z+xy\end{aligned} \tag{1.6}$$

With $\sigma = 10$ and $r = 28$ the solution is chaotic. Similarly, if we couple Maxwell equations with Schrödinger equations for N atoms confined in a cavity and we expand the field in cavity modes, keeping only the first mode E which goes unstable, this is coupled with the collective variables P and Δ, describing the atomic polarization and population inversion as follows [2, 3]

$$\begin{aligned}\dot{E} &= kE+gP\\ \dot{P} &= -\gamma_\perp P+gE\Delta\\ \dot{\Delta} &= -\gamma_\parallel(\Delta-\bar{\Delta})-2g(P^*E+PE^*)\end{aligned} \tag{1.7}$$

where k, $\gamma_\perp$, $\gamma_\parallel$ are the loss rates for field, polarization and population respectively, g is a coupling constant and $\bar{\Delta}$ is the population inversion which would be established by the pump mechanism in the atomic medium in the absence of coupling. While the first of (1.7) comes from Maxwell equations, the two others imply the reduction of each atom to the levels which are resonantly coupled with the field, that is, a description of each atom in a isospin space of spin 1/2. The last two equations are Bloch equations which describe the spin precession. Therefore (1.7) are called Maxwell-Bloch equations.

Eqs. (1.6) and (1.7) are phenomenological. The presence of loss rates means that three relevant degrees of freedom are in contact with a "sea" of other degrees of freedom. In principle, they could be deduced from microscopic equations by statistical reduction techniques. The fluctuation-dissipation theorem would impose the addition of stochastic forces.

However, we show that for $N \geq 3$ degrees of freedom, deterministic chaos may be reached in nonlinear eqs. (1.6) or (1.7) without consideration of stochastic forces. These latter ones would modify some details of the phenomena, without relevant changes in the qualitative picture. In a dissipative system there is a contraction of the phase space volume. If at time t = 0 the ensemble of initial conditions is confined in a hypersphere of radius ϵ, (initial volume $V_o = \epsilon^N$) at time t the volume will be (referring to the principal axes)

$$V(t) = \epsilon^N \cdot e^{\sum_1^N \lambda_i t}, \tag{1.8}$$

where the growth rates λ_i are the Lyapunov exponents. The contraction requirement means

$$\sum_1^N \lambda_i < 0 . \tag{1.9}$$

Now, if we start from a single point at t = 0 (well determined initial condition) a single trajectory emerges, and on it we have obviously $\lambda = 0$. For N = 1, (1.9) imposes $\lambda_1 < 0$, hence there is no trajectory. Indeed, the asymptotic volume can only be 0-dimensional, that is, a single point; hence we have only a stationary solution. For N = 2, if $\lambda_1 < 0$, it may be that $\lambda_2 < 0$ (no trajectory) but also $\lambda_2 = 0$, that is, the final volume is 1-dimensional (limit cycle) and we have a periodic oscillation. For N = 3, besides fixed points ($\lambda_1, \lambda_2, \lambda_3$ all negative) and limit cycles (λ_1, λ_2 negative, $\lambda_3 = 0$) we may have $\lambda_1 < 0$, $\lambda_2 = 0$, $\lambda_3 > 0$, with $\lambda_3 < |\lambda_1|$ to satisfy (1.9). This means that in direction 3 we have a stretching from ϵ to $\epsilon e^{\lambda_3 t}$. Even if two initial conditions are very close, the representative points after a long time will be largely distant. This sensitive dependence on the initial conditions is the indicator of deterministic chaos.

The asymptotic phase space locus (after a long transient) for Lorenz equations (1.6) is well known. That locus attracts all neighboring initial conditions because of the compression of the phase volume peculiar of dissipative systems. It is then an <u>attractor</u>, as the fixed point for N = 1 or the limit cycle for N = 2. But nearby points at a given time must diverge after a long interval, because of $\lambda_3 > 0$. Hence the attractor will never close on itself and is called <u>strange</u>. The unpredictable behavior of paths started from initial conditions specified with an arbitrary (but finite) precision is a fundamental obstacle to long-term nonprobabilistic forecasting. This is why computers can produce only a sufficiently short realization of the path of a dynamic system with one (or more) $\lambda > 0$ (of course, the machine may go on computing, but, for large t, the path is no longer related to the initial conditions).

As for the physical realizability of deterministic chaos, (1.7) indicate that lasers are candidates for large varieties of situations as one changes the atomic species (g, $\gamma_\perp$, $\gamma_\parallel$), the pump rate ($\bar{\Delta}$) or the losses of the e.m. cavity (k) [6]. On the other hand, it is well known that commercial laser sources are good examples of stable dynamical systems. The main reason is that of time scales. As shown by the coefficients of (1.6), a strange attractor is obtained when the damping rates of x,y,z, are comparable.

On the contrary, for noble gas lasers (Ne, Ar), the atomic damping rates are much faster ($\gamma_\perp \sim \gamma_\parallel \sim 10^8 \div 10^9\ s^{-1}$) than the field loss rates ($K \sim 10^6 \div 10^7\ s^{-1}$). Hence the second and third of (1.7) can be solved at equilibrium ($\dot{P} = \dot{\Delta} = 0$) with respect to the rather slow variations of E, and substitution into the first yields a single nonlinear dissipative equation $\dot{E} = f(E)$ which allows only for a fixed point. The procedure is called adiabatic elimination [2] of the fast variables, which are slaved by the slow variable. The latter can be considered as the only relevant dynamical variable (order parameter), as was implicit in the heuristic considerations of Fig.3. We call these lasers Class A.

As we scan most atomic or molecular species radiating in the visible or in the infra-red, it is practically impossible to find one whose rates satisfy Lorenz conditions. Hence we must enhance the topological complexity by adding extra-degrees of freedom in different ways. This will be the subject of the last part of these lectures, the plan of which is as follows: Sec. 2 will give the "numerology", to get acquainted with lasers; Sec. 3 is a derivation of the fundamental Maxwell-Bloch equations and the practical application of them to specific situations; Sec. 4 discusses the onset of chaotic phenomena in quantum optical systems including the generalized multistability and the noise induced jumps among attractors, which give rise to long time phenomena (low frequency tails in the power spectra).

2 Physics of stimulated emission processes

In a rectangular e.m. cavity of volume $V = X_1 X_2 X_3$ the plane wave expansion for the field is

$$E(x,y,z,t) = \sum_{\vec{k}} E(\vec{k},t) \exp i(k_1 x + k_2 y + k_3 z) \tag{2.1}$$

where $k_i = n_i \cdot 2\pi / X_i$ $(i=1,2,3;\ n_i=1,2,\ \ldots)$.

For each set of k_i we have a different field configuration, or mode. The dispersion relation imposes a constraint between frequency ω_k and amplitude of the $\vec{k}$ vector: $\omega_k = c\,k$. In $\vec{k}$ space each mode occupies an elementary volume $\delta^3 k = (2\pi)^3/V$. In a spherical shell of radius k and thickness $\Delta k = 2\pi\,\Delta\nu/c$ there are

$$\Delta M = 2\,\frac{4\pi k^2 \Delta k}{\delta^3 k} = 8\pi\,\frac{\nu^2}{c^3}\,\Delta\nu\,V \tag{2.2}$$

modes. The extra factor 2 accounts for the two possible polarizations for each vector. Let us assume the cavity contains radiators (atoms on the walls or inside).

The distinction between spontaneous and stimulated emission came in 1917 in Einstein's derivation of the black body formula. In the interaction between an atom and a field mode with n photons, the spontaneous rate A and the stimulated emission rate B are relates by

$$A/B = \Delta M \tag{2.3}$$

This can be integrated by representing the field modes as boxes and the excited atom as being linked to all of them as in Fig.4. With the probabilities there indicated, the stimulated emission probability into the mode with n photons is larger than the total spontaneous emission over the empty ΔM modes (all those within the line-width of the atomic emission) when

$$n > \Delta M. \tag{2.4}$$

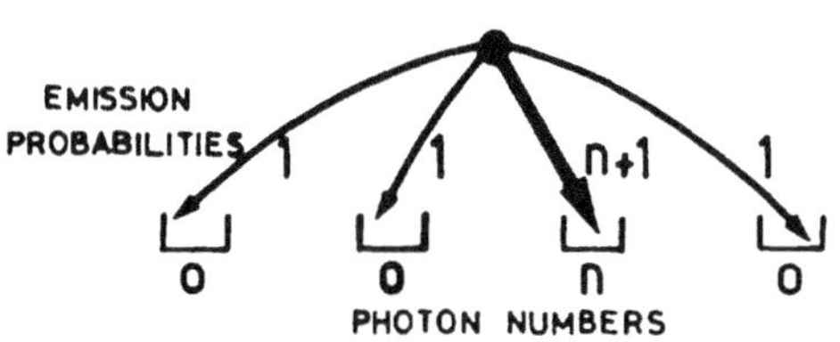

Figure 4 Decay channels of an excited atom into different field modes

Let us call ΔN the atomic population difference between upper and lower states, P the rate of excitation (pump), n the photon number in the laser mode and

$$T_c = \frac{L}{c}\frac{1}{\Theta} \tag{2.5}$$

the decay time of photons in the cavity made by two facing mirrors separated by a length L. T_c is equal to $1/\Theta$ transit times, since the limited mirror transmittivity $\Theta < 1$ increases the number of transits. Condition (2.5) stems from considering photons as particles. It is a necessary, but not sufficient condition. Indeed, if we account for phase matching between forward and backard waves, the cavity is resonant for those frequencies corresponding to the standing wave condition

$$m\frac{\lambda}{2} = L \qquad \text{(m integer)} ,$$

which amounts to a minimum frequency separation

$$\Delta \nu_{m,m+1} = \frac{c}{2L} . \tag{2.6}$$

Only for these resonances is the escape time given by (2.5), otherwise it is much faster (just one transit time, L/c). The rate equations for photons and population inversion are then

$$\begin{aligned} \dot{n} &= Bn\,\Delta N - n/T_c \\ \Delta\dot{N} &= P - Bn\,\Delta N \end{aligned} \tag{2.7}$$

where we have neglected spontaneous processes. Solving them at equilibrium, the first gives

$$\Delta N = \frac{1}{BT_c} \tag{2.8}$$

and the second

$$P = Bn\,\Delta N . \tag{2.9}$$

Combining the two with (2.4), the pump rate must be

$$P > \Delta M / T_c . \tag{2.10}$$

Let us introduce the concept of atomic cross section σ. The stimulated emission rate Bn can be written as $Bn = \sigma\Phi$ where Φ = cn/v is the photon flux and hence $\sigma = BV/c$. When the atomic line is broadened only by spontaneous emission processes, put $\Delta\nu$ = A in (2.2) and have

$$\Delta M = \frac{8\pi}{\lambda^2}\frac{VB}{c}\Delta M$$

Hence

$$\sigma = \frac{\lambda^2}{8\pi} \qquad (\sim 10^{-10}\ \text{cm}^2 \text{ in the visible}) \tag{2.11}$$

If there is an extra broadening $\Delta\nu_a > A$ for collision processes or other decays, σ reduces as

$$\sigma = \frac{\lambda^2}{8\pi}\frac{A}{\Delta\nu_a} . \tag{2.12}$$

This cross section holds for a bound electron, while for a free electron σ is the square of the classical electron radius $r_o \sim 10^{-13}$ cm:

$$\sigma_o = \frac{8\pi}{3} r_o^2 \sim 10^{-25}\ \text{cm}^2 . \tag{2.13}$$

On the other hand, writing the volume as V = SL, condition (2.10) can be rewritten as

$$P > \Theta \frac{S}{\sigma} \Delta\nu , \tag{2.14}$$

which shows that the pump rate required for lasing is proportional to the ratio between the laser beam cross section S and the atomic cross section. We see that a bound cross section is many orders of magnitude bigger than that for free electrons, hence bound transitions, whenever available, should be preferred for the coherent generation of e.m. waves. However, nowadays, using high energy ($>$10 MeV) free electrons in an accelerator one can produce laser action down at wavelengths less than λ = 1 μm (FEL = free electron lasers). Figure 5 summarizes the different interactions and the spectral regions covered by lasers.

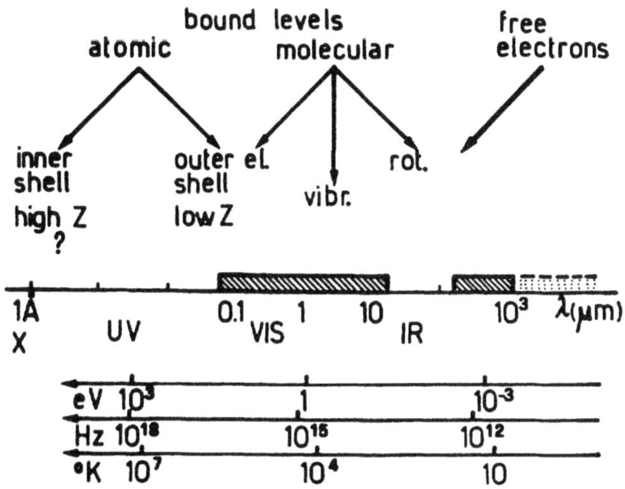

Figure 5 Map of available lasers

Once $n > \Delta M$ is fulfilled, that is, once the privileged mode has enough photons to neglect spontaneous decay channels, we must also take care of the cavity losses, and by (2.8) require that $B\,\Delta N \gtrsim 1/T_c$. This condition is represented in Fig. 6 for two different Δ N. In the first case only one mode is above threshold, hence we have a single monochromatic frequency. In the second case we may have emission at three frequencies. Here we must introduce the fundamental difference: homo - inhomogeneous linewidth. In the former case a monochromatic transition is broadened by circumstances which are equal for all atoms in the cavity (as spontaneous lifetime broadening in a gas, as photon interaction in a solid matrix). All atoms can contribute over the whole linewidth. Hence, once the mode nearest to the peak has been excited, as the associated field "sweeps" the cavity, it will "eat" all atomic contributions, forbidding the other modes from going above threshold.

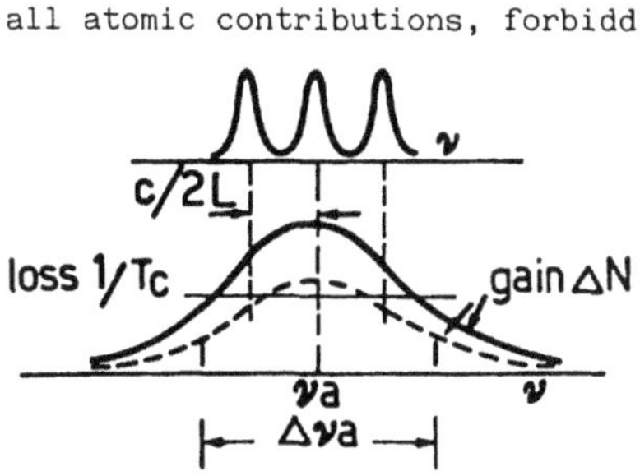

Figure 6 One and three mode laser (dashed and solid gain respectively)

The inhomogeneous line broadening corresponds to different frequency locations of different atoms. This can be due, e.g., to Doppler shift in a gas where thermal agitation gives a distribution of velocities. Another inhomogenneity occurs in a crystal where active ions are exposed to a crystal field which changes from site to site. For an inhomogeneous line, different modes can go above threshold even without a standing wave pattern.

In general, if c/2 L is much smaller than the atomic linewidth there are many independent laser lines without phase relations. If the different laser fields have fixed phase relations (mode locking), they act as the different Fourier components of a train of pulses, each lasting $1/\Delta\nu_a$ and separated by 2L/c (Fig.7).

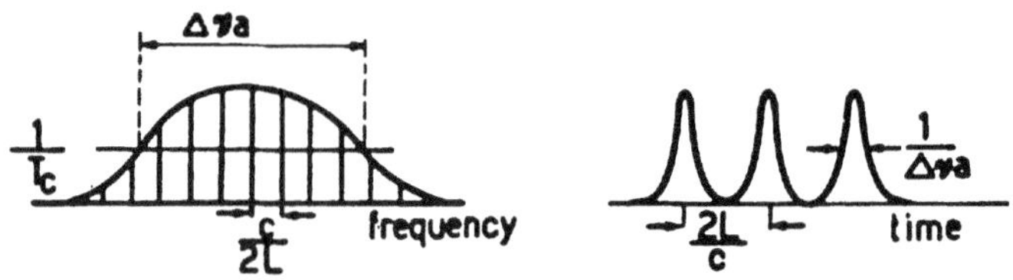

Figure 7
Mode locking operation

By using a Doppler broadened atomic line in a gas (like He-Ne, or A^+ laser) then

$$\Delta\nu_a \sim \frac{\nu}{c}\sqrt{\frac{KT}{M}} \sim 10^9\ Hz$$

hence $t_{pulse} \sim 1$ ns .

Using ions of a transition element embedded in a crystal or glass matrix, such as Cr^{3+} in Al_2O_3 (ruby), or Nd^{3+} in glass, one has large $\Delta\nu_a$. A large $\Delta\nu_a$ is also obtained in the case of complex dye molecules in a liquid solution because of the overlapping among many vibrational and rotational levels. Nowadays it is easy to have $\Delta\nu_a > 10^{13}\ Hz$ and hence $t_{pulse} \sim 100$ fs (1 femtosec = 10^{-15} s).

Eq. (2.12) gives the radiative cross section for a material with spontaneous lifetime $\tau_{sp} = 1/A = 1/\Delta\nu_0$ and total linewidth $\Delta\nu_a$. The total gain per unit length α (cm^{-1}) is given by

$$\alpha = \rho\,\sigma \tag{2.15}$$

where ρ is the density of inverted atoms. The minimal length ℓ of a lasing medium has to be of the order of $1/\alpha$. Once $\alpha\ell \sim 1$, laser action can be obtained even with a small feedback at the end of the rod.

In order to compare different media, we must briefly review the origins of the line broadening $\Delta\nu_a$. For sake of simplicity let us take a classical field, and evaluate the dipole source due to a radiating atom. It will be the expectation value of the dipole operator over the atomic state $|\psi\rangle$

$$\langle d\rangle = e\langle r\rangle = e\langle\psi|r|\psi\rangle . \tag{2.16}$$

Now, let us consider an atomic transition from an s to a p state. It is easily seen from symmetry arguments that $\langle d\rangle = 0$ when $|\psi\rangle$ is the pure s or p state. When interacting with a resonant E field, an atom is driven back and forth in a reversible way as shown in Fig. 8 at a rate (Rabi frequency)

$$\Omega = \frac{\hbar}{\mu E} \tag{2.17}$$

where

$$\mu = e\langle s|r|p\rangle \tag{2.18}$$

is the transition matrix element (taken for simplicity as a real number) and Ω is called the Rabi frequency. The instantaneous wavefunction ψ (t) is given by

$$\psi(t) = a(t)\,|s\rangle + b(t)\,|p\rangle \tag{2.19}$$

where a(t) and b(t) have sinusoidal variations as shown in the figure. Correspondingly the induced dipole evaluated from (2.16)

$$\langle d\rangle = \mu\,(a^*b + ab^*)$$

is shown by the dashed line in Fig.8. It depends on the phase of the wave function, whereas the population inversion $|b|^2 - |a|^2$ is phase independent. Both $|a|$ and $|b|$ return to the equilibrium values $|a| = 1$, $b = 0$ by spontaneous emission processes. On a faster scale, phase destroying processes, as collisions, interrupt the coherent Rabi precession as shown in Fig. 9. The average interruption time is called T_2, and its reciprocal is the homogeneous linewidth $1/T_2 = \Delta\nu_a \gtrsim \Delta\nu_n$.

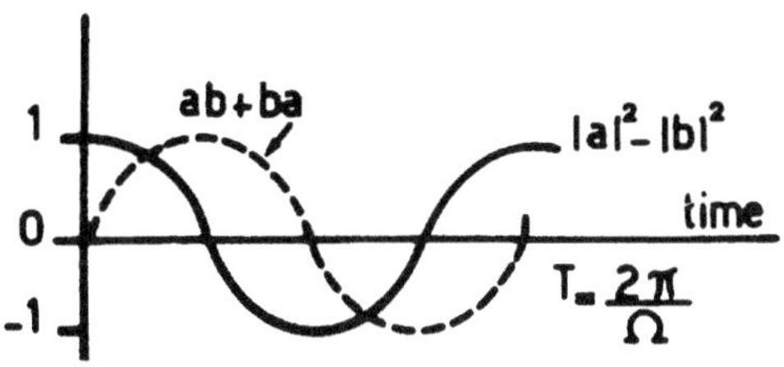

Figure 8 Population difference (solid line) and induced dipole (dashed line)

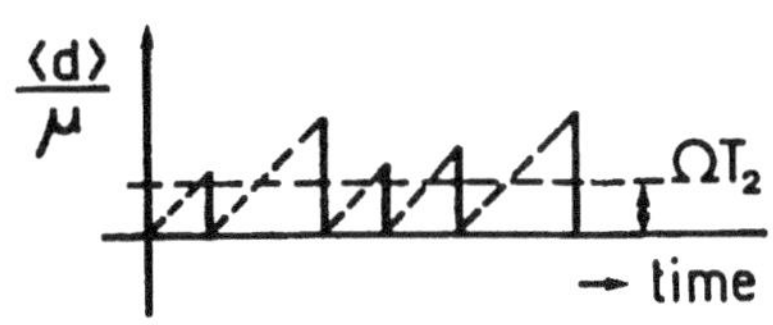

Figure 9 Interruptions due to decay processes, giving a non zero polarization

Since $\langle d\rangle/\mu = \Omega T_2$ then the average dipole is

$$\langle d\rangle = \mu^2 E T_2/\hbar = \mu^2 E/(\hbar\Delta\nu_a)$$

and the polarization (ρ being the atomic density)

$$P = \rho\langle d\rangle = \frac{\rho\mu^2}{\hbar\Delta\nu_a}E \,. \tag{2.20}$$

We have thus shown a simple derivation of the susceptibility χ . If however there is an inhomogeneous broadening (as Doppler broadening in a gas) whose linewidth is bigger than $1/T_2$ (that is, $1/\Delta\nu_{inh} \equiv T_2^* < T_2$), then σ is evaluated with $\Delta\nu_{inh}$. The Table summarizes the parameters of some lasers.

	Ruby Nd	Dye	He-Ne	Ar^+	CO_2	Semiconductor
τ_{sp}(sec)	10^{-3}	10^{-8}	10^{-8}	10^{-8}	0.2	10^{-9}
T_2(sec)	10^{-12}	10^{-13}	10^{-8}	10^{-8}	$(1.1\text{x}10^7/\text{torr})^{-1}$	10^{-14}
T_2^*(sec)	10^{-10}	10^{-13}	10^{-10}	$\frac{1}{3}\,10^{-10}$	$(3.14\text{x}10^8)^{-1}$	--
$\sigma(cm^2)$	$3\text{x}10^{-20}$	$3\text{x}10^{-16}$	10^{-12}	10^{-12}	10^{-15}	10^{-15}
$\rho(cm^{-3})$	10^{19}	10^{15}	10^{9}	10^{10}	10^{14}	10^{18}
$1/\alpha=\ell$ (cm)	10	0.1	10^{3}(*)	10^{2}(*)	10	$3\text{x}10^{-2}$

(*) Good mirrors reduce the cavity length to 10 - 100 cm

3 The Maxwell - Bloch equations

Let us consider an e.m. field and a set of atoms confined in a cavity. The interaction Hamiltonian is

$$H_{f-a} = -\frac{e}{m}\sum_i \vec{A}(\vec{x}_i)\cdot\vec{p}_i$$

$\vec{x}_i$ being the position of atom i, $\vec{A}(\vec{x}_i)$ the value of the vector potential at $\vec{x}_i$, and $\vec{p}_i$ the momentum of the i-th electron.

Expand the field in plane waves

$$\vec{A}(x,t) = \sum_k \left(\frac{\hbar}{2\epsilon_0\omega_k V}\right)^{1/2} \left(a_k e^{-i(\omega_k t - \vec{k}\cdot\vec{x})} + c.c.\right) \tag{3.1}$$

(V = cavity volume; mks system used) and consider a_k and the conjugate a_k^+ as Bose operators:

$$\left[a_k, a_{k'}^+\right] = \delta_{kk'} . \tag{3.2}$$

For simplicity, we skip vector relations and give a scalar theory.

We consider the atoms as two-level atoms, so that the Hilbert space of a single atom is fully described by the identity operator I plus the three Pauli operators $\sigma^{\pm}$, σ_3

$$\left[\sigma^+, \sigma^-\right] = \sigma_3 \quad , \quad \left[\sigma^{\pm}, \sigma_3\right] = \mp 2\sigma^{\pm} . \tag{3.3}$$

It is then a straight-forward matter to obtain the following model Hamiltonian

$$\frac{H}{\hbar} = \sum_k \omega_k a_k^+ a_k + \frac{\omega_0}{2}\sum_i \sigma_{3(i)} + \sum_{i,k} g_k \left(a_k \sigma_i^+ e^{ikx_i} + h.c.\right) \tag{3.4}$$

where

$$g_k = \left(\frac{\omega_k}{\hbar\varepsilon_0 V}\right)^{1/2}\mu \tag{3.4'}$$

$\mu = \langle 1|ex|2\rangle$ being the dipole moment matrix element between the two atomic states coupled by the radiation at frequency ω_k. We introduce the collective operators

$$J^{\pm} = \sum_i \sigma_i^{\pm} e^{\pm ikx_i} \quad , \quad J_3 = \sum_i \sigma_{3(i)} \tag{3.5}$$

Only when the space dependence can be neglected, that is, for three cases

(a) point laser (cavity length $\ll \lambda$),

(b) single mode case (one $k = k_0$ value),

(c) travelling plane wave field (again one k_0 value),

one has closed commutation relation for $J^{\pm}$, J_z, and the Heisenberg equations of motion are obtained:

$$\begin{aligned} \dot{a} &= -i\omega a - igJ^- \\ \dot{J}^- &= -i\omega_0 J^- + igaJ_3 \\ \dot{J}_3 &= ig\left(a^+J^- - aJ^+\right) . \end{aligned} \tag{3.6}$$

They do not form a closed set, because one would also need an equation for the motion of bilinear operators such as Ja^+, etc .. We study here the self-consistent field approximation (SCFA):

$$\langle a J^+ \rangle = \langle a \rangle \langle J^+ \rangle . \tag{3.7}$$

Thus, once the quantum commutation relations have led us to (3.6), a suitable approximation consists in considering them as C-number equations. In such a case, the first is equivalent to Maxwell equations for a single mode in Hamiltonian form, the other two are analogous to Bloch equations in nuclear magnetic resonance, and in vector form can be summarized as

$$\vec{J} = \vec{\Omega} \times \vec{J} \tag{3.8}$$

where $\vec{\Omega} = (g\langle a \rangle, 0, \omega_0)$.

The semiclassical approach (only field or only atoms quantized) is summarized in the following scheme:

Classical current $\xrightarrow{\text{Maxwell}}$ field state $\langle a \rangle \neq 0$ ↖

coherent states

Classical field $\xrightarrow{\text{Bloch}}$ atomic state $\langle J^{\pm} \rangle \neq 0$ ↙

Let $|\psi\rangle = a\,|s\rangle + b\,|p\rangle$ be the wave function of a two level atom. We set up the linear combination

$$\vec{J} = \begin{pmatrix} a^*b + ab^* \\ i(ab^* - a^*b) \\ |b|^2 - |a|^2 \end{pmatrix} \tag{3.9}$$

of the two complex amplitudes a and b which weight the eigenstates $|s\rangle$ and $|p\rangle$ of the free-atom Hamiltonian. By evaluating $\langle\psi|er|\psi\rangle$, it is easy to see that the three components of $\vec{J}$ have the following meaning:

J_1 is proportional to the in-phase polarization (dispersion),

J_2 is proportional to the out-of-phase polarization (absorption or gain),

J_3 is proportional to the population inversion.

Let us now introduce phenomenological lifetimes T_2 for polarization components, T_1 for inversion ($T_2 \leq T_1$), and $1/K = T_c$ for the field. Further, let us introduce, besides the atoms and field, a third partner, the 'pump', which in the absence of field yields a steady-state value of J_3. For simplicity we refer to resonance ($\omega - \omega_0 = 0$), so that we have real field amplitude and one polarization component. Then MB equations become

$$\begin{aligned} \dot{a} &= gS - ka \\ \dot{S} &= 2ga\,\Delta N - \gamma_\perp S \\ \Delta\dot{N} &= -2gaS - \gamma_\parallel(\Delta N - N/2) \end{aligned} \tag{3.10}$$

where g was given in (3.4) and $k = 1/T_c$, $\gamma_\perp = 1/T_2$, $\gamma_\parallel = 1/T_1$ are field and atom loss rates. The normalization is such that $a^2 = n$ is the photon number, and ΔN the number of the inverted atoms in volume V. Also, N is a source term; for an absorbing medium in the ground state (no pumping) it must be changed into $-N_{tot}$.

The order of magnitude, for dilute gas of atoms with allowed transitions in the visible, and for $V = 1\ \text{cm}^3$, is

$$g \sim 10^4 s^{-1}, \qquad \gamma_\perp \sim \gamma_\parallel = \gamma \sim 10^8 s^{-1},$$

and

$$k \sim 10^7 s^{-1} \qquad \text{or } 10^{10} s^{-1}$$

depending on whether the gas is in a laser cavity or is distributed over a length of some centimeters, without mirrors at the end.

The numbers given above suggest two distinct time scales, when $k \ll \gamma$. The fast atomic variables S, Δ N relax toward their equilibrium values, while the slow variable a changes little. Hence a is the order parameter in terms of which the coupled system can be described.

Single Mode Laser

When $\gamma \gg k$, equations for S and ΔN can be solved at steady state in terms of field amplitude, giving

$$S = \frac{gN}{\gamma}\frac{a}{1+x^2} \sim \frac{gN}{\gamma} a(1-x^2) \tag{3.11}$$

where we have introduced the ratio of Rabi frequency to atomic loss rate

$$x = 2ga/\gamma$$

Since g/γ 10^{-4}, it takes $a^2 \sim 10^8$ photons to have $x = 1$. Hence for $n < 10^8$ photons the saturation term in the polarization can be approximated as a cubic correction. Replacing S in the first of eqs. (3.10), the field equation becomes

$$\dot{x} = k\left(C\frac{x}{1+x^2} - x\right) \simeq k\left[(C-1)x - Cx^3\right] \tag{3.12}$$

where C is a cooperation parameter given by

$$C = g^2 N/(\gamma K). \tag{3.13}$$

The steady, stable solution x^2 goes smoothly from zero to a nonzero value as C goes from below to above 1 (Fig. 10).

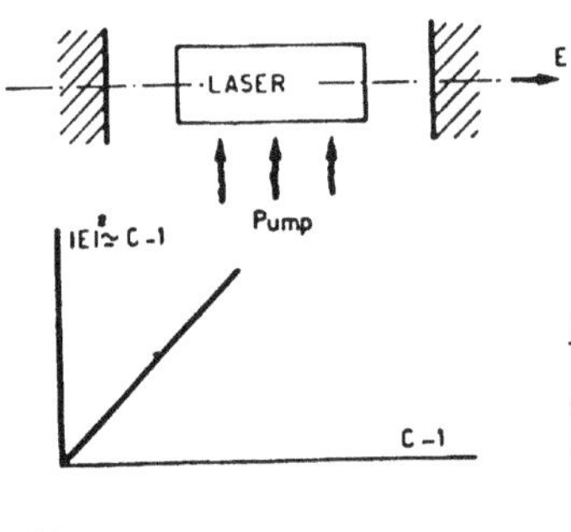

Figure 10 Plot of the square of the order parameter (intensity of the laser field) versus the cooperation number (C-1) is proportional to the difference between gains and losses

Optical Bistability

Take the same configuration (atoms within a cavity), leaving the atoms in the ground state (this changes the sign of S) and injecting an external field $\underline{a}$, which amounts to changing the first of eqs. (3.10) into

$$\dot{a} = gS - k(a - \underline{a}) \tag{3.14}$$

Introducing $y = 2g\underline{a}/\gamma$ gives the steady solution

$$y = x + \frac{C\,x}{1 + x^2} \tag{3.15}$$

Plots of transmitted versus injected field are given in Fig. 11 for increasing atomic densities, that is, for increasing C values. For C = 0 the cavity is tuned so that trasmitted and incident fields are equal. For $C < 1$ the system is in the linear absorption regime. For $C \gg 1$ the system jumps into the saturated regime, where atoms become transparent, and hence there is a return to full transmission. By decreasing the impinging field, one gets a hysteresis cycle peculiar of a first-order phase transition.

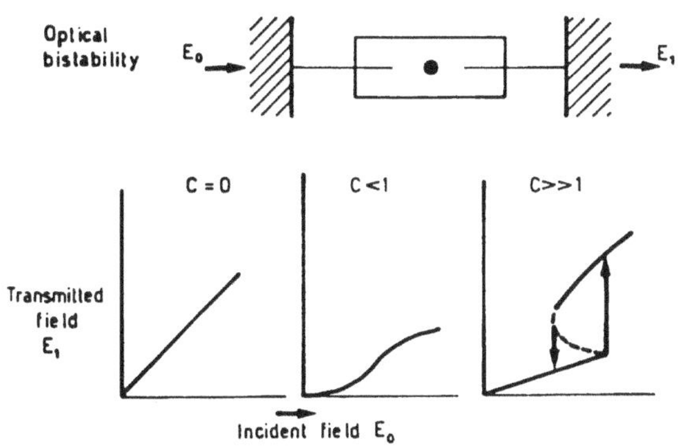

Figure 11 Plots of transmitted field versus impinging field in an optical bistable device for three different values of cooperation number C

4 Onset of chaos, generalized multistability and low frequency spectra in Quantum Optics

As discussed in Sec. 1, even though Maxwell-Bloch equations are like Lorenz equations, in some laser systems (those called class A) time scale considerations reduce the dynamics to one single variable, or order parameter. In some other systems (class B, e.g. ruby, and CO_2 lasers) the population decays slowly, so that the dynamics are described by the two coupled rate equations, which are still insufficient to yield dynamic instabilities. We call class C systems those ones for which the three decay rates for the polarization, population and field are of the same order of magnitude. Thus far, no instabilities have been reported for simple class C systems.

In order to increase by at least 1 the number of the degrees of freedom of class B systems, we must either

i) make the systems non-autonomous, by a time dependent parameter, or

ii) increase the number of lasing modes, or

iii) increase the number of independent gain packets, that is, using an inhomogeneous line.

Point i) was shown in [7]. Point ii) has been shown for two counter-rotating modes in a ring cavity [8]. Point iii) was treated in experimental and theoretical papers [9,10], without, however, a simple analysis in terms of a compact set of equations. In such a case, due to the large number of available packets, distinction in class A and B is immaterial. On the contrary, as pointed out in [11], the beauty of a single mode laser with a homogeneous line is that the relevant instabilities can be modelled with the small number of equations resembling the Lorenz model.

Let us first describe case i). We have chosen a CO_2 laser system because the relaxation time $1/\gamma_{\parallel}$ of the population inversion is much larger ($1/\gamma_{\parallel} = 0.4$ ms) than the memory time $1/\gamma_{\perp}$ of the induced dipole ($1/\gamma_{\perp} = 10^{-8}$ s), thus reducing the single-mode dynamics to the coupling between two degrees of freedom, namely photon population n and molecular population inversion Δ. Introducing within the cavity a time-dependent perturbation by an electro-optical modulator driven by a sinusoisal frequency, we have a non autonomous differential system in n and Δ, amounting to the crucial three degrees of freedom necessary to obtain chaos. As shown later, the relevant range of modulation frequencies is correlated to the value $\sqrt{K\gamma_{\parallel}}$; hence the choice of a CO_2 laser has put the working frequency range in the easily accessible 50-130kHz region.

The coupled field-molecules equations for a single-mode laser, after adiabatic elimination of the polarization, reduce to the following rate equations:

$$\dot{\Delta} = R - Gn - \gamma_{\parallel}\Delta \quad , \quad \dot{n} = 2Gn\Delta - K(t)n, \tag{4.1}$$

where

$$G = \omega\mu^2/\hbar\,\varepsilon_0\gamma_{\perp} V = 0.25 \times 10^{-4} s^{-1} \tag{4.2}$$

is the coupling constant (SI units) including the frequency ω and the dipole matrix element μ of the transition and the collisional broadening rate $\gamma_{\perp}$; R is the pump rate; V, the cavity volume, and

$$K(t) = K_1\ (1+m\cos\Omega t) \tag{4.3}$$

is the cavity damping rate, modulated by the inserted electro-optical device. A linear perturbation analysis around the steady-state volume $\bar{\Delta}$, $\bar{n}$, yields a linearized frequency value $\Omega_0/2\pi \sim 43$kHz for the following typical parameters: $K_1 = 3 \times 10^7 s^{-1}$ (corresponding to a cavity length of 2 m with losses of 20% per pass), and pumping rate R such that $\bar{n} \sim \gamma_{\parallel}/G \sim 10^8$ (corresponding to a dc power output of 50 μW). We select for the two modulation parameters m and Ω as follows. We set K_1 consistently below the maximum damping rate $K_0 = GR/\gamma_{\parallel}$ compatible with the fixed pump rate R. It is easily shown that the linearized eigenfrequency is

$$\Omega_0 \sim (\gamma_{\parallel} k)^{1/2} \tag{4.4}$$

provided we choose $K_1 \sim K_0/2$ and a modulation depth m sufficiently small to have

$$(1/K_1)(dK/dt) = \Omega\, m < \gamma_{\parallel} \tag{4.5}$$

The driving frequency $f = \Omega/2\pi$ is chosen to vary in the region from $\Omega_0/2\pi$ on, that is, from 40 to 150 kHz. As a consequence, $m < \gamma_{\parallel}/\Omega \sim 10^{-2}$. We have explored modulation values between 1% and 5%. The range m = 1% - 5% does not display m dependence; therefore we limit ourselves to giving experimental results at m = 1% for various Ω values.

In Fig.12 we show experimental data in a narrow region between 62.7 and 64.25kHz where various bifurcations occur. This region is limited above and below by wide

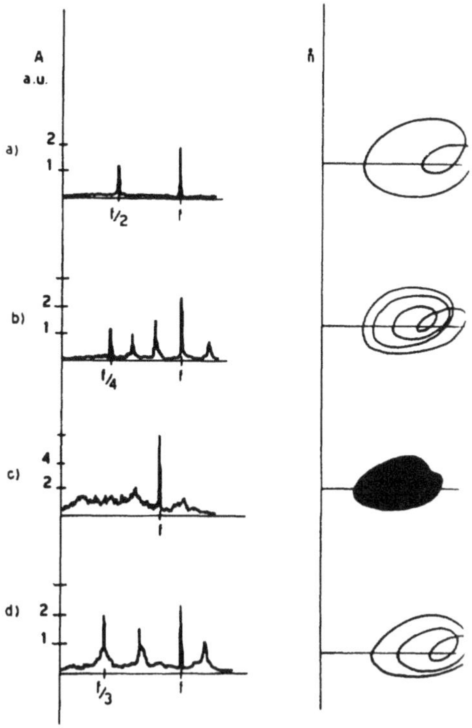

Figure 12 Experimental phase-space portraits ($\dot{n}$ - n) (right side) and the corresponding frequency spectra (left side) for different modulation frequencies f. (a). Period-two limit cycle and corresponding f/2 subharmonic. (b) f/4 subharmonic. (c) strange attractor (the oscilloscope spot could not resolve single windings). The power spectrum is a quasicontinuous one with a small peak at the modulation frequency. (d) f/3 subharmonic

intervals with stable single-period limit cycles. Fig. 12(a) shows the f/2 bifurcation at f = 62.7 kHz; Fig. 12(b) the f/4 case for f = 63.8 kHz; Fig. 12(c) shows the strange attractor and a broad-band spectrum for f = 64.0 kHz; and Fig. 12(d) shows the f/3 case for f = 64.2 kHz.

As for case ii), we show in Fig. 13 the experimental set up allowing for independent detection of forward and backward waves in a ring cavity [8]. The coupling

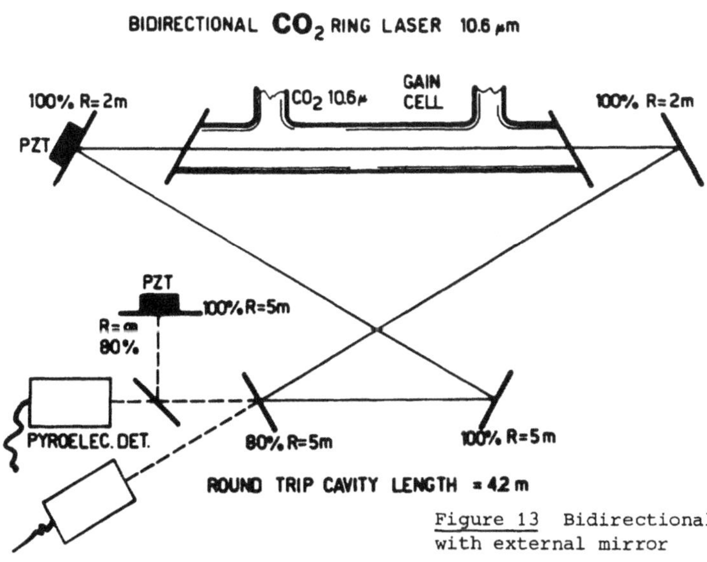

Figure 13 Bidirectional CO_2 ring laser system with external mirror

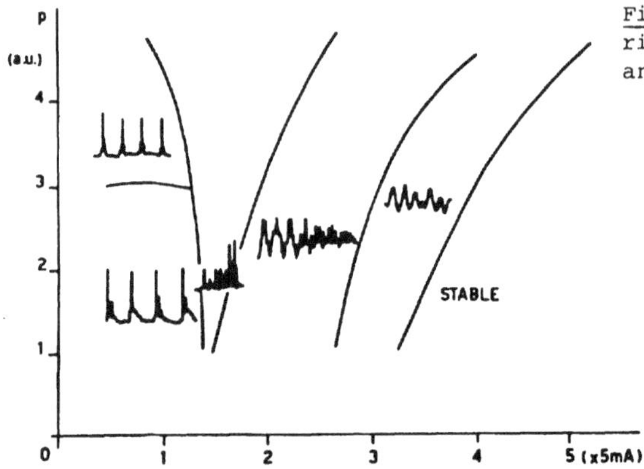

Figure 14 Phase space diagram for CO_2 ring laser. p and I are gas pressure and discharge current

of the two waves via the nonlinear gain medium gives rise to competition, as shown in the phase diagram of Fig.14 (p and I being the pressure and current of the gas discharge in the gain cell: increasing p and I corresponds roughly to increasing the density of excited atoms and hence the gain). Fig.14 shows 5 different regions for increasing current. At low currents we have a regular spiking on both directions with a spike repetition given by $\gamma_{\parallel}^{-1}$. The spikes are synchronous as both fields Q-switch. As we reduce the pressure, each spike develops a damped oscillation with a frequency $\Omega_0 \sim (k\gamma_{\parallel})^{1/2}$ as for eq. (4.4). This oscillation is 180° out of phase for the two fields, showing a coupling of one mode with the other. Increasing the current, we have an irregular transition region followed by a chaotic oscillation where the two modes alternate in a bistable fashion. This chaotic oscillation has a low frequency $\gamma_{\parallel}$ plus a superposed high frequency $(\gamma_{\parallel}\cdot k)^{1/2}$. Measurement of the Grassberger-Procaccia correlation exponent shows a fractal dimension increasing from 2.5 to 5 in regions 2 to 4. Eventually in the fifth region one mode quenches the other and we have stable behavior. The experimental situations of Fig. 14 have been reproduced by a seven equation model, which couple the two complex field amplitudes (E_+ forward and E_- backward: 4 dynamical variables) via the complex second harmonic N_{2k} of the population inversion (2 more variables). The 7th real variable is the uniform component N_o of population inversion N(x). More explicitly let us Fourier expand the population N(x):

$$N_k = \frac{1}{L}\int_0^L dx\, e^{-ikx} N(x). \tag{4.6}$$

The 0th component is that considered in the elementary single wave approach leading to Maxwell equations and to the rate equations (4.3) (where it was called Δ). The higher components act as diffraction grating scattering photons from the forward field into the backward and viceversa, thus providing a coupling. Our numerical solutions of the seven coupled equations for E_+, E_-, N_{2k} (complex) and N_o (real) show that a truncation to N_{2k} is sufficient for a qualitative explanation of the experimental chaos, without having to introduce higher order components of N_k.

The physics of this chaotic ring configuration has implied a reinjection of one field into the other. We have studied specifically the single mode laser with an injected signal [11]. The injected signal has a frequency ω_1 , different from the free running laser frequency ω_L . In a reference frame rotating at the frequency ω_1 , the dynamical system is autonomous, but the field amplitude is made of two relevant variables, one in phase, the other in quadrature with the external signal. Adding the population difference (we refer to a CO_2 laser, hence a class B system) we have the three necessary degrees of freedom to have deterministic chaos. Indeed in Ref. 11 we reported a quasi-periodic route to chaos, which for a limited parameter range is described by a circle map. Further work is in progress on this point.

Generalized multistability

Nonlinear dissipative systems can be many simultaneously coexisting basins of attraction (generalized multistability - GM) [8,11,13]. This situation can be destabilized by changes of the control parameters, merging two independent attractors into a single one via an intermediate region which is only sporadically visited near the transition. The associated dynamics imply a low frequency tail (deterministic diffusion [14, 15]. Viceversa, when the above coexistence is stable, application of external noise may induce jumps between two otherwise disjoint regions of phase space.

For some simple systems we study the two different phenomenologies near their borderline, giving a scaling relation for the escape rates in terms of control parameter and noise amplitude.

Clear evidence of generalized multistability was first shown in an electronic oscillator [13] and then in the modulated laser [7]. Fig. 15 and 16 show the coexistence of two attractors and the corresponding low frequency tails for an electronic nonlinearity obeying the Duffing equation and for the laser with modulated losses, respectively. In both cases, besides the qualitative appearance of different attractors in phase space, there was a low frequency spectral component due to noise induced jumps among different attractors.

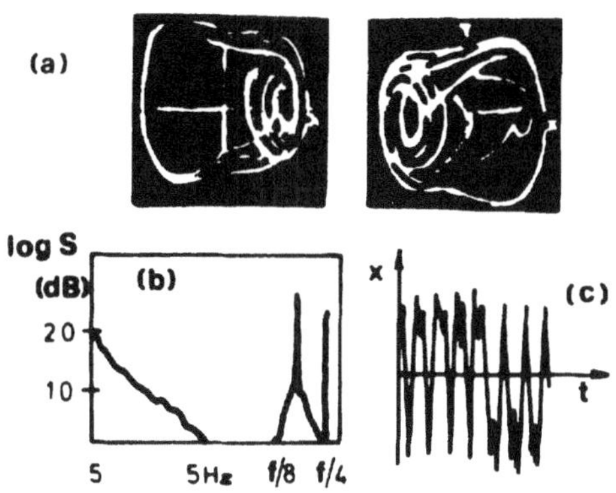

Figure 15 Hopping between two attractors and associated low frequency spectrum in the purely bistable case. (a) Symmetric phase-space plots; (b) log-log spectrum showing the low frequency branch, a broadened f/8 line, and a narrow f/4 line; (c) a sample of the x(t) plot

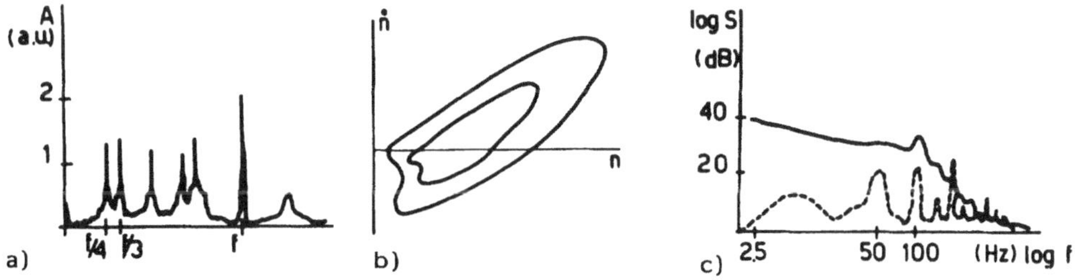

Figure 16 Bistability in a CO_2 laser with loss modulation a-b) coexistence of two attractors (period 3 and 4 respectively); c) comparison between the low frequency cut-off when the two attractors are stable and the low frequency divergence when the attractors are strange

A clear analysis of GM was first given for a cubic recursive map, allowing for two simultaneous attractors [13]. Here I review results of a numerical study [16] of a forced Duffing oscillator with a double well potential, whose equation is

$$\ddot{x} + \gamma \dot{x} - x + 4x^3 = A \cos \omega t. \qquad (4.7)$$

Numerical solution of this equation, for $\gamma = 0.154$ and for different choices of the parameters A, ω, yields a particularly interesting region of the parameter space displayed in Fig. 17. There we have reported the boundaries of the regions of existence of some attractors, corresponding to subharmonics of order from 4 to 7, as denoted by the associated numbers. The meaning of the two lines C and D may be explained with reference to the two valleys of the potential of eq.(4.7). Line C is the borderline above which the motion is no longer confined in one single valley. Below line C, there is a manifold of lines approximately parallel to it, corresponding to a sequence of subharmonic bifurcations, with mutual distances ruled by Feigenbaum universal rate δ. These have been omitted in Fig. 17 for clarity. They have already been observed experimentally [17] (see Fig. 1 of Ref. 17) in an electronic oscillator ruled by (4.7), which is, however, affected by too large a noise to be able to display the other details reported in Fig. 17.

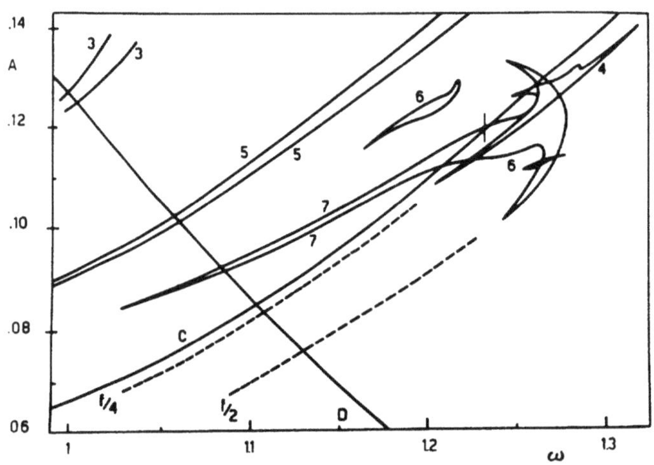

Figure 17 Phase diagram of Duffing equation showing the type of solutions for each pair of driving parameters (A, ω). The borderlines are the frontiers of parameter regions corresponding to ordered motion, and the associated numbers denote the periodicity. Curves C and D show the upper limit (in the amplitude A) for the stability region of solutions confined in one valley. The vertical bar at $\omega = 1.22$ indicates the region where the escape times have been measured

We measure the mean escape time from the period-7 attractor versus the amplitude σ of an applied noise in the region denoted by a vertical bar in Fig. 17. Specifically, at each integration step (which is 10^{-2} the forcing period) we shift $\dot{x}$ by a random number sorted from a rectangular distribution with zero mean and r.m.s. σ.

Fig. 18 shows log T vs. $1/\sigma$ for different values (A_1, A_2, A_3, A_4) of the driving amplitude A at fixed frequency ω. $A_3 \sim A_c$ is the parameter at which the period-7 undergoes crisis [18], that is, representative points "escape" from the attractor after an infinite time in the absence of noise. This is equivalent to the definition of crisis, i.e. A_c is the value for which the attractor of period-7 collides with the unstable period-7 solution.

For $A_4 > A_c$ even a zero noise ($1/\sigma \to \infty$) yields a finite escape time as is evident from the horizontal asymptote of the plot. For $A < A_c$ and as $\sigma \to 0$, $T \to \infty$ faster as A is reduced, as shown by comparison between A_1 and A_2. The finite escape time across a bounded region for $A > A_c$ corresponds to the deterministic (noise-free) diffusion discussed elsewhere. On the other hand, for $A < A_c$, the attractor is structurally stable, therefore, the phase point cannot escape unless we apply external noise. This is the phenomenon of noise induced jumps. Notice that, both for $A < A_c$ and $A > A_c$, around the crisis region the large escape times give low frequency power spectra which are qualitatively similar. The essential difference is that for $A < A_c$ no jumps occur in the absence of noise.

It is apparent from Fig. 17 that one can have the same T for different A's, adjusting the noise amplitude. This may be expressed in terms of a scaling relation as given for other chaotic scenarios [19]. For simplicity we refer to the logistic map

$$x_{n+1} = \mu x_n (1-x_n) \tag{4.8}$$

with μ around the crisis value $\mu_c = 4$. The invariant density for noise-free map is given by [20]

$$\rho_c(x) = \frac{1}{\pi} \frac{1}{\sqrt{x(1-x)}} . \tag{4.9}$$

The divergence of ρ_c at x = 1 is due to the flat behavior of the map around the pre-image point x = 1/2.

We first evaluate the mean escape rate R from the right for $\mu \sim \mu_c$. If either a small noise is added, or μ is set slightly above μ_c, there is no longer an invariant density; however, we can reasonably assume that the distribution of points relaxes to a pseudo-invariant $\rho(x)$ on a time scale much shorter than the

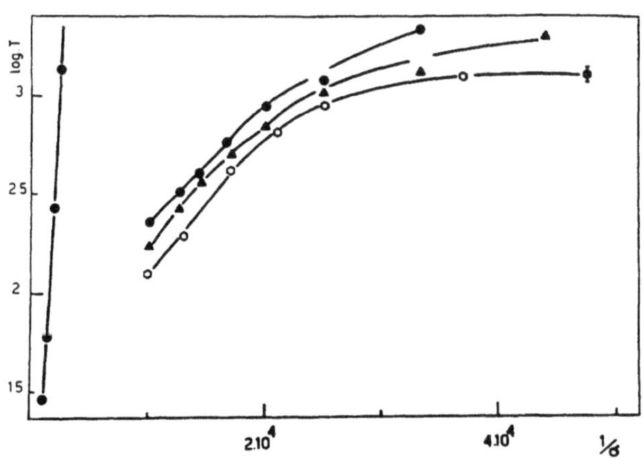

Figure 18 The mean escape time from the period-7 region vs. the inverse of the noise amplitude. All the curves refer to the same frequency, but with different A's. Namely the symbols ⊗ • ▲ o represent respectively: A=.1165, .117280, .117285. The error bar is drawn for only one measurement since it is always the same

mean escape time. In particular, for $\mu < \mu_c$ and without noise, $\rho(x)$ may be a highly singular distribution, however an additional random term smooths out almost all of the peaks over a continuous $\rho(x)$, whose rightmost part is a scaled version of the corresponding part of $\rho_c(x)$, that is $\sim 1/\pi\sqrt{1-\epsilon-x}$, with the asymptote shifted from x = 1 to $x = 1 - \epsilon$ ($\epsilon = 1 - \mu/4$).

When applying a gaussian noise of r.m.s. σ, the escape rate R at the point x is equal to the area of the Gaussian centered at x, outside the segment (0,1). Hence, for $\epsilon \ll 1$, we have the escape rate from the right given by ($y = 1-x-\epsilon$)

$$R(\epsilon,\sigma) = \frac{1}{\pi}\int_0^\infty \frac{dy}{\sqrt{y}} \int_y^\infty \frac{e^{-\frac{(z+\epsilon)^2}{2\sigma^2}}}{\sqrt{2\pi\sigma^2}}\, dz . \tag{4.10}$$

Integrating by parts we have

$$R(\epsilon,\sigma) = \frac{1}{\pi}\sqrt{\frac{\sigma}{2}}\; e^{-\epsilon^2/4\sigma^2} D_{-3/2}(\epsilon/\sigma) , \tag{4.11}$$

where D is the parabolic cylinder function.

It is easily seen that the function $R(\epsilon,\sigma)$ satisfies the general scaling law

$$R(\epsilon,\sigma) = \sigma^\alpha \, \phi(\epsilon, \sigma^\beta) \tag{4.12}$$

with the exponents $\alpha = 1/2$ and $\beta = 1$. Similar scaling laws have already been given for the period doubling cascade and for intermittency [19], but with different exponents, as well as a different physical meaning for R.

At the crisis ($\epsilon = 0$), R scales as the square root of σ, namely,

$$R(0,\sigma) \sim \sigma^{1/2} \tag{4.13}$$

Below the crisis ($\epsilon > 0$) and for $\epsilon \gg \sigma$, we have the asymptotic relation

$$R(\epsilon,\sigma) = \frac{\sqrt{2}}{\pi}\,\frac{\sigma^2}{\epsilon^{3/2}}\, e^{-\epsilon^2/2\sigma^2} . \tag{4.14}$$

Above the crisis ($\epsilon < 0$) we can escape even without noise: indeed the asymptotic expansion of (4.10) in the limit of large ϵ/σ yields $R \sim (2/\pi)|\epsilon|^{1/2}$.
We have shown that the left escape rate L is much smaller (around 10%) than R at $\mu = \mu_c$ and even less important for $\mu < \mu_c$ for noise induced jumps [16].

In conclusion these long escape times provide low frequency tails (as shown in the above reported experiments) which are fully distinct from the high frequency spectra. The latter correspond to the decay of correlations within one attractor, while the former are a marker of noise induced jumps. In Ref. [17] there is a discussion of the conditions under which these tails display a 1/f behavior.

References

1 F.T. Arecchi: "Photocount distributions and field statistics" (Proc.E. Fermi School 1967) in Quantum Optics ed. by R.J. Glauber (Academic Press, 1969)

2 H. Haken: Synergetics (Springer Verlag, 1977)

3 F.T. Arecchi: "Experimental Aspects of Transition Phenomena in Quantum Optics" (Proc. XVIIth International Solvay Conference on Physics, 1978) in Order and Fluctuations in Equilibrium and Nonequilibrium Statistical Mechanics, ed. by G. Nicolis et al. (J. Wiley, 1981)

4 S. Chandrasekhar: Hydrodynamics and Hydromagnetic Stability (Oxford U., London 1961)
5 E. Lorenz: J. Atmos. Scie. 20, 130 (1963)
6 H. Haken: Phys. Lett. 53A, 77 (1975)
7 F.T. Arecchi, R. Meucci, G. Puccioni, J. Tredicce: Phys. Rev. Lett. 49, 1217 (1982)
8 J. Tredicce, G.L. Lippi, F.T. Arecchi and N.B. Abraham: Proc. Royal Soc. (London 1984); N.B. Abraham, F.T. Arecchi, G.L. Lippi and J. Tredicce Infra-red Physics, 1984
9 R.S. Gioggia and N.B. Abraham: Phys. Rev. Lett. 51, 650 (1983)
10 R. Graham, Y. Cho: Opt. Commun. 47, 52 (1983)
11 F.T. Arecchi, G.L. Lippi, G.P. Puccioni and J. Tredicce: in Coherence and Quantum Optics V (Proc. Rochester Conference 1983) ed. L. Mandel and E. Wolf, 1984; Optics Commun. 1984 (to be published)
12 F.T. Arecchi and F. Lisi: Phys. Rev. Lett. 49, 94 (1982); and 50, 1328 (1983)
13 F.T. Arecchi, R. Badii and A. Politi: Phys. Rev. A29, 1006 (1984)
14 T. Geisel and J. Nierwetberg: Phys. Rev. Lett. 48, 7 (1982); S. Grossman and H. Fujisaka Phys. Rev. A26, 504 (1982)
15 Y. Aizawa: Prog. Theor. Phys. 68, 64 (1982)
16 F.T. Arecchi, R. Badii, and A. Politi: Phys. Lett. 103A, 3 (1984)
17 F.T. Arecchi and A. Califano: Phys. Lett. 101A, 443 (1984)
18 C. Grebogi, E. Ott and J.A. Yorke: Phys. Lett. 49, 1507 (1982)
19 J. Crutchfield, M. Nauenberg, J. Rudnick: Phys. Rev. Lett. 46, 933 (1981); R. Shraiman, C.E. Wayne and P.C. Martin Phys. Rev. Lett. 46, 935 (1981); J.G. Hirsch, M. Nauenberg, D.J. Scalapino Phys. Lett. 87A, 391 (1982)
20 P. Collet and J.P. Eckmann: Iterated Maps on the Interval as Dynamical Systems (Birkhäuser, Boston 1980)

Furthermore, we suggest as a general reference to collective effects in quantum optics:
Part I ("Basic theory and laser physics") of Laser Handbook, ed. by F.T. Arecchi and E.O. Schulz-Du Bois (North Holland, 1972)

From Optical Bistability to Chaos

H.M. Nussenzveig

Departamento de Física, Pontifícia Universidade Católica, 22453 Rio de Janeiro, Brasil

1. Introduction

Important advances have recently taken place in our understanding of nonlinear phenomena, in two essentially opposite directions. On the one hand, a wide class of completely integrable systems, associated with solitons, has been found. On the other hand, some universal features have been detected in the approach of a fully deterministic system to chaotic behavior.

This progress arises from the confluence of parallel developments in several different areas, going back to Poincaré's studies of qualitative properties of dynamical systems, inspired by celestial mechanics, and to the investigations on ergodic theory and the foundations of statistical mechanics. Developments in quantum optics and in the theory of critical phenomena, as well as in fluid dynamics, have played a significant role.

Common features have been identified in the behavior of a variety of nonlinear dissipative systems far from thermodynamic equilibrium. When a control parameter is varied, such systems undergo qualitative changes of behavior known as bifurcations, that may be regarded as generalizations of the concept of phase-transitions in equilibrium thermodynamics.

Quantum optics has been a source of contributions to the recent advances in both directions. Self-induced transparency is a beautiful illustration of soliton-type effects. The approach to chaos through a sequence of bifurcations is illustrated by optical bistability.

The same theoretical model may be employed in both cases. This model, which was developed in connection with the theory of the laser, shows a remarkable richness of behavior, ranging from complete integrability to chaos, depending on the excitation conditions and the choice of parameters. The laser threshold is analogous to a second-order phase-transition far from equilibrium. The bifurcations found in optical bistability are generalizations of first-order phase-transitions.

The advantage of quantum optics over fluid dynamics, where similar phenomena are found, is that the underlying theoretical model is much simpler and better understood, while still remaining in good agreement with experiment.

In the present lectures, we are concerned with optical bistability. Besides its theoretical interest, this area has great potential promise for practical application, e. g., in terms of optical transistors, very fast miniature optical switches, optical memory elements. Our emphasis is on theory.

A *multistable* element (bistability is a particular case) is one that may yield a variety of different outputs for a given input. Some of them may actually be unstable and others metastable, so that a detailed discussion of stability is essential. Typically, the operating point will depend on the previous history, i.e., we have hysteresis.

Elements of this type have long been known in electronics, resulting from a combination of nonlinear characteristics and feedback. We consider only intrinsic bistability, arising from purely optical elements, excluding hybrid systems, where optical components are combined with electronic circuits to obtain bistable behavior. Furthermore, we restrict our consideration to *passive* systems, where no pumping is present, thus excluding the study of laser instabilities.

We assume a general background in quantum optics [1]. More detailed discussions of optical bistability, including the experimental aspects, and more complete lists of references, may be found in several reviews and proceedings [2 -5].

We begin with a formulation and discussion of the theoretical model. This is followed by a heuristic explanation of the two types of optical bistability: absorptive and dispersive. We concentrate mainly on the simplest model, the semiclassical theory of absorptive optical bistability. After a brief review of the quantum treatment, we discuss side-mode instabilities and self-pulsing. Finally, we deal with multistability and the approach to chaos.

2. The ring cavity

The simplest optical system for the analysis of optical bistability, as was recognized by BONIFACIO and LUGIATO [6], is a ring cavity (Fig.1). The advantage is that one can assume unidirectional propagation, whereas in a Fabry-Perot one must deal with both forward and backward travelling waves. However, the treatment has been extended to the Fabry-Perot cavity as well.

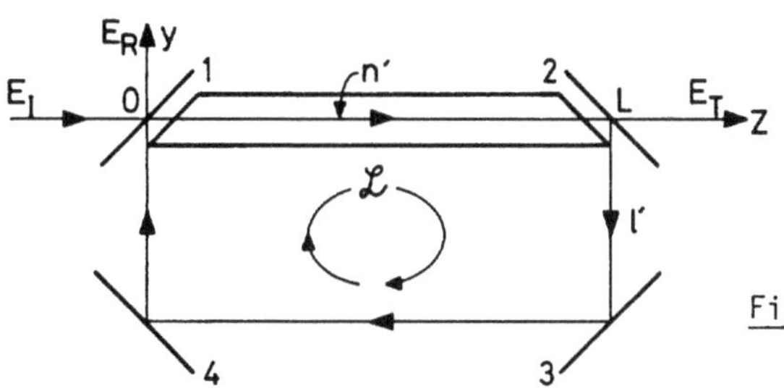

Fig. 1 The ring cavity

Mirrors 1 and 2 have reflectivity R, whereas 3 and 4 are totally reflecting. The active medium, with complex refractive index n', is inserted in a sample of length L between mirrors 1 and 2. The round-trip length of the cavity is $\mathcal{L} = 2(L + \ell')$.

We take perpendicular polarization, so that, with the geometry of Fig. 1, the electric field along the z axis is of the form

$$\vec{E} = E(z,t) \exp[i(kz-\omega t)]\vec{i} \equiv E_x\vec{i} \tag{2.1}$$

where $k = \omega/c$ and we are making the following assumptions: (i) *Plane-wave approximation*, i.e., we neglect transverse beam effects, such as a Gaussian profile; (ii) *Slowly-varying amplitude and phase approximation* [1]:

$$|\partial E/\partial z| \ll k|E|, \quad |\partial E/\partial t| \ll \omega|E|. \tag{2.2}$$

We employ the notations E_I and E_T for the incident and transmitted fields, respectively,

$$E_I(t) = E(0-,t), \qquad E_T(t) = E(L+,t), \tag{2.3}$$

so that $E(z,t)$ in (2.1) will denote the field inside the sample; thus, $E(0,t) \equiv E(0+,t)$; $E(L,t) \equiv E(L-,t)$.

If the active medium is treated in the linear approximation,

$$E(z,t) = E_0 \exp[\,i(n'-1)kz\,], \tag{2.4}$$

and (2.2) requires

$$|n' - 1| \ll 1, \tag{2.5}$$

corresponding to a rarefied active medium. The presence of a host (background) medium of refractive index $\tilde{n}$ can be taken into account by substituting $k \to \tilde{n}k$ in (2.1).

Let the complex reflection and transmission amplitudes for perpendicular polarization on mirrors 1 and 2 be given by $\sqrt{R}\exp(i\psi_R)$ and $\sqrt{T}\exp(i\psi_T)$, respectively (assuming the same angle of incidence), where

$$R + T = 1. \tag{2.6}$$

Let $\exp(i\psi'_R)$ be the complex reflection amplitude on the totally reflecting mirrors 3 and 4.

The boundary condition at mirror 2 is

$$E_T(t) = \sqrt{T}\exp(i\psi_T)E(L,t). \tag{2.7}$$

The boundary condition at mirror 1 is obtained by taking into account, besides the transmission of the incident field, the feedback term due to reflection of the field at the end of the sample on all four mirrors:

$$E(0,t) = \sqrt{T}\exp(i\psi_T)E_I(t) + R\exp(i\psi)\exp(ik\mathcal{L})E\left(L,t-\frac{\ell}{c}\right), \tag{2.8}$$

where $\psi = 2(\psi_R+\psi'_R)$ is the total phase-shift due to reflection and

$$\ell = \mathcal{L} - L \tag{2.9}$$

is the length of the free-space path outside of the sample. The phase factor $\exp(ik\mathcal{L})$ arises from the rapidly-varying factor in (2.1) followed around the ring.

Since the factor multiplying the field at the end of the sample in the feedback term arises only from free-space propagation, the boundary condition remains valid when one takes into account the nonlinearity of the active medium.

In order to simplify the boundary conditions, let us redefine the incident and transmitted amplitudes, as well as the round-trip phase factor, so as to incorporate transmission and reflection phase-shifts:

$$\exp(i\psi_T)E_I(t) \to E_I(t),$$

$$\exp(-i\psi_T)E_T(t) \to E_T(t),$$

$$k\mathcal{L} + \psi \to k\mathcal{L}. \tag{2.10}$$

With these redefinitions, boundary conditions (2.7) and (2.8) become

$$\text{(I)}\quad E_T(t) = \sqrt{T}\,E(L,t), \tag{2.11}$$

$$\text{(II)}\quad E(0,t) = \sqrt{T}\,E_I(t) + R\exp(ik\mathcal{L})\,E\left(L,t-\frac{\ell}{c}\right). \tag{2.12}$$

2.1. Solution for a Linear Medium

In the linear approximation, $E(z,t)$ is given by (2.4), where E_0 is a constant and

$$n' = n + i\frac{\alpha}{2k}, \tag{2.13}$$

n being the real refractive index and α the linear absorption coefficient (by (2.4), the intensity is damped as $\exp(-\alpha z)$).

Substituting (2.4) in (2.11) and (2.12), we find

$$E_0 = \frac{E_T}{\sqrt{T}} \exp[-i(n'-1)kL] = \sqrt{T}\, E_I/\{1-R\exp[ikL + i(n'-1)kL]\}, \tag{2.14}$$

so that the output is related to the input by

$$\frac{E_T}{E_I} = \frac{T\exp[i(n'-1)kL]}{1-R\exp[ikL+i(n'-1)kL]}. \tag{2.15}$$

This is identical to the relationship between transmitted and incident amplitudes in a Fabry-Perot interferometer [7], where T and R correspond to the transmissivity and reflectivity of the plate surfaces; the phase-shift δ associated with double traversal of the Fabry-Perot plate corresponds in (2.15) to the round-trip phase-shift

$$\delta = k[L+(n-1)L] = k(\ell + nL) \equiv kL', \tag{2.16}$$

where L' is the optical path associated with a round trip (cf.(2.9); in (2.10), ψ gives rise to a negligible contribution, of the order of a wavelength).

By (2.13) and (2.16), the last term in the denominator of (2.15) can be rewritten as $R\exp(-\frac{1}{2}\alpha L + ikL')$, so that we may expand (2.15) as a geometric series:

$$E_T/E_I = T\,[1 + R\exp(-\tfrac{1}{2}\alpha L + i\delta) + R^2\exp(-\alpha L + 2i\delta) + \ldots]\exp[i(n'-1)kL]. \tag{2.17}$$

This has an immediate physical interpretation in terms of multiple internal reflections, which provide an alternative method of solution [7].

The intensity I is proportional to $|E|^2$, so that, defining the overall transmissivity $\mathcal{T}$ as

$$\mathcal{T} \equiv I_T/I_I = |E_T|^2/|E_I|^2, \tag{2.18}$$

we find, from (2.15) and (2.16),

$$\mathcal{T} = \frac{T^2\exp(-\alpha L)}{|1-R\exp(-\frac{1}{2}\alpha L + i\delta)|^2} = \frac{T^2}{\left[\exp(\frac{\alpha L}{2}) - R\right]^2 + 4R\exp(\frac{\alpha L}{2})\sin^2\frac{\delta}{2}}, \tag{2.19}$$

which corresponds to the result for a lossy Fabry-Perot.

As is well known [7], there are *transmission resonances* when the successive terms of (2.17) interfere constructively,

$$\delta = 2m\pi \Rightarrow k = k_m = 2m\pi/L' \quad (m = 1,2,3,\ldots), \tag{2.20}$$

corresponding to a maximum of the transmissivity,

$$\mathcal{T}_{max} = T^2 \Big/ \left[\exp\left(\frac{\alpha L}{2}\right) - R\right]^2. \tag{2.21}$$

and there are transmission *antiresonances* when the multiple-beam interference is destructive,

$$\delta = (2m-1)\pi, \qquad m = 1,2,3,\ldots \tag{2.22}$$

corresponding to a minimum of the transmissivity,

$$\mathcal{T}_{min} = T^2 \Big/ \left[\exp\left(\frac{\alpha L}{2}\right) + R\right]^2 . \tag{2.23}$$

For a thick absorber, $\alpha L \gg 1$, there is very little contrast between maxima and minima: we have $\mathcal{T} \approx T^2 \exp(-\alpha L) \ll 1$. We will usually consider only thin absorbers, with $\alpha L \ll 1$. In order to obtain very sharp resonance fringes, we also need very small transmissivity of mirrors 1 and 2, $T \ll 1$. Under these conditions, taking into account (2.6), we see from (2.21) and (2.23) that

$$\mathcal{T}_{max} \approx 1, \quad \mathcal{T}_{min} \approx T^2/4 \ll 1. \tag{2.24}$$

Furthermore, from (2.14), still for $\alpha L \ll 1$, $T \ll 1$,

$$E_0 \approx \frac{E_T}{\sqrt{T}} \approx \frac{E_I}{\sqrt{T}} \gg E_I \text{ (at resonance)},$$

$$E_0 \approx \frac{E_T}{\sqrt{T}} \approx \frac{\sqrt{T}}{2} E_I \ll E_I \text{ (at antiresonance)}, \tag{2.25}$$

so that the large overall transmissivity at resonance, in spite of having $T \ll 1$, is due to the buildup of a large field within the sample, resulting from constructive interference.

Taking into account (2.21), we can rewrite (2.19) as

$$\mathcal{T}/\mathcal{T}_{max} = \left\{1 + F \sin^2 \frac{\delta}{2}\right\}^{-1} , \tag{2.26}$$

where

$$F = 4R \exp\left(-\frac{\alpha L}{2}\right) \Big/ \left[1 - R \exp\left(-\frac{\alpha L}{2}\right)\right]^2 \approx 4R/T^2 \quad \text{if} \quad \alpha L \ll 1. \tag{2.27}$$

The halfwidth ε of a resonance fringe, as a function of δ, is obtained by setting $\mathcal{T} = \mathcal{T}_{max}/2$ for $\delta = 2m\pi \pm \varepsilon/2$, yielding

$$\varepsilon \approx 4/\sqrt{F}, \tag{2.28}$$

and the cavity *finesse* $\mathcal{F}$ is defined as

$$\mathcal{F} = \frac{\text{fringe separation}}{\text{halfwidth}} = \frac{2\pi}{\varepsilon} \approx \frac{\pi}{2}\sqrt{F}. \tag{2.29}$$

For $\alpha L \ll 1$, $T \ll 1$, we have

$$F \approx 4/T^2 \approx \mathcal{T}_{max}/\mathcal{T}_{min} \gg 1, \quad \mathcal{F} \gg 1, \tag{2.30}$$

so that the fringes are very sharp, with strong contrast between maxima and minima.

2.2. Natural modes

The natural modes of oscillation (free oscillations) of the ring cavity are associate with values of k such that $E_0 \neq 0$ with $E_I = 0$ (no input) in (2.14), i.e., such that

$$1 - R \exp\{ik[\mathcal{L} + (n'-1)L]\} = 0, \tag{2.31}$$

so that they are poles of the transfer function (2.15).

Let

$$n' = n + iK \tag{2.32}$$

where $K \geq 0$ (cf. (2.13)). In order to solve (2.31), we neglect dispersion, assuming that n and K do not depend on k within the range of interest. Substituting (2.16) and (2.32) in (2.31), we find the roots

$$i\tilde{k}_m(L' + iKL) = -\ln R + 2im\pi, \tag{2.33}$$

where m is an integer, which will usually be large at optical frequencies (we take $m > 0$).

Setting

$$\eta = KL/L', \tag{2.34}$$

we find from (2.33)

$$\tilde{k}_m = k'_m - i\beta_m, \tag{2.35}$$

$$k'_m = k_m [1 + (\eta \ln R/2m\pi)]/(1 + \eta^2), \tag{2.36}$$

$$\beta_m = k_m [\eta - (\ln R/2m\pi)]/(1 + \eta^2), \tag{2.37}$$

where k_m are the resonance wave numbers (2.20).

The natural modes are therefore associated with complex poles in the lower half of the k-plane ($\beta_m > 0$). The corresponding frequencies are $\tilde{\omega}_m = c\tilde{k}_m$, so that

$$\exp(-i\tilde{\omega}_m t) = \exp(-ick'_m t) \exp(-c\beta_m t), \tag{2.38}$$

showing that, as expected, the modes are exponentially damped in time. If (2.1) remained valid, they would also blow up exponentially as $z \to \infty$, but the excitation conditions must be taken into account. For further discussion of natural modes and their physical interpretation, see [8].

If $K \ll 1$, $T \ll 1$, the poles (2.35) are located very close to the positions of transmission resonances on the real axis. For $K = 0$ (no absorption), we find

$$\tilde{k}_m = k_m + i\ln R/L' \quad (K=0), \tag{2.39}$$

and the damping factor (2.38) becomes

$$\exp(-c\beta_m t) = \exp(ct\ln R/L') = R^{t/t_c}, \tag{2.40}$$

where (cf. (2.16))

$$t_c = \frac{L'}{c} = \frac{\ell}{c} + \frac{L}{v}, \quad v = c/n, \tag{2.41}$$

is the *time of flight* around the cavity.

The physical interpretation of (2.40) is obvious: for each round trip, the reflectivity of mirrors 1 and 2 damps the field amplitude by R, and t/t_c is the number of round trips up to time t. Rewriting (2.40) as $\exp(-t/\tau_c)$, we see that the *mode lifetime* τ_c is given by

$$\tau_c = 1/c\beta_m = L'/c|\ln R| \quad (K=0) \tag{2.42}$$

which approaches infinity as $T \to 0$ ($R \to 1$).

In the limiting case R=1, the damping arises entirely from $K > 0$. Still for $K \ll 1$, (2.37) now yields

$$\exp(-c\beta_m t) \approx \exp(-k_m KLt/t_c), \tag{2.43}$$

so that the amplitude is damped by $\exp(-k_m KL)$ per round trip, as it should be.

3. The Maxwell-Bloch equations

As a model for the nonlinear interaction between the field and the active medium in the sample, we employ the Maxwell-Bloch equations [1,9-12]. The coupling takes place through the medium polarization $\vec{P}$, which acts as a source term for the field:

$$\frac{\partial^2 \vec{E}}{\partial z^2} - \frac{1}{c^2} \frac{\partial^2 \vec{E}}{\partial t^2} = \frac{4\pi}{c^2} \frac{\partial^2 \vec{P}}{\partial t^2} , \tag{3.1}$$

where, as before, we assume no host medium (sample in vacuum apart from the active medium).

The polarization is written as

$$\vec{P} = P(z,t) \exp[i(kz - \omega t)] \vec{i} \equiv P_x \vec{i}, \tag{3.2}$$

where, as in (2.1), it is understood that we must take the real part (care must be taken when nonlinear terms arise). The slowly-varying amplitude and phase approximation (2.2) is also assumed to hold with $E(z,t)$ replaced by $P(z,t)$.

Substituting (2.1) and (3.2) in (3.1), and neglecting terms that are small according to the slowly-varying amplitude and phase approximation, we find that Maxwell's equation (3.1) reduces to

$$\frac{\partial E}{\partial z} + \frac{1}{c}\frac{\partial E}{\partial t} = 2\pi i k P. \tag{3.3}$$

The active medium consists of two-level atoms, with upper level a (above) and lower level b (below), with energy separation $\hbar\omega_0$ (the zero level of energy is taken halfway between a and b). We make the following assumptions:

(i) The atoms are coupled only through their interaction with the common radiation field.

(ii) Only *homogeneous broadening* will be taken into account (all atoms see the same field at a given location; there are no Doppler shifts).

(iii) The *dipole approximation* holds. In particular, since we have electric dipole transitions induced by linearly polarized light, the levels must satisfy the selection rule

$$\Delta m = 0. \tag{3.4}$$

The single-atom Hamiltonian is of the form

$$H = H_A - exE_x, \tag{3.5}$$

where H_A is the unperturbed atomic Hamiltonian and E_x is given by (2.1). The transition dipole moment is

$$\mu = ex_{ab} = e(\psi_a, x\psi_b), \tag{3.6}$$

which can be taken as real, without loss of generality, by employing real bound-state wave functions ψ_a and ψ_b.

A general pure state

$$\psi = a(t)\psi_a + b(t)\psi_b \qquad (|a|^2 + |b|^2 = 1) \tag{3.7}$$

may be represented by the density operator

$$\rho(t) = |\psi\rangle\langle\psi| = \begin{pmatrix} a \\ b \end{pmatrix} (a^* b^*) = \begin{pmatrix} |a|^2 & ab^* \\ a^*b & |b|^2 \end{pmatrix} , \tag{3.8}$$

which obeys the equation of motion

$$i\hbar \frac{\partial\rho}{\partial t} = [H, \rho] , \quad H = \begin{pmatrix} \hbar\omega_0/2 & -\mu E_x \\ -\mu E_x & -\hbar\omega_0/2 \end{pmatrix} . \tag{3.9}$$

This is equivalent to

$$\dot{\rho}_{aa} = -i \frac{\mu}{\hbar} E_x(\rho_{ab} - \rho_{ba}) = -\dot{\rho}_{bb}, \tag{3.10}$$

$$\dot{\rho}_{ab} = -i\omega_0\rho_{ab} - i \frac{\mu}{\hbar} E_x(\rho_{aa} - \rho_{bb}) = \dot{\rho}^*_{ba}. \tag{3.11}$$

Let N be the number of (identical) two-level atoms per unit volume. The polarization P_x (cf. (3.2)) is given by

$$P_x = N\langle ex\rangle = N\mathrm{Tr}(\rho ex) = N\mathrm{Tr}\left[\rho\begin{pmatrix} 0 & \mu \\ \mu & 0 \end{pmatrix}\right] = N\mu(\rho_{ab} + \rho_{ba}) = 2N\mu\mathrm{Re}\rho_{ab}. \tag{3.12}$$

Comparing (3.12) with (3.2), we can identify

$$P(z,t) = 2N\mu\rho_{ab} \exp[-i(kz - \omega t)] . \tag{3.13}$$

We set

$$\rho_{ab} \equiv p \exp[i(kz - \omega t)] \tag{3.14}$$

where p, which is dimensionless (like ρ), is called the *complex dimensionless polarization* of the two-level atom. From (3.13) and (3.14),

$$P(z,t) = 2N\mu p. \tag{3.15}$$

Let

$$w = \frac{1}{2}(\rho_{aa} - \rho_{bb}) = \frac{1}{2}(|a|^2 - |b|^2), \tag{3.16}$$

so that Nw is one-half of the *population inversion density*. The equilibrium value at zero temperature is

$$w_{eq} = -1/2. \tag{3.17}$$

By (3.10), (3.14) and (3.16),

$$\dot{w} = -i \frac{\mu}{\hbar} E_x \{p \exp[i(kz - \omega t)] - p^* \exp[-i(kz - \omega t)]\}. \tag{3.18}$$

Similarly, by (3.11), (3.14) and (3.16),

$$\dot{p} = i\Delta\omega p - \frac{2i\mu}{\hbar} E_x\, w \exp[-i(kz - \omega t)] , \tag{3.19}$$

where

$$\Delta\omega = \omega - \omega_0 \tag{3.20}$$

is the *atomic detuning* (detuning from atomic resonance).

In the nonlinear equations (3.18) and (3.19), we must take the real representation for E_x, i.e., by (2.1),

$$E_x = \frac{1}{2} E \exp[i(kz-\omega t)] + \frac{1}{2} E^* \exp[-i(kz-\omega t)] , \tag{3.21}$$

where, by (3.3) and (3.15), E verifies Maxwell's equation

$$\text{(a)} \quad \frac{\partial E}{\partial z} + \frac{1}{c}\frac{\partial E}{\partial t} = 4\pi i N\mu k p . \tag{3.22}$$

In (3.18) and (3.19), we need an additional approximation:

(iv) *Rotating wave approximation*: We drop terms varying like $\exp(\pm 2i\omega t)$ in quantities that were assumed to be slowly-varying. This is needed for consistency with the slowly-varying amplitude and phase approximation (other such terms have been dropped before).

Substituting (3.21) in (3.18) and (3.19) and making this approximation, we get

$$\dot{p} = i\Delta\omega p - i\frac{\mu}{\hbar} E w , \tag{3.23}$$

$$\dot{w} = -\frac{i\mu}{2\hbar}(pE^* - p^* E) . \tag{3.24}$$

In order to convert (3.23) and (3.24) into the Bloch equations, we must take into account, besides these "coherent" contributions to $\dot{p}$ and $\dot{w}$, the "incoherent" contributions, that lead to damping terms [1,9-12].

For w, the damping (e.g., by spontaneous emission) implies a relaxation to the equilibrium value (3.17), so that

$$(\dot{w})_{\text{damping}} = -\gamma_\parallel (w - w_{eq}) = -\gamma_\parallel \left(w + \frac{1}{2}\right) , \tag{3.25}$$

where the longitudinal width $\gamma_\parallel = 1/T_1$ (T_1 is the energy-damping lifetime). For the dimensionless polarization,

$$(\dot{p})_{\text{damping}} = -\gamma_\perp p , \tag{3.26}$$

where the transverse width $\gamma_\perp = 1/T_2'$, and T_2' is the *homogeneous* contribution to the dipole moment damping lifetime. It includes, besides the effects of T_1, also the effects due to phase interruptions (collisional broadening).

Adding the incoherent contributions (3.25), (3.26) to (3.23) and (3.24), we get

$$\text{(b)} \quad \frac{\partial p}{\partial t} = (i\Delta\omega - \gamma_\perp)p - \frac{i\mu}{\hbar} w E , \tag{3.27}$$

$$\text{(c)} \quad \frac{\partial w}{\partial t} = -\gamma_\parallel \left(w + \frac{1}{2}\right) + \frac{i\mu}{2\hbar}(p^* E - pE^*) . \tag{3.28}$$

Equations (3.22), (3.27) and (3.28), which we will refer to as (a), (b), (c), respectively, are the Maxwell-Bloch equations, in the form employed by IKEDA [13]. Together with the boundary conditions (2.11) and (2.12), which we refer to as (I) and (II), respectively, they are the basic equations of the model.

In their papers on optical bistability, BONIFACIO and LUGIATO [5,6] employ a different notation, which we will also adopt in this connection, so that we now show how to go over to their notation.

What they call "polarization" is

$$\mathcal{P} = -iNp, \quad \text{where} \tag{3.29}$$

$$N = \mathcal{N}V \tag{3.30}$$

is the total number of two-level atoms, V being the volume of the sample. According to (3.15), the slowly-varying envelope of the actual polarization that appears in

Maxwell's equations is related to $\mathcal{P}$ by

$$P = 2i\mu\mathcal{P}/V. \tag{3.31}$$

Their variable corresponding to the population inversion is (cf. (3.16))

$$D = -Nw = \frac{N}{2}(\rho_{bb} - \rho_{aa}) = \frac{1}{2}(N_b - N_a), \tag{3.32}$$

which is half the population difference between the *lower* and the *upper* level; thus, by (3.17), its equilibrium value is

$$D_{eq} = N/2 \tag{3.33}$$

They also define the dimensionless atomic detuning parameter (cf. (3.20))

$$\Delta = -\Delta\omega/\gamma_\perp = (\omega_0 - \omega)/\gamma_\perp. \tag{3.34}$$

Substituting (3.29) and (3.30) in (3.22), we get their version of Maxwell's equation (a):

$$\text{(A)}\quad \frac{\partial E}{\partial t} + c\frac{\partial E}{\partial z} = -g\mathcal{P}, \tag{3.35}$$

where the "coupling constant" g is given by

$$g = 4\pi\omega\mu/V. \tag{3.36}$$

Multiplying (3.27) by -iN and taking into account (3.29), (3.32) and (3.34), we get

$$\text{(B)}\quad \frac{\partial \mathcal{P}}{\partial t} = \frac{\mu}{\hbar}ED - \gamma_\perp(1 + i\Delta)\mathcal{P}. \tag{3.37}$$

Similarly, multiplying (3.28) by -N, we get

$$\text{(C)}\quad \frac{\partial D}{\partial t} = -\frac{\mu}{2\hbar}(E\mathcal{P}^* + E^*\mathcal{P}) - \gamma_\parallel\left(D - \frac{N}{2}\right). \tag{3.38}$$

Equations (3.35), (3.37), (3.38), respectively referred to as (A), (B), (C), are the Bonifacio-Lugiato version of the Maxwell-Bloch equations. The boundary conditions (I) and (II) remain unchanged. Their definition of the linear absorption coefficient (cf. (2.13)) also differs from ours: $\alpha_{BL} = \alpha/2$, but we will stick to our definition, which is the conventional one.

4. The linear approximation

Because of the product terms wE, p^*E, pE^* in (3.27) and (3.28), the Maxwell-Bloch equations (a), (b), (c) are a coupled set of *nonlinear* partial differential equations. For sufficiently weak fields E, the active medium behaves like a linear medium. Under these conditions, the polarization is proportional to the field, and we can solve the equations in this linear approximation, neglecting nonlinear terms in E.

Since $p \propto E$, the last term in (3.28) is $\mathcal{O}(|E|^2)$ and may be neglected, so that setting (cf. (3.17))

$$w = -\frac{1}{2} + \zeta = w_{eq} + \zeta, \tag{4.1}$$

we get from (c)

$$\frac{\partial\zeta}{\partial t} = -\gamma_\parallel\zeta \qquad \zeta = \zeta_0\exp(-\gamma_\parallel t), \tag{4.2}$$

i.e., the inversion relaxes to zero with lifetime T_1. Similarly, from (b), we would get a transient expressing the relaxation of p with lifetime T_2'. We are only interested in the steady-state behavior ($t \gg T_1, T_2'$), so that we set $\partial/\partial t = 0$ in (a), (b), (c). Therefore,

$$w = w_{eq} = -1/2. \tag{4.3}$$

Substituting this in (3.27), with $\partial p/\partial t = 0$, we get

$$p = -\frac{\mu E}{2\hbar(\Delta\omega + i\gamma_\perp)}. \tag{4.4}$$

By (3.15), this confirms the linear dependence of P on E in this approximation:

$$P = \frac{N\mu^2 E}{\hbar(\omega_0 - \omega - i\gamma_\perp)} = \chi'(\omega)E, \tag{4.5}$$

where $\chi'(\omega)$ is the complex linear susceptibility of the active medium.

Replacing (4.4) in (3.22) with $\partial E/\partial t = 0$, and taking into account (4.5), we get

$$dE/dz = 2\pi i\chi' kE. \tag{4.6}$$

Recalling the Maxwell relationship

$$4\pi\chi'(\omega) = \varepsilon'(\omega) - 1 \approx n'^2(\omega) - 1 \approx 2[n'(\omega) - 1], \tag{4.7}$$

where $\varepsilon'(\omega)$ is the complex dielectric constant, and the last form follows from the dilute character of the medium (cf. (2.5)), (4.6) becomes

$$dE/dz = i(n' - 1)kE \;\{\; E = E_0 \exp[i(n' - 1)kz], \tag{4.8}$$

so that we recover the solution (2.4) for a linear medium.

From (4.5) and (4.7), we get the linear complex refractive index of our medium of two-level atoms:

$$n'(\omega) = 1 + \frac{2\pi N\mu^2}{\hbar(\omega_0 - \omega - i\gamma_\perp)}. \tag{4.9}$$

This has a single complex pole at $\omega = \omega_0 - i\gamma_\perp$ in the lower half-plane, in agreement with causality. However, it does not satisfy the crossing symmetry relation $n'(-\omega) = n'^*(\omega)$; this is due to our having used the rotating-wave approximation.

From (4.9) and (2.13), we get the real refractive index,

$$n(\omega) = 1 - \frac{2\pi N\mu^2}{\hbar}\frac{\Delta\omega}{[(\Delta\omega)^2 + \gamma_\perp^2]}, \tag{4.10}$$

as well as the linear absorption coefficient,

$$\alpha(\omega) = \frac{4\pi N\mu^2 k}{\hbar}\frac{\gamma_\perp}{[(\Delta\omega)^2 + \gamma_\perp^2]}. \tag{4.11}$$

The well-known Kramers-Heisenberg dispersion formula gives

$$n'^2(\omega) - 1 = 4\pi N\frac{e^2}{m}\sum_j \frac{f_{jo}}{(\omega_{jo}^2 - \omega^2 - i\omega\gamma_{jo})}, \tag{4.12}$$

where the oscillator strength f_{jo} is given by

$$f_{jo} = \frac{2m}{\hbar e^2} \omega_{jo} |ex_{jo}|^2 . \tag{4.13}$$

Specializing (4.12) to a dilute medium of two-level atoms, sufficiently close to resonance, so that $\omega_{jo}^2 - \omega^2 \approx 2\omega_{jo}(\omega_{jo} - \omega)$ and $\omega\gamma_{jo} \approx \omega_{jo}\gamma_{jo}$, it is readily verified that the result coincides with (4.9) (where $\gamma_\perp \equiv \gamma_{jo}/2$).

4.1. Transition to the nonlinear regime

Why does a medium of two-level atoms, in the linear regime, behave like a collection of harmonic oscillators? As is well known [1, 12], one may associate with the Maxwell-Bloch equations a sort of classical analogue, representing the motion of a spherical pendulum ("Bloch pendulum"), with "horizontal" components given by the absorptive and dispersive parts of the transition dipole moment and "vertical" component determined by the degree of excitation of the two-level atoms.

The *Rabi frequency*

$$\Omega = [(\mu E/\hbar)^2 + (\Delta\omega)^2]^{1/2} \tag{4.14}$$

measures the rate at which the electric field E coherently induces transitions between the levels b and a. At resonance, (4.14) becomes

$$\Omega = \mu E/\hbar \qquad (\Delta\omega = 0). \tag{4.15}$$

In a time t_R such that $\Omega t_R = \pi$, the Bloch pendulum (if it were subject only to E, neglecting damping effects) would flip over completely. The *Rabi flipping time* t_R is therefore given by

$$t_R = \pi/\Omega = \pi\hbar/\mu E . \tag{4.16}$$

The damping effects bring the pendulum back to its stable equilibrium position (atomic ground state). If this happens before it can swing far away from equilibrium, i.e., if T_1, T_2' are $\ll t_R$, the pendulum will perform small damped oscillations around equilibrium, with $|\zeta| \ll 1$ in (4.1). We are then in the linear regime, and we see why one gets harmonic oscillator-like behavior.

If E becomes strong in (4.16) so that

$$t_R \lesssim T_1, T_2' , \quad \text{i.e.,} \quad \Omega \gtrsim \gamma_\parallel, \gamma_\perp , \tag{4.17}$$

the Bloch pendulum will swing around, tending to equalize the atomic populations in the upper and lower states, which is a typical *saturation* effect. This characterizes the *nonlinear regime*, so that (4.17) is the condition for the appearence of nonlinearity. For typical values of the atomic lifetimes, this condition is readily attained with focussed beams from lasers of moderate power.

5. Optical bistability

We now give a heuristic discussion of optical bistability in the ring cavity system. As in (2.18), we denote by I_I and I_T the incident and transmitted intensities, respectively, and by $I_0 \propto |E_0|^2$ (cf. (2.4)) a typical intensity within the sample. We assume that $T \ll 1$.

Depending on the dominant nonlinear feature of the active medium that is responsible for bistability, one usually considers as extreme cases (purely) *absorptive* optical bistability and (purely) *dispersive* optical bistability. In practice, one usually has a mixture of dispersive and absorptive features, but it is useful to discuss the extremes.

5.1. Absorptive optical bistability

The nonlinear feature of the active medium that is relevant in this case is the *saturation of the absorption* at a typical value I_S (determined by (4.17)) of the intensity I_0: for $I_0 \gtrsim I_S$, the medium gets bleached and the absorption decreases significantly.

Let us start with the cavity tuned at a transmission resonance, so that, in the absence of absorption, one would have (cf. (2.24)) $I_T = I_I$, and let us gradually increase I_I from zero. In the linear regime, for small I_I, the linear absorption destroys the total transmission, yielding instead (cf. (2.21)), if $\alpha L \gtrsim 1$,

$$I_T \sim T^2 I_I \exp(-\alpha L) \ll I_I. \tag{5.1}$$

This is valid for sample intensities below saturation, i.e. (cf. (2.14)), for

$$I_0 \sim TI_I < I_S. \tag{5.2}$$

On the other hand, for high enough I_I, the intensity within the sample can go above I_S,

$$I_0 > I_S. \tag{5.3}$$

Since the absorption is bleached, this restores the total transmission,

$$I_T = I_I. \tag{5.4}$$

As we saw in (2.25), this is possible, even though $T \ll 1$, because the sample intensity builds up to a large value,

$$I_0 \sim I_T/T = I_I/T, \tag{5.5}$$

so that, by (5.3),

$$I_I/T > I_S. \tag{5.6}$$

Since $T \ll 1$, there is a *range* of values of I_I, within the nonlinear regime, in which *both* inequalities (5.2) and (5.6) are simultaneously satisfied (e.g., this happens for $I_I \sim I_S$). Within this range, one can have *bistability*, with a *low-transmission branch* (5.1) and a *high-transmission branch* (5.4).

Actually, instead of taking $\alpha L \gtrsim 1$, we will consider the case of weak linear absorption,

$$\alpha L \ll 1, \tag{5.7}$$

so that, according to (2.21), we must replace (5.1) by

$$\mathcal{T}_{max} = (I_T/I_I)_{max} = T^2/[\exp(\alpha L/2) - R]^2 \approx T^2/(1 - R + \alpha L/2)^2,$$

which, taking into account (2.6), may be rewritten as (at resonance)

$$I_T \approx I_I/(2C + 1)^2, \tag{5.8}$$

where

$$C = \frac{\alpha L}{4T} \tag{5.9}$$

is the relevant parameter in this situation.

In order to retain the low-transmission branch ($I_T \ll I_I$) characteristic of bistability, we need, according to (5.8),

$$C \gg 1. \tag{5.10}$$

The parameter C, which plays a fundamental role in the theory, is called the *cooperation parameter*, for reasons to be discussed further on. Note that

C ~ absorption damping/transmission damping.

Thus, (5.10) means that the field gets reflected many times around the cavity before linear absorption gets a chance to act.

5.2. Dispersive optical bistability

We now start with the cavity tuned away from resonance, still for $T \ll 1$, $\alpha L \ll 1$, so that (cf. (2.28), (2.30), (2.24))

$$I_T/I_I \sim T_{min} \approx T^2/4 \ll 1 \tag{5.11}$$

in the linear regime.

Furthermore (cf. (3.34)), we take

$$\Delta \gg 1, \tag{5.12}$$

so that (cf. (4.10), (4.11)) absorptive effects in the complex refractive index are much smaller than dispersive ones. The relevant nonlinear property of the medium now is the *intensity dependence of the real refractive index* n. The simplest example is a Kerr medium, for which

$$n(I) = n_0 + n_2 |E|^2 , \tag{5.13}$$

where n_0 is the linear refractive index and n_2 is the quadratic coefficient of the nonlinear refractive index.

For a high enough sample intensity, say $I_0 \sim I_r$, the *change* in n is sufficient to restore the transmission resonance (cf. (2.16)). The result (5.11) is then valid only for small sample intensity,

$$I_0 \sim TI_I < I_r \Rightarrow I_T \ll I_I. \tag{5.14}$$

On the other hand, if $I_0 > I_r$, the transmission resonance is restored, so that, as in (5.5),

$$I_T \approx I_I \quad \text{for} \quad I_0 = \frac{I_T}{T} \approx \frac{I_I}{T} > I_r. \tag{5.15}$$

Again, for $T \ll 1$, there is a range of values of I_I, in the nonlinear regime, such that $TI_I < I_r$, but $I_I/T > I_r$ ($I_I = I_r$ belongs to this range). Within this range, both responses, $I_T \ll I_I$ and $I_T \approx I_I$, are possible, and we have *dispersive* optical bistability, again with a low-transmission branch and a high-transmission branch.

5.3. Remarks on history and applications

The first proposals concerning absorptive optical bistability were made by SZÖKE, DANEU, GOLDHAR and KURNIT [14], who outlined an argument similar to the above one, and by SEIDEL [15].

The first experimental observation of optical bistability was made by GIBBS, McCALL and VENKATESAN [16]. They employed a cell filled with sodium vapor at 10^{-4} - 10^{-5} Torr pressure, placed within a Fabry-Perot interferometer with mirror transmissivity T ~ 10% (the cooperation parameter was C ~ 14), and a 50 mW dye laser tuned within the sodium D lines. They were looking for absorptive bistability, but they actually found dispersive bistability, which is easier to observe (Na vapor has a large nonlinear coefficient of refractive index); what they observed was a hysteresis cycle of I_T as a function of I_I. Since that time, optical bistability has been observed in many systems [4,5], including semiconductors like GaAs and InSb, which are important for applications.

Possible applications for optically bistable devices are as optical circuit elements (optical transistors, limiters, switches) in light pulse communications, and as optical memory elements in computing. Technical developments are in the direction of miniaturization (sizes of a few microns), fast switching times (of the order of picoseconds) and small switching energies (fractions of picojoules). For up-to-date achievements in this rapidly developing field, one must refer to the current literature.

6. Semiclassical theory: steady state

The semiclassical theory of optical bistability [6] is based on the Maxwell-Bloch equations, which we will employ in the Bonifacio-Lugiato version (A), (B), (C), together with the boundary conditions (I) and (II). We now look for the steady-state solution of the Maxwell-Bloch equations, by setting $\partial/\partial t = 0$.

6.1. Steady-state solution of the Maxwell-Bloch equations

In the steady state, E, P and D become functions only of z. In order not to encumber the notation, we do not indicate the z dependence explicity.

From (b), with $\partial P/\partial t = 0$, we get

$$P = \mu ED/[\hbar\gamma_\perp(1 + i\Delta)] . \tag{6.1}$$

Substituting this result in (C), with $\partial D/\partial t = 0$, and solving for D, we get

$$D = \frac{N/2}{1 + \dfrac{|F|^2}{1 + \Delta^2}} , \tag{6.2}$$

where we have introduced the *dimensionless electric field* variable

$$F = \mu E/\hbar\sqrt{\gamma_\| \gamma_\perp} = \Omega/\sqrt{\gamma_\| \gamma_\perp} , \tag{6.3}$$

and Ω is the *resonant Rabi frequency* (4.15) associated with E.

The linear regime corresponds to neglecting the intensity-dependence (dependence on $|F|^2$) in (6.2). This holds for $|F|^2 \ll 1 + \Delta^2$, leading to the replacement of (6.2) by (3.33) (cf. also (4.3)).

Substituting (6.2) back into (6.1), we get

$$P = \mu NE/\left\{ 2\hbar\gamma_\perp(1 + i\Delta) \left[1 + \frac{|F|^2}{1 + \Delta^2}\right]\right\} , \tag{6.4}$$

which can be reexpressed in terms of (6.3) as

$$P = \sqrt{\frac{\gamma_\|}{\gamma_\perp}} \, \frac{N}{2} \, \frac{(1 - i\Delta)F}{(1 + \Delta^2 + |F|^2)} . \tag{6.5}$$

Equations (6.2) and (6.5) express the *stationary solution of Bloch's equations* (B) and (C) in terms of the local electric field variable $F(z)$. Maxwell's equation (A), with $\partial E/\partial t = 0$, yields

$$dE/dz = -\, gP/c,$$

which, in terms of F, reads

$$dF/dz = -\,(\mu g/\hbar c\,\sqrt{\gamma_{\|}\gamma_{\perp}})\,P(z). \qquad (6.6)$$

Substituting $P(z)$ by (6.5), we obtain

$$\frac{dF}{dz} = -\frac{\mu g N}{2\hbar c\gamma_{\perp}}\,\frac{[(1-i\Delta)F(z)]}{[1+\Delta^2+|F(z)|^2]} = -\frac{2\pi\mu^2 Nk}{\hbar\gamma_{\perp}}\,\frac{[(1-i\Delta)F(z)]}{[1+\Delta^2+|F(z)|^2]}, \qquad (6.7)$$

where we have employed (3.30) and (3.36). This nonlinear differential equation must be solved subject to the boundary conditions (I) and (II) in order to obtain the stationary solution for the field. However, before doing this, it is worth while to pause for a physical discussion of the results so far obtained.

6.2. Nonlinear refractive index of the Bloch medium

From (3.30), (3.31) and (6.4), we obtain the relationship between P and E for a Bloch medium,

$$P = \frac{i\mu^2 N}{\hbar\gamma_{\perp}}\,\frac{(1-i\Delta)}{(1+\Delta^2+|F|^2)}\,E. \qquad (6.8)$$

The coefficient of E in (6.8) is the nonlinear susceptibility $\chi'(\omega)$ of the Bloch medium, and (4.7) now yields the complex nonlinear refractive index of the Bloch medium,

$$n' - 1 = 2\pi\chi' = \frac{2\pi N\mu^2(i+\Delta)}{\hbar\gamma_{\perp}(1+\Delta^2+|F|^2)}, \qquad (6.9)$$

which now depends both on the frequency ω (through Δ) and on the intensity I (through $|F|^2$).

Comparing (6.7) with (6.9), we see that (6.7) may be rewritten as

$$\frac{dF}{dz} = i(n'-1)kF, \qquad (6.10)$$

which is a generalization of (4.8), with

$$n' = n'(\omega, I). \qquad (6.11)$$

In the limit as $I \to 0$ ($|F| \to 0$), (6.9) goes over into the linear complex refractive index (4.9). Separating the real and imaginary parts of (6.9), we get (cf. (2.13)) the nonlinear real index and absorption coefficient,

$$n(\omega,I) - 1 = \frac{2\pi N\mu^2\Delta}{\hbar\gamma_{\perp}(1+\Delta^2+|F|^2)}, \qquad (6.12)$$

$$\alpha(\omega,I) = \frac{4\pi N\mu^2 k}{\hbar\gamma_{\perp}(1+\Delta^2+|F|^2)}, \qquad (6.13)$$

which reduce to (4.10) and (4.11), respectively, for $I \to 0$.

From these formulas we see that the Bloch medium shows both types of nonlinear features needed in connection with absorptive or dispersive optical bistability. Indeed, from (6.13) at resonance ($\Delta = 0$), we get

$$\alpha(\omega_0, I) = \frac{\alpha_0}{1 + I/I_s}, \tag{6.14}$$

where α_0 is the resonant linear absorption coefficient,

$$\alpha_0 = \alpha(\Delta = 0, I = 0) = 4\pi N\mu^2 k_0/\hbar\gamma_\perp \quad \text{and} \tag{6.15}$$

$$I/I_s = |F|^2, \quad \text{i.e.,} \quad |F_s|^2 = 1, \tag{6.16}$$

where I_s is the *saturation intensity* of (5.2). Note that $|F_s| = 1$ is equivalent to (4.17). For $I \gg I_s$, we see that α decreases like $1/I$ (by (6.2), the same happens with D).

Similarly, (6.12) gives the intensity-dependence of the real refractive index. For

$$|F|^2 \ll 1 + \Delta^2 \tag{6.17}$$

we can expand (6.12) in powers of $|F|^2/(1 + \Delta^2)$, keeping only the linear term. By (6.3), we get

$$n - 1 \approx n_0 - 1 + n_2|E|^2, \tag{6.18}$$

where n_0 is the linear index, given by (4.10), and

$$n_2 = -\frac{2\pi N\mu^4\Delta}{\hbar^3\gamma_\parallel\gamma_\perp^2(1 + \Delta^2)^2} = \frac{-\alpha(\omega, 0)\mu^2\Delta}{2k\hbar^2\gamma_\parallel\gamma_\perp(1 + \Delta^2)}. \tag{6.19}$$

Thus, in this approximation, the medium of two-level atoms behaves like a Kerr medium (cf. (5.13)), with quadratic coefficient of refractive index n_2 given by (6.19).

6.2. State equation for absorptive bistability

We consider *purely absorptive* bistability, for which the atomic detuning vanishes (cf. (3.34)):

$$\Delta = 0. \tag{6.20}$$

According to (4.10) and (4.11), this has the following implications for the linear optical constants:

$$n = 1, \qquad \alpha = \alpha_0 \tag{6.21}$$

where α_0 is given by (6.15). Thus, when the field frequency is resonant with the atomic transition frequency, the medium becomes purely absorptive, its real refractive index going through unity.

From (6.7) and (6.21), we get

$$\frac{dF}{dz} = -\frac{\alpha_0}{2}\frac{F(z)}{[1 + |F(z)|^2]}. \tag{6.22}$$

Since the coefficients are real, we can take the dimensionless field F (as well as E, by (6.3)) to be real, without loss of generality. In fact, we see directly from (A), (B), (C) that, for $\Delta = 0$, we can take *both E and P as real* (D is always real).

The same is true for boundary conditions (I) and (II), where E_T and E_I can be taken as real. In fact, by (2.16) and (2.20),

$$n = 1 \Rightarrow \delta = kL = 2m\pi \Rightarrow e^{ikL} = 1, \tag{6.23}$$

where we must remember that, in absorptive bistability, the cavity is tuned to a transmission resonance. Thus, (2.12) also has real coefficients.

The fact that E, P, D are all real in purely absorptive bistability is a great theoretical simplification, because we have to deal only with three real variables. In the general case of dispersive bistability, E and P are both complex, so that one must deal with the equivalent of five real variables. For this reason, it is much simpler to deal with purely absorptive bistability, and we devote most of our attention to this case.

Taking into account (6.23) and the time-independence of the fields, the boundary conditions (I) and (II) become

$$\text{(I)} \quad E_T = \sqrt{T}\, E(L),$$

$$\text{(II)} \quad E(0) = \sqrt{T}\, E_I + RE(L).$$

Multiplying both sides by $\mu/\hbar\sqrt{\gamma_{\parallel}\gamma_{\perp}}$ to express the results in terms of (6.3), we get

$$\text{(I)} \quad F(L) = x, \tag{6.24}$$

$$\text{(II)} \quad F(0) = Ty + Rx = x + T(y - x), \tag{6.25}$$

where we have introduced the dimensionless input and output electric field amplitudes

$$x = \mu E_T/(\hbar\sqrt{\gamma_{\parallel}\gamma_{\perp} T}) \qquad \text{(output)}, \tag{6.26}$$

$$y = \mu E_I/(\hbar\sqrt{\gamma_{\parallel}\gamma_{\perp} T}) \qquad \text{(input)}. \tag{6.27}$$

Note the extra factor $1/\sqrt{T}$ in these definitions as compared with (6.3). It will turn out that in the bistable situation y(x) remains single-valued, whereas x(y) is not, which is the reason why they are defined in this way.

Since F(z) is real, (6.22) may be rewritten as (omitting the index in α_0 for simplicity)

$$\frac{dF}{F} + F dF = -\frac{\alpha}{2} dz.$$

Integrating both sides between z = 0 and z = L, we get

$$\ln[F(L)/F(0)] + \frac{1}{2}[F^2(L) - F^2(0)] = -\frac{1}{2}\alpha L. \tag{6.28}$$

Substituting (6.24) and (6.25) into (6.28), we finally get the connection between output and input in the steady state [6]:

$$\ln\left[1 + T\left(\frac{y}{x} - 1\right)\right] + \frac{x^2}{2}\left\{\left[1 + T\left(\frac{y}{x} - 1\right)\right]^2 - 1\right\} = \frac{1}{2}\alpha L. \tag{6.29}$$

By analogy with equilibrium thermodynamics, we will call this the *exact state equation* for purely absorptive optical bistability.

6.3. Mean field theory

The state equation (6.29) depends only on two parameters, T and αL. For $T \ll 1$ and $\left|\frac{y}{x} - 1\right|$ not too large, so that

$$\left|T\left(\frac{y}{x} - 1\right)\right| \ll 1, \tag{6.30}$$

we can expand (6.29) in powers of this small parameter, yielding, to first order,

$$T\left(\frac{y}{x} - 1\right)(1 + x^2) \approx \frac{\alpha L}{2}.$$

Remembering the definition (5.9), this becomes

$$y = x + \frac{2Cx}{1 + x^2}, \tag{6.31}$$

which now depends only on the cooperation parameter.

The simplified state equation (6.31) follows *exactly* from (6.29) if we perform the *double limit*

$$T \to 0, \quad \alpha L \to 0, \quad \alpha L/4T = C \quad \text{(constant)}, \tag{6.32}$$

but (6.31) remains a good approximation to (6.29) as long as (6.30) is satisfied, even if αL does not approach zero simultaneously with T.

This is confirmed by plotting the graph of (6.31) and comparing it with the exact curves obtained from (6.29) for various values of T and αL (or C). It is found [6] that (6.31) is still a reasonably good approximation, for T = 0.1, even if $\alpha L = 4 (C = 10)$.

In the limit as $\alpha L \to 0$, we may integrate (6.22) as follows:

$$F(L) - F(0) = -\frac{\alpha}{2}\int_0^L \frac{F(z)}{1 + F^2(z)}\,dz \approx -\frac{\alpha L}{2}\,\frac{F(L)}{1 + F^2(L)}, \tag{6.33}$$

which amounts to recognizing that F(z) almost does not vary across the sample and may be replaced by a *mean field*

$$\langle F(z) \rangle \approx F(L), \tag{6.34}$$

to first order in αL. Indeed, since the variation of F(z) across the sample is $\mathcal{O}(\alpha L)$, replacing L by any other value of z in (6.34) and in the r.h.s. of (6.33) only leads to a second-order correction, $\mathcal{O}(\alpha^2 L^2)$. Substituting (6.24) and (6.25) into (6.33), we are led again to (6.31), which may therefore be interpreted as the *mean field theory* state equation.

A more systematic derivation of mean field theory directly from the Maxwell-Bloch equations may given. For this purpose, we assume that, in the double limit (6.32), we may replace in the equations E(z, t), $\mathcal{P}$(z, t) and D(z, t) by the *mean field variables*

$$E(t) = \langle E(z,t) \rangle, \quad \mathcal{P}(t) = \langle \mathcal{P}(z,t) \rangle, \quad D(t) = \langle D(z,t) \rangle, \tag{6.35}$$

where

$$f(t) \equiv \langle f(z,t) \rangle \equiv \frac{1}{L}\int_0^L f(z,t)\,dz. \tag{6.36}$$

We also assume that, in this limit,

$$< ED > = < E > < D >, \quad < E P > = < E > < P >, \tag{6.37}$$

i.e., that correlations between the field and atomic variables can be neglected.

Remembering that $E = E^*$, $P = P^*$ for $\Delta = 0$, and taking the space averages of (3.37) and (3.38), we then get, for the mean field variables,

$$\text{(B)} \quad \dot{P} = \frac{\mu}{\hbar} ED - \gamma_\perp P, \tag{6.38}$$

$$\text{(C)} \quad \dot{D} = -\frac{\mu}{\hbar} EP - \gamma_\parallel \left(D - \frac{N}{2} \right), \tag{6.39}$$

where dots denote time derivatives.

On the other hand, the space average of Maxwell's equation (3.35) yields

$$\text{(A)} \quad \dot{E} + \frac{c}{L} [E(L,t) - E(0,t)] = -gP, \tag{6.40}$$

where, according to (2.11), (2.12), (6.23) and (6.34), the boundary conditions are

$$\text{(I)} \quad E_T/\sqrt{T} = E(L,t) \approx < E(z,t) > = E(t), \tag{6.41}$$

$$\text{(II)} \quad E(0,t) = \sqrt{T}\, E_I + R\, E \left(L, t - \frac{\ell}{c} \right). \tag{6.42}$$

Let us assume that $E(t)$ varies sufficiently little during the delay time ℓ/c so that

$$E\left(L, t - \frac{\ell}{c} \right) \approx E\left(t - \frac{\ell}{c} \right) \approx E(t) - \frac{\ell}{c} \dot{E}(t). \tag{6.43}$$

Then, (6.42) becomes

$$E(0,t) - E(L,t) \approx T \left(\frac{E_I}{\sqrt{T}} - E \right) - R \frac{\ell}{c} \dot{E}, \tag{6.44}$$

where we can take $R = 1$ in the last term, in the limit (6.32). Substituting this result in (6.40) and taking into account (2.9), we get

$$\dot{E} = -g \frac{L}{\mathcal{L}} P - \frac{cT}{\mathcal{L}} \left(E - \frac{E_I}{\sqrt{T}} \right). \tag{6.45}$$

From (2.42), (2.16) and (6.21), we find that the *cavity mode damping constant* γ_c, for $T \ll 1$, is given by

$$\gamma_c = 1/\tau_c = c|\ln(1 - T)|/\mathcal{L} \approx cT/\mathcal{L}, \tag{6.46}$$

which (since n=1) is not affected by the active medium. Thus, (6.45) may be rewritten as

$$\text{(A)+(II)} \quad \dot{E} = -g \frac{L}{\mathcal{L}} P - \gamma_c \left(E - \frac{E_I}{\sqrt{T}} \right), \tag{6.47}$$

which is the mean field theory version of Maxwell's equation, with boundary condition (II) already incorporated. Boundary condition (I), by (6.41), gives the transmitted field in terms of the solution of (6.47): $E_T = \sqrt{T}E$.

The stationary solution of the mean field Bloch equations (6.38) and (6.39) is still given by (6.2) and (6.5) (with $\Delta = 0$ in the present case), where E, F now represent the mean field. Substituting these results in (6.47), with $\dot{E} = 0$, we get the stationary mean field theory solution

$$E - \frac{E_I}{\sqrt{T}} = -\frac{gL}{cT} P = -2\pi k_0 \mu N \frac{L}{T}\sqrt{\frac{\gamma_\parallel}{\gamma_\perp}} \frac{F}{(1+F^2)} = -\hbar\sqrt{\gamma_\parallel \gamma_\perp} \frac{\alpha L}{2\mu T} \frac{F}{(1+F^2)} , \tag{6.48}$$

where we have employed (3.36) and (6.15). Taking into account (6.41), (5.9), (6.26) and (6.27), we see that this result is indeed equivalent to the state equation (6.31).

Note that, for an empty cavity ($P=0$) with no input field ($E_I=0$), (6.47) becomes $\dot{E}=-\gamma_c E$, with the solution $E(t) = E(0)\exp(-\gamma_c t)$, as expected (empty cavity mode damping). At the same time, we see that, as a result of the spatial averaging (6.36), only the resonant cavity mode is taken into account, i.e., *mean field theory is a one-mode theory*.

On the other hand, the steady-state solution of (6.47) for an empty cavity ($P=0$) with a nonvanishing input field E_I is $E = E_I/\sqrt{T}$, which is just the resonant Fabry-Perot field within the sample (cf. (2.25)), corresponding, by (6.41), to the transmission resonance $E_T = E_I$ ($x = y$) to which the cavity is tuned. Therefore, we can interpret

$$E_r = E - \frac{E_I}{\sqrt{T}} \tag{6.49}$$

as the "*reaction field*" arising from the polarization of the active medium, which, in the steady state, is given by the r.h.s. of (6.48).

7. Discussion of the state equation

The mean field theory state equation (6.31) defines a family of response curves dependent on the parameter C. By (6.24) and (6.16), we may take x as a measure of the dimensionless field inside the sample, and $x \sim 1$ is the threshold of nonlinearity, so that $x \ll 1$ in the linear regime and $x \gg 1$ in the extreme nonlinear range.

From (6.31), we have, in the linear range,

$$x \ll 1 \Rightarrow y \approx (2C + 1)\, x, \tag{7.1}$$

a straight line through the origin with slope $(2C+1)$, in exact agreement with (5.8). On the other hand, in the extreme nonlinear range,

$$x \gg 1 \Rightarrow y \approx x + \frac{2C}{x} , \tag{7.2}$$

so that we are always in the high-transmission situation (transmission resonance) in this case, due to the bleaching effects of saturation.

7.1. Optical transistor effect

Differentiating (6.31), we get

$$\frac{dy}{dx} = 1 + \frac{2C(1-x^2)}{(1+x^2)^2} . \tag{7.3}$$

Thus, for any $C > 0$,

$$dy/dx < 1 \quad \text{for} \quad x > 1, \tag{7.4}$$

i.e., as soon as we enter the nonlinear range, the slope, which is > 1 for $x < 1$, becomes < 1.

By (6.26) and (6.27), this is equivalent to

$$dx/dy = dE_T/dE_I > 1 \qquad (x > 1), \tag{7.5}$$

i.e., to having *differential gain*. This amounts to the *amplification of small signals*: a small input signal increment Δy gives rise to an output signal increment $\Delta x > \Delta y$ (Fig. 2). Therefore, the system behaves like an *optical transistor*.

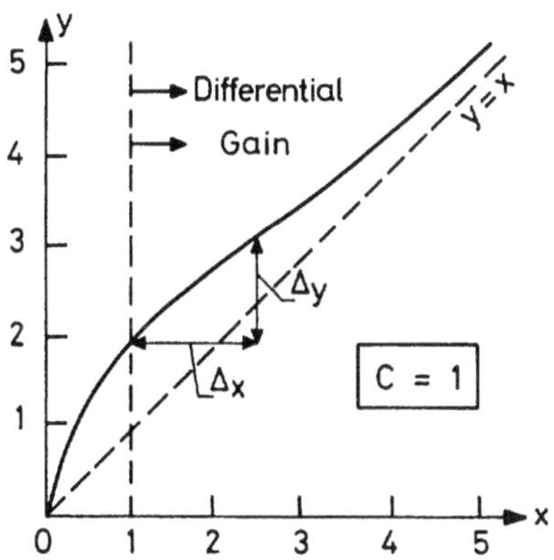

Fig. 2 Plot of (6.31) for C = 1, showing differential gain for x > 1

The fact that *a resonant saturable absorber can amplify small signals* was already noted by MOLLOW [17]. At first, this looks paradoxical, because the medium is *passive* (no pumping!) and strongly absorbing in the linear regime. Mollow discussed the response of two-level atoms driven by a strong pump field together with a weak signal field (perturbation). He showed that the weak field can induce *stimulated emission* rather than absorption. There is no problem with the energy balance, because the amplification of the small signal occurs at the expense of the pump field, which suffers an increased rate of attenuation to compensate for the amplification of the signal field.

Similarly, in the present case, ΔE_I may be amplified, even though the *total* input field $E_I + \Delta E_I$ is attenuated. Indeed, the state equation (6.31) shows that, for $x > 0$, we always have $y > x$, i.e., $E_T < E_I$ (the curve in Fig. 2 lies above the bisector).

A general argument to explain this effect was given by McCALL [18]. Adapted to the present case, it is based on the fact that, due to saturation, the rate of energy-loss S within the sample must approach a constant S_0 as the input intensity I_I increases (by (6.2), $D \to 0$). The energy-loss rate is proportional to

$$\vec{J}_P \cdot \vec{E} = \vec{E} \cdot \partial \vec{P}/\partial t, \tag{7.6}$$

where $\vec{J}_P$ is the polarization current. Therefore, J_P must behave as S_0/E for large E, yielding a negative "differential conductivity"

$$\partial J_P/\partial E \sim - S_0/E^2,$$

which is equivalent to amplification in the nonlinear range.

In the present case, the average of (7.6) over a period would be proportional to (cf. (3.31))

$$\omega \operatorname{Im}(PE^*) \propto \omega \operatorname{Re}(PE^*), \tag{7.7}$$

and (6.5) indeed shows that $P \propto 1/F$ for large F, in agreement with the above argument.

7.2. Bistability

In order to reach bistability, the response curve must fold over, so that the condition for bistability is

$$\text{Bistability} \Longleftrightarrow dy/dx \;\; \text{changes sign.} \tag{7.8}$$

Thus, dy/dx must go through zero, which, by (7.3), happens if and only if

$$C \geqslant 4, \tag{7.9}$$

and the zeros of dy/dx are then given by

$$x_{\pm}^{\;2} = C - 1 \pm \sqrt{C(C-4)}. \tag{7.10}$$

Therefore, *C = 4 is the threshold for optical bistability*. For C = 4, the response curve has an inflection point with horizontal tangent, at $x_+ = x_- = \sqrt{3}$. We will usually assume that

$$C \gg 1, \tag{7.11}$$

in which case

$$x_+ \equiv x_m \approx \sqrt{2C} \Rightarrow y_m \approx 2\sqrt{2C} \quad \text{(minimum)},$$

$$x_- \equiv x_M \approx 1 \Rightarrow y_M \approx C \quad \text{(maximum)}. \tag{7.12}$$

For $x \gg 1$, the response curve may be approximated by (7.2).

The response curve (6.31) for C = 10 is plotted in Fig. 3(a), and again in Fig. 3(b) with x = x(y), i.e., output as a function of input. In the range $y_m < y < y_M$, there are three branches x_1, x_2, x_3 corresponding to a given $y = y_0$, but the branch x_2, for which $dy/dx < 0$, is expected to be unstable (this will be confirmed below).

The upper branch in Fig. 3(b) is the high-transmission branch and the lower one is the low-transmission branch. For reasons to be explained below, the high-transmission branch will be called *one-atom branch* and the low-transmission one *cooper-*

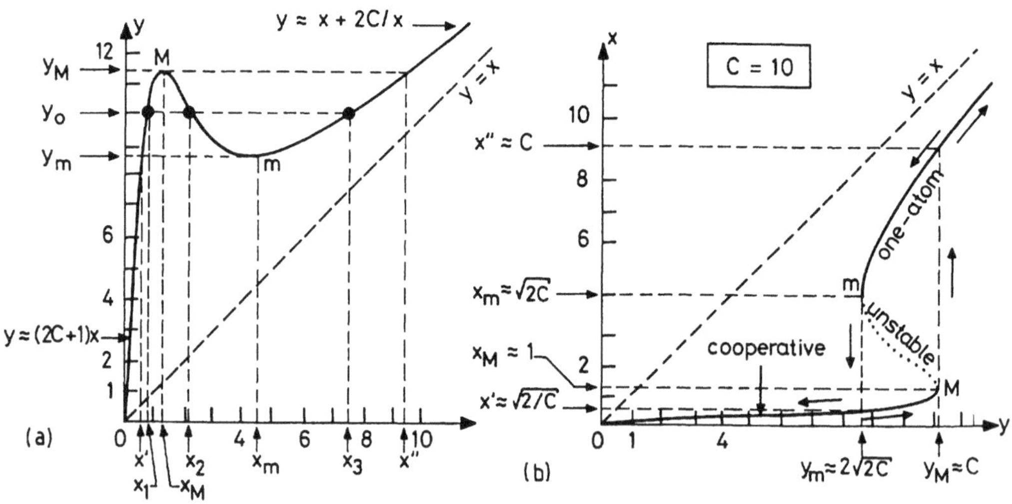

Fig. 3 Response curve for C = 10; (a) y = y(x); (b) x = x(y), showing also the hysteresis cycle

ative branch. Bistability occurs for incident fields in the range $y_m \leqslant y \leqslant y_M$. The family of response curves $y = y(x,C)$ may be roughly compared with the isotherms of a Van der Waals gas, with (y,x,C) playing roles somewhat analogous to (P,V,T), respectively. The curve for $C=4$ would then be analogous to the critical isotherm.

We also expect to find the analogues of the hysteresis effects characteristic of first-order phase-transitions. Thus, if the dimensionless input field y is increased from zero, we expect the system to follow the path indicated by arrows in Fig. 3(b), with x first increasing along the cooperative branch up to $x_M \approx 1$, then discontinuously jumping up to $x'' \approx C \approx Cx_M$ in the one-atom branch and continuing to increase along it. If we now decrease y, the response x will decrease past x'', continuing along the one-atom branch, until it reaches $x_m \approx \sqrt{2C}$, when it jumps down to $x' \approx \sqrt{2/C} \approx x_m/C$ in the cooperative branch, and then continues to decrease along it. Notice that both jumps are by a factor of C. The actual behavior, as will be seen later, is more complicated than this "naïve" expectation.

7.3. Fluorescent Intensity

The behavior of the *excited state population* (cf. (3.32))

$$N_a = N - N_b = \frac{N}{2} - D \tag{7.13}$$

may be monitored by looking at the intensity I_F of the *fluorescent light* emitted at 90° from the direction of propagation, because I_F should be proportional to N_a. By (6.2), in the steady state,

$$N_a = \frac{N}{2}\left(1 - \frac{1}{1+x^2}\right) = \frac{N}{2}\,\frac{x^2}{(1+x^2)} \quad , \tag{7.14}$$

where we have also employed (6.34) and (6.24).

The hysteresis cycle for N_a ($\propto I_F$) that corresponds to Fig. 3(b) is shown in Fig. 4. Along the one-atom branch, $x \gg 1$, so that (7.14) yields $N_a \approx N/2$, i.e., *the fluorescent intensity is proportional to the number of atoms*. This is just what one would expect from *incoherent emission*: each atom behaves as if the others were not present, justifying the name "one-atom branch". Along the cooperative branch, typically, $x = \mathcal{O}(1/\sqrt{C})$, so that, by (7.14), N_a is of order N/C. Since C is proportional to N (cf. (5.9), (6.15)), the fluorescence is *reduced* by a factor proportional to the number of atoms. This can only arise from coherent, cooperative behavior.

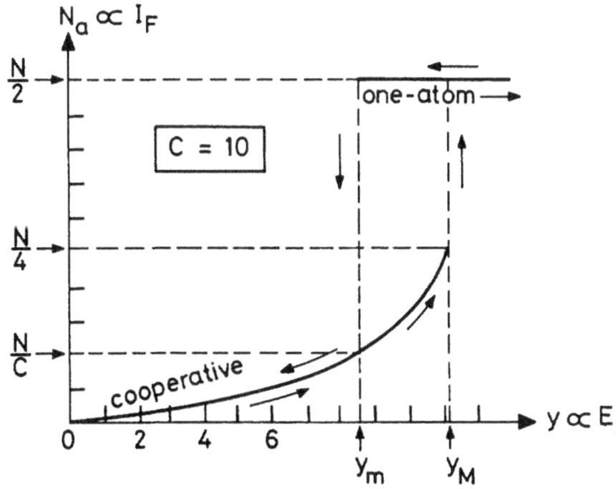

Fig. 4 Hysteresis cycle of the fluorescent intensity

We already know an effect where atomic cooperation leads to an *enhancement* of fluorescence by a superradiant factor proportional to N, namely, *superfluorescence* [3], the cooperative emission by an initially fully inverted system of two-level atoms. One usually considers a pencil-shaped sample (length L, cross-sectional area A) with Fresnel number $A/\lambda_0 L \sim 1$ (λ_0 = resonant wavelength), so as to concentrate the emission in a single mode, with small diffraction losses.

The cooperative decay rate γ_R for *pure superfluorescence* is the inverse temporal width of the hyperbolic secant pulse emitted in this situation, given by [19]

$$\gamma_R \sim \gamma N \frac{\lambda_0^2/A}{4\pi} , \tag{7.15}$$

where γ is the single-atom spontaneous emission rate, and λ_0^2/A is the diffraction solid angle, so that the last factor in (7.15) is the fraction of the total solid angle 4π within which we get the superradiant enhancement by N[1]. Since

$$\gamma = 4\mu^2 k_0^3/3\hbar , \tag{7.16}$$

we have, replacing N by (3.30), where V = AL,

$$\gamma_R \sim \alpha_0 L \gamma_\perp , \tag{7.17}$$

where we have employed (6.15).

Comparing (7.17) and (5.9), we get

$$C \sim \gamma_R/(\gamma_\perp T). \tag{7.18}$$

In pure superfluorescence, there are no end mirrors, so that T = 1. The condition for cooperative decay then is that the cooperative rate γ_R be much faster than the incoherent rate $\gamma_\perp$, so that $C \sim \gamma_R/\gamma_\perp \gg 1$. The extra factor 1/T in (7.18) represents the enhancement due to the number of reflections at the end mirrors. The condition $C \gg 1$ (cf. (7.11)) still stands for strong cooperation, justifying the name "cooperation parameter" given to C.

An alternative way to look at the results is in terms of the reaction field (cf. (6.48), (6.49))

$$E_r = E - \frac{E_I}{\sqrt{T}} = \frac{\hbar}{\mu} \sqrt{\gamma_\parallel \gamma_\perp}\, (x - y), \tag{7.19}$$

which is represented graphically by the vertical distance from the bisector to the response curve in Fig. 3(b), and, by (6.48), is proportional to the polarization of the active medium.

The reaction field is negligible for $x \gg x'' \approx C$, i.e., well along the one-atom branch, where each atom is just "doing its own thing" in its interaction with the incident field. On the other hand, as we come down this branch, the sudden precipitation to the lower branch at $y = y_m$ may be regarded as a "polarization catastrophe" associated with the internal field, where the field polarizes the medium, the medium polarization further increases the field, and so on, which is obviously a cooperative effect.

Conversely, as we increase the field from zero along the cooperative branch, the transmission is small in the linear range because of the strong polarization set up in the absorber (cf. (5.8)). As we reach the nonlinear range, the absorption bleaching by saturation increases the internal field, which in its turn increases the bleaching, and so on, until we switch over to the high-transmission branch. On the way down, the internal field remains strong enough for saturation, so that precipitation to the lower branch takes place at a lower value of the incident intensity than that required for the upwards transition, thus explaining the hysteresis effect [5].

The steady state hysteresis cycle of the fluorescent intensity has recently been experimentally observed [20].

8. Linear Stability Analysis

The steady states found in Sect. 7 represent equilibrium points: in order to find out whether they are physically realized, we must discuss the *stability* of equilibrium. For this purpose, we must consider displacements from the stationary solutions and the *transient behavior* of the system following such displacements. If the displacements are *damped out* in time, returning the system to the steady state, we have *stable equilibrium;* if they *grow*, taking the system further and further away from the stationary solution, we have *unstable equilibrium*, which is not physically realizable, because there always exist fluctuations, externally or internally generated.

One can imagine, for instance, that the input field is suddenly shifted away from a steady value E_I to $E_I + \delta E_I$: the transient approach of the system to a new steady operating point may in principle be experimentally observed.

We begin by considering only *small displacements* from equilibrium, so that one can *linearize* the equations of motion for the displacements; this is called *linear stability analysis*. It is important to realize that, *if the system is found to be unstable for small displacements, it is absolutely unstable,* because it will not return to the starting point. However, *stability against small displacements does not prove absolute stability,* because the system may still be found to be unstable under the action of larger displacements. Therefore, a complete discussion requires also a fully nonlinear stability analysis, which will be performed later.

Let δE, δP, δD be the small displacements from the stationary solutions (E, P, D) of (3.35), (3.37), (3.38), together with boundary conditions (2.11) and (2.12), for $\Delta = 0$. The linearized set of equations of motion which they satisfy is readily obtained from the "parent" equations by differentiation (this is equivalent to neglecting nonlinear terms in the displacements):

$$(A') \qquad \delta\dot{E} + c\frac{\partial}{\partial z}(\delta E) = -g\delta P, \tag{8.1}$$

$$(B') \qquad \delta\dot{P} = \frac{\mu}{\hbar}(E\delta D + D\delta E) - \gamma_\perp \delta P, \tag{8.2}$$

$$(C') \qquad \delta\dot{D} = -\frac{\mu}{\hbar}(E\delta P + P\delta E) - \gamma_\parallel \delta D, \tag{8.3}$$

$$(II') \qquad \delta E(0,t) = R\delta E\left(L, t - \frac{\ell}{c}\right), \tag{8.4}$$

where E, P, D are the stationary solutions found in Sect. 6, $\delta\dot{E} = \partial(\delta E)/\partial t$ (similarly for δP, δD) and (6.23) has also been employed.

As is well known, a linear differential system of this type has solutions of the form

$$\delta f(z,t) = \delta f_\lambda(z) \exp(\lambda t), \tag{8.5}$$

where f stands for E, P or D and λ is a root of the characteristic equation, obtained by substituting the "Ansatz" (8.5) in the system. Note that, if λ is complex, it is understood that we must take the real part of the r.h.s. of (8.5). The roots λ will be called the *eigenvalues* for the linear stability problem.

If we find *one* eigenvalue with $\mathrm{Re}\lambda > 0$, the corresponding perturbations (8.5) grow exponentially in time (until the linear approximation breaks down), showing that the stationary solution (E, P, D) is *unstable*. If *all* the eigenvalues have $\mathrm{Re}\lambda < 0$, the stationary solution is *linearly stable*, but, as remarked above, this does not guarantee absolute stability.

Substituting (8.5) in (8.4), we find

$$\delta E_\lambda(0) = R\delta E_\lambda(L)\exp(-\lambda\ell/c), \tag{8.6}$$

which will play the role of a boundary condition for (8.1)-(8.3).

Substituting D and P in (8.2), (8.3) by the stationary solutions (6.2), (6.5) (with $\Delta = 0$), and expressing E in terms of F with the help of (6.3), we get (with $\dot{\delta f} = \lambda\delta f$ by (8.5))

$$\delta P = \frac{\sqrt{\gamma_\parallel\gamma_\perp}}{\gamma_\perp + \lambda}F\delta D + \frac{N\mu}{2\hbar(\gamma_\perp + \lambda)}\frac{\delta E}{(1 + F^2)}, \tag{8.7}$$

$$\delta D = -\frac{\sqrt{\gamma_\parallel\gamma_\perp}}{\gamma_\parallel + \lambda}F\delta P - \frac{\mu}{\hbar}\sqrt{\frac{\gamma_\parallel}{\gamma_\perp}}\frac{N}{2(\gamma_\parallel + \lambda)}\frac{F\delta E}{(1 + F^2)}, \tag{8.8}$$

where we have dropped the index λ and the explicit z dependence.

Substituting (8.8) in (8.7) and solving for δP as a function of δE, and replacing the result, as well as (8.5), in (8.1), we find

$$\frac{\frac{\partial}{\partial z}(\delta E)}{\delta E} = \frac{\frac{\partial}{\partial z}(\delta F)}{\delta F} = -\frac{\lambda}{c} - \frac{\mu N g}{2\hbar c}\frac{\{\gamma_\parallel[1 - F^2(z)] + \lambda\}}{[(\gamma_\parallel + \lambda)(\gamma_\perp + \lambda) + \gamma_\parallel\gamma_\perp F^2(z)][1 + F^2(z)]}. \tag{8.9}$$

8.1 The Characteristic Equation

Before proceeding with the determination of the eigenvalues, it is worth while to consider the physical meaning of (8.9) in the special case when we set $\lambda = 0$, so that (8.5) becomes a *time-independent* perturbation. We can then interpret the solution as describing the *small-signal response* of our system. Taking into account (6.7) and (6.15), we see that (8.9) then becomes

$$\frac{\frac{d}{dz}(\delta F)}{\delta F} = -\frac{\alpha_0}{2}\left\{\frac{[1 - F^2(z)]}{[1 + F^2(z)]^2}\right\} \quad (\lambda = 0). \tag{8.10}$$

In the linear regime, $F^2 \ll 1$, this just describes the absorption of the small signal: $\delta F(z) = \delta F(0)\exp(-\alpha_0 z/2)$. However, in the nonlinear regime, the effective "absorption coefficient" for small signals is multiplied by the factor within curly brackets in (8.10). Thus, for $F > 1$, we get *gain* rather than absorption, i.e., small signals are amplified, in agreement with (7.3) and with the same effective gain factor. This is again the optical transistor effect, but now applied locally, at each z.

Getting back to (8.9), let us now integrate both sides over z from 0 to L; we get

$$\ln[\delta E(L)/\delta E(0)] = -\lambda L/c - \frac{\alpha}{2}\gamma_\perp\int_0^L \frac{\{\gamma_\parallel[1 - F^2(z)] + \lambda\}\,dz}{[(\gamma_\parallel + \lambda)(\gamma_\perp + \lambda) + \gamma_\parallel\gamma_\perp F^2(z)][1 + F^2(z)]}, \tag{8.11}$$

where $F(z)$ is the stationary solution, given (implicitly) by (6.28) with L replaced by z.

On the other hand, the boundary condition (8.6) yields

$$\ln[\delta E(L)/\delta E(0)] = \ln\left[\frac{1}{R}\exp(\lambda\ell/c)\right] = -\ln R + \frac{\lambda\ell}{c} + 2in\pi \quad (n=0, \pm 1,\dots). \tag{8.12}$$

The eigenvalues λ_n are obtained by identifying (8.12) with (8.11); this yields (remembering (2.9))

$$\lambda_n = -i\nu_n + \frac{c}{L}\ln R - \frac{\alpha c\gamma_\perp}{2L}\int_0^L \frac{\{\gamma_\parallel[1 - F^2(z)] + \lambda_n\}dz}{[(\gamma_\parallel + \lambda_n)(\gamma_\perp + \lambda_n) + \gamma_\parallel\gamma_\perp F^2(z)][1 + F^2(z)]}, \tag{8.13}$$

where

$$\nu_n \equiv 2n\pi c/L \qquad (n = 0, \pm1, \pm2,\dots). \tag{8.14}$$

Eq. (8.13) is the *characteristic equation*, an implicit equation for the eigenvalues λ_n. By (2.1), where (cf. (2.20), (6.23))

$$k = k_m = 2m\pi/L \Rightarrow \omega = \omega_m = 2m\pi c/L \tag{8.15}$$

are associated with the resonant cavity mode, we find, taking into account (8.5), that the electric field perturbation associated with the eigenvalues λ_n is of the form

$$\delta\vec{E}_n = \delta E_{\lambda_n}(z)\exp(ik_m z)\exp[-i(\omega_m + i\lambda_n)t]\,\vec{i}, \tag{8.16}$$

where $\delta E_{\lambda_n}(z)$ is the eigensolution of (8.9), (8.6) corresponding to the eigenvalue λ_n.

In order to get further insight into the physical interpretation of the eigenvalues and eigensolutions, let us consider them first in the *empty cavity limit* ($g = 0 = \alpha$). In this limit, the last term of (8.13) vanishes, so that

$$\lambda_n = -i\nu_n + \frac{c}{L}\ln R \tag{8.17}$$

and

$$\omega_m + i\lambda_n = \omega_m + \nu_n + i\frac{c}{L}\ln R = c\left[\frac{2(m+n)\pi}{L} + i\frac{\ln R}{L}\right] = c\,\tilde{k}_{m+n} \qquad (g = 0) \tag{8.18}$$

is the complex frequency associated with the natural mode $m + n$ ($n = 0, \pm1,\dots$) according to (2.39) (where, by (2.16), $L' = L$, since the refractive index $n = 1$).

Also in this limit, (8.9) yields

$$\delta E_{\lambda_n}(z) = \delta E_{\lambda_n}(0)\exp(\lambda_n z/c), \tag{8.19}$$

so that, by (8.18), the associated eigensolution (8.16) becomes

$$\delta\vec{E}_n = \delta E_{\lambda_n}(0)\exp[i\tilde{k}_{m+n}(z-ct)]\,\vec{i} \quad (g = 0), \tag{8.20}$$

which describes the propagation of the *natural mode* $m+n$ of the empty cavity: the eigenvalue $n = 0$ corresponds to the *resonant cavity mode*, and $n = \pm1, \pm2,\dots$ are the *side modes* on either side of it. Actually, n should not be too large, in order not to violate the slowly-varying amplitude and phase approximation, but we will need only the lowest values of n.

We see, therefore, that the eigenvalue equation (8.13) defines the *natural modes for the coupled system cavity field + active atoms in the linearized approximation.* This may also be seen directly from (8.1)-(8.4), which are the linearized Maxwell-Bloch equations for the ring cavity with no input field ($E_I = 0$; cf. Sect. 2.2).

Let us now go over to the *mean field approximation*, by making the replacement (cf. (6.34), (6.24))

$$F(z) \to \langle F(z) \rangle = F(L) = x \tag{8.21}$$

in the integral (8.13). We also note that (since $T \ll 1$)

$$\frac{c}{L} \ln R = \frac{c}{L} \ln(1 - T) \approx - \frac{cT}{L} = - \gamma_c , \tag{8.22}$$

where γ_c is the cavity mode damping constant (6.46). The result is

$$\lambda_n = -i\nu_n - \gamma_c - \frac{2C}{(1 + x^2)} \frac{\gamma_c \gamma_\perp [\gamma_\parallel (1 - x^2) + \lambda_n]}{[(\gamma_\parallel + \lambda_n)(\gamma_\perp + \lambda_n) + \gamma_\parallel \gamma_\perp x^2]} . \tag{8.23}$$

Eliminating the denominator, we find that the characteristic equation becomes the following cubic equation for λ_n:

$$\begin{aligned} &\lambda_n^3 + (\gamma_\parallel + \gamma_\perp + i\nu_n + \gamma_c)\,\lambda_n^2 + [\,(\gamma_\parallel + \gamma_\perp)\, i\nu_n \\ &+ \gamma_\parallel \gamma_\perp (1 + x^2) + \gamma_c \,(\gamma_\parallel + \gamma_\perp + \frac{2C\gamma_\perp}{1 + x^2})]\,\lambda_n \\ &+ i\nu_n \gamma_\parallel \gamma_\perp (1 + x^2) + \gamma_c \gamma_\parallel \gamma_\perp \left[1 + x^2 + \frac{2C(1 - x^2)}{1 + x^2} \right] = 0 . \end{aligned} \tag{8.24}$$

For each value of n, this equation will yield three eigenvalues λ_{nj} ($j = 1,2,3$), corresponding to the three degrees of freedom (real E, P, D) of the Maxwell-Bloch equations for absorptive optical bistability. In order that the stationary solution (x,y) be stable, (8.24) must yield, for this value of x,

$$\mathrm{Re}\, \lambda_{nj}(x) < 0 \quad \text{for all } (n,j.) . \tag{8.25}$$

This *linear stability criterion* is a *necessary* condition, but, as was emphasized at the beginning of this Section, it is not a sufficient one: it requires confirmation by the nonlinear stability analysis.

In particular, setting n=0 (i.e., by (8.14), $\nu_n = 0$) in (8.24), we get the characteristic equation for mean field theory, where only the resonant mode is considered. This can also be derived [21] directly from the mean field theory equations (6.38), (6.39), (6.47). However, a complete linear stability analysis requires considering also the other cavity modes ($n \neq 0$), which is the reason why we employed (8.1)-(8.4). In this respect, though we employ (8.21) to approximate the integral in (8.13), we go beyond mean field theory.

8.2. Critical Slowing Down

Writing (8.24) as $(\lambda_n - \lambda_{n1})(\lambda_n - \lambda_{n2})(\lambda_n - \lambda_{n3}) = 0$, we see that the constant term, taken with opposite sign, is the product of the three eigenvalues. In particular, for $n = 0$,

$$-\lambda_{01}\lambda_{02}\lambda_{03} = \gamma_c\gamma_{\parallel}\gamma_{\perp}(1+x^2)\left[1+\frac{2C(1-x^2)}{(1+x^2)^2}\right]$$

$$= \gamma_c\gamma_{\parallel}\gamma_{\perp}(1+x^2)\frac{dy}{dx}, \quad (8.26)$$

where the last equality follows from (7.3).

The r.h.s. of (8.26) vanishes at the two turning points M and m of Fig. 3(a). Therefore, there must be an eigenvalue that vanishes at these points. This corresponds to what is called a *marginal mode* or *soft mode* at these points, where the unstable branch begins or ends, and the jumps in the hysteresis cycle are expected to occur.

Therefore, the closer the system is to one of these turning points, the longer the regression to equilibrium must take following a small departure from the steady state. This *critical slowing down* effect in absorptive optical stability has been directly observed in a recent experiment [22].

8.3. Good Cavity Limit

We discuss the roots of (8.24) first in the limit of a "good cavity", defined as the situation in which the cavity mode lifetime τ_c (cf. (6.46)) is much greater than the atomic lifetimes T_1, T_2', i.e.,

$$\gamma_c \ll \gamma_{\parallel}, \gamma_{\perp}. \quad (8.27)$$

In this limit, we can, as a first approximation, neglect the terms proportional to γ_c in (8.23) (after eliminating the denominator), yielding

$$(\lambda_n + i\nu_n)[\lambda_n^2 + (\gamma_{\parallel} + \gamma_{\perp})\lambda_n + \gamma_{\parallel}\gamma_{\perp}(1+x^2)] = 0, \quad (8.28)$$

with the roots

$$\lambda_{n1} = -i\nu_n, \quad (8.29)$$

$$\lambda_{n2,3} = -\frac{1}{2}(\gamma_{\parallel} + \gamma_{\perp}) \pm \left[\frac{1}{4}(\gamma_{\parallel} - \gamma_{\perp})^2 - \gamma_{\parallel}\gamma_{\perp}x^2\right]^{1/2} \quad (8.30)$$

to zeroth order in γ_c.

This justifies associating λ_{n1} with *cavity natural modes* and $\lambda_{n2,3}$ with *atomic natural modes*. In fact, in the range

$$x \ll |\gamma_{\parallel} - \gamma_{\perp}| / \sqrt{2\gamma_{\parallel}\gamma_{\perp}}, \quad (8.31)$$

we may approximate (8.30) by

$$\lambda_{n2} \approx -\gamma_{\parallel} - \frac{\gamma_{\parallel}\gamma_{\perp}x^2}{\gamma_{\perp} - \gamma_{\parallel}}, \qquad \lambda_{n3} \approx -\gamma_{\perp} + \frac{\gamma_{\parallel}\gamma_{\perp}x^2}{\gamma_{\perp} - \gamma_{\parallel}}, \quad (8.32)$$

so that λ_{n2} is associated with D and λ_{n3} with P (cf. (3.37), (3.38)). In any case,

$$\mathrm{Re}\,\lambda_{n2,3} < 0. \quad (8.33)$$

To compute $\mathrm{Re}\lambda_{n1}$, we must take account of the first-order correction in γ_c, which (by first-order perturbation theory) is obtained by replacing the zero-order result

(8.29) back in (8.23):

$$\lambda_{n1} = -i\nu_n - \gamma_c \left\{ 1 + \frac{2C\gamma_\perp[\gamma_\parallel (1 - x^2) - i\nu_n]}{(1 + x^2)[(\gamma_\parallel - i\nu_n)(\gamma_\perp - i\nu_n) + \gamma_\parallel \gamma_\perp x^2]} \right\}. \qquad (8.34)$$

In particular, for the resonant mode n = 0, the only one taken into account in mean field theory,

$$\lambda_{01} = -\gamma_c \left[1 + \frac{2C(1 - x^2)}{(1 + x^2)^2} \right] = -\gamma_c \frac{dy}{dx} , \qquad (8.35)$$

so that λ_{01} is the eigenvalue responsible for critical slowing down in (8.26). Furthermore, from (8.35),

$$\frac{dy}{dx} < 0 \Rightarrow \mathrm{Re}\lambda_{01} > 0, \qquad (8.36)$$

thus proving what had only been assumed so far, namely, that the negative-slope portion of the response curve (shown dotted in Fig. 3(b)) is indeed *unstable*.

Taking the real part of (8.34), we get

$$\mathrm{Re}\lambda_{n1} = G_n - \gamma_c, \qquad (8.37)$$

where

$$G_n(x,C,\nu_n) = -\frac{2C\gamma_c\gamma_\perp\{\gamma_\parallel(1 - x^2)[\gamma_\parallel\gamma_\perp(1 + x^2) - \nu_n^2] + \nu_n^2(\gamma_\parallel + \gamma_\perp)\}}{(1 + x^2)\{[\gamma_\parallel\gamma_\perp(1 + x^2) - \nu_n^2]^2 + \nu_n^2(\gamma_\parallel + \gamma_\perp)^2\}}. \qquad (8.38)$$

For $n \neq 0$, further discussion is required to find out whether $\mathrm{Re}\lambda_{n1}$ can become positive (this would lead to *side-mode instabilities*). We postpone this discussion to Sect. 13.

8.4. Adiabatic Elimination

In the linear approximation (cf. (3.33), (4.3)) and at resonance ($\Delta = 0$), (3.37) becomes

$$\frac{\partial \mathcal{P}}{\partial t} = -\gamma_\perp \mathcal{P} + \frac{\mu N}{2\hbar} E. \qquad (8.39)$$

The solution of this equation, assuming no polarization at $t \to -\infty$, is

$$\mathcal{P}(t) = \frac{\mu N}{2\hbar} \int_{-\infty}^{t} \exp[-\gamma_\perp(t - t')]\, E(t')dt' . \qquad (8.40)$$

In the good-cavity limit (8.27), the field E(t') varies very slowly during the time interval $0 \leq t - t' \lesssim 1/\gamma_\perp$ that contributes significantly to the integral, so that we may approximate $E(t') \approx E(t)$ and take it out of the integral:

$$P(t) \approx \frac{\mu N}{2\hbar} E(t) \int_{-\infty}^{t} \exp[-\gamma_\perp(t - t')]\, dt' = \frac{\mu N}{2\hbar\gamma_\perp} E(t). \qquad (8.41)$$

The same result follows directly from (8.39) by setting $\partial P/\partial t = 0$. Under these conditions, we say that the polarization *follows the field adiabatically*, adjusting "immediately" (viewed from the slow time scale of the field) to the instantaneous value of the field. The solution of (8.39) by this procedure (taking $\partial P/\partial t = 0$ and expressing P in terms of E) is called *adiabatic elimination* of the polarization variable.

The linear approximation in this example was taken only for illustrative purposes. Still in the good-cavity limit (8.27), we may adiabatically eliminate both atomic variables P and D from the mean field Bloch equations (6.38), (6.39):

$$P = \frac{\mu}{\hbar\gamma_\perp} ED, \qquad D = \frac{N}{2} - \frac{\mu}{\hbar\gamma_\parallel} E P. \tag{8.42}$$

Solving for P as a function of E, we get

$$P(t) = \frac{\mu N}{2\hbar\gamma_\perp} E(t) \Bigg/ \left[1 + \frac{\mu^2 E^2(t)}{\hbar^2\gamma_\parallel\gamma_\perp}\right], \tag{8.43}$$

which reduces to (8.41) in the linear approximation and, apart from the time dependence, coincides with (6.5) (for $\Delta = 0$), as it must, since it was obtained by setting the l.h.s. of (6.38), (6.39) equal to zero.

Substituting (8.43) in the mean field theory Maxwell equation (6.47), and expressing the results in terms of the variables (6.24), (6.27), we get

$$\dot{x} = \gamma_c \left[y - x - \frac{2Cx}{(1 + x^2)}\right]. \tag{8.44}$$

The stationary solution of (8.44) is the state equation (6.31). For a small displacement δx around the stationary solution, we find from (8.44) the linearized equation

$$\delta\dot{x} = -\gamma_c \delta x \left[1 + \frac{2C(1 - x^2)}{(1 + x^2)^2}\right] = -\gamma_c \frac{dy}{dx}\delta x = \lambda_{01}\delta x, \tag{8.45}$$

where we have employed (8.35). This agrees with (8.5), showing that adiabatic elimination is a simple way to obtain the most relevant eigenvalue λ_{01} (associated with the slowest decay) in this situation.

8.5. Bad Cavity Limit

This is the opposite extreme with respect to (8.27), i.e.,

$$\gamma_c \gg \gamma_\parallel, \gamma_\perp, \tag{8.46}$$

so that the field now decays much faster than the atomic variables (as it would in superfluorescence). The relevant eigenvalues should therefore be associated with atomic modes; the simplest way to obtain them is through adiabatic elimination, which should now be applied to the field.

In order to perform the adiabatic elimination of E, we set $\partial E/\partial t = 0$ in (6.47) and solve for E:

$$E(t) = \frac{E_I}{\sqrt{T}} - \frac{gL}{\gamma_c \mathcal{L}} P(t). \tag{8.47}$$

Substituting this result in (6.38), (6.39), and taking into account (6.7) and (6.15), we get

$$\dot{P} = \sqrt{\gamma_\| \gamma_\perp}\; yD - \frac{4C}{N}\gamma_\perp PD - \gamma_\perp P, \tag{8.48}$$

$$\dot{D} = -\sqrt{\gamma_\| \gamma_\perp}\; yP + \frac{4C}{N}\gamma_\perp P^2 - \gamma_\| \left(D - \frac{N}{2}\right). \tag{8.49}$$

Linearizing around the stationary solution and substituting the "Ansatz" (8.5), we get

$$\delta\dot{P} = \lambda\delta P = \sqrt{\gamma_\| \gamma_\perp}\; y\delta D - \frac{4C}{N}\gamma_\perp (P\delta D + D\delta P) - \gamma_\perp \delta P, \tag{8.50}$$

$$\delta\dot{D} = \lambda\delta D = -\sqrt{\gamma_\| \gamma_\perp}\; y\delta P + \frac{4C}{N}\gamma_\perp \cdot 2P\delta P - \gamma_\| \delta D. \tag{8.51}$$

Substituting P and D by the stationary solutions (6.2) and (6.5) (expressed in terms of x), and employing (6.31), this becomes

$$(\lambda + \gamma_\perp y/x)\delta P - \sqrt{\gamma_\| \gamma_\perp}\; x\delta D = 0, \tag{8.52}$$

$$\sqrt{\gamma_\| \gamma_\perp}\; (2x - y)\delta P + (\lambda + \gamma_\|)\delta D = 0. \tag{8.53}$$

The characteristic equation is obtained by equating to zero the determinant of this set of equations:

$$\lambda^2 + \gamma_\perp \left(\frac{\gamma_\|}{\gamma_\perp} + \frac{y}{x}\right)\lambda + \gamma_\| \gamma_\perp \left[2x^2 + \frac{y}{x}(1 - x^2)\right] = 0. \tag{8.54}$$

This equation can also be obtained from (8.24) by dropping the cubic term and retaining only the terms proportional to $\gamma_\mathfrak{r}$ (cf. (8.46)) in the coefficients of other powers of λ.

The roots of (8.54) are

$$\lambda_\pm = -\frac{\gamma_\perp}{2}\left\{\left(\frac{\gamma_\|}{\gamma_\perp} + \frac{y}{x}\right) \pm \left[\left(\frac{\gamma_\|}{\gamma_\perp} - \frac{y}{x}\right)^2 - 4\frac{\gamma_\|}{\gamma_\perp}x(2x - y)\right]^{1/2}\right\}, \tag{8.55}$$

which correspond to λ_{02} and λ_{03}. It suffices to consider the *slowest*-decaying mode, which is associated with λ_-.

For a point on the *cooperative branch* of Fig. 3(b), we have, for $C \gg 1$, that $x \lesssim x_M \approx 1$ and $y/x \gtrsim C \gg 1$, so that we can expand the square root in (8.55) into a power series, with the following result:

$$\lambda_- \approx -\gamma_\| \frac{x}{y}(1 + x^2)\frac{dy}{dx} \approx -2\gamma_\| \frac{(1 - y^2/C^2)^{1/2}}{[1 + (1 - y^2/C^2)^{1/2}]}. \tag{8.56}$$

For $y \ll y_M \approx C$, (8.56) yields $\lambda_- \approx -\gamma_\|$. As $y \to y_M$, we see that $\lambda_- \to 0$, so that it is the eigenvalue responsible for *critical slowing down* in this case.

On the other hand, for a point on the *one-atom branch* in Fig. 3(b), with $y \gtrsim y_M$, we have, by (7.2),

$$\frac{y}{x} \approx 1 + \frac{2C}{x^2} = 1 + \mathcal{O}\left(\frac{1}{C}\right). \tag{8.57}$$

Neglecting the $\mathcal{O}(1/C)$ correction, (8.54) becomes

$$\lambda^2 + (\gamma_{\|} + \gamma_{\perp})\lambda + \gamma_{\|}\gamma_{\perp}(1 + x^2) = 0, \quad (8.58)$$

so that

$$\lambda_{\pm} = -\frac{1}{2}(\gamma_{\|} + \gamma_{\perp}) \pm \left[\frac{1}{4}(\gamma_{\|} + \gamma_{\perp})^2 - \gamma_{\|}\gamma_{\perp}(1 + x^2)\right]^{1/2} . \quad (8.59)$$

Since $x \sim y \gtrsim y_M \approx C \gg 1$, (8.59) becomes

$$\lambda_{\pm} \approx -\frac{1}{2}(\gamma_{\|} + \gamma_{\perp}) \pm i\sqrt{\gamma_{\|}\gamma_{\perp}}\, x . \quad (8.60)$$

By (6.26) and (4.15),

$$\sqrt{\gamma_{\|}\gamma_{\perp}}\, x = \frac{\mu}{\hbar}\frac{E_T}{\sqrt{T}} = \frac{\mu E}{\hbar} = \Omega, \quad (8.61)$$

where Ω is the *Rabi frequency* associated with the mean field inside the sample. Thus, (8.60) becomes

$$\lambda_{\pm} \approx -\frac{1}{2}(\gamma_{\|} + \gamma_{\perp}) \pm i\Omega \qquad (y \gtrsim y_M). \quad (8.62)$$

Under these conditions, therefore, the regression towards equilibrium is *oscillatory*, with frequency of oscillation given by the Rabi frequency. According to (8.59) and (8.61), the eigenvalues become complex when

$$\Omega \gtrsim \frac{1}{2}(\gamma_{\|} + \gamma_{\perp}), \quad (8.63)$$

i.e., the Rabi frequency must be larger than an atomic linewidth. Some experimental observations of transients in optical bistability, including Rabi oscillations for a bad cavity, have been made [23].

9. Dispersive Optical Bistability

In dispersive optical bistability (cf. Sect. 5.2), we have both atomic detuning ($\Delta \neq 0$) and cavity detuning, i.e., we start with the cavity tuned away from resonance, so that (6.23) is no longer valid. By analogy with (3.34), we define a *dimensionless cavity detuning parameter*

$$\theta = (\omega_c - \omega)/\gamma_c, \quad \text{where} \quad (9.1)$$

$$\omega_c = ck_m = 2m\pi c/L \quad (9.2)$$

is the frequency of the cavity mode closest to the field frequency ω (resonant mode), and γ_c is the associated mode damping constant (6.46). It follows from these definitions that

$$kL = 2m\pi - \theta T. \quad (9.3)$$

From (6.7) and (6.15), Maxwell's equation in the steady state becomes (we can set $k = k_0$ in the coefficient of (6.7), since $\gamma_{\perp}/\omega_0$ is very small)

$$\frac{dF}{dz} = -\frac{\alpha_0}{2}\frac{(1 - i\Delta)F(z)}{[1 + \Delta^2 + |F(z)|^2]} . \quad (9.4)$$

On the other hand, taking into account (6.3), (6.27) and (9.3), boundary condition (2.12) becomes, in the steady state,

$$F(0) = Ty + R\exp(-i\theta T)F(L). \tag{9.5}$$

Since $F(z)$ is now complex, the exact integration of (9.4), (9.5) is more involved than that for the absorptive case [5], and it does not lead to an explicit expression like (6.29). We confine our discussion to the *mean field limit state equation*, defined by (6.32) together with the condition

$$\theta = (\omega_c - \omega)\, L/cT = \text{constant}, \tag{9.6}$$

so that, while T becomes very small, the cavity detuning remains comparable with the mode width, though much smaller than the mode spacing; thus, the mean field limit remains a one-mode approximation.

Making the mean field approximation (8.21) in the r.h.s. of (9.4) and integrating between $z = 0$ and $z = L$, we get

$$x - F(0) = -\frac{\alpha_0 L}{2}\,\frac{(1 - i\Delta)x}{(1 + \Delta^2 + |x|^2)}\,. \tag{9.7}$$

On the other hand, by (6.32) and (9.6), (9.5) becomes

$$F(0) = Ty + (R - i\theta T)x. \tag{9.8}$$

Substituting (9.8) into (9.7) and taking into account (5.9), we get

$$\frac{y}{x} = 1 + \frac{2C}{1 + \Delta^2 + |x|^2} + i\left(\theta - \frac{2C\Delta}{1 + \Delta^2 + |x|^2}\right), \tag{9.9}$$

which reduces to the mean field state equation (6.31) of purely absorptive optical bistability for

$$\theta = \Delta = 0 \qquad \text{(absorptive)}, \tag{9.10}$$

as it should.

From (6.9) and (6.15) we see that (9.9) may be rewritten as

$$E_I/E_T = 1 + i\theta - i(n' - 1)kL/T, \tag{9.11}$$

where n' is the complex nonlinear refractive index of the Bloch medium. It can readily be verified that (9.11) also follows directly from (2.15) with the additional conditions

$$|n' - 1|kL \ll 1, \qquad T \ll 1, \tag{9.12}$$

with C and θ fixed, corresponding to the mean field approximation. Thus, (9.11) is just the expression for the *inverse transmission amplitude of a nonlinear Fabry-Perot* in this approximation.

To write the state equation, we define the dimensionless output and input intensities associated with (6.26), (6.27),

$$X = |x|^2 \qquad \text{(output)}, \tag{9.13}$$

$$Y = y^2 \qquad \text{(input)}, \tag{9.14}$$

where E_I is taken to be real. Equating the squared moduli of both sides of (9.9), we get

$$\frac{I_I}{I_T} = \frac{Y}{X} = \left(1 + \frac{2C}{1 + \Delta^2 + X}\right)^2 + \left(\theta - \frac{2C\Delta}{1 + \Delta^2 + X}\right)^2, \tag{9.15}$$

wich is the *mean field state equation for mixed absorptive-dispersive optical bistability*, for a homogeneously broadened medium [24, 25].

Note from (9.15) that the first parenthesis contains the absorptive part of the complex refractive index, whereas the second one contains the dispersive part. We have *purely dispersive* optical bistability when the atomic detuning (cf. (3.34)) is so large that the absorptive part becomes negligible ($\Delta^2 \gg 1$), so that (9.1) becomes

$$\frac{Y}{X} = 1 + \left(\theta - \frac{2C\Delta}{\Delta^2 + X}\right)^2 \qquad \text{(dispersive)} \tag{9.16}$$

The mechanism for dispersive bistability discussed in Sect. 5.2 now becomes apparent: for

$$X = X_r \equiv \Delta^2\left(\frac{2C}{\theta\Delta} - 1\right), \tag{9.17}$$

the nonlinear change in the real refractive index drives the cavity back to resonance: $Y_r = X_r$ by (9.16). This corresponds to the intensity I_r in Sect. 5.2. By (9.11), we see that (9.17) corresponds to

$$(n - 1)kL = \theta T, \tag{9.18}$$

and, by (9.3) and (2.16), this indeed leads back to $\delta = 2m\pi$.

From (9.13) and (9.17) we get the condition

$$2C > \theta\Delta, \tag{9.19}$$

and, by requiring that (9.16) be a good approximation to (9.15) also for X around X_r, we find the conditions for the validity of (9.16) [24]:

$$\Delta^2 \gg 1, \qquad \theta\Delta \gg 1, \qquad \theta/\Delta \ll 1. \tag{9.20}$$

For $X \ll \Delta^2$ (cf. (6.17)), we can approximate

$$(\Delta^2 + X)^{-1} \approx \Delta^{-2} - \Delta^{-4}X \qquad (X \ll \Delta^2), \tag{9.21}$$

so that the state equation for Y becomes a cubic in X, corresponding to the Kerr medium approximation (6.18).

In order to have bistability, the response curve must have an *inflection point* $X_0 > 0$ with negative slope at X_0, i.e.,

$$X_0 > 0, \qquad (dY/dX)_0 < 0. \tag{9.22}$$

By a careful discussion of these conditions [24], the following results can be shown to hold:

(a) For $C < 4$, no optical bistability (absorptive, dispersive or mixed) can exist.

(b) For fixed $C > 4$, the largest hysteresis cycle is obtained in the purely absorptive case $\theta = \Delta = 0$. However, for $\theta \neq 0$, $\Delta \neq 0$, the switching from low to high transmission occurs for lower values of the input field, which may be a practical advantage. Similarly, we see from (9.21) and (6.2) that dispersive bistability can be achieved with input intensities well below saturation.

The effect of *inhomogeneous broadening*, which we have excluded from our treatment, can also be taken into account. It is found [24, 25] that, if the inhomogeneous linewidth $1/T_2^*$ is much larger than the homogeneous one $\gamma_\perp$ (i.e., $\gamma_\perp T_2^* \ll 1$), there are values of C such that purely absorptive bistability ($\theta = \Delta = 0$) cannot occur, but the system is bistable for large enough θ and Δ.

Thus, dispersive bistability is easier to observe, explaining why it was observed earlier [16]. A very nice experimental observation of absorptive optical bistability has been made recently [26].

The linear stability analysis in the dispersive case is more difficult, because both E and P are complex, so that there are five degrees of freedom instead of three. We refer to the literature [5] for further discussion.

10. Quantum Theory of Optical Bistability

So far, the quantized field treatment of optical bistability has been confined almost exclusively to the mean field model, i.e., to the *single-mode* situation; we consider only the purely absorptive case. We omit detailed derivations, giving only an outline of the main results.

The model employed is a straightforward generalization of the quantum theory of the laser in single-mode operation; we assume familiarity with the laser model [1,9-11]. The main difference is that no pumping is present; on the other hand, we have an incident laser field, which we describe by a pure coherent state.

Let a and a^+ be the annihilation and creation operators of photons in the resonant cavity mode. The two-level atoms are described by the operators

$$r^\pm = \frac{1}{2}(\sigma_x \pm i\sigma_y) \quad , \qquad r_3 = \frac{1}{2}\sigma_z , \tag{10.1}$$

where σ_x, σ_y, σ_z are the Pauli spin matrices; $r^+(r^-)$ is a raising (lowering) operator. We need a triplet of operators for each two-level atom: $r_i^\pm$, r_{3i} ($i = 1,2,\ldots, N$; operators associated with different atoms commute).

Let z_i be the z coordinate of the ith two-level atom along the sample. It is convenient to introduce collective dipole operators

$$R^\pm = \sum_{i=1}^{N} r_i^\pm \exp(\pm i k_0 z_i), \tag{10.2}$$

together with the operator associated with one half of the population inversion

$$R_3 = \sum_{i=1}^{N} r_{3i}. \tag{10.3}$$

These operators obey the angular momentum commutation rules

$$[R^+, R^-] = 2R_3,$$

$$[R_3, R^\pm] = \pm R^\pm. \tag{10.4}$$

Let ρ_{AF} be the density operator for the combined system two-level atoms (A) + field (F). Omitting the indices in ρ_{AF} for brevity, we take it to obey the following master equation in the *interaction picture*:

$$d\rho/dt = -iL_{AF}\rho + \Lambda_F\rho + \Lambda_A\rho. \tag{10.5}$$

Here, L_{AF} is the Liouville operator which describes the atom-field interaction in the electric dipole and rotating-wave approximations,

$$L_{AF}\rho = [H_{AF}, \rho]/\hbar, \qquad (10.6)$$

where H_{AF} is the interaction Hamiltonian

$$H_{AF} = i\hbar f(a^+R^- - aR^+), \qquad (10.7)$$

and the coupling constant f is defined as

$$f = \left(\frac{2\pi\omega_0}{\hbar V}\frac{L}{L}\right)^{1/2} \mu. \qquad (10.8)$$

This interaction is the quantized version of the last term in (3.5) summed over all the atoms in the rotating-wave approximation (cf. (3.24)). The field propagation factor is absorbed into (10.2). All normalization factors, such as that in (10.8), are so chosen as to recover the semiclassical mean field theory equations of Sect. 6.3 from the expectation values of the quantized equations of motion, as will be shown below.

The remaining terms in (10.5) represent the incoherent contributions arising from the interactions with the field and atomic reservoirs. The general form for such a term, obtained from the quantum theory of damping [9, 10, 27], involves the following steps: the system is linearly coupled with the reservoir and the equation of motion for the combined system is solved in second-order perturbation theory; one then goes back to the reduced density operator for the system by taking the trace with respect to the reservoir variables, assuming no initial correlation and no memory effects (Markoff approximation).

For the single-mode laser, the field loss term resulting from the interaction with the field loss reservoir, found by this procedure, is [1]

$$\Lambda_F\rho = -\Gamma(a^+a\rho + \rho a^+a - 2a\rho a^+) = \Gamma([a, \rho a^+] + [a\rho, a^+]) \quad \text{(laser)}, \qquad (10.9)$$

where Γ is the field damping constant, which, in the present case, is given by γ_c (cf. (6.46)).

However, (10.9) cannot be the correct expression for optical bistability, because it does not take into account the incident field E_I (which is not present in the laser problem). Indeed, in the absence of active atoms ($H_{AF} = \Lambda_A = 0$), the steady-state solution of (10.5), $d\rho/dt = \Lambda_F\rho = 0$, is the vacuum state $\rho = |0><0|$ (as we see by inserting this in (10.9)), which is the correct solution for the laser.

For the bistability problem, the steady state of the internal field for the empty cavity must correspond to the classical field amplitude $E_I/\sqrt{T}$ at the transmission resonance (cf. (2.25)). Since we represent the incident laser field by a pure coherent state [1], we must have

$$\rho_{steady} = |\alpha><\alpha| \qquad \text{for } H_{AF} = \Lambda_A = 0, \qquad (10.10)$$

where (the normalization will be checked below)

$$\alpha = \left(\frac{VL}{8\pi\hbar\omega_0 L}\right)^{1/2} \frac{E_I}{\sqrt{T}} = \alpha^*. \qquad (10.11)$$

Since $(a - \alpha)|\alpha> = <\alpha|(a^+ - \alpha) = 0$, it suffices to replace a by $a - \alpha$ in (10.9), so that, for bistability,

$$\Lambda_F\rho = \gamma_c\{[(a - \alpha), \rho(a^+ - \alpha)] + [(a - \alpha)\rho, (a^+ - \alpha)]\}. \qquad (10.12)$$

Analogously to (10.9), the energy damping contribution to atomic loss (cf. (3.25)) corresponds to

$$(\Lambda_A\rho)_\parallel = \frac{\gamma_\parallel}{2}\sum_{i=1}^{N} ([r_i^-, \rho r_i^+] + [r_i^-\rho, r_i^+]), \tag{10.13}$$

where the damping mechanism corresponds to the deexcitation of an atom accompanied by the emission of a reservoir photon.

However, as was mentioned in connection with (3.26), the homogeneous transverse linewidth $\gamma_\perp$ includes, besides the effects arising from (10.13), also the effects of *dephasing* processes (due, e.g., to atomic collisions), that broaden the line by interrupting the emission. Since the atom is left in the same state, these terms contain r_{3i} rather than $r_i^\pm$; they are given by (note that $\gamma_\perp \geqslant \gamma_\parallel/2$)

$$(\Lambda_A\rho)_{\text{dephasing}} = \left(\gamma_\perp - \frac{\gamma_\parallel}{2}\right)\sum_{i=1}^{N} ([r_{3i}, \rho r_{3i}] + [r_{3i}\rho, r_{3i}]), \tag{10.14}$$

with $\gamma_\parallel = 2\gamma_\perp$ for purely radiative damping. Finally,

$$\Lambda_A\rho = (\Lambda_A\rho)_\parallel + (\Lambda_A\rho)_{\text{dephasing}}. \tag{10.15}$$

The master equation (10.5), with L_{AF}, Λ_F and Λ_A given by (10.6), (10.12) and (10.15), respectively, was formulated by BONIFACIO and LUGIATO [24]. Let us now verify that in the semiclassical approximation it indeed corresponds to the mean field model of Sect. 6.3.

For this purpose, we find the equations of motion for the time-dependent expectation values $\langle R^- \rangle(t)$, $\langle R_3 \rangle(t)$, $\langle a \rangle(t)$, where, for a general interaction picture operator A,

$$\langle A \rangle(t) \equiv \mathrm{Tr}\,[\rho(t)A] \tag{10.16}$$

is the expectation value of A at time t. It follows that

$$\frac{d}{dt}[\langle A \rangle(t)] \equiv \langle \dot{A} \rangle = \mathrm{Tr}\left[\frac{d\rho}{dt}(t)A\right], \tag{10.17}$$

where $d\rho/dt$ is given by (10.5).

For the evaluation of the time derivatives, it suffices to replace A by R^-, R_3 and a, to employ (10.5), (10.6), (10.12) and (10.15), and to make use of the commutation relations, as well as of the cyclic invariance of the trace. The results are

$$\langle \dot{R}^- \rangle = 2f\langle aR_3 \rangle - \gamma_\perp \langle R^- \rangle, \tag{10.18}$$

$$\langle \dot{R}_3 \rangle = -f(\langle a^+R^- \rangle + \langle aR^+ \rangle) - \gamma_\parallel\left(\langle R_3 \rangle + \frac{N}{2}\right), \tag{10.19}$$

$$\langle \dot{a} \rangle = f\langle R^- \rangle - \gamma_c(\langle a \rangle - \alpha). \tag{10.20}$$

The semiclassical approximation is obtained, as usual, by setting

$$\langle aR_3 \rangle = \langle a \rangle\langle R_3 \rangle, \qquad \langle a^\pm R^\mp \rangle = \langle a^\pm \rangle\langle R^\mp \rangle, \tag{10.21}$$

where, all coefficients being real (cf. (10.11)), we can take

$$\langle a \rangle = \langle a^+ \rangle, \qquad \langle R^- \rangle = \langle R^+ \rangle. \tag{10.22}$$

Substituting these results in (10.18) - (10.20), we find that the resulting equations become identical to the mean field theory equations (6.38), (6.39) and (6.45), with the following identifications (cf. (10.3), (3.32))

$$< R_3 > = -D, \quad < R^{\pm} > = -P, \quad 2f < a > = \mu E/\hbar, \tag{10.23}$$

where we employ the relations

$$2\hbar f^2/\mu = gL/L, \quad 2\hbar f\alpha/\mu = E_I/\sqrt{T}, \tag{10.24}$$

which follow from (10.8), (10.11), and (3.36). This justifies regarding (10.5) as the quantized version of the semiclassical mean field theory of purely absorptive optical bistability.

The last equation in (10.23) gives the semiclassical analogue of the photon number:

$$< n > = < a^+ a > \to < a >^2 = \mu^2 E^2/4f^2\hbar^2. \tag{10.25}$$

Taking into account (6.3) and (6.16), we see that the *saturation photon number* n_S, corresponding to the saturation intensity, is given by

$$n_s = \gamma_{\parallel}\gamma_{\perp}/4f^2. \tag{10.26}$$

Furthermore, (6.24) and (6.27) become

$$x = < a > / \sqrt{n_s}, \quad y = \alpha/\sqrt{n_s} . \tag{10.27}$$

Finally, taking into account (5.9), (6.46), (6.15), (6.7) and (10.24), we find that the cooperation parameter can be expressed as

$$C = \frac{\mu g N L}{4\hbar c \gamma_{\perp} T} = \frac{f^2}{2\gamma_c \gamma_{\perp}} N. \tag{10.28}$$

10.1 The Fokker-Planck Equation

It is usually easier to work with a c-number equation, rather than with the operator master equation (10.5). For this purpose, one can make use of the well-known "quantum-classical correspondence" that was first formulated by WIGNER [28], and that has been extended to "generalized phase-space representations". We just outline the procedure, referring to quantum optics texts [1, 9-10] for further discussion.

The basic idea is to associate a "quasiprobability distribution" P, which is a c-number function, with the density operator ρ. The master equation is translated into a partial differential equation for P, which is usually easier to handle. In particular, quantum statistical averages can be computed from P just as classical statistical averages, in the generalized phase space associated with P.

Since the quantum effects must be contained in P, it cannot behave like a purely classical object, and this becomes apparent in the following respects: (a) P is not a true probability density, but only a "quasiprobability", because it can take negative values. (b) In associating an operator expression with a c-number function, there is an *ordering ambiguity*, so that the association is not unique: in order to get uniqueness, we must adopt an *ordering prescription*.

The prescription most commonly adopted in quantum optics for the field creation and annihilation operators is *normal ordering*, which leads to the "optical equivalence theorem" [1] between quantum and classical coherence functions. The corresponding phase-space representation is then the *diagonal coherent-state representation* [1], which is employed, in particular, in the quantum theory of the laser. For the laser problem, the master equation is translated into a *Fokker-Planck equation* for P.

The Fokker-Planck equation for a one-dimensional probability density $P(x,t)$ is

$$\frac{\partial P}{\partial t} = -\frac{\partial}{\partial x}[A(x)P] + \frac{1}{2}\frac{\partial^2}{\partial x^2}[D(x)P] . \qquad (10.29)$$

The expectation value of x at time t is

$$\langle x\rangle(t) = \int_{-\infty}^{\infty} P(x,t)x\, dx. \qquad (10.30)$$

Assuming P to be of sufficiently fast decrease as $|x| \to \infty$, we readily find from (10.29) and (10.30), with the help of partial integration,

$$\frac{d}{dt}\langle x \rangle = \langle A(x)\rangle. \qquad (10.31)$$

so that $\langle A(x)\rangle$ governs the time evolution of $\langle x\rangle$. For this reason, $A(x)$ is called the *drift coefficient*. Similarly, (10.29) yields

$$\frac{d}{dt}(\langle x^2\rangle - \langle x\rangle^2) = \langle D(x)\rangle + 2(\langle xA\rangle - \langle x\rangle\langle A\rangle), \qquad (10.32)$$

so that $\langle D(x)\rangle$, under these conditions, governs the time dependence of the mean squared fluctuation: $D(x)$ is known as the *diffusion coefficient*.

Note also that (10.29) may be written in the form of a continuity equation,

$$\partial P/\partial t + \partial j/\partial x = 0, \qquad (10.33)$$

where the probability current density is

$$j(x,t) = A(x)P(x,t) - \frac{1}{2}\frac{\partial}{\partial x}[D(x)P(x,t)] . \qquad (10.34)$$

In particular, for $D = 0$, we see that $A(x)$ plays the role of a local drift velocity. On the other hand, if $A = 0$ and D is a constant, (10.29) reduces to the diffusion equation.

For the optical bistability problem, unlike the ordinary laser problem, the normal-ordering prescription is not compatible with the Fokker-Planck approximation. The reason is [29] that the resulting Fokker-Planck equation does not have a positive definite diffusion matrix (higher-dimensional generalization of D).

However, this problem does not arise [29] if one adopts as the ordering prescription *symmetrical (Weyl) ordering*, which corresponds to the original WIGNER [28] choice. For this purpose, we begin by defining the *characteristic function* associated with $\rho(t)$ in (10.5):

$$C(\xi,\xi^*, \eta, \zeta,\zeta^*,t) = \mathrm{Tr}\{\rho(t) \exp[i(\xi^* R^+ + \xi R^- + \eta R_3 + \zeta^* a^+ + \zeta a)]\}. \qquad (10.35)$$

The corresponding (generalized) Wigner distribution is the Fourier transform of C:

$$P(\tilde{v},\tilde{v}^*, \tilde{m}, \tilde{\beta},\tilde{\beta}^*,t) = (2\pi)^{-5}\int C(\xi,\xi^*, \eta,\zeta,\zeta^*,t).$$
$$\times \exp[-i(\xi^*\tilde{v}^* + \xi\tilde{v} + \eta\tilde{m} + \zeta\tilde{\beta} + \zeta^*\tilde{\beta}^*)]\, d_2\xi d\eta d_2\zeta, \qquad (10.36)$$

where, e.g., $d_2\xi = d(\mathrm{Re}\xi)d(\mathrm{Im}\xi)$. This function generates symmetrized (Weyl-ordered) expectation values; e.g.,

$$\int\tilde{\beta}^*\tilde{\beta}\, P(\tilde{v},\tilde{v}^*, \tilde{m},\tilde{\beta},\tilde{\beta}^*,t)d_2\tilde{v}d\tilde{m}d_2\tilde{\beta} = \langle a^+a + aa^+\rangle/2. \qquad (10.37)$$

Taking into account (6.2), (6.5) and (10.27), it is convenient to define the normalized variables v, m, x, y by

$$\tilde{v} = -\frac{N}{2}\sqrt{\delta}\, v, \quad \tilde{m} = -\frac{N}{2} m, \quad x = \frac{\tilde{\beta}}{\sqrt{n_s}}, \quad y = \frac{\alpha}{\sqrt{n_s}}, \tag{10.38}$$

where n_s is the saturation photon number (10.26) and

$$\delta = \gamma_{\parallel}/\gamma_{\perp}. \tag{10.39}$$

The equation for $\partial P/\partial t$ that corresponds to the master equation (10.5) can be obtained either by laborious disentangling techniques [9, 10] or by a less rigorous, but more direct moment evaluation method [29]. The equation can be expanded in powers of the small parameter $(n_s)^{-1/2}$, coupled with derivatives of correspondingly increasing orders. Since $n_s \gg 1$ in practice, one can stop the expansion at the second order (neglecting $\mathcal{O}(n_s^{-3/2})$). This yields the *Fokker-Planck approximation* [29, 5]

$$\partial P/\partial t = \Gamma P(v, v^*, m, x, x^*, t), \tag{10.40}$$

where the Fokker-Planck operator Γ is given by

$$\begin{aligned}\Gamma = {} & \left\{-\frac{\partial}{\partial v}[-\gamma_{\perp}(v - mx)] + \text{c.c.}\right\} \\ & - \frac{\partial}{\partial m}\left\{-\gamma_{\parallel}\left[m - 1 - \frac{1}{2}(v^* x + v x^*)\right]\right\} \\ & + \left\{-\frac{\partial}{\partial x}[-\gamma_c(x - y + 2Cv)] + \text{c.c.}\right\} \\ & + \frac{\gamma_{\perp}^2}{2\gamma_c C n_s}\left[\frac{\partial^2}{\partial v^* \partial v} - \frac{\delta^2}{4}\frac{\partial}{\partial m}\left(\frac{\partial}{\partial v} v + \frac{\partial}{\partial v^*} v^*\right)\right. \\ & \left. + \frac{\delta^2}{4}\frac{\partial^2}{\partial m^2}(1 - m)\right] + \frac{\gamma_c}{n_s}\frac{\partial^2}{\partial x \partial x^*}.\end{aligned} \tag{10.41}$$

It is readily verified that the equations of type (10.31) associated with the drift terms in (10.41) reproduce the equations of motion (10.18) - (10.20), as they should. For instance, we find

$$\frac{d}{dt}\langle v\rangle = -\gamma_{\perp}(\langle v\rangle - \langle m x\rangle), \tag{10.42}$$

which is equivalent to (10.18). On the other hand, the matrix of diffusion coefficients in (10.41) is positive definite.

The Fokker-Planck approximation (10.40) (neglect of higher-order derivatives in Γ) is justified by the smallness of $(n_s)^{-1/2}$. On the other hand, while the Fokker-Planck equation can be employed to compute mean values and second-order fluctuations, it would not in general yield correct values for higher-order moments [30]. However, we will only apply it for moments up to second order.

The diffusion terms in (10.41), which determine the fluctuations, contain the various damping parameters $\gamma_{\perp}$, $\gamma_{\parallel}$ and γ_c, so that the fluctuations are connected with dissipation, as expected. The scaling of diffusion terms by $\mathcal{O}(n_s^{-1})$ relative to drift terms implies that the fluctuations will usually be very small.

11. Spectrum of the Transmitted Light

By the Wiener-Khintchine theorem [1], the spectrum of the transmitted light (cf. (6.26), (10.27)) is the Fourier transform of the temporal coherence function,

$$S(\omega) = \frac{1}{\pi} \mathrm{Re} \int_0^\infty \exp(-i\Delta\omega t) < a^+(t)a(0) >_{st} dt, \tag{11.1}$$

where the subscript "st" implies evaluation in the steady state, and $\Delta\omega = \omega - \omega_0$, as in (3.20) (the frequency difference arises from our use of the interaction picture).

Let x_{st} be the steady-state output, chosen either in the cooperative or in the one-atom branch, associated with the input y by (6.31), so that, by (10.27),

$$< a >_{st} = \sqrt{n_s}\, x_{st} = < a^+ >_{st}, \tag{11.2}$$

and let

$$\delta a(t) = a(t) - < a >_{st} \tag{11.3}$$

be the deviation from the steady state. Then,

$$< a^+(t)a(0) >_{st} = < a^+ >_{st} < a >_{st} + <\delta a^+(t)\delta a(0) >_{st}, \tag{11.4}$$

and, taking into account (11.2), (11.1) becomes

$$S(\omega) = S_{coh}(\omega) + S_{inc}(\omega), \quad \text{where} \tag{11.5}$$

$$S_{coh}(\omega) = < a >^2_{st} \cdot \frac{1}{\pi} \mathrm{Re} \int_0^\infty \exp(-i\Delta\omega t)dt = n_s x^2_{st}\, \delta(\omega - \omega_0) \tag{11.6}$$

and

$$S_{inc}(\omega) = \frac{1}{\pi} \mathrm{Re} \int_0^\infty \exp(-i\Delta\omega t) < \delta a^+(t)\delta a(0) >_{st} dt. \tag{11.7}$$

The term $S_{coh}(\omega)$ is just the monochromatic classical intensity (coherent part). The effects of quantum fluctuations are contained in the incoherent contribution (11.7). This contribution arises from the diffusion terms in the Fokker-Planck equation (cf. (10.32), (10.37)).

We briefly outline the derivation [31] for the case of a good cavity (cf. (8.27)). For the semiclassical theory, we saw in Sect. 8.4 that this allowed adiabatic elimination of the atomic variables, leading to the equation of motion (8.44) for the field variable alone.

The adiabatic elimination of atomic variables from the Fokker-Planck equation (10.41) can be performed by a procedure due to GORDON [32]. The result [31, 5] is

$$\frac{1}{\gamma_c} \frac{\partial}{\partial t} P(x,x^*,t) = \Bigg\{ - \frac{\partial}{\partial x} \left(y - x - \frac{2Cx}{1 + |x|^2} \right) + \text{c.c.}$$

$$- \frac{C}{2n_s} \frac{\partial^2}{\partial x^2}\, x^2\, \frac{(1 + |x|^2)^2 + 1 + \delta}{(1 + |x|^2)^3} + \text{c.c.}$$

$$+ \frac{1}{n_s} \frac{\partial^2}{\partial x^* \partial x} \left[1 + C\, \frac{(1 + |x|^2)^2 (2 + |x|^2) - (1 + \delta)|x|^2}{(1 + |x|^2)^3} \right] \Bigg\} P, \tag{11.8}$$

where $P(x,x^*,t)$ is the reduced quasiprobability distribution for the field variable alone, obtained from the full distribution by integrating over the atomic variables. Note that the equation of motion (10.31) associated with (11.8) coincides with the semiclassical equation (8.44).

Since the fluctuations are small for $n_s \gg 1$, we can linearize the Fokker-Planck equation (11.8) around the steady state, taking small deviations

$$x' = x - x_{st}, \qquad x'^* = x^* - x_{st}, \tag{11.9}$$

which we assume to be independent, so that, in contrast with (8.45), we vary both amplitude and phase. It is therefore convenient to introduce the "real and imaginary part" perturbations,

$$x_1' = \frac{1}{2}(x' + x^{*'}), \qquad x_2' = \frac{1}{2i}(x' - x^{*'}), \tag{11.10}$$

which are associated with amplitude and phase fluctuations.

Substituting (11.9) in (11.8) and linearizing, one finds [31, 33]

$$\frac{\partial}{\partial t} P(x_1', x_2', t) = \left\{ \lambda_{01} \frac{\partial}{\partial x_1'} x_1' + \lambda_\phi \frac{\partial}{\partial x_2'} x_2' + \frac{\gamma_c}{4n_s}\left[1 + 2C \frac{x_{st}^4 + (1-\delta)x_{st}^2 + 1}{(1 + x_{st}^2)^3}\right] \frac{\partial^2}{\partial x_1'^2} + \frac{\gamma_c(1+2C)}{4n_s} \frac{\partial^2}{\partial x_2'^2} \right\} P, \tag{11.11}$$

where λ_{01} is given by (8.35) and

$$\lambda_\phi = -\gamma_c \frac{y}{x_{st}} = -\gamma_c \left(1 + \frac{2C}{1 + x_{st}^2}\right). \tag{11.12}$$

Since $\delta \leqslant 2$ by (10.14) and (10.39), the positive definiteness of the diffusion matrix is apparent in (11.11).

The associated equations of motion (10.31) yield

$$\frac{d}{dt} < \delta x_1 > (t) = \lambda_{01} < \delta x_1 > ,$$

$$\frac{d}{dt} < \delta x_2 > (t) = \lambda_\phi < \delta x_2 >, \tag{11.13}$$

where x_1 and x_2 are the hermitian and antihermitian parts of the operator $\hat{x} = a/\sqrt{n_s}$ (cf. (10.27), (11.10)). The first equation is the same as (8.45), yielding the regression to the steady state of small amplitude perturbations (cf. (8.5))

$$< \delta x_1 > (t) = < \delta x_1 > (0) \exp(\lambda_{01} t). \tag{11.14}$$

Similarly, the second equation shows that $-\lambda_\phi$ is the damping rate for phase perturbations.

To compute the temporal coherence function in (11.7), one can now apply the *quantum regression theorem*, due to LAX [34], according to which the regression of fluctuations from the steady state obeys, on the average, the same laws as the linear deviations, so that $< \delta x_1(t) \delta x_j(0) >_{st}$ ($j = 1,2$) behaves like (11.14), and

$<\delta x_2(t)\,\delta x_j(0)>_{st}$ behaves like $<\delta x_2>(t)$. The initial values of these quantities are the stationary fluctuations $<\delta x_i(0)\delta x_j(0)>_{st}$, which can be obtained from (10.37) and (11.11) in the steady state by relations similar to (10.32). For instance, it follows from (10.32), when $A(x) = -\lambda x$ and D = constant (cf. (11.11)), that $<(\delta x)^2>_{st} = D/2\lambda$.

Therefore, the coherence function in (11.7) becomes a linear combination of exponentials such as (11.14) with constant coefficients, so that its Fourier transform, the incoherent spectrum, is a superposition of Lorentzians, given by [31]

$$S_{inc}(\omega) = \frac{\gamma_c C}{2\pi}\,\frac{x_{st}^2}{1+x_{st}^2}\left\{\frac{(2x_{st}^2+1-\delta)}{(1+x_{st}^2)^2\,[(\Delta\omega)^2+\lambda_{01}^2]}+\frac{1}{(\Delta\omega)^2+\lambda_{\phi}^2]}\right\}, \tag{11.15}$$

where the first Lorentzian is associated with amplitude fluctuations and the second one with phase fluctuations.

Let us follow the behavior of $S_{inc}(\omega)$ around the hysteresis cycle of Fig.3(b), assuming for definiteness that there is no collisional damping, so that (cf. (10.14), (10.39)) $\delta = 2$ in (11.15). Starting on the lower branch near point M, we see from (8.35) that, as M is approached, $\lambda_{01} \to 0$, so that we have *line narrowing* as a manifestation of the critical slowing down.

As one jumps to the upper branch, we have $x_{st} \gg 1$, so that (cf. (8.35), (11.12)) we have $\lambda_{01} \approx \lambda_{\phi} \approx -\gamma_c$, and the spectrum changes discontinuosly from a narrow line to a line whose width coincides with the empty cavity linewidth $2\gamma_c$. This is just what one should expect for a good cavity along the one-atom branch: all atoms emit incoherently, following the field adiabatically, so that the linewidth is determined by the cavity.

Moving downwards along the upper branch, another line narrowing occurs as we approach point m in Fig. 3(b). Jumping down to the lower branch, x_{st} becomes $\ll 1$, so that, from (8.35) and (11.12), we get $\lambda_{01} \approx \lambda_{\phi} \approx -2C\gamma_c (C \gg 1)$. Thus, the spectrum suddenly broadens to a characteristic width $4C\gamma_c \gg 2\gamma_c$ along the low transmission branch. This linewidth is proportional to the number of atoms, i.e., we have *cooperative line broadening* (cf. the discussion following (7.15)).

For $\delta = 2$ and $x_{st} \ll 1$, we also see that the weight factor in the first Lorentzian of (11.15) becomes negative, so that it interferes destructively with the second one and the line gets a hole in the middle. This is a purely quantum effect, associated with the fact that, as was remarked above, the quasiprobability distribution can take on negative values when quantum effects are dominant. It has been shown([33], [35]), that two other manifestations of purely quantum effects, known as *photon antibunching* and *squeezing*, also appear in the present situation.

11.1 Bad Cavity

One can also compute the spectrum in the bad cavity limit (8.46) by similar techniques [31]. As was done in the semiclassical treatment of Sect. 8.5, one now applies adiabatic elimination to the field, so that instead of (11.7) one finds that $S_{inc}(\omega)$ is the Fourier transform of an *atomic* correlation function.

The calculations [31] show that atom-atom correlations are generally negligible along the one-atom branch, as one would expect, but they become large (divergent in the linear approximation) as one approaches the transition points in the hysteresis cycle, so that one can think of the transitions as being driven by atom-atom correlations, in agreement with the discussion given in Sect. 7.

The spectrum is again a superposition of Lorentzians. On the cooperative branch, one of them is associated with the eigenvalue (8.56), which again gives rise to line narrowing as the transition point M in Fig. 3(b) is approached.

As one jumps to the one-atom branch, the eigenvalues (8.62) become relevant, and their imaginary part gives rise to Lorentzian peaks shifted by $\pm\,\Omega$ from the resonance frequency. The spectrum becomes suddently three-peaked, and, for $\gamma_{\parallel} = 2\gamma_{\perp} = \gamma$ ($\delta = 2$ in (10.39)) and $y \gg y_M$ in Fig. 3(b), it is given by [31]

$$S_{inc}(\omega) \propto \frac{\gamma/2}{(\Delta\omega)^2 + (\gamma/2)^2} + \frac{1}{2}\left[\frac{3\gamma/2}{(\Delta\omega - \Omega)^2 + (3\gamma/2)^2} + \frac{3\gamma/2}{(\Delta\omega + \Omega)^2 + (3\gamma/2)^2}\right], \quad (11.16)$$

which is identical with the spectrum of one-atom resonance fluorescence at high intensity [36]. Thus we have the *discontinuous appearance of a dynamic* (AC) *Stark shift*.

12. Photon Statistics

In order to determine the photon statistics of the transmitted light in the steady state, one must find the steady-state solution of the Fokker-Planck equation (11.8) (only the good cavity case is treated).

For this purpose, we rewrite (11.8) in polar coordinates, setting

$$x = re^{i\phi}. \quad (12.1)$$

The result is

$$\frac{1}{\gamma_c}\frac{\partial}{\partial t}P(r,\phi,t) = \left\{\frac{1}{r}\left[\frac{\partial}{\partial r}\,r\left(r + \frac{2Cr}{1+r^2} - y\cos\phi\right) + \frac{1}{r}\frac{\partial}{\partial\phi}\,y\sin\phi\right] + \frac{1}{4n_s}\left[\frac{1}{r}\frac{\partial}{\partial r}\,r\,\frac{\partial}{\partial r}\left(1 + 2C\,\frac{r^4 + (1-\delta)r^2 + 1}{(1+r^2)^3}\right) + (1+2C)\,\frac{1}{r^2}\frac{\partial^2}{\partial\phi^2}\right]\right\}P(r,\phi,t). \quad (12.2)$$

For the steady state solution of the one-dimensional Fokker-Planck equation (10.29), we see from (10.33) with $\partial P/\partial t = 0$ that $\partial j_{st}/\partial x = 0$, so that the steady-state current $j_{st}(x)$ must be constant. Since P is of rapid decrease as $|x| \to \infty$, the constant must be zero:

$$j_{st}(x) = 0 \quad \text{(steady state)}, \quad (12.3)$$

which is the *detailed balance* condition.

For a constant diffusion coefficient D, (10.34) and (12.3) yield

$$\frac{D}{2}\frac{dP_{st}}{dx} = A(x)P_{st}(x), \quad (12.4)$$

which has the solution

$$P_{st}(x) = C\exp\left[-\frac{V(x)}{D/2}\right], \quad (12.5)$$

where

$$A(x) = -dV/dx \quad (12.6)$$

and C is a normalization constant ($\int P dx = 1$).

If we interpret V(x) as the analogue of a mechanical potential, (10.31) can be thought of as the average equation of motion for an overdamped oscillator [37], and (12.5) shows that V(x) plays the role of a generalized free energy, the minima of which yield the most probable values of x.

It has already been mentioned that the equation of motion (10.31) associated with (11.8) is the semiclassical equation (8.44), which can be written as

$$\dot{x}/\gamma_c = -dV_y/dx, \quad V_y(x) = -yx + \frac{x^2}{2} + C\ln(1 + x^2). \tag{12.7}$$

According to the discussion given in Sect. 7, when $C > 4$, $V_y(x)$ has a single minimum x_{st} for $y < y_m$ in Fig. 3(b) (cooperative branch), as well as for $y > y_M$ (one-atom branch). However, in the bistable region $y_m < y < y_M$, $V_y(x)$ develops two minima which, for $y = y_0$ in Fig. 3(a), would correspond to the points x_1 (cooperative) and x_3 (one-atom), with a maximum in between, corresponding to the unstable point x_2. Thus, $V_y(x)$ becomes a (generally asymmetric) double well. One would expect, by analogy with (12.5), that the deeper well would yield the absolutely stable solution, while the shallower one would yield a metastable solution.

Unfortunately, the discussion that led to (12.5) does not apply to (12.2), for several reasons: (i) Recasting the two-dimensional equation (12.2) in the form of a continuity equation, one finds [38] that the associated components of the current cannot vanish simultaneously, so that *there is no detailed balance.* (ii) The amplitude diffusion coefficient in (12.2) is not constant: it is intensity-dependent.

On the other hand, the phase diffusion coefficient in (12.2) is essentially ($C \gg 1$)

$$q = C/2n_s \tag{12.8}$$

and this is usually extremely small, due to the large value of n_s. Hence, it is a good approximation to neglect phase fluctuations, setting $\phi = 0$, which is the stationary value (x_{st} is real in purely absorptive bistability). This amounts to phase locking with the phase of the incident field (α is real in (10.11)).

Actually, the assumption that the incident laser field can be represented by the pure coherent state (10.10) is unrealistic: there will generally be fluctuations associated with the incident field, as well as thermal fluctuations. Taking into account these additional fluctuations [39] and setting $\cos\phi = 1$ in (12.2), one finds [5] that

$$P(r,t) = \int_0^{2\pi} rP(r,\phi,t)d\phi \qquad \left(\int_0^{\infty} P(r,t)dr = 1\right) \tag{12.9}$$

obeys the following Fokker-Planck equation:

$$\frac{1}{\gamma_c}\frac{\partial}{\partial t}P(r,t) = -\frac{\partial}{\partial r}\left[A_y(r) - q\frac{\partial}{\partial r}\mathcal{D}(r)\right]P, \tag{12.10}$$

where q is given by (12.8),

$$A_y(r) = y - r - \frac{2Cr}{1 + r^2}, \quad \text{and} \tag{12.11}$$

$$\mathcal{D}(r) = \frac{1 + 2\bar{n}}{2C} + \frac{r^4 + (1 - \delta)r^2 + 1}{(1 + r^2)^3}, \tag{12.12}$$

with $\bar{n}$ corresponding to the external (incident and thermal) fluctuations.

We see that (12.10) is of the form (10.29), and its stationary solution is

$$P_{st}(r) = \frac{C}{\mathcal{D}(r)} \exp\left[-\frac{1}{q} U_y(r)\right] , \tag{12.13}$$

where C is a normalization constant (cf. (12.9)) and $U_y(r)$ is given by the indefinite integral

$$U_y(r) = -\int^r \frac{A_y(r')}{\mathcal{D}(r')} dr' . \tag{12.14}$$

If the external term $\bar{n}$ is large enough to dominate all fluctuations, we can replace $\mathcal{D}(r)$ in (12.12) by the constant $\bar{n}/C$, and we get

$$\mathcal{D}(r) \to \frac{\bar{n}}{C} \Rightarrow \frac{1}{q} U_y(r) \to \frac{1}{\bar{q}} V_y(r), \tag{12.15}$$

where V_y is given by (12.7) and

$$\bar{q} = \bar{n}/2n_s . \tag{12.16}$$

Thus, in this limit of constant diffusion, (12.13) goes over into the form (12.5) and the discussion following (12.7) applies: V_y plays the role of generalized free energy. In the bistable region, P_{st} has two peaks, one around each of the minima of V_y (steady states x_1, x_3 of Fig. 3(a)). However, due to the smallness of $\bar{q}$, and to the exponential dependence of $P_{st} = C\exp(-V_y/\bar{q})$, even a slight asymmetry of $V_y(x)$ implies that the area of the absolutely stable peak of P_{st} (around the bottom of the deeper well) is much greater than that of the metastable one, so that the absolutely stable steady state is strongly dominant.

Therefore, it is only in a very narrow range of incident fields y, centered around the value $\bar{y}$ for which $V_y(x)$ becomes symmetrical,

$$V_{\bar{y}}(x_1) = V_{\bar{y}}(x_3) \qquad (x_1, x_3 = \text{minima}), \tag{12.17}$$

that the two peaks have comparable areas, so that both states are metastable. This narrow range is the transition region between the one-atom and the cooperative branch.

In the thermodynamic limit, $\bar{q} \to 0$, and the transition becomes discontinuous at $\bar{y}$, just like a first-order phase transition. In terms of the Van der Waals analogy (Sect. 7.2), x_1 and x_3 are like liquid and vapour specific volumes at coexistence. On the other hand, (12.17) is equivalent to

$$-\int_{x_2}^{x_1} A_{\bar{y}}(x)dx = -\int_{x_2}^{x_3} A_{\bar{y}}(x)dx, \tag{12.18}$$

which (cf. Fig. 3(a)) corresponds to the *Maxwell rule* of equilibrium thermodynamics: the straight line $y = \bar{y}$ intersects the response curves so as to form equal areas above and below.

If the intrinsic quantum fluctuations are comparable with the external ones, however, the approximation (12.15) is no longer valid, and one must employ (12.13), (12.14), so that the role of generalized free energy is now played by U_y. Since $\mathcal{D}(r)$ is not constant (cf. (12.12)), we have $U_y \neq V_y$, but (12.14) shows that the extrema of U_y coincide with those of V_y. Since $q \ll 1$ and $\mathcal{D}(r)$ is slowly-varying, these extrema practically coincide with those of P_{st}, so that the steady states still corresponds to x_1, x_3, as before.

Again, due to the smallness of q, we have a very small transition region. Outside of this region, one of the double peaks of P_{st} around the steady states is strongly dominant, and it can be shown that in this situation the peaks are approximately Gaussian ([5], [38]), and the relative fluctuation $\langle (\delta\hat{x})^2 \rangle_{st} / \langle \hat{x} \rangle^2_{st}$ is very small.

The transition region is again centered around the value of y at which the two peaks of P_{st} have equal areas, which implies

$$\langle \hat{x} \rangle_{st} = \frac{1}{2}(x_1 + x_3), \tag{12.19}$$

and in the thermodynamic limit $q \to 0$ and the transition at this point is again discontinuous. However, the transition now occurs at the value $\tilde{y}$ of the incident field corresponding to (12.19), i.e., by (12.13), such that

$$U_{\tilde{y}}(x_1) = U_{\tilde{y}}(x_3) \qquad (x_1, x_3 = \text{minima}), \tag{12.20}$$

and, since $U_y \neq V_y$, *the Maxwell rule is no longer obeyed*. We have therefore a clear difference between this bifurcation far from equilibrium and a first-order phase transition in equilibrium thermodynamics (cf. Sect. 1).

The fluctuations of the transmitted field in the steady state (in particular, the photon statistics)can be obtained from the stationary solution (12.13). The relative fluctuation $\langle (\delta\hat{x})^2 \rangle_{st} / \langle \hat{x} \rangle^2_{st}$ only becomes large in the narrow transition region ($q \ll 1$) where the two peaks of P_{st} have comparable areas and the field is "undecided" about where to go. Numerical examples and figures can be found in the literature ([5], [38]).

The expectation value (12.19) should be interpreted with care: it does *not* coincide with the expected result of individual field measurements. In the transition region, such measurements should yield the values x_1 and x_3 with comparable probabilities, and $\langle \hat{x} \rangle_{st}$ would emerge as the statistical average over many such measurements.

In fact, the above stationary treatment says nothing about the *time scale* over which large fluctuations, such as those between x_1 and x_3, may be expected to occur. Such fluctuations correspond to *tunnelling* over the central hump in the double well, and they may be treated [40] by a generalization of a method due to KRAMERS [41]. One finds that the transition times are proportional to $\exp(\Delta U_y/q)$, where ΔU_y is the height of the generalized free energy barrier that has to be surmounted. In practice, therefore, the probability of such "spontaneous switching" is extremely small, thus reconciling the above treatment with the actual occurrence of bistability (long metastable lifetime) and observation of the hysteresis loop.

13. Side Mode Instabilities

We now go back to the semiclassical theory in order to extend the discussion of stability, which has been confined so far only to the resonant mode (n = 0 in (8.14)), also to the side modes (n = ±1, ±2,... in (8.14)).

13.1 Linear Analysis

We note from (8.23) and (8.14) that

$$\lambda_{-n,j} = \lambda^*_{n,j} \qquad (j = 1, 2, 3), \tag{13.1}$$

so that modes symmetrically placed with respect to the resonant mode are either simultaneously stable or simultaneously unstable, and it suffices to consider $n > 0$.

We confine our discussion to the good cavity case, in which the relevant eigenvalue λ_{n1} is given by (8.34), with $\mathrm{Re}\lambda_{n1}$ given by (8.37), where γ_c is the passive cavity loss. On the other hand, G_n is the effective saturated absorption for small signals, which, as we saw in (8.10) and (7.5), can become a gain in the nonlinear regime, for n = 0. Thus, (8.37) is of the form "gain"-loss.

To simplify the discussion, let us set

$$\gamma_{\parallel} = \gamma_{\perp} = \gamma. \tag{13.2}$$

Then, (8.37) and (8.38) become

$$-\frac{\mathrm{Re}\lambda_{n1}}{\gamma_c} = \frac{\nu_n^4 - 2\gamma^2(x^2 - C - 1)\nu_n^2 + \gamma^4[(1 + x^2)^2 + 2C(1 - x^2)]}{[\gamma^2(1 + x^2) - \nu_n^2]^2 + 4\nu_n^2\gamma^2}. \tag{13.3}$$

The roots of the numerator of this expression are

$$\nu_\pm^2 = S \pm \sqrt{R}, \quad \text{where} \tag{13.4}$$

$$S = \gamma^2(x^2 - C - 1), \quad R = \gamma^4(C^2 - 4x^2). \tag{13.5}$$

In order to have instability for side mode n, ν_n must be such that the numerator of (13.3) is < 0, which requires the following conditions: (i) *The roots are real*, so that

$$R \geqslant 0, \qquad S \pm \sqrt{R} \geqslant 0. \tag{13.6}$$

(ii) *ν_n is between the roots*, i.e.,

$$\nu_- < \nu_n < \nu_+ \tag{13.7}$$

On the cooperative branch, we have (cf. Fig. 3(b)) $x \leqslant x_M \approx 1$, so that $S \pm \sqrt{R} < 0$. Thus, *all side modes are stable on the cooperative branch.* On the one-atom branch, R > 0 requires

$$x < C/2. \tag{13.8}$$

In order that the above conditions be satisfied, one must have [42]

$$C > 2(1 + \sqrt{2}), \tag{13.9}$$

which is somewhat above the bistability threshold (7.9).

We conclude that, *if (13.9) is satisfied, the points on the one-atom branch of Fig. 3(b) lying below x = C/2 (cf. (13.8)) are unstable (in spite of the positive slope) provided that at least one side mode $\nu_n (n > 0)$ lies in the range (13.7).*

The instability range (13.7) for the first side mode n = 1 when C = 20 is the region ABE of the plane $(\nu_1/\gamma, x)$ shown in Fig. 5: it is bounded by the curves $\nu_\pm(x)/\gamma$ defined by (13.4) (of course, by (13.1), *both* of the side modes n = ±1 are unstable within this region). The shaded region is the domain where the modes n = ±2 become unstable, and there are others below it (not shown) for n = ±3, etc. Note (cf. (8.14)) that $\nu_1 = 2\pi c/L$ is the spacing between the modes, so that the instability domains for the higher modes are just scaled-down versions of the domain ABE.

What happens to a small initial perturbation around the stationary solution in the one-atom branch within the instability domain, e.g., at points like 1 and 2 in Fig. 5, where only the side modes n = ±1 are unstable? According to linear stability theory, the perturbations start growing exponentially, like $\exp(\mathrm{Re}\lambda_{11}t)$. However, as soon as they get large, the linear approximation breaks down and we must resort to other methods.

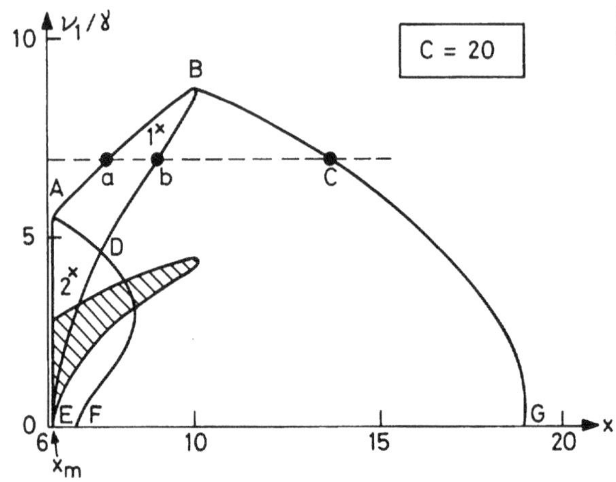

Fig. 5 Instability boundaries for $C = 20$, $\gamma_\parallel = \gamma_\perp = \gamma$ (after [44])

One way to find out what happens is to perform a numerical integration of the complete Maxwell-Bloch equations [43]. The results begin by showing, as expected, an initial exponentially growing oscillation with frequency (cf. (8.34)) $|\mathrm{Im}\lambda_{11}| \approx \nu_1$, i.e., with period $t_c = L/c$ equal to the time of flight around the cavity. The nonlinear effects stabilize this in two different ways, depending on the location of the unstable stationary state in Fig. 5. (i) For a point in the upper portion of the instability domain, such as 1 in Fig. 5, the amplitude of oscillation stabilizes and the oscillation goes on. The corresponding envelope is shown in Fig. 6(a). (ii) For a point in the lower portion of the instability domain, such as 2 in Fig. 5, the system makes an abrupt transition to the lower (cooperative) branch, as shown by the envelope in Fig. 6(b).

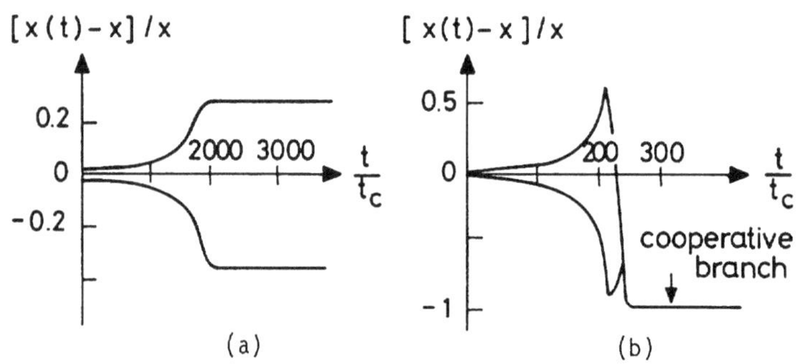

Fig. 6 Envelope of relative deviation from unstable stationary solution as a function of t/t_c (a) For point 1 in Fig. 5 (b) For point 2 (after [43])

Behavior of type (i) is called *self-pulsing*, and that of type (ii) is called *precipitation*. Note in Fig. 6(a) that

$$\langle x(t) \rangle - x < 0 \tag{13.10}$$

in the steady self-pulsing regime. This indicates an attraction by the lower branch that increases as one moves down from 1 towards 2 in Fig. 5, until we go over from self-pulsing to precipitation.

The physical mechanism responsible for these side-mode instabilities is similar to that discussed in Sect. 7.1, the small-signal gain for a resonant saturable absorber, combined with the feedback provided by the boundary conditions. The nonlinear medium plays the role of a parametric oscillator that transfers energy from the incident field to the side modes, which grow at the expense of the resonant mode. Analogous mode-mode coupling effects are found in mode competition in a laser ([1], [11]).

In the self-pulsing regime, the system converts cw light to pulsed light, with period of the order of the cavity time of flight t_c. The effect has not yet been observed: the experimental conditions required are difficult to achieve in an all-optical system [5].

13.2 Nonlinear Analysis

The numerical work described above shows that the exponential growth obtained in the linear approximation is stabilized by nonlinear effects. Therefore, to find the long-time behavior, we must take into account nonlinear terms, that were neglected in the linear stability analysis. This can be done by a procedure that was developed by BENZA and LUGIATO [44].

In order to go over from the Maxwell-Bloch equations (3.35), (3.37), (3.38) (for $\Delta = 0$) to a perturbed set, one replaces the stationary solution (E, $\mathcal{P}$, D) by (E + δE, $\mathcal{P}$ + $\delta\mathcal{P}$, D + δD). The corresponding variation of the nonlinear terms ED and E$\mathcal{P}$ is

$$(E + \delta E)(D + \delta D) - ED = E\delta D + D\delta E + (\delta E\delta D),$$

$$(E + \delta E)(\mathcal{P} + \delta\mathcal{P}) - E\mathcal{P} = E\delta\mathcal{P} + \mathcal{P}\delta E + (\delta E\delta\mathcal{P}). \tag{13.11}$$

In the linearized variational equations (8.1)-(8.3), the *only* terms that were neglected are those within parenthesis on the r.h.s. of (13.11). Thus, to get the *exact* variational equations, it suffices to add these terms to the r.h.s. of (8.1)-(8.3), which may then be written in matrix form as

$$\frac{\partial\vec{q}}{\partial t} = \underline{\underline{L}}\,\vec{q} + \vec{\Psi}_{NL}, \qquad \vec{q}(z, t) = \begin{pmatrix} \delta E \\ \delta\mathcal{P} \\ \delta D \end{pmatrix}, \tag{13.12}$$

where $\underline{\underline{L}}$ is the linear differential (with respect to z) operator defined by the r.h.s. of (8.1)-(8.3), and

$$\vec{\Psi}_{NL} = \frac{\mu}{\hbar}\begin{pmatrix} 0 \\ \delta E\delta\mathcal{P} \\ -\delta E\delta D \end{pmatrix}. \tag{13.13}$$

In the linear stability analysis of Sect. 8, we found the eigenvalues λ_{nj} of $\underline{\underline{L}}$. As was mentioned following (8.20), the corresponding *eigenvectors* $\vec{O}_{nj}(z)$,

$$\underline{\underline{L}}\,\vec{O}_{nj}(z) = \lambda_{nj}\vec{O}_{nj}(z), \tag{13.14}$$

where n = 0, ±1,..., j = 1,2,3, are the *natural modes of the coupled system* in the linearized approximation. We saw that, in the good cavity limit we are considering, j = 1 is a predominantly "field" (cavity) mode, while j = 2,3 are predominantly "atomic" modes, although they already include the atom-field interaction in linear approximation (for this reason, BENZA and LUGIATO [44] call them "dressed modes").

If one now expands the vector $\vec{q}$ in terms of the natural modes,

$$\vec{q}(z, t) = \sum_{n,j} \exp(-i\nu_n t)\, S_{nj}(t)\vec{O}_{nj}(z), \tag{13.15}$$

and one substitutes this expansion in (13.12), one finds a set of coupled nonlinear differential equations for the mode amplitudes $S_{nj}(t)$,

$$\dot{S}_{nj} = (\lambda_{nj} + i\nu_n)S_{nj} + \sum_{n'j'}\sum_{n''j''} \Gamma(nj, n'j', n''j'')S_{n'j'}S_{n''j''}, \tag{13.16}$$

where the explicit expressions of the coefficients Γ can be evaluated [44]. In the linear approximation to (13.16), we recover the exponential behavior in time.

In the situation considered above, where only the modes $n = \pm 1$ become unstable, the only relevant amplitudes should be those for $n = 0$ and $n = \pm 1$. Furthermore, for a good cavity, as we saw in Sects. 8.3, 8.4, we expect that the "atomic" modes $j = 2,3$ can be adiabatically eliminated, so that only the "field" modes $j = 1$ should be relevant.

To perform the adiabatic elimination in (13.16), one makes use of (8.27), according to which the atomic modes settle down very fast in the characteristic "long" time scale of the field, given by $\tau_c = \gamma_c^{-1}$. Therefore, we can set $\dot{S}_{n2} = \dot{S}_{n3} = 0$ in (13.16). Furthermore, one finds that the coefficients $\Gamma(nj, n'j', n''j'')$ are of different orders in the smallness parameter T (like $\mathcal{O}(1)$, $\mathcal{O}(T)$ $\mathcal{O}(T^2)$) depending on the values of $(nj, n'j', n''j'')$. Keeping only the leading order $\mathcal{O}(1)$ (which would be exact in the limit $T \to 0$), the system (13.16) becomes extremely simple (it becomes linear in S_{n2} and S_{n3}), so that *the adiabatic elimination can be performed explicitly* (and *exactly*, as $T \to 0$)

The only remaining amplitudes after the adiabatic elimination (keeping only the relevant field modes) are $S_{01}(t)$, $S_{11}(t)$ and $S_{-1,1}(t) = S_{11}^*(t)$. From (13.15) and (13.12), denoting by $O_{nj}^{(1)}$ the first component of the vector $\vec{O}_{nj}$, the perturbed electric field amplitude becomes

$$
\begin{aligned}
E(z,t) &= E(z) + \delta E(z,t) = E(z) + S_{01}(t) O_{01}^{(1)}(z) \\
&+ [S_{11}(t) O_{11}^{(1)}(z) \exp(-i\nu_1 t) + \text{c.c.}] .
\end{aligned} \tag{13.17}
$$

Correspondingly, the transmitted field (6.41) becomes

$$
x(t) = x + \sigma(t) + \rho(t) \cos[\nu_1 t - \phi(t)] , \quad \text{where} \tag{13.18}
$$

$$
\sigma(t) \equiv A O_{01}^{(1)}(L) S_{01}(t), \quad \frac{1}{2} \rho(t) e^{i\phi(t)} = A O_{11}^{(1)}(L) S_{11}(t),
$$

with $A = \mu/\hbar\sqrt{\gamma_\| \gamma_\perp}$ (cf. (6.3), (6.24)). Thus, $\rho(t)$ is the amplitude associated with the side-mode oscillations, and $\langle \sigma(t) \rangle = \langle x(t) \rangle - x$ is the mean deviation from the stationary solution.

The system (13.16) is reduced to the equations

$$
\frac{d\rho}{dt} = f(\rho, \sigma) , \quad \frac{d\sigma}{dt} = g(\rho, \sigma), \tag{13.19}
$$

where the explicit form of the functions f and g is known ($\phi(t)$ is determined by ρ and σ). The nonlinear stability analysis can therefore be performed just by discussing this two-dimensional system. Note that $x + \sigma \pm \rho$ define the upper and lower envelopes shown in Fig. 6. In particular, we have

$$
\begin{aligned}
\rho(\infty) &\neq 0 \equiv \text{self-pulsing}, \\
\rho(\infty) &= 0 \equiv \text{precipitation}.
\end{aligned} \tag{13.20}
$$

The detailed stability analysis [44] shows that self-pulsing solutions can exist outside of the domain ABE of Fig. 5 obtained in the linear analysis. One finds that the self-pulsing range is the region ABCGFDA in Fig. 5, and the precipitation range is ADE, excluding the domains where higher-order lateral modes become unstable (the shaded domain in Fig. 5 for $n = \pm 2$ and smaller ones below it).

If one increases the field along the one-atom branch, following the horizontal line abC in Fig. 5, self-pulsing appears at a, beginning with zero amplitude (2nd-order transition). The amplitude of oscillation ρ grows until we reach the nonlinear instability boundary at point C, where the system suddenly jumps to the one-atom branch (1st-order transition). If one now decreases the field, the system stays on the one-atom branch until point b in the linear instability boundary is reached, when it suddenly begins to self-pulse with finite amplitude ρ (1st-order transition).

Thus, we have a new type of hysteresis cycle, involving the steady bistable states and the self-pulsing regime. Within the nonlinear instability domain BGFD in Fig. 5, the system requires a finite-amplitude excitation (*hard excitation*) to go into self-pulsing; for a small perturbation (*soft excitation*), it returns to the one-atom branch.

The transition from self-pulsing to precipitation takes place along the line AD of Fig. 5. It can be shown [44] that it corresponds to a line of Hopf bifurcations into unstable limit cycles.

14. Multistability

We now go back to the situation where dispersive effects are present ($\Delta \neq 0$) and no longer restrict our consideration to $\alpha L \ll 1$. This situation has been analysed by IKEDA [13], and it will be convenient to switch back to his version of the Maxwell-Bloch equations, namely, (3.22), (3.27) and (3.28).

14.1. Retarded-Time Integration

By changing variables to the "light frame"

$$z' = z, \qquad \tau = t - \frac{z}{c}, \tag{14.1}$$

we can rewrite (3.22), (3.27) and (3.28) as (cf. (3.34))

$$\text{(a)} \quad \frac{\partial E}{\partial z} = 4\pi i N\mu k p, \tag{14.2}$$

$$\text{(b)} \quad \frac{\partial p}{\partial \tau} = -\gamma_\perp(1 + i\Delta)p - \frac{i\mu}{\hbar} wE, \tag{14.3}$$

$$\text{(c)} \quad \frac{\partial w}{\partial \tau} = -\gamma_\parallel\left(w + \frac{1}{2}\right) + \frac{i\mu}{2\hbar}(p^*E - pE^*), \tag{14.4}$$

where we have dropped the prime on z.

Contrary to the mean field situation, we now want to consider what happens when propagation effects become important. However, we still want the atomic relaxation to be fast in the time scale of field variations. We begin by assuming *fast transverse relaxation*, namely, that

$$\gamma_\perp^{-1} \ll t_c \sim L/c, \tag{14.5}$$

where t_c is the cavity time of flight (this is more restrictive than the "good cavity" condition $\gamma_\perp^{-1} \ll \tau_c$).

This condition allows the adiabatic elimination of the polarization (cf. Sect. 8.4). From (14.3), setting $\partial p/\partial\tau = 0$, we find

$$p = -\frac{i\mu}{\hbar\gamma_\perp(1 + i\Delta)} wE. \tag{14.6}$$

Substituting in (14.2), and taking into account (4.9), we get

$$\frac{1}{E}\frac{\partial E}{\partial z}(z,\tau) = \frac{4\pi N\mu^2 k}{\hbar\gamma_\perp(1 + i\Delta)} w = -2i(n' - 1)kw(z, \tau), \tag{14.7}$$

where n' is the *linear* refractive index of the Bloch medium. At equilibrium (cf. (3.17)), $w = -1/2$, and we recover (4.8).

We now integrate both sides of (14.7) over z, for *fixed* τ (retarded-time integration):

$$\ln E(z',\tau)\Big|_0^z = -2i(n'-1)k\int_0^z w\left(z',\ \tau = t - \frac{z'}{c}\right)dz'.$$

Going back to "laboratory time", this is equivalent to

$$E\left(z,\ \tau + \frac{z}{c}\right) = E(0,\ \tau)\exp\left[-2i(n'-1)kW(z,\ \tau)\right], \tag{14.8}$$

where $\tau + z/c = t$ (τ fixed), and

$$\frac{1}{z}W(z,\ \tau) = \frac{1}{z}\int_0^z w\left(z',\ \tau + \frac{z'}{c}\right)dz' \tag{14.9}$$

is the *average atomic inversion* felt by a wavefront as it traverses the path (0,z) (from $z' = 0$ at $t - z/c$ to $z' = z$ at t). From (4.9),

$$-2i(n' - 1)k = \alpha(1 - i\Delta), \tag{14.10}$$

where α is given by (4.11).

Substituting (14.6) and (14.8) in (14.4), we get, taking into account (14.10),

$$\frac{\partial}{\partial\tau}w\left(z',\ \tau + \frac{z'}{c}\right) = -\gamma_{\|}\left[w\left(z',\ \tau + \frac{z'}{c}\right) + \frac{1}{2}\right]$$
$$- \frac{\mu^2}{\hbar^2\gamma_{\perp}(1 + \Delta^2)}w\left(z',\ \tau + \frac{z'}{c}\right)|E(0,\ \tau)|^2\exp\left[2\alpha W(z',\ \tau)\right].$$

Integrating both sides over z' from 0 to z, we get, by (14.9),

$$\frac{\partial}{\partial\tau}W(z,\ \tau) = -\gamma_{\|}\left[W(z,\ \tau) + \frac{z}{2}\right]$$
$$- \frac{\mu^2}{\hbar^2\gamma_{\perp}(1 + \Delta^2)}|E(0,\ \tau)|^2\int_0^z \exp\left[2\alpha W(z',\ \tau)\right]\frac{d}{dz'}W(z',\ \tau)dz';$$

so that, finally,

$$\frac{\partial}{\partial\tau}W(z,\ \tau) = -\gamma_{\|}\left[W(z,\ \tau) + \frac{z}{2}\right]$$
$$- \frac{\mu^2|E(0,\ \tau)|^2}{2\alpha\hbar^2\gamma_{\perp}(1 + \Delta^2)}\left\{\exp\left[2\alpha W(z,\ \tau)\right] - 1\right\}. \tag{14.11}$$

The coupled equations (14.8) and (14.11) represent the result of the retarded-time integration of (a), (b), (c).

However, we must still satisfy the boundary conditions (2.11) and (2.12). For (2.12), we need the expression of $E\left(L,\ t - \frac{\ell}{c}\right)$. Setting

$$z = L,\ \tau = t - \frac{L}{c} \Rightarrow \tau + \frac{z}{c} = t - \frac{L - L}{c} = t - \frac{\ell}{c} \tag{14.12}$$

in (14.8), we get

$$E\left(L, t - \frac{\ell}{c}\right) = E\left(0, t - \frac{L}{c}\right) \cdot$$

$$\times \exp\left[\alpha(1 - i\Delta)\, W\left(L, t - \frac{L}{c}\right)\right] . \qquad (14.13)$$

Let us introduce the dimensionless variables

$$\tilde{t} = \gamma_{\|} t, \qquad (14.14)$$

$$\phi(\tilde{t}) = \frac{1}{L} W\left(L, t - \frac{L}{c}\right) , \qquad (14.15)$$

which is the "light frame" average atomic inversion over the *whole sample*, and

$$\epsilon(z, \tilde{t}) = \frac{\mu E(z, t)}{2\hbar\sqrt{\gamma_{\|}\gamma_{\perp}(1 + \Delta^2)}} , \qquad (14.16)$$

analogous to F in (6.3).

Then, substituting (14.12) in (14.11) and dividing both sides by $\gamma_{\|} L$, we get

$$d\phi(\tilde{t})/d\tilde{t} = -\left[\phi(\tilde{t}) + \frac{1}{2}\right]$$

$$-2|\epsilon(0, \tilde{t} - \tilde{t}_c)|^2 \left[e^{2\alpha L \phi(\tilde{t})} - 1\right] / \alpha L, \quad \text{where} \qquad (14.17)$$

$$\tilde{t}_c = \gamma_{\|} L/c = \gamma_{\|} t_c \qquad (14.18)$$

is the dimensionless round-trip time.

By (14.13) and (14.15),

$$e^{ikL} E\left(L, t - \frac{\ell}{c}\right) = E(0, t- t_c) \exp[ikL + \alpha L(1 - i\Delta)\phi]$$

$$= E(0, t - t_c) \exp\left[\alpha L\phi - i\alpha L\Delta\left(\phi + \frac{1}{2}\right) + i\left(kL + \alpha L\frac{\Delta}{2}\right)\right] .$$

From (14.10) and (2.16),

$$kL + \alpha L\frac{\Delta}{2} = kL + (n - 1)kL = \delta = 2m\pi + \delta_0, \qquad (14.19)$$

where δ_0 is the *linear Fabry-Perot round-trip phase-shift* (reduced modulo 2π).

Therefore, (2.12) becomes

$$\epsilon(0,\tilde{t}) = \sqrt{T}\,\epsilon_I(\tilde{t}) + R\epsilon(0,\tilde{t}-\tilde{t}_c)\exp(i\delta_0) \cdot$$

$$\times \exp[\alpha L\phi(\tilde{t})] \exp\left\{-i\alpha L\Delta\left[\phi(\tilde{t}) + \frac{1}{2}\right]\right\} , \qquad (14.20)$$

where (cf. (14.16), (6.27))

$$\epsilon_I(\tilde{t}) = \frac{\mu E_I(t)}{2\hbar\sqrt{\gamma_{\|}\gamma_{\perp}(1 + \Delta^2)}} . \qquad (14.21)$$

The other boundary condition (2.11) can be written as $E_T(t - \ell/c) = \sqrt{T}\, E(L, t - \ell/c)$, so that it becomes

$$\epsilon_T(\tilde{t}) = \sqrt{T}\, \epsilon(0, \tilde{t} - \tilde{t}_c) e^{i(\delta_0 - kL)} \cdot$$
$$\times \exp[\alpha L \phi(\tilde{t})] \exp\left\{-i\alpha L \Delta \left[\phi(\tilde{t}) + \frac{1}{2}\right]\right\} , \tag{14.22}$$

where we have defined

$$\epsilon_T(\tilde{t}) = \frac{\mu E_T\left(t - \frac{\ell}{c}\right)}{2\hbar\sqrt{\gamma_{\parallel}\gamma_{\perp}(1 + \Delta^2)}} . \tag{14.23}$$

Equations (14.17) and (14.20) can be regarded as a *set of coupled differential-diference equations* in the variables $\phi(\tilde{t})$, $\epsilon(0, \tilde{t})$, with driving term $\epsilon_I(\tilde{t})$. Equation (14.22) just serves to determine the output in terms of the solutions of this set.

In the *linear approximation*, by (4.3), (14.9) and (14.15),

$$\phi = \phi_{eq} = -1/2 \qquad \text{(linear)}, \tag{14.24}$$

and (14.20), (14.22) reduce to (2.14). Therefore, in (14.20), we can write

$$\tilde{\delta} = \delta_0 - \alpha L \Delta \left(\phi + \frac{1}{2}\right) \equiv \delta_0 + \Psi, \tag{14.25}$$

where Ψ is the *nonlinear contribution* to the total roundtrip phase-shift $\tilde{\delta}$, to be obtained from the solution of (14.17), the second term of which represents the nonlinear effect of the field on the average atomic inversion.

Similarly, the last term in (14.20) contains the effects of the nonlinear contribution to the phase-shift as well as those of the nonlinear absorption $\alpha L\left(\phi + \frac{1}{2}\right)$. Equations (14.17) and (14.20) are due to IKEDA [13]. To determine a time-dependent solution, one would need to specify $\phi(0)$ and $\epsilon(0, \tilde{t})$ for $-\tilde{t}_c \leq \tilde{t} \leq 0$.

14.2 Stationary Solution. Multistability

The stationary solution is obtained by setting $d/d\tilde{t} = 0$, so that all quantities become time-independent. We can then eliminate ϵ between (14.20) and (14.22), yielding (cf. (14.25))

$$\frac{\epsilon_T}{\epsilon_I} = \frac{T \exp[\alpha L \phi + i(\tilde{\delta} - kL)]}{1 - R \exp(\alpha L \phi + i\tilde{\delta})} , \tag{14.26}$$

where ϕ is given by the implicit equation (cf. (14.17), (14.22))

$$\frac{\alpha L}{2}\left(\phi + \frac{1}{2}\right) \Big/ \left(1 - e^{2\alpha L \phi}\right) = |\epsilon|^2 = e^{-2\alpha L \phi} |\epsilon_T|^2 / T , \quad \text{i.e.,}$$

$$\frac{\alpha L}{2} \frac{\left(\phi + \frac{1}{2}\right)}{(e^{-2\alpha L \phi} - 1)} = \frac{|\epsilon_T|^2}{T} . \tag{14.27}$$

The result (14.26) reduces to (2.15) in the linear approximation (14.24). It yields the *state equation*

$$\frac{|\epsilon_T|^2}{|\epsilon_I|^2} = \frac{T^2}{(e^{-\alpha L\phi} - R)^2 + 4Re^{-\alpha L\phi}\sin^2(\tilde{\delta}/2)} , \tag{14.28}$$

which is just the *nonlinear Fabry-Perot transmissivity*, reducing to (2.19) in the linear case.

In the *mean field limit* (6.32), (9.3), (9.6), we see that (14.28) reduces to

$$\frac{X}{Y} = \left[\left(1 - \frac{\alpha L\phi}{T}\right)^2 + \left(\theta - \frac{\alpha L\phi}{T}\Delta\right)^2\right]^{-1} , \tag{14.29}$$

in terms of the variables (9.13), (9.14), whereas (14.27), taking into account (4.11) and (6.15), becomes

$$-\frac{\phi + \frac{1}{2}}{4\phi} = \frac{X}{1 + \Delta^2} \Rightarrow \frac{\alpha L\phi}{T} = -\frac{2C}{1 + \Delta^2 + X} . \tag{14.30}$$

Substituting (14.30) into (14.29), we recover the mean field state equation (9.15), which is therefore a limiting form of (14.28).

Since the r.h.s. of (14.27) is $\geqslant 0$, ϕ cannot become positive: it ranges from $\phi_{eq} = -1/2$ at $\epsilon_T = 0$ to the saturation value $\phi = 0$ as $|\epsilon_T| \to \infty$, increasing monotonically with $|\epsilon_T|^2$. Thus, from (14.25),

$$\tilde{\delta}(\epsilon_T = 0) = \delta_0 ; \quad \tilde{\delta}(|\epsilon_T| \to \infty) = \delta_0 - \frac{\alpha L\Delta}{2} . \tag{14.31}$$

It follows that, *if* $\alpha L\Delta$ *is large enough* (cf. the beginning of Sect. 14), the factor $\sin^2(\tilde{\delta}/2)$ in (14.28) will oscillate as the field intensity increases (due to the intensity-dependent nonlinear phase-shift in (14.25)), and one can go over from bistability to *multistability*. Physically, this corresponds to scanning over several Fabry-Perot resonances due to the nonlinear changes in refractive index.

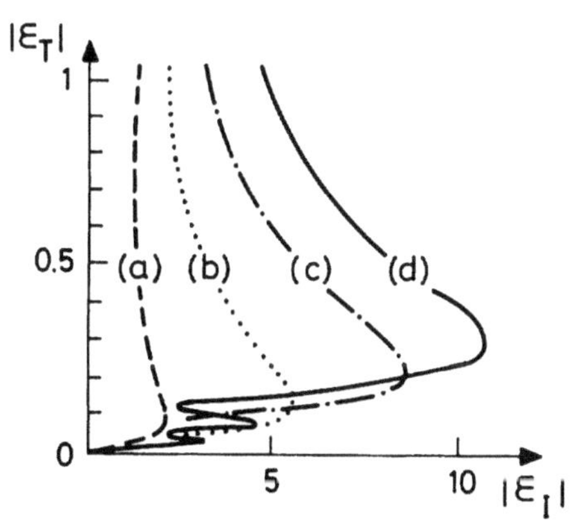

Fig. 7 Plot of the state equation (14.28) for: (a) $\alpha L\Delta$=0; (b) $\alpha L\Delta=2\pi$; (c) $\alpha L\Delta=4\pi$; (d) $\alpha L\Delta=6\pi$ (after [13])

Fig. 7 (cf. [13]) shows plots of the state equation (14.28) for various values of $\alpha L\Delta$, with $\delta_0 = 0$, $T = 0.05$ and $\alpha L = 4$ (so that $C = 20$). For $\alpha L\Delta = 0$, we have a curve similar to that for absorptive bistability in the mean field limit. Note that the scales for $|\epsilon_T|$ and $|\epsilon_I|$ are different, and that $|\epsilon_T| = \sqrt{T}\, x/2$ and $|\epsilon_I| = \sqrt{T}\, y/2$, where $\sqrt{T}/2 \approx 0.1$ for the values of the parameters chosen. As $\alpha L\Delta$ increases, we get more and more peaks (distorted Fabry-Perot resonances) in the plot, so that multiple stationary states can exist. However, we must discuss the question of their stability. This will be done in the following Section.

15. Ikeda Instability. Approach to Chaos

We now assume that, in addition to (14.5), we also have *fast longitudinal relaxation*, so that

$$\gamma_{\|}^{-1} \ll t_c \Rightarrow \tilde{t}_c = \gamma_{\|} t_c \gg 1, \tag{15.1}$$

and the population inversion can also be adiabatically eliminated, setting $d/d\tilde{t} = 0$ in (14.17).

With

$$\tilde{t}_n = \tilde{t}_o + n\tilde{t}_c, \quad \epsilon_n = \epsilon(0, \tilde{t}_n), \quad \phi_n = \phi(\tilde{t}_n), \tag{15.2}$$

equations (14.17), (14.20) and (14.22) simplify to

$$\epsilon_n = \sqrt{T}\,\epsilon_{In} + R\epsilon_{n-1}\exp(\alpha L\phi_n).\quad \exp\{-i[\alpha L\Delta\left(\phi_n + \tfrac{1}{2}\right) - \delta_o]\}, \tag{15.3}$$

$$\epsilon_{Tn} = \sqrt{T}\,\epsilon_{n-1}\exp(\alpha L\phi_n)\exp\{-i[\alpha L\Delta\left(\phi_n + \tfrac{1}{2}\right) - \delta_o + kL]\}, \tag{15.4}$$

$$\left(\phi_n + \tfrac{1}{2}\right)\Big/[1 - \exp(2\alpha L\phi_n)] = 2|\epsilon_{n-1}|^2\Big/\alpha L, \tag{15.5}$$

which is a set of *finite-difference equations* at intervals corresponding to the cavity roundtrip time.

One can now discuss the stability of the multiple stationary states by linear stability analysis ([13], [45]). The results for curve (d) ($\alpha L\Delta = 6\pi$) of Fig. 7 are shown in Fig. 8, where unstable branches are represented in dotted line (note the breaks and changes in the scales). As expected, all branches with negative slope are unstable, but *instability is also found along large portions of branches with positive slope*. This is known as the *Ikeda instability*.

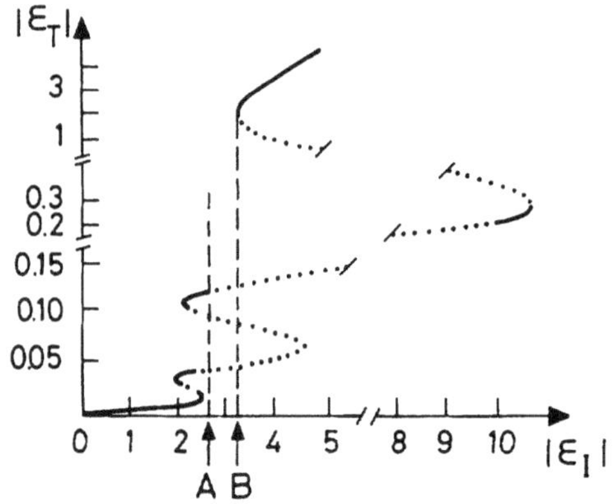

Fig. 8 Stability for $\alpha L\Delta = 6\pi$ (after [13])

Note that the stable portions (shown in solid line in Fig. 8) are located next to the turning points, the analogues of m and M in the bistable case. In the fast limit of longitudinal relaxation, $\gamma_{\|} t_c \to \infty$, the eigenvalues for *all field modes* become unstable simultaneously, rather than just for a finite number of them. Indeed, the condition (15.1) is equivalent to $c/L \ll \gamma_{\|}$, so that many cavity modes fall within the atomic linewidth (cf. Sect. 13).

In Fig. 8, one can see a range of incident fields $|\epsilon_I|$, between points A and B, where *all stationary solutions are unstable*. What happens within this range? In order to find out, IKEDA [13] started out from an initial value ϵ_o located inside it, and he iterated (15.2)-(15.4) numerically. The resulting output is a series of points that wander *chaotically* with (discrete) time. He also discussed what happens when $|\epsilon_I|$ is varied around this region: hysteresis effects and period-doubling bifurcations are found [13].

15.1 The Dispersive Limit

The discussion can be simplified [46] by going over to the *dispersive limit* [45]

$$|\Delta| \gg 1, \qquad 0 < \phi + \frac{1}{2} \ll 1, \tag{15.6}$$

but still allowing (cf. (14.25))

$$|\Psi| = \alpha L|\Delta| \left(\phi + \frac{1}{2}\right) \gtrsim 1. \tag{15.7}$$

The second condition (15.6) implies that the average atomic deviation from the ground state is still small, so that we can replace

$$\exp[\alpha L\phi(\tilde{t})] \rightarrow \exp(-\alpha L/2) \tag{15.8}$$

in (14.17), (14.20) and (14.22). We get from (14.17)

$$d\phi(\tilde{t})/d\tilde{t} = -\left[\phi(\tilde{t}) + \frac{1}{2}\right] - 2|\epsilon(0, \tilde{t} - \tilde{t}_c)|^2 (e^{-\alpha L} - 1)/\alpha L. \tag{15.9}$$

Let us introduce a new rescaling of the field

$$E = [2|\Delta|(1 - e^{-\alpha L})]^{1/2} \epsilon. \tag{15.10}$$

Then, multiplying both sides of (15.9) by $(-\alpha L\Delta)$, recalling the expression (14.25) of the nonlinear phase-shift and going back to t from $\tilde{t} = \gamma_{\parallel} t$, we get

$$\gamma_{\parallel}^{-1}\dot{\Psi}(t) = -\Psi(t) + \zeta|E(t - t_c)|^2, \quad \text{where} \tag{15.11}$$

$$\zeta = \operatorname{sign}(\Delta). \tag{15.12}$$

Also (14.20) becomes much simpler:

$$E(t) = A + BE(t - t_c) \exp\{i[\Psi(t) + \delta_0]\}, \tag{15.13}$$

where we write $E(t)$ for $E(0, t)$, and

$$A = \sqrt{T}\, E_I, \qquad B = Re^{-\alpha L/2}. \tag{15.14}$$

The parameter A corresponds to the incident field and B is an attenuation factor. Finally from (14.22),

$$|E_T|^2 = Te^{-\alpha L} |E(t - t_c)|^2 \tag{15.15}$$

measures the transmitted intensity in the same dimensionless scale (15.10).

From (15.10) and (14.16), we see that

$$E = \left[\frac{|\Delta|\mu^2(1 - e^{-\alpha L})}{2\hbar^2\gamma_{\parallel}\gamma_{\perp}(1 + \Delta^2)}\right]^{1/2} E, \tag{15.16}$$

which may be rewritten as

$$\mathcal{E} = \left[k|n_2| \frac{(1 - e^{-\alpha L})}{\alpha} \right]^{1/2} E \tag{15.17}$$

where n_2 is the quadratic coefficient of refractive index (Kerr effect), given by (6.19). From now on, for definiteness, we choose $\zeta = +1$ in (15.12); for $\zeta = -1$, it suffices to change $\mathcal{E} \to \mathcal{E}^*$, $\delta_0 \to -\delta_0$.

The *stationary solution* of (15.11), (15.13) is

$$\Psi = |\mathcal{E}|^2, \tag{15.18}$$

$$\mathcal{E} = A + B\mathcal{E} \exp[\, i(|\mathcal{E}|^2 + \delta_0)] \,, \tag{15.19}$$

yielding the state equation

$$|\mathcal{E}|^2 \, [1 - 2B \cos(|\mathcal{E}|^2 + \delta_0) + B^2] = A^2. \tag{15.20}$$

This is represented by the curve in full line in Fig. 9, which corresponds to $B = 0.4$ and $\delta_0 = 0$. The curve is a plot of A (which is proportional to $\mathcal{E}_I$, by (15.14)) as a function of $|\mathcal{E}|$, which is proportional to $|\mathcal{E}_T|$, by (15.15). The stable portions of this curve, derived by linear stability analysis ([46]), are shown in thick line: they are adjacent to the turning points and get narrower as A increases, so that most of the stationary solutions become unstable.

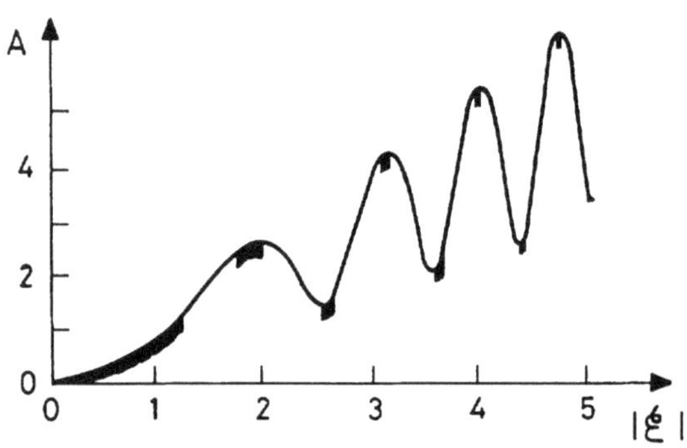

Fig. 9 Plot of (15.20) (after [46])

In the fast limit of longitudinal relaxation (15.1), one gets from (15.11) and (15.13), in a notation similar to (15.2),

$$\Psi_n = |\mathcal{E}_{n-1}|^2, \tag{15.21}$$

$$\mathcal{E}_n = A + B\,\mathcal{E}_{n-1} \exp[\, i(|\mathcal{E}_{n-1}|^2 + \delta_0)] \,. \tag{15.22}$$

Note that (15.22) is now a difference equation involving only the field, which can also be regarded as a *two-dimensional mapping* (since $\mathcal{E}_n$ is complex). The stationary solutions are fixed points of this mapping, that lie on the curve (15.19). From linear stability analysis ([46]), one finds, as mentioned above, that most stationary solutions are unstable when $A^2B \gg 1$.

To find out what happens when the stationary solution becomes unstable, it suffices to plot the time series $\mathcal{E}_0$, $\mathcal{E}_1$, $\mathcal{E}_2$,... on a computer. One finds that, as the instability threshold is crossed, the solution first bifurcates into a self-pulsing one, with period $2t_c$. Then, as one continues to increase A ($\propto \mathcal{E}_I$), it undergoes a sequence of period-doubling bifurcations, leading into the chaotic regime, where the points $\mathcal{E}_n$ wander erratically in the complex plane. For $B \ll 1$, they tend to cluster around a roughly spiral-shaped trajectory that may be regarded as a strange attractor ([46]).

When the time variations or fluctuations become very fast, approaching the atomic time scale $\gamma_{\parallel}^{-1}$, the adiabatic approximation $\gamma_{\parallel}^{-1}d/dt = d/d\tilde{t}_c \to 0$ is no longer valid, and one has to go back to (15.11), (15.13). These equations simplify still further when

$$B \ll 1, \tag{15.23}$$

which does not prevent A^2B from being $\gtrsim 1$.

In this case, iterating (15.13) once, we get

$$E(t) = A + B\,[A + \mathcal{O}(B)]\,\exp\{i[\Psi(t) + \delta_0]\},$$

which may be approximated by

$$E(t) \approx A + AB\exp\{i[\Psi(t) + \delta_0]\}\ , \tag{15.24}$$

so that $E(t)$ is determined by the *instantaneous* value of $\Psi(t)$ in this approximation. Thus,

$$|E(t)|^2 \approx A^2\{1 + 2B\cos[\Psi(t) + \delta_0]\}, \tag{15.25}$$

again neglecting terms of order B^2.

Substituting (15.25) in (15.11), we get

$$\gamma_{\parallel}^{-1}\dot{\Psi}(t) = -\Psi(t) + A^2\{1 + 2B\cos[\Psi(t - t_c) + \delta_0]\}, \tag{15.26}$$

which is a *differential-difference equation* for the nonlinear phase-shift $\Psi(t)$. In this approximation, all the dynamics is contained in this equation; once we solve it, the field is determined by (15.24).

For the stationary solution, (15.18) holds and (15.25) is then equivalent to (15.20) for $B \ll 1$. To perform a linear stability analysis, we go back to the dimensionless variable $\tilde{t} = \gamma_{\parallel}t$, setting

$$\Psi(\tilde{t}) = \Psi + \delta\Psi(\tilde{t}) = \Psi + \delta\Psi_\lambda\exp(\lambda\tilde{t}). \tag{15.27}$$

Substituting in (15.26), we get

$$(\lambda + 1)\exp(\lambda\tilde{t}_c) = -2A^2B\sin(\Psi + \delta_0). \tag{15.28}$$

From (15.25) and (15.18),

$$G \equiv \frac{d}{dA^2}|E|^2 = -2A^2B\sin(\Psi + \delta_0)G + 1, \tag{15.29}$$

so that (15.28) becomes

$$\lambda + 1 = \left(1 - \frac{1}{G}\right)\exp(-\lambda\tilde{t}_c), \tag{15.30}$$

a complex transcendental equation that may be solved by asymptotic methods [47].

By (15.29), we see that G measures the slope of the steady-state characteristic plotted in Fig. 9. It is readily verified that one must have $\mathrm{Re}\lambda < 0$ when $|1 - 1/G| < 1$, so that all modes are (linearly) stable for

$$0 < \frac{1}{G} = \frac{dA^2}{d|E|^2} < 2, \tag{15.31}$$

which corresponds to the positive-slope neighborhood of the peaks shown in Fig. 9 (a smaller and smaller neighborhood as the peaks get sharper).

The transition from period-doubling bifurcations to optical chaos is directly reflected in the power spectrum of $E(t)$ [46]. When only one pair of modes is unstable, corresponding to the first bifurcation, the power spectrum shows sharp peaks at the fundamental frequency and smaller peaks at its harmonics (due to the nonlinearity). As A increases, so does the number of unstable modes, and when the chaotic regime is reached, one finds the characteristic transition to a broad, continuous power spectrum, on top of which one can still see peaks close to eigenmode frequencies, including higher-frequency modes that remain stable. This suggests that energy transfer from unstable lower-frequency modes to stable higher-frequency ones takes place through mode-mode coupling.

The above discussion allows us to form a picture of the physical mechanism that leads to chaos in the Ikeda instability. The Fabry-Perot mechanism (feedback) makes the field in the sample strongly dependent on the total phase-shift. In the intense-field regime, however, the dispersive nonlinearity renders the phase-shift field-dependent. Therefore, for each turn around the cavity (delayed feedback), one gets erratic wandering of the field: this *sensitive dependence on initial conditions* is characteristic of chaos.

The first observation of the Ikeda instability was made in a hybrid system [48], where the predicted period-doubling bifurcations to chaos were seen; more recently [49], the bifurcations to chaos were observed in an all-optical bistable system.

To show that life imitates art, and that the concept of pitchfork bifurcations to chaos was intuitively grasped long ago by a great Latin American writer, let me finish with a quotation:

"Me detuve, como es natural, en la frase: *Dejo a los varios porvenires (no a todos) mi jardín de senderos que se bifurcan*. Casi en el acto comprendí; *el jardín de los senderos que se bifurcan* era la novela caótica; la frase *varios porvenires (no a todos)* me sugirió la imagen de la bifurcación en el tiempo, no en el espacio... he conjeturado el plan de ese caos... Esa trama de tiempos que se aproximan, se bifurcan, se cortan o que secularmente se ignoran, abarca *todas* las posibilidades."

(Jorge Luis Borges, "El Jardín de Senderos que se Bifurcan").

Acknowledgments
I would like to thank Marlan Scully for his hospitality at the Institute for Modern Optics in Alburquerque, New Mexico, where a preliminary version of these lectures was given. This work was supported by CNPq and FINEP.

References

1. H.M. Nussenzveig: Introduction to Quantum Optics (Gordon and Breach, New York, 1973).
2. C.M. Bowden, M. Ciftan and H.R. Robl, editors: Optical Bistability (Plenum, New York, 1981)
3. R. Bonifacio, editor: Dissipative Systems in Quantum Optics (Springer-Verlag, Berlin, 1982)
4. E. Abraham and S.D. Smith: Rep. Prog. Phys. 45, 815 (1982)
5. L.A. Lugiato: "Theory of Optical Bistability", in Progress in Optics, vol. 21, edited by E. Wolf (North-Holland, Amsterdam, 1984)
6. R. Bonifacio and L. Lugiato: Lett. Nuovo Cimento 21, 505 (1978)
7. M. Born and E. Wolf: Principles of Optics (Pergamon Press, London, 1959)
8. H.M. Nussenzveig: "Causality and Analyticity in Optics", in Optics in Four Dimensions, edited by M. Machado and L.A. Narducci (American Institute of Physics, New York, 1980)
9. H. Haken: "Laser Theory", in Handbuch der Physik, vol. XXV/2c (Springer-Verlag, Berlin, 1970)

10. W.H. Louisell: Quantum Statistical Properties of Radiation (Wiley, New York, 1973)
11. M. Sargent III, M.O. Scully and W.E. Lamb, Jr.: Laser Physics (Addison-Wesley, Reading, 1974)
12. L. Allen and J.H. Eberly: Optical Resonance and Two-Level Atoms (Wiley, New York, 1975)
13. K. Ikeda: Opt. Commun. 30, 257 (1979)
14. A. Szöke, V. Daneu, J. Goldhar and N.A. Kurnit: Appl. Phys. Letters 15, 376 (1969)
15. H. Seidel: U.S. Patent 3,610,731 (1971; filed in 1969)
16. H.M. Gibbs, S.L. McCall and T.N.C. Venkatesan: Phys. Rev. Lett. 36, 1135 (1976)
17. B.R. Mollow: Phys. Rev. A5, 2217 (1972)
18. S.L. McCall: Phys. Rev. A9, 1515 (1974)
19. R. Bonifacio and L.A. Lugiato: Phys. Rev. A11, 1507 (1975)
20. D.E. Grant and H.J. Kimble: Opt. Lett. 7, 353 (1982); F.T. Arecchi, G. Giusfredi, E. Petriella and P. Salieri: J. Appl. Phys. 29, 79 (1982)
21. R. Bonifacio and L.A. Lugiato: Phys. Rev. A18, 1129 (1978)
22. D.E. Grant and H.J. Kimble: Opt. Commun. 44, 415 (1983)
23. F. Rohart and B. Macke: J. Physique 41, 837 (1980); G. Grynberg, F. Biraben and E. Giacobino: Appl. Phys. B26, 155 (1981)
24. R. Bonifacio and L.A. Lugiato: Lett. Nuovo Cimento 21, 517 (1978); R. Bonifacio, M. Gronchi and L.A. Lugiato: Nuovo Cimento B53, 311 (1979)
25. S.S. Hassan, P.D. Drummond and D.F. Walls, Opt. Commun: 27, 480 (1978)
26. A.T. Rosenberger, L.A. Orozco and H.J. Kimble: Phys. Rev. A28, 2569 (1983)
27. F. Haake: Statistical Treatment of Open Systems by Generalized Master Equations, Springer Tracts in Modern Physics, vol. 66 (Springer-Verlag, Berlin, 1973)
28. E.P. Wigner: Phys. Rev. 40, 749 (1932). Cf. also J.E. Moyal: Proc. Camb. Phil. Soc. 45, 99 (1939)
29. M. Gronchi and L.A. Lugiato: Lett.Nuovo Cim. 23, 593 (1978)
30. N.G. Van Kampen: Adv. Chem. Phys. 15, 65 (1969)
31. L.A. Lugiato: Nuovo Cimento B50, 89 (1979)
32. J.P. Gordon: Phys. Rev. 161, 367 (1967)
33. F. Casagrande and L.A. Lugiato: Nuovo Cimento B55, 173 (1980)
34. M. Lax: Phys. Rev. 129, 2342 (1963); 157, 213 (1967)
35. P.D. Drummond and D.F. Walls: J. Phys. A13, 725 (1980); Phys. Rev. A23, 2563 (1981)
36. B.R. Mollow: Phys. Rev. 188, 1969 (1969)
37. H. Haken: Rev. Mod. Phys. 47, 67 (1975)
38. R. Bonifacio, M. Gronchi and L.A. Lugiato: Phys. Rev. A18, 2266 (1978)
39. A. Schenzle and H. Brand: Opt. Commun. 23, 151 (1978); 27, 485 (1978); 31, 401 (1979)
40. R. Bonifacio, L.A. Lugiato, J.D. Farina and L.M. Narducci: IEEE J. Quantum Electron. QE-17, 357 (1981)
41. H.A. Kramers: Physica 7, 284 (1940)
42. R. Bonifacio and L.A. Lugiato: Lett. Nuovo Cimento 21, 510 (1978)
43. R. Bonifacio, M. Gronchi and L. A. Lugiato: Opt. Commun. 30, 129 (1979); M. Gronchi, V. Benza, L.A. Lugiato, P. Meystre and M. Sargent III, Phys. Rev. A. 24, 1419 (1981)
44. V. Benza and L.A. Lugiato: Z. Phys. B35, 383 (1979); 47,79 (1982); L.A. Lugiato, V. Benza, L.M. Narducci and J.D. Farina: Z. Phys. B49, 351 (1983)
45. H.J. Carmichael, R.R. Snapp and W.C. Schieve: Phys. Rev. A26, 3408 (1982)
46. K. Ikeda, H. Daido and O. Akimoto: Phys. Rev. Lett. 45, 709 (1980)
47. H.M. Nussenzveig: Nucl. Phys. 11, 499 (1959); Nuovo Cimento 20, 694 (1961)
48. H.M. Gibbs, F.A. Hopf, D.L. Kaplan and R.L. Shoemaker: Phys. Rev. Lett. 46, 474 (1981); F.A. Hopf, D.L. Kaplan, H.M. Gibbs and R.L. Shoemaker: Phys. Rev. A25, 2172 (1982)
49. H. Nakatsuka, S. Asaka, H. Itoh, K. Ikeda and M. Matsuoka: Phys. Rev. Lett. 50, 109 (1983)

Part III **Fluids**

Recent Developments in Rayleigh-Benard Convection and Interface Dynamics

A. Libchaber

The James Franck Institute, The Enrico Fermi Institute, The University of Chicago
Chicago, IL 60637, USA

We present three topics in fluid dynamics which have recently sparkled the interest of the physics community:

1) Rayleigh-Benard convection
2) Saffman Taylor Instability
3) Solid liquid interface dynamics.

The common ground is that in all cases, non-linear dynamics is the relevant basic ingredient. One finds always a marginal stability analysis for the onset of instabilities. In the nonlinear regime, the wavelength selection sits as an essentially unanswered problem. Finally, the transition to disordered pattern in time or space leads to relevant questions.

I will try to give a simple, often experimental, introduction to those problems.

I The Rayleigh Benard Instability

Since Benard early experiments [1] in 1900, thermal convection has been one of the main paradigms for the study of fluid instabilities. The theoretical approach has also moved a long way since the early work of Rayleigh [2] and JEFFREYS [3]. The main architect for the general framework of thermal convection is Busse, whose review article [4] is the best general reference. A simple introduction can be found in a recent Scientific American article [5]. For the evolution of the experimental scene the articles by KRISHNAMURTI [6], GOLLUB [7], RAYLEIGH [8] and AHLERS [9] are good references.

Why are Rayleigh-Benard experiments such a paradigm for the observation of hydrodynamic instabilities?

The primary instability, transition from heat diffusion to heat convection, is well understood and documented. It is a second-order mean-field phase-transition. The convective pattern may develop as an ordered two-dimensional flow, for low-aspect-ratio cell, or when an ordering field is applied [10,11]. Cylindrical rolls, parallel to each other, will then define a new ordered state. Building upon it, and increasing the control parameter (Rayleigh number), a sequence of bifurcations will evolve up to a weak turbulent state. Let us emphasize that, in the more general case of a large-aspect-ratio cell and with no ordering field, the convective state above the transition will be disordered, and the route to weak turbulence more complicated, and not adequately approached by a dynamical-system theory [9].

It is also one of the rare cases, in nonlinear physics, where a detailed analysis of the secondary instabilities has been successfully performed [4], as a function of the Rayleigh number and of the Prandtl number. Finally, from a pure experimental point of view, the localization of the fluid flow in a small and closed vessel, with excellent thermal regulation of the boundaries, leads to very high signal-to-noise performance in the measurements [12]. In the oscillating state of the fluid, signal-to-noise ratio can be of the order of 60 dB.

Ia Basics of Thermal Convection

The Oberbeck-Boussinesq approximation is used; this means that all fluid properties are assumed constant, with the exception of the temperature-dependence of the density, taken into account in the gravity term. To get dimensionless numbers one used d as a length scale, d^2/κ as a time scale, $(T_2-T_1)/R$ as a temperature scale.

The governing equations become then

$$P^{-1}(\frac{\partial}{\partial t}\vec{v} + (v\cdot\nabla)\vec{v}) = -\vec{\nabla}p + \theta\vec{k} + \nabla^2\vec{v}, \tag{1}$$

$$\vec{\nabla}\cdot\vec{v} = 0, \tag{2}$$

$$\frac{\partial}{\partial t}\theta + (v\cdot\nabla)\theta = R\vec{k}\cdot\vec{v} + \nabla^2\theta, \tag{3}$$

where $\vec{k}$ is a vertical unit vector opposed to gravity, θ the deviation from the static temperature distribution, ∇p all terms that can be written in gradient form, and where R and P are the Rayleigh number and the Prandtl number.

The onset of convection defines a critical Rayleigh number, R_c, and a critical wave number, α_c, for a fluid layer of infinite horizontal extent:

$R_c = 1707, \quad \alpha_c = 3.117.$

Whereas for the onset of convection the Rayleigh number is the only relevant parameter, the Prandtl number enters insofar as the nonlinear properties of convection are concerned. Depending on the type of fluids used, i.e. the Prandtl number, various types of instabilities are observed as one increases the Rayleigh number. The main ones are called oscillatory instability (OS), skewed varicose (SV), zigzag (ZZ), cross rolls (CR). The so-called <<Busse balloon>> is shown in Fig. 1: it indicates the region of stable rolls and the marginal stability curves.

The Prandtl number is the ratio between the kinematic viscosity ν (viscosity divided by density) and the thermal diffusivity. In the following table we list the P number of the usual fluids used.

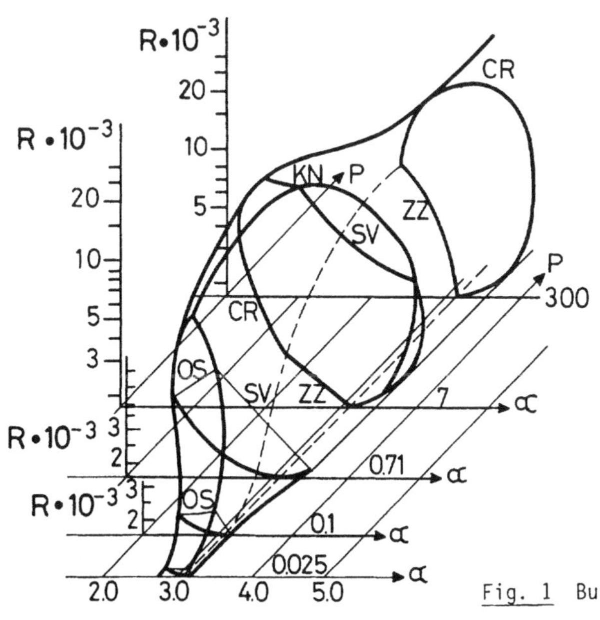

Fig. 1 Busse stability analysis

	K	Hg	He	Gas	H_2O	Oils	
P	5×10^{-3}	2×10^{-2}	0.5	0.7	7	100	and up

Looking at the Busse balloon, some experimental behavior appear immediately. For oils, in one extreme case, the convective state will be stable up to high Rayleigh numbers where Cross Rolls will destabilize. In the other limit, like mercury, the region of stability is very small, and one reaches a time-dependent instability, called the Oscillatory instability. For intermediate P the Skewed varicose instability is the main one.

Ib The Order Parameter Equations

Without entering into the detail of the analysis, let us define the two order parameters relevant for the problem of thermal convection: the vertical velocity W and the vertical vorticity ζ. They are enough to define a divergence-free velocity field. The two equations for those order parameters are:

$$\partial_t \nabla^2 W - \vec{k}.\vec{rot}.\vec{rot}\,[(\vec{v}.\vec{\nabla})\vec{v}\,] = P\nabla^4 W + RP\Delta\,\theta \tag{4}$$

$$\partial_t \zeta + \vec{k}.\vec{rot}[(\vec{v}.\vec{\nabla})\vec{v}] = P\nabla^2\zeta \tag{5}$$

Note that the vertical velocity W and the temperature θ are linearly coupled whereas the vertical vorticity simply obeys a diffusion equation, in the linear approximation. Vertical vorticity modes of wavenumber k are thus damped, proportionally to $\exp(-Pk^2t)$, and do not appear in the linear approximation. In the nonlinear regime, and for small P, they will be relevant.

One can show that for the onset of convection, the marginal stability analysis leads to the following Landau equation, for the amplitude A of the vertical velocity:

$$\partial_t A = \eta A - \gamma A^3 \tag{6}$$

where $\eta = (k_c^{\,2} + \pi^2)\,\frac{P}{1+P}\,r$

$$\gamma = \frac{k_c^{\,2}}{8}\,\frac{P}{1+P}$$

$$r = \frac{R-R_c}{R_c}$$

R_c and k_c being the critical Rayleigh number and wavenumber for the onset of convection.

This analysis is valid only at the onset of the convective state, where only one mode, of wavevector k_c, is unstable. As one increases the temperature difference, a band of possible wavenumbers exists. The way to treat the problem is to allow a slow variation of the amplitude of the velocity field along the convective planform. One finds then the SEGEL NEWELL WHITEHEAD [13] equation:

$$\partial_t A = \eta_0[rA + \ell_0^{\,2}(\partial_x - \frac{i}{2k_c}\partial^2_{yy})^2 A] - \gamma A^3 \tag{7}$$

This equation takes into account the space variation of the order parameter. It is a type of Ginzburg-Landau equation.

Up to now we left aside equation (5) for the vertical vorticity. But for higher Rayleigh numbers the nonlinear coupling of it to the velocity field will become relevant. For low P and small k values, long wave-length, the vertical vorticity term is almost marginally stable. The problem has been treated by SIGGIA and ZIPPELUIS[14]. The vertical vorticity term will lead to the oscillatory instability, the new order parameter equations being

$$\partial_t A = \eta_o [rA + \ell_o^2 (\partial_x - \frac{i}{2k_c} \partial^2_{yy}) A - \gamma A^3 - i k_c B_x A] \tag{8}$$

$$\zeta_t = P \partial^2_{yy} \zeta + 2 \partial_y [A^* (\partial_x - \frac{i}{2k_c} \partial^2_{yy}) A] \tag{9}$$

$$\zeta = - \partial_y B_x$$

Ic Experimental Situation

In convective boxes wide enough to contain many rolls, complicated patterns are common, involving curved rolls, defects in the pattern such as dislocations, and sometimes messy regions. We refer to CROSS and NEWELL analysis [15] of those complicated patterns. There are two methods to simplify the problem.

The first one consists in reducing the number of rolls in a small aspect ratio cell, thus in a way truncating the possible solutions. In doing such experiments one finds that the dynamical behavior follows qualitatively dynamical system theory approach. Let us refer the reader to papers by RUELLE [16] and the author for the analysis.

The second one, in large aspect ratio cells, consists in imposing well ordered initial conditions and then following the pattern evolution with Rayleigh number. WHITEHEAD and BUSSE [4] have then been able to observe all the various bifurcations predicted by the theory. One can also in this context follow the evolution of defects imposed as initial conditions, such as a dislocation or a grain boundary. CROQUETTE and POCHEAU [17] have by this technique studied the dynamics of defects and infer from it a wavelength selection.

In this very short presentation we have tried to sum up the basis of convection. More complex problems appear in binary mixtures [18] or in problems where two control parameters can act on the flow, like a magnetic field [19] or rotation. Also the problem of the very small Prandtl number is not completely solved, the nonlinear terms being important, as close as one wants from the onset.

II Fluid - Fluid Interface, the Saffman Taylor Problem

Sharp interfaces occur frequently in fluids. The equations describing the interface dynamics are generally nonlinear. One simple case is represented by the interface between two immiscible fluids, in a quasi two-dimensional cell; what is called the Hele Shaw cell. It consists of two plates whose interspacing distance d is smaller than the fluids interface possible wavelength. (See Fig. 2).

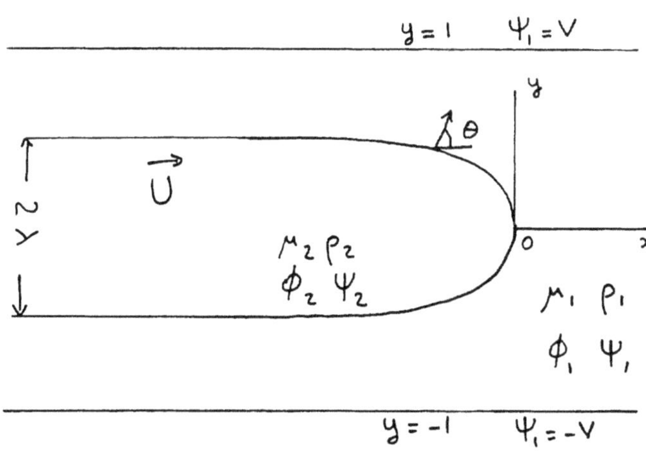

Fig.2 Finger in a channel

A remarkable analytical solution for this problem was found by SAFFMAN and TAYLOR [20], for the case of no surface tension of the interface and in the limit of zero viscosity for one of the fluids. It is one of the rare cases in nonlinear dynamics where an analytic solution exists. In fact Saffman and Taylor results show that there is a family of possible shapes for the interface, and again the question is posed about the selection mechanism, for the experiments show only one interfacial shape. Surface tension is certainly relevant, but no general results have been obtained up to now.

This problem has stimulated many studies because of its practical implications. In oil fields, for example, water fingering through the oil trapped in porous rocks, leads to a large amount of unpumped oil.

We will now present the problem and elaborate somewhat on various interesting aspects. For this we shall follow Saffman and Taylor presentation. A very good up-to-date article on the question, with complete references, is given by TRYGGVASON and HASSAN AREF [21].

IIa Interface Stability

Let us take two fluids, immiscible, of viscosities μ_1 and μ_2 and densities ρ_1 and ρ_2, in a two-dimensional cell. We now apply a pressure on one side of the cell (Fig. 2). We will first show that in order for the interface to be unstable, the low viscosity fluid has to push the high viscosity one.

We will develop a marginal stability analysis, supposing a disturbance from the flat interface

$$x = ae^{\sigma t + iny} \tag{10}$$

We are interested in the sign of σ, the growing rate.

For a thin cell the motion of the fluids is governed by DARCY law [20]

$$\vec{u} = -\frac{k}{\mu}\,\vec{\text{grad}}\,P = \text{grad}\,\phi \tag{11}$$

u is the velocity, k the permeability of the medium to the fluid and ϕ the velocity potential.

The continuity of normal velocity at the interface leads to

$$\frac{\partial\phi_1}{\partial x} = \frac{\partial\phi_2}{\partial x} = V + a\sigma e^{\sigma t+iny} \tag{12}$$

where V is the interface uniform motion. The fluids being incompressible ϕ follows Laplace equation $\nabla^2\phi=0$.

Hence

$$\phi_1 = Vx-\frac{a\sigma}{n}e^{iny-nx+\sigma t} \tag{13}$$

$$\phi_2 = Vx+\frac{a\sigma}{n}e^{iny+nx+\sigma t} .$$

Equating the values of pressure at the interface leads to

$$\frac{\sigma}{n}\frac{(\mu_1+\mu_2)}{k} = \frac{\mu_1-\mu_2}{k}V \tag{14}$$

A positive growing rate, $\sigma>0$, leads to $\mu_1>\mu_2$. Thus the interface is unstable if the low viscosity fluid pushes the high viscosity one.

IIb Shape of the Interface

The elegance of the Saffman Taylor analysis comes from using a conformal mapping which leads simply to the interface shape.

One defines a stream function for this two-dimensional geometry. The velocities along the x and y directions are thus

$$u = \frac{\partial \phi}{\partial x} = \frac{\partial \psi}{\partial y}$$

$$v = \frac{\partial \phi}{\partial y} = -\frac{\partial \psi}{\partial x}$$

and $\phi + i\,\psi$ is an analytic function of $x + i\,y$. The continuity of the normal component of the velocity at the interface leads to:

$$\frac{\partial \phi_1}{\partial n} = \frac{\partial \phi_2}{\partial n} = U\cos\theta$$

But $\frac{\partial \phi}{\partial n} = \frac{\partial \psi}{\partial s}$ and $\cos\theta = \frac{\partial y}{\partial s}$. So $\psi_1 = \psi_2 = Uy$. The boundary conditions are

$$\phi_1 = Vx \qquad x \to \infty$$

$$\phi_2 = Ux \qquad x \to -\infty$$

$$\psi_1 = \pm V \qquad y = \pm 1 \quad \text{(walls are streamlines)}$$

Let us now suppose that fluid 2 has zero viscosity, $\mu_2 = o$. Ignoring surface tension one gets for the continuity of pressure:

$$\mu_1 \phi_1 = \mu_2 \phi_2$$

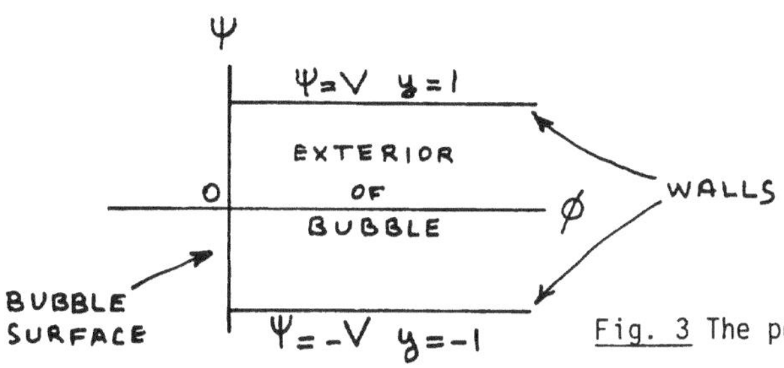

Fig. 3 The potential plane for channel motion

Thus $\phi_1 = 0$ at the surface. In the $\phi\psi$ plane the bubble looks as shown in Fig. 3. Now $x + iy$ is an analytic function of $\phi + i\psi$. A simple expansion will lead to the family of solutions

$$x = \frac{1-\lambda}{\pi} \text{Ln} \frac{1}{2} (1 + \cos\frac{\pi y}{\lambda})$$

IIc Interesting Aspects of the Problem

The shape of the finger is shown in Fig. 2. The experimental results show that one always reaches asymptotically a finger of universal shape, with $\lambda = \frac{1}{2}$. Then the equation becomes

$$2\,x = \text{Ln} \frac{1}{2} (1 + \cos 2\,y)$$

As yet there is no general explanation for this shape selection.

If one includes surface tension the problem becomes more complex. One gets then a discontinuity in the tangential components of the velocity at the interface, which imposes in this Laplacian field to introduce a vortex sheet along the interface boundary. An experimental study of this situation is far from complete, but computer simulations [21] indicate a very rich behavior for the interface dynamics.

III Solid-Liquid Interface, From Cells to Dendrites

Let us now introduce another boundary problem of interest, a moving solid-liquid interface, in a temperature gradient. In this problem, the motion of the interface is dominated by the rejection of impurities into the liquid side, from the solid. Like in the Saffman Taylor instability, only one direction of motion is unstable, the one where the solid is growing. Again here, to simplify the problem we will tend to reduce the space dimensions by working in a two-dimensional geometry, as shown on Fig. 4.

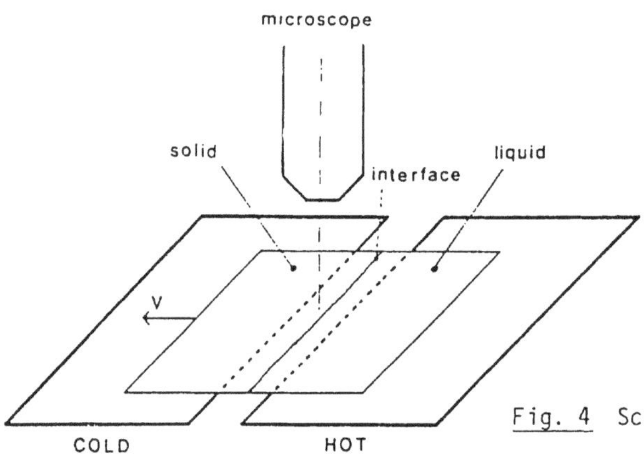

Fig. 4 Schematic of crystal growth

Finally, the material used is relevant. We want our moving interface to be smooth, non-faceted. This means that our material should have a small entropy of melting, as was shown by JACKSON and HUNT [22]. In a more modern terminology, we would say that the interface should be at a temperature above the roughening transition. Plastic crystals usually fall in that range, and one of the commonly used is Succinonitrile, $NC(CH_2)_2CN$.

The following is observed in this case. At low translational velocity the interface is flat. For a critical velocity, of the order of 1μm/sec, the interface becomes unstable, with a characteristic wavy shape. For larger velocities, the cells will bifurcate to various shapes, but essentially solid fingers will develop. Strictly speaking, in this problem the thermal and chemical diffusion fields are coupled. One can simplify the problem, noticing that the thermal diffusion constant is much larger than the chemical diffusivity. The latent heat being small in the material used, we can ignore the heat generated at the interface.

Assuming equal thermal conductivities for the solid and the liquid, the temperature profile between the hot and cold plates (Fig. 4) is

$$T = T_0 + Gz \tag{15}$$

Let us now look at the phase diagram in Fig. 5. Because of (15) the temperature axis is a z axis. Let us start with an interface at rest, equal impurity concentration C_0 in the solid and in the liquid, at infinity. At the interface there is a jump in concentration, C_0 for the liquid, κC_0 for the solid, where κ is the ratio of the slopes of the solidus and of the liquidus.

Let us now move the interface. Impurities will be rejected towards the liquid side, building up a front of impurities as the solid is advancing. The interface will slowly move to its new equilibrium position where the solid concentration becomes C_0 and the liquid one larger C_0/k. This transient may take an hour, $\tau \sim D/V^2$, where D is the chemical diffusivity, and V the translational velocity.

In front of the interface there will be a spike in solute concentration, decreasing exponentially as shown in Fig. 5. This concentration gradient will increase with velocity, and may then enter in the two-phase region. The front becomes then unstable, it is a chemical version of the MULLINS-SEKERKA [23] instability.

The slope of the spike at the interface is of order $\frac{V\Delta C}{GD}$, where ΔC is the jump of impurity concentration at the interface. This has to be compared with the liquidus slope $\frac{dc}{dT}$, and thus the control parameter will be:

$$\nu = \frac{v\Delta c}{GD\frac{dc}{dt}}$$

It is about equal to one at the instability threshold.

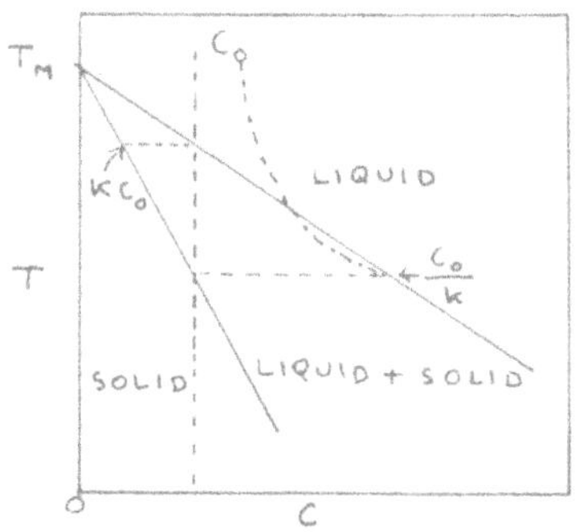

Fig. 5 Portion of the phase diagram

As one increases the velocity further, the interface becomes wavylike, but seems to stabilize into a shape with fingers and grooves (Fig.6). Then, as in our preceeding problems, we will find a wavelength selection, and in this case the wavelength increases with velocity. This will be true until we reach a higher velocity regime, where the shapes become dendritic, the wavelength will then decrease with velocity. This wavelength selection is well documented in TRIVEDI articles [24].

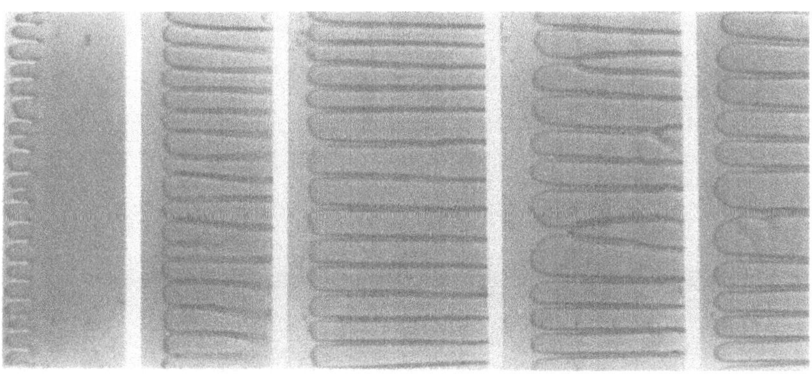

Fig. 6 Cellular growth along the <100> direction for increasing velocities. Note the wavelength selection

We have found recently that the crystal orientation is very important to understand the shapes appearing [25]. The marginal stability theory is quite elaborate, and the nonlinear study in its infancy, so that we shall not elaborate on it in this article. Let us refer the reader to the excellent review by J. LANGER [26], and a recent article by DEE and MATHUR [27] which should lead to the relevant publications.

Acknowledgements

This work was supported by the Materials Research Lab at The University of Chicago under Grant NSF DMR 82-16892 and also NSF Grant DMR 83-16204.

REFERENCES

[1] H. Benard: Rev. Gen. Sci., 1, 1261 (1900).
[2] Lord Rayleigh: Philos. Mag., 32, 529 (1916).
[3] H. Jeffreys: Philos. Mag., 2, 823 (1926); Proc. R. Soc. London, Ser. A, 118, 195 (1928).
[4] F. H. Busse: Rep. Prog. Phys., 41, 1929 (1978).
[5] M. Velarde and C. Normand: Sci, Am., 243, 78 (1980).
[6] R. Krishnamurti: J. Fluid Mech., 42, 295 (1970); 60, 285 (1973).
[7] J. P. Gollub and S.V.Benson: J. Fluid Mech., 100, 449 (1980); J.P.Gollub, A.R. McCarriar and J.F. Steinman: J. Fluid Mech., 125, 259 (1982).
[8] P. Berge: Chaos and Order in Nature, edited by H. Haken (Berlin,1981).
[9] G. Ahlers and R.W.Walden: Phys. Rev. Lett. 44, 445 (1980).
[10] H. Benard and D. Avsec: J. Phys.(Paris), 11, 486 (1938).
[11] S. Fauve, C.Laroche and A.Libchaber: J. Phys.(Paris) Lett., 42,L455(1981).
[12] A. Libchaber and J. Maurer: Non Linear Phenomena at Phase Transitions and Instabilities, edited by T. Riste (New York, N.Y., 1981).
[13] A.C.Newell, J.A. Whitehead: J. Fluid Mech. 38,279 (1969).
[14] E. Siggia, A. Zippelius: Phys. Rev. A26, 1788 (1982).
[15] M.C.Cross, A.C.Newell: Preprint.
[16] D. Ruelle: E. Fermi School of Physics, Varenna (1983).
[17] A. Pocheau, V. Croquette: J. Physique (to be published).
[18] G. Ahlers and coworkers: Preprint.
[19] S. Chandrasekhar: Hydrodynamic and hydramagnetic stability, Dover(New York 1981).
[20] P. G. Saffman, G. I. Taylor: Proc. R. Soc. London A245, 312 (1958).
[21] G. Tryggvason, Hassan Aref: J. Fluid Mech, 136, 1 (1983).
[22] K. A. Jackson, J. D. Hunt: Acta Metallurgica 13, 1212 (1965).
[23] W. W. Mullins, R. F. Sekerka: J.A.P. 35, 444 (1964).
[24] R. Trivedi: Metal. Trans. 15A, 977 (1984).
[25] F. Heslot, A. Libchaber: to appear Physica Scripta (1984).
[26] J. S. Langer: Rev. Mod. Phys. 52, 1 (1980).
[27] G. Dee, R. Mathur: Phys. Rev. B27, 7073 (1983).

Part IV Astrophysics and General Relativity

Nonlinear Problems in Stellar Dynamics

G. Contopoulos

European Southern Observatory, D-8046 Garching bei München, Fed. Rep. of Germany

We deal with orbits of stars in galaxies (1) on the plane of symmetry of a spiral galaxy (2) on the meridian plane of an axisymmetric galaxy and (3) three-dimensional orbits in non-axisymmetric galaxies. The theory of the "third integral" explains the ordered motions. However, there are also stochastic motions that play an important role near resonances. The transition to stochasticity is studied. Finally some new types of phenomena that appear in 3-dimensional systems are discussed.

1. Introduction

Stellar systems are among the most difficult objects in science; their complete understanding implies the solution of the N-body problem under the action of Newtonian forces, and this is a really tremendous task. However some progress has been made in this area, especially in recent years. There are three approaches to this problem:

(1) An observational approach, based on the study of the observed stellar systems, clusters, galaxies, clusters of galaxies, and the whole Universe;
(2) An experimental approach, based on computer simulations of various N-body systems, and
(3) A theoretical approach based on analytical studies of relatively simple models.

The most spectacular types of stellar systems in the sky are galaxies. They contain 10^{10}-10^{12} stars and most of them have typical forms; they are elliptical, spiral, or barred (Fig. 1). Many spiral galaxies have a characteristic two-armed structure. This characteristic form is reproduced both by N-body calculations and by theoretical models that deal with stars and gas in a galaxy.

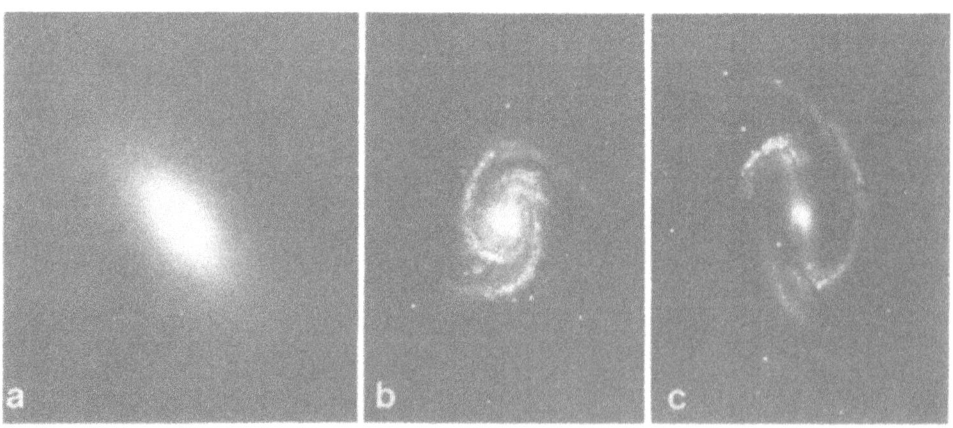

Fig. 1. The main types of regular galaxies (a) elliptical, (b) spiral, (c) barred spiral.

Stellar dynamics studies the behaviour of stars in stellar systems. In the case of large systems, like galaxies, the study is simplified by considering the motions of stars under the influence of the smooth average field, and then finding the statistical deviations due to the graininess of the system. Thus we distinguish between collisionless and collisional dynamics (by "collisions" we mean close approaches of stars; actual collisions are extremely rare). Deviations from the collisionless dynamics are important only after times of the order of the "relaxation time", which, in the case of galaxies, is very long, larger than the age of the Universe. Thus a good approach for the study of galaxies is collisionless dynamics. Under this heading we distinguish three main problems:

(1) The study of orbits in various galactic potentials
(2) The density produced by the stars in such models (this is called the response density), and
(3) Assuming that the potential of a galaxy is due to its stars only (this is found by Poisson's equation and is called the imposed density) we formulate the self-consistency equation

$$\text{response density} = \text{imposed density}. \tag{1}$$

The solution of this equation under various conditions (e.g. various symmetries) is the main goal of galactic dynamics.

The basic theory that seems to explain the spiral and barred galaxies is the density wave theory. The spiral arms are not composed always of the same stars, but as the stars move through the spiral arms they stay longer there, producing a maximum of density. The density wave theory developed by LIN and SHU [1], KALNAJS [2], TOOMRE [3], and many others has had much success up to now in explaining the main features of galaxies. The present version of this theory is a linear one. It deals with weak spirals, or bars, and it explains the onset and maintenance of these features. However, as soon as the amplitude of the spirals or bars becomes appreciable (say more than 10% of the axisymmetric background), nonlinear effects become important. Such nonlinear effects appear mainly at resonances. In resonances the basic assumptions of the linear theory are no more valid, and a new study of the problem is necessary ab initio.

In the present lectures we will deal with some of the main nonlinear problems that we encounter in orbit theory, like:

(a) The various types of orbits on the plane of symmetry of a galaxy, especially near each resonance,
(b) The types of orbits on a meridian plane of an axisymmetric galaxy
(c) The theory of the third integral of motion (besides the energy and angular momentum)
(d) The distinction between ordered and stochastic orbits and especially the transition between these two types, and
(e) The types of three-dimensional orbits and the new effects that appear in the dynamics of 3-dimensional systems.

These problems have many points in common with many other non-linear problems appearing in Physics, Chemistry or even Biology. The basic equations of stellar dynamics have a structure similar to that of equations used in many branches of science. For this reason, one can consider "non-linear problems" as a new field of science, that does not depend strongly on the particular subject of interest.

2. Stellar Motions on the Plane of Symmetry of a Spiral Galaxy

The potential on the plane of symmetry is of the form

$$V = V_0(r) + V_1(r,\theta) \ , \tag{2}$$

where $V_0(r)$ is the axisymmetric background and $V_1(r)$ the spiral perturbation. The spiral is assumed to rotate with constant angular velocity Ω_s. If θ is the azimuth in the rotating frame, the potential of a two-armed spiral can be written

$$V_1 = A(r)\cos[\Phi(r)-2\theta] \; . \tag{3}$$

The minimum of the potential is along the two symmetric spiral arms

$$\theta = \frac{1}{2}\,\Phi(r) \pm \frac{\pi}{2} \; . \tag{4}$$

If $\Phi(r)$ = const. the spiral is reduced to a bar.

The equations of motion are derived from the Hamiltonian

$$H \equiv \frac{1}{2}\left(\dot{r}^2 + \frac{J_o^{\,2}}{r^2}\right) + V_o(r) + V_1(r,\theta) - \Omega_s J_o = h \; , \tag{5}$$

where $\dot{r}$ is the radial velocity, J_o the angular momentum, and h the (constant) value of the Hamiltonian. If the transverse velocity in the rotating frame is v_θ, we have

$$J_o/r = \Omega_s r + v_\theta \tag{6}$$

and Eq. (5) becomes

$$H \equiv \frac{1}{2}\left(\dot{r}^2 + v_\theta^{\,2}\right) - \frac{1}{2}\Omega_s^{\,2} r^2 + V_o(r) + V_1(r,\theta) = h \; . \tag{7}$$

Thus the equation

$$-\frac{1}{2}\Omega_s^{\,2} r^2 + V_o(r) + V_1(r,\theta) = h \tag{8}$$

represents a "curve of zero velocity". If h is smaller than a limiting value h_1 there are two closed curves of zero velocity (Fig. 2). The motion takes place either inside the inner curve, or outside the outer curve. If $h=h_1$ the two curves join at the points L_1, L_2 (called the "unstable Lagrangian points") and for a little larger h the curves of zero velocity close around L_4, L_5 (the "stable Lagrangian points"). Finally, if h is larger than another limiting value h_4 ($>h_1$) the curves of zero velocity disappear. The points L_1, L_2, L_4, L_5 are near "corotation" where the mean angular velocity of the stars around the center, Ω, is equal to Ω_s. When $h>h_1$ the integral H=h provides no restriction on the motion of a star, i.e. a star from the central region of a galaxy may escape to infinity. However, this does not happen necessarily. E.g. in an axisymmetric galaxy, the motion with energy smaller than the "escape energy" remains restricted. This restriction is due to a second integral of motion, which is the angular momentum.

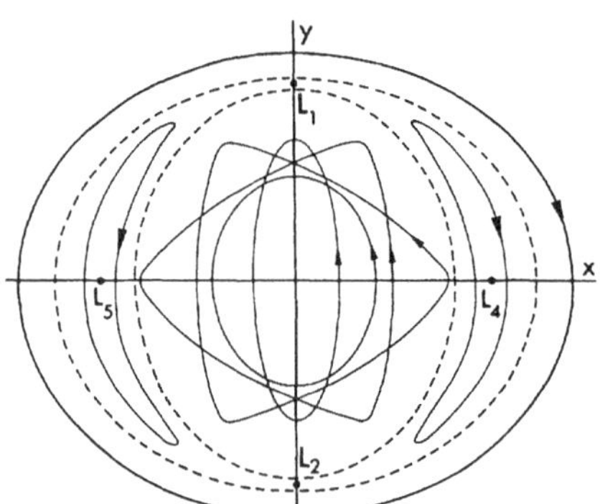

Fig. 2. Curves of zero velocity (----) and periodic orbits (——) around the center and around the Lagrangian points L_4 and L_5 (in a frame rotating with angular velocity Ω_s).

On the other hand, if the spiral field is relatively strong, the stellar orbits near corotation become stochastic, and eventually the stars escape to infinity.

A star with angular momentum J_o describes a circular orbit of radius r_o in the unperturbed Hamiltonian

$$H_o \equiv \frac{1}{2}\left(\dot{r}^2 + \frac{J_o^2}{r^2}\right) + V_o(r) - \Omega_s J_o \ . \tag{9}$$

In this case

$$V_o'(r_o) = J_o^2/r_o^3 \tag{10}$$

and the angular velocity is

$$\Omega = J_o/r_o = (V_o'/r_o)^{1/2} \ . \tag{11}$$

If we develop H in powers of $r-r_o$ we find

$$H \equiv V_o(r_o) + \frac{J_o^2}{2r_o^2} + \frac{1}{2}\left[\dot{r}^2 + \kappa^2(r-r_o)^2\right] \ \ldots \ + V_1(r,\theta) = h \tag{12}$$

where

$$\kappa = \left[V_o''(r_o) + 3V_o'(r_o)/r_o\right]^{1/2} \tag{13}$$

is called the "epicyclic frequency". Thus we can consider the motion of a star as consisting of a circular motion with frequency $\Omega-\Omega_s$ (in the rotating frame) plus an epicyclic motion with frequency κ.

If the ratio

$$\text{rot} = \frac{\kappa}{\Omega-\Omega_s} \tag{14}$$

is a rational number we have a resonance. The most important resonances in a galaxy are the inner and outer Lindblad resonances, where rot = ±2, and the corotation resonance where $\Omega=\Omega_s$.

We can now express the Hamiltonian (12) in action angle variables. We define the radial action

$$I_1 = \frac{1}{2\pi}\oint \dot{r}\,dr \tag{15}$$

and the transverse action

$$I_2 = J_0 - J_1 \ , \tag{16}$$

where J_1 is the angular momentum of a circular orbit with energy h. Then we can develop the Hamiltonian (12) in the form

$$H = h + \omega_1 I_1 + \omega_2 I_2 + aI_1^2 + 2bI_1I_2 + cI_2^2 + V_1 \ , \tag{17}$$

where the quantities $\omega_1 = \kappa_c$, $\omega_2 = \Omega_c-\Omega_s$ and a,b,c are calculated at $r=r_c$. The corresponding angles θ_1, θ_2 are the "epicyclic angle", and the azimuth of the center of the epicycle. In the lowest approximation

$$r-r_o = \left(\frac{2I_1}{\omega_1}\right)^{1/2}\cos\theta_1 \ , \quad \dot{r} = -(2\omega_1 I_1)^{1/2}\sin\theta_1 \ . \tag{18}$$

More accurately, the values of r, $\dot{r}$ and θ are expressed as trigonometric series in the angles θ_1 and θ_2, with coefficients functions of I_1, I_2 [4]. Thus we find

$$V_1 = A_c\cos 2\theta_2 + \varepsilon_1\cos(\theta_1-2\theta_2) + \varepsilon_2\cos(\theta_1+2\theta_2) + \ldots \tag{19}$$

(A_c, in particular, is independent of the actions.) The Hamiltonian (17) with V_1 given by (19) represents in general a nonintegrable dynamical system.

If we are not close to a resonance, we can make a change of variables that brings H into an integrable form

$$H = h + \omega_1\tilde{I}_1 + \omega_2\tilde{I}_2 + a\tilde{I}_1^2 + 2b\tilde{I}_1\tilde{I}_2 + c\tilde{I}_2^2 + \dots , \qquad (20)$$

i.e. a form independent of the angles, in which case the new actions $\tilde{I}_1$, $\tilde{I}_2$ are integrals of motion. This transformation is in general only formal, but it gives very good numerical results if the spiral perturbation is not large. The trigonometric terms (19) are eliminated by using a generating function

$$G = \theta_1\tilde{I}_1 + \theta_2\tilde{I}_2 - \varepsilon_1 \frac{\sin(\theta_1-2\theta_2)}{\omega_1-2\omega_2} - \varepsilon_2 \frac{\sin(\theta_1-2\theta_2)}{\omega_1+2\omega_2} - \frac{A_c\sin 2\theta_2}{2\omega_2} \qquad (21)$$

and connecting the old and new variables (I_i, θ_i) and $(\tilde{I}_i, \tilde{\theta}_i)$ by the relations

$$I_i = \frac{\partial G}{\partial \theta_i}, \quad \tilde{\theta}_i = \frac{\partial G}{\partial \tilde{I}_i} . \qquad (22)$$

It is obvious that the eliminating transformation (22) cannot apply if one of the quantities $\omega_1-2\omega_2$, $\omega_1+2\omega_2$, ω_2 is zero or close to zero. These conditions in fact represent the inner and outer Lindblad resonances and the corotation resonance.

If we are close to a resonance, it is not possible to eliminate all trigonometric terms as above because one denominator is zero or close to zero. This is the problem of "small divisors" of celestial mechanics. However, we can eliminate all angle combinations except the critical one. E.g. if $\omega_2 = 0$ (corotation) we use a generating function like (21) but without its last term. Then we find the new Hamiltonian in the form

$$H \equiv h_* + \omega_1 I_1 + aI_1^2 + 2bI_1I_2 + cI_2^2 + A_*\cos 2\theta_2 = h , \qquad (23)$$

where we have dropped the tildes. The action I_1 is now an integral of motion (because the angle θ_1 is ignorable) but I_2 is not. Equation (23) represents orbits in the (I_2,θ_2) plane for given values of I_1 and h. If we solve this equation for $I_1 = 0$ we find

$$I_2 = \pm\left[\frac{1}{c}(h-h_*-A_* \cos 2\theta_2)\right]^{1/2} = \pm\left[\frac{2A_*}{c}\left(\frac{h-h_*+A_*}{2A_*} - \cos 2\theta_2\right)\right]^{1/2} . \qquad (24)$$

Here we have $A_*>0$ and $c<o$. Therefore we have real solutions only if

$$\cos 2\theta_2 > \frac{h-h_*+A_*}{2A_*} . \qquad (25)$$

The angle θ_2 is zero at the Lagrangian point L_4 (Fig. 2). It can take all possible values from 0 to 2π if $h < h_1 = h_*-A_*$; then the orbits surround the centre (Fig. 2). But if

$$1 > \frac{h-h_*+A_*}{2A_*} > -1 \qquad (26)$$

the angle θ_2 cannot become absolutely larger than $\cos^{-1}[(h-h_*+A_*)/2A_*]$.

In this case, the orbits librate around L_4 (or L_5) (Fig. 2). Finally, if $(h-h_*+A_*)/2A_* > 1$ (hence $h > h_4 = h_*+A_*$) there is no periodic orbit for this value of I_1 (I_1=0). (See further details in [5].)

This example shows that the orbits may completely change their character near a resonance. In a similar way, we have resonant orbits of type 2/1 (near the inner Lindblad resonance), 3/1, etc. (Fig. 2).

3. Stellar Motions on the Meridian Plane of an Axisymmetric Galaxy

If the galaxy is axisymmetric around the z-axis, the z-component of the angular momentum J_o is a constant. Then the Hamiltonian is written in the form

$$H = \frac{1}{2}\left(\dot{r}^2 + \frac{J_o^2}{r^2} + \dot{z}^2\right) + V(r,z) \ . \tag{27}$$

This Hamiltonian represents the motions of stars on a meridian plane (r,z) passing through the star and the axis of symmetry of the galaxy.

We assume now that the galaxy has also a plane of symmetry ($z=0$) and we develop the Hamiltonian around the point ($r=r_o$, $z=0$), where r_o is the radius of a circular orbit with angular momentum J_o, as given by Eq. (10). Writing also

$$r = r_o + x, \quad z = y, \quad \dot{z} = \dot{x}, \quad \dot{z} = \dot{y} \tag{28}$$

we find finally

$$H = \text{const.} + \frac{1}{2}\left(\dot{x}^2 + \dot{y}^2 + \omega_1^2 x^2 + \omega_2^2 y^2\right) + \varepsilon x y^2 + \varepsilon' x^3 + \dots \tag{29}$$

where ω_1 is equal to κ given by Eq. (13), $\omega_2^2 = (\partial^2 V/\partial z^2)_o$, and ε, ε' depend also on r_o.

If now $\varepsilon = \varepsilon' = 0$ and all higher order terms are zero, this problem is separable, and the energies along the x and y axes

$$\Phi_{10} = \frac{1}{2}\left(\dot{x}^2 + \omega_1^2 x^2\right) , \quad \Phi_{20} = \frac{1}{2}\left(\dot{y}^2 + \omega_2^2 y^2\right) \tag{30}$$

are separately conserved. If ε, ε' are small, we try to find new integrals of motion, that are generalizations of Φ_{10} and Φ_{20}. E.g. we write

$$\Phi_1 = \Phi_{10} + \Phi_{11} + \dots \tag{31}$$

and we require that $d\Phi_1/dt = 0$ along an orbit. This condition is written in the form of Poisson brackets

$$(\Phi_1, H) = \frac{\partial\Phi_1}{\partial x}\frac{\partial H}{\partial \dot{x}} + \frac{\partial\Phi_1}{\partial y}\frac{\partial H}{\partial \dot{y}} - \frac{\partial\Phi_1}{\partial \dot{x}}\frac{\partial H}{\partial x} - \frac{\partial\Phi_1}{\partial \dot{y}}\frac{\partial H}{\partial y} = 0 \tag{32}$$

and can be solved as a series in ε, ε' [6]. E.g.

$$\Phi_{11} = \frac{\varepsilon}{\omega_1^2 - 4\omega_2^2}\left[(\omega_1^2 - 2\omega_2^2)xy^2 - 2x\dot{y}^2 + 2\dot{x}y\dot{y}\right] + \varepsilon' x^3 \ . \tag{33}$$

The higher order terms are given by computer algebra; the first expansions of this form were given by CONTOPOULOS and MOUTSOULAS [7] and GUSTAVSON [8]. The series Φ_1 is called in Astronomy "the third integral", because it comes after the energy and the angular momentum. It can be calculated for any Hamiltonian in the form of a polynomial of any degree, starting with second order terms. The series (31) takes a different form near each resonance. In fact the Hamiltonian (29) can be written in action-angle variables in the form (17), where V_1 is a sum of terms of the form

$$f_{m,n}(I_1,I_2)\cos(m\theta_1 - n\theta_2) \ . \tag{34}$$

The computer program eliminates all trigonometric terms except the resonant one, and gives the result in Cartesian coordinates.

The applicability of the third integral can be checked mainly by two methods:
(a) By calculating how constant it is along a given orbit, and
(b) By finding the theoretical invariant curves and comparing them with the empirical ones.

If we take a given surface of section in phase-space (e.g. $y=0$) we eliminate $\dot{y}$ between Eqs. (29) and (31) and find the equation of an invariant curve

$$f(x, \dot{x}) = \text{const.} \tag{35}$$

If we numerically calculate an orbit, its successive intersections by the surface of section ("consequents" according to Poincaré) are on the curve (35). This is true if the system is integrable, in which case the invariant curves are, in general, smooth and closed. If, however, the system is non-integrable then there are two types of orbits:

(1) ordered orbits, which define closed invariant curves (Fig. 3a), and
(2) stochastic orbits, in which case the successive consequents appear scattered in a random way (Figs. 3d,e).

In fact, the consequents lie on open invariant curves that make infinite oscillations. A particular case of open invariant curves are the asymptotic curves emanating from an unstable invariant point that represents an unstable periodic orbit (Fig. 3d,e).

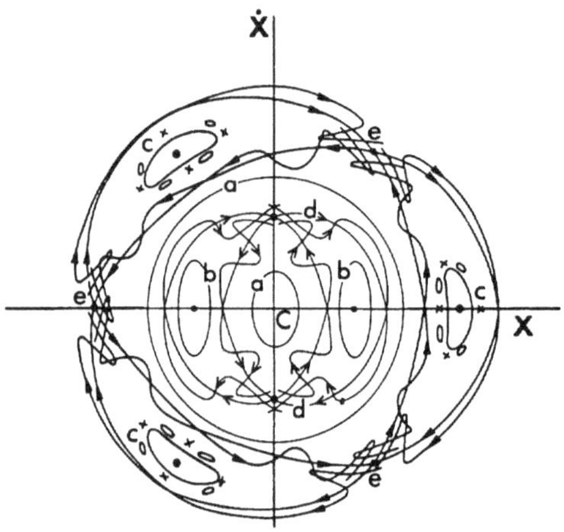

Fig. 3. Invariant curves of ordered (a,b,c) and stochastic orbits (d,e).

Extensive numerical calculations have shown that, in the case of regular orbits the third integral is better and better conserved if we add higher and higher order terms in it. Also, the agreement between theoretical and empirical invariant curves becomes better and better. On the other hand, if an orbit is stochastic the function Φ_1, truncated at any level, is varying considerably, and its variation does not decrease by adding higher-order terms. Furthermore, the theoretical invariant curves cannot represent the apparently stochastic scattering of the consequents on a surface of section.

4. Transition from Ordered to Stochastic Motion

The transition between ordered and stochastic orbits is rather abrupt. The critical transition parameter (e.g. critical energy) can be found approximately by the so-called criterion of overlapping of resonances. This criterion applies essentially as follows. If the energy is small, we have mainly ordered motions that produce closed invariant curves (Fig. 3a,b,c). The invariant curves close either around the "central" invariant point, which represents the main periodic orbit of the system, or around some secondary invariant points, representing resonant periodic orbits; in the latter case, the invariant curves are composed of one or more islands.

At the same time, one sees unstable invariant points between the islands, and around such points there are orbits with some degree of stochasticity (Fig. 3d,e). But these stochastic regions are well separated from each other by invariant curves closing around the origin. E.g. in Fig. 3 the stochastic regions of multiplicity 3 do not communicate with the stochastic regions of multiplicity 2. Only if the perturbation becomes large enough do the various stochastic regions communicate, and a large stochastic "sea" is formed. This is what we call a resonance overlap. This phenomenon was described by ROSENBLUTH et al. [9], CONTOPOULOS [10] and CHIRIKOV [11]. One can calculate theoretically the size of the various resonant regions as a function of the perturbation, and thus estimate the critical perturbation needed to have this interaction of resonances.

This phenomenon is related to the famous KAM theorem (after Kalmogorov, Arnold and Moser), that proves the existence of invariant surfaces in phase-space (and invariant curves on a surface of section). The theorem requires that the ratio of frequencies ω_1/ω_2 is far from all resonances. This means that for all n and m the difference $(\omega_1/\omega_2 - n/m)$ should be absolutely larger than a quantity ε_0/m^3, where ε_0 is a small quantity related to the perturbation.

Although near every irrational number ω_1/ω_2 there are infinite rational numbers n/m, it is easy to prove that the set of irrational numbers that are far from rationals is not small, if ε_0 is small. In fact, if we exclude regions of order $2\varepsilon_0/m^3$ centered on each rational $n/m < 1$ we find that the sum total of these regions is

$$\sum_{m=1}^{\infty} \sum_{n=1}^{m-1} \frac{2\varepsilon_0}{m^3} < 2\varepsilon_0 \sum_{m=1}^{\infty} \frac{1}{m^2} \simeq 3.3\varepsilon_0 \; . \qquad (36)$$

Therefore, if ε_0 is small the total measure of the stochastic regions is small. But if ε_0 goes beyond a critical value, the stochastic regions increase rather abruptly and a large degree of stochasticity appears.

The KAM theorem gives an exact proof of the existence of closed invariant curves. The set of invariant curves is nowhere continuous, because between any two invariant curves there is always an infinity of resonances. However, this set has a positive measure if the perturbation is small enough. The original KAM theorem gives a very small estimate for the perturbation ε. As an example, let us consider the perturbation of the moon on a satellite of the earth. The earth is considered as a point mass. The KAM theorem proves the stability of a nearly circular satellite orbit if its distance from the earth is sufficiently small. This "sufficiently small" turns out to be of the order of 1 cm, therefore not quite realistic! On the other hand, we have numerical evidence that there are stable satellite orbits even if their distances are larger than one half the distance of the moon. Thus in recent years there have been several extensions of the KAM theorem, with very wide applications.

The main question that has been explored in recent years is how the "last KAM invariant curve" is destroyed as the perturbation increases. It is known that between any two rational rotation numbers (like 1/3 and 1/2 in Fig. 3) there is an irrational rotation number, which is the most remote from all rationals. This number can be written as a continuous fraction (GREENE [12], MACKAY [13]):

$$\text{rot} = a_0 + \cfrac{1}{a_1 + \cfrac{1}{a_2 + \dots}} \; , \qquad (37)$$

where $a_i = 1$ for i larger than a certain number N. This is called a "golden number". If ε is smaller than a critical value ε_* there is a closed KAM curve with rotation number rot. For every truncated fraction P/Q of rot there corresponds a pair of periodic orbits with rotation number P/Q. One of these orbits is stable for small ε but becomes unstable if ε goes beyond a value ε_Q. As ε tends to the critical value ε_* $(>\varepsilon_Q)$ all the periodic orbits above a certain order become unstable, and for $\varepsilon > \varepsilon_*$ there is no closed invariant curve any more. SHENKER and KADANOFF [14] have demonstrated that the invariant curve with rotation number rot becomes more and more corrugated as ε approaches ε_*. For $\varepsilon = \varepsilon_*$ this is the "last KAM curve", which has corrugations on all scales, and is, in fact, a fractal. The details of this curve are repeated on smaller and smaller scales i.e. we have here a renormalization phenomenon. Furthermore, it seems that this renormalization is universal, i.e. it has the same scales for any "golden" rotation number.

The problem that remains to be examined is how the transition from stability to instability of periodic orbits leads to stochasticity. It is well known that in nonintegrable systems there is some stochasticity near every unstable periodic orbit. The question is whether a large degree of stochasticity is produced when some periodic orbits become unstable.

In many cases of systems of two degrees of freedom, when a stable orbit becomes unstable a new family of stable periodic orbits is formed. Let this happen for $\varepsilon = \varepsilon_1$. The new family in general has double period. This family becomes also unstable for $\varepsilon = \varepsilon_2 > \varepsilon_1$, giving its stability to a quadruple period family, and so on. This procedure converges approximately geometrically. Namely, the intervals between successive bifurcations have almost constant ratios. More accurately the limit

$$\delta = \lim_{n\to\infty} \frac{\varepsilon_{n+1} - \varepsilon_n}{\varepsilon_n - \varepsilon_{n-1}} \tag{38}$$

is a constant. It was found by FEIGENBAUM [15] and by COULLET and TRESSER [16], both theoretically and by numerical experiments, that this number is universal for all dissipative systems[1], namely it is

$$\delta_d = 4.67 \; . \tag{39}$$

This means that the period doubling is independent of the particular system we deal with. Therefore the sequence $\varepsilon_1, \varepsilon_2, \varepsilon_3, \ldots$ converges to a critical value ε_∞. Beyond that value all the infinite orbits produced by bifurcations are unstable. Thus, in a large region of phase-space there is no stable periodic orbit produced by period doubling, that could trap non-periodic orbits around it, and a large degree of stochasticity follows.

In conservative (non-dissipative) systems of two degrees of freedom the bifurcation ratio is equal to

$$\delta_c = 8.72 \; . \tag{40}$$

This was found in the case of area-preserving maps by BENETTIN et al. [17] and later by GREENE et al. [18], BOUNTIS [19] and others. In Hamiltonian systems several period doubling sequences with the ratio (40) were found by CONTOPOULOS [20, 21].[2]

We also found cases of infinite bifurcations along the same family. Namely, a family that is initially stable becomes unstable, then stable again, and so on. At every change of stability type, another family, of equal or double period, bifurcates from the original family. In a particular case that we studied numerically [23] we found that the bifurcation ratio for these bifurcations is $\delta = 9.22$. But it was proved later by HEGGIE [24] that such bifurcations are not universal. In our particular model the value of δ is $\delta = \exp(\sqrt{\pi}/2) = 9.22$, but in other models one finds different bifurcation ratios.

---

1) There are some exceptions to this rule that we will not discuss here.

2) We found a sequence of pitchfork bifurcations in a Hamiltonian system already in 1970 [22]. However, we did not calculate the bifurcation ratio at that time. In retrospect it is found consistent with the value (40).

In all cases of infinite bifurcations we have the formation of infinite unstable families of periodic orbits, and these lead to a large degree of stochasticity. In particular, in the case of infinite pitchfork bifurcations all the families produced by successive period doublings are unstable for ε somewhat larger than ε_∞.

On the other hand, it is possible that for a range of values of ε larger than ε_∞ we have another region of stability. One common case is the appearance of "irregular" families of periodic orbits. These are families not related to the original family. Namely, their characteristic curves do not cross the characteristics of the original family or of the families bifurcating from it. Such families have a stable region close to their minimum $\varepsilon = \varepsilon_{min}$. In fact there are two branches joining at such a minimum, one stable and one unstable. The stable branch becomes unstable for some $\varepsilon > \varepsilon_{min}$ and produces a cascade of pitchfork bifurcations, in the same way as the original family. Several irregular families were found in galactic models [22], but they appear also in mappings (e.g. the period-3 family in the standard mapping; MAY [25]).

Another way of producing stability for large value of a parameter is by an inverse sequence of bifurcations. Namely, we have found several cases where the infinite families produced by pitchfork bifurcations join in the opposite sequence and are reduced finally to only one family (the continuation of the original family), which is then stable (Fig. 4). This is expected in realistic models representing rotating galaxies, because far from a galaxy, its field is approximately that of a point mass and the Hamiltonian is approximately integrable. Therefore orbits that remain far from the galaxy are approximately Keplerian, and they do not show any stochasticity. The main stable periodic orbits in this case are the direct and retrograde circular orbits, and these trap most of the non-periodic orbits around them.

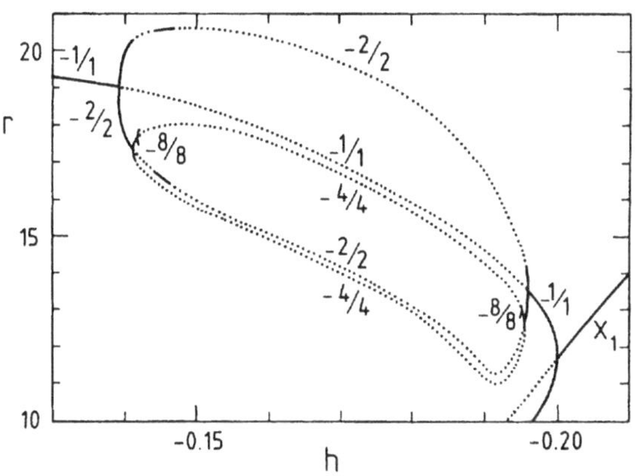

Fig. 4. A sequence of infinite pitchfork bifurcations followed by an inverse sequence of bifurcations in a rotating galactic model. The bifurcation ratio has the same universal value δ = 8.72 on both ends.

5. Bifurcations and Gaps

Let us study a simple case, where bifurcations and gaps appear in the characteristics of periodic orbits. We consider the Hamiltonian [20]

$$H = \omega_1 I_1 + \omega_2 I_2 + aI_1^2 + 2bI_1 I_2 + cI_2^2 + \varepsilon(2I_1)^{m/2}(2I_2)\cos(m\theta_1 - 2\theta_2), \quad (41)$$

near the resonance $m\omega_1 - 2\omega_2 = 0$. We may consider this Hamiltonian as derived from the original Hamiltonian (17) after the non-resonant combinations of angles have been removed by a canonical transformation. The Hamiltonian (41) can be further simplified by another canonical transformation

$$\psi_1 = \theta_1 - \frac{2}{m}\theta_2 , \quad J_1 = I_1$$
$$\psi_2 = \theta_2 \quad , \quad J_2 = I_2 + \frac{2}{m} I_1 \tag{42}$$

and takes the form

$$H = \omega_2 J_2 + cJ_2^2 + [(\omega_1 - \frac{2}{m}\omega_2) + 2(b - \frac{2}{m}c)J_2]J_1 + (a - \frac{4}{m}b + \frac{4}{m^2}c)J_1^2$$
$$+ \varepsilon(2J_1)^{m/2}(2J_2 - \frac{4}{m}J_1)\cos m\psi_1 . \tag{43}$$

It is evident that the angle ψ_2 is ignorable, therefore J_2 is an integral of motion. We consider it as a parameter and set

$$x = \sqrt{2J_1}\cos\psi_1 , \quad \dot{x} = -\sqrt{2J_1}\sin\psi . \tag{44}$$

Then we find

$$\frac{H'}{Q} = q(J_{20}-J_2)\frac{1}{2}(x^2 + \dot{x}^2) + \frac{1}{4}(x^2 + \dot{x}^2)^2$$
$$+ \varepsilon_1[2J_2 - \frac{2}{m}(x^2 + \dot{x}^2)][x^m - \frac{m(m-1)}{2}x^{m-2}\dot{x}^2 + \ldots] , \tag{45}$$

where

$$Q = a - \frac{4}{m}b + \frac{4}{m^2}c , \quad qQ = -2(b - \frac{2}{m}c) ,$$
$$qQJ_{20} = \omega_1 - \frac{2}{m}\omega_2 , \quad \varepsilon_1 Q = \varepsilon , \tag{46}$$

and we omitted in H' the constant term $\omega_1 J_2 + cJ_2^2$.

Equation (45) gives the form of the invariant curves on the plane $(x,\dot{x})$. The periodic orbits are found if

$$\frac{\partial H'}{\partial \dot{x}} = \frac{\partial H'}{\partial x} = 0 . \tag{47}$$

Hence

$$\dot{x}\{q(J_{20}-J_2) + x^2+\dot{x}^2 + \varepsilon_1[-\frac{4}{m}x^m$$
$$- m(m-1)x^{m-2}(2J_2 - \frac{2}{m}x^2) + O(\dot{x}^2) + \ldots]\} = 0 \quad \text{and} \tag{48}$$

$$x[q(J_{20}-J_2) + x^2+\dot{x}^2] + \varepsilon_1[-\frac{2}{m}(m+2)x^{m+1} + 2mJ_2x^{m-1} + O(\dot{x}^2) + \ldots] = 0 . \tag{49}$$

After finding the periodic orbits, we can find the invariant curves close to them and decide whether a given periodic orbit is stable or unstable. One solution of Eq. (48) is always $\dot{x} = 0$. Then the solutions for x are of the form $x = x(J_2; m,\varepsilon_1)$. E.g. if $\varepsilon_1 = 0$ we have the solutions $x = 0$ and

$$x^2 = q(J_2-J_{20}) . \tag{50}$$

If $\varepsilon_1 \neq 0$ we keep the lowest order terms in x and ε_1 and we distinguish various cases. If $m > 4$ we have $x = 0$ and

$$x^2 = q(J_2-J_{20}) - 2J_2m\varepsilon_1x^{m-2} , \tag{51}$$

which is close to the solution (50) for small x. The family $x = 0$ is always stable. If $\varepsilon_1 < 0$ the bifurcating family is stable if m = even and stable only for $x > 0$ if m = odd (Fig. 5a). If $\varepsilon_1 > 0$ the stable bifurcating families become unstable and vice-versa.

Also in these cases we have one more solution

$$x = 0 , \quad \dot{x}^2 = q(J_2-J_{20}) , \tag{52}$$

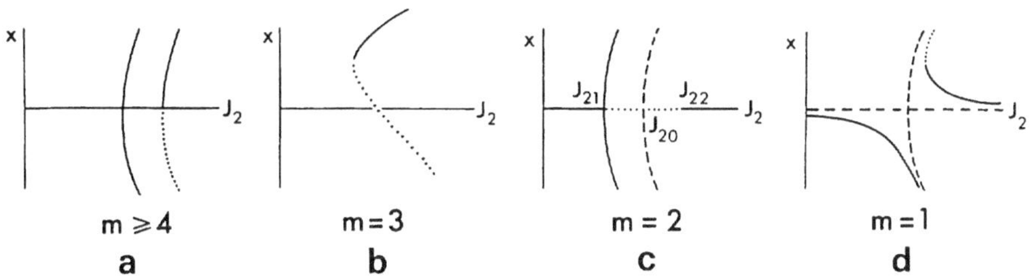

Fig. 5. Bifurcations for Hamiltonians of the form (43) for various values of m; (——) stable (····) unstable families of periodic orbits; (----) unperturbed cases.

which bifurcates from the point $J_2 = J_{20}$ on the plane $(\dot{x}, J_2)$.

If m = 3 we have

$$x^2 + 6J_2\varepsilon_1 x + q(J_{20}-J_2) = 0 \ . \tag{53}$$

Then the minimum J_2 is not on the J_2-axis (Fig. 5b). In all the above cases the bifurcating family intersects the J_2 axis at the point $J_2 = J_{20}$.

If m = 2 we have two solutions

$$\dot{x} = 0 \ , \quad x^2(1-4\varepsilon_1) = q(J_2-J_{20}) - 4\varepsilon_1 J_2 \quad \text{and} \tag{54}$$

$$x = 0 \ , \quad \dot{x}^2(1+2\varepsilon_1) = q(J_2-J_{20}) + 4\varepsilon_1 J_2 \ . \tag{55}$$

The two solutions are parabolae on the planes (x, J_2) and $(\dot{x}, J_2)$ respectively. They intersect the J_2-axis at the points

$$J_{21} = qJ_{20}/(q - 4\varepsilon_1) \quad \text{and} \quad J_{22} = qJ_{20}/(q + 4\varepsilon_1) \tag{56}$$

respectively (Fig. 5c). A simple stability analysis shows that the family $x = \dot{x} = 0$ is unstable between the points J_{21} and J_{22} (Fig. 5c). This is the first case where we encounter a loss of stability of the main family.

Finally, if m = 1 we have $\dot{x} = 0$, and

$$2J_2\varepsilon_1 + xq(J_{20}-J_2) - 6\varepsilon_1 x^2 + x^3 = 0 \ . \tag{57}$$

In this case the axis $x = 0$ does <u>not</u> represent a family of periodic orbits. Equation (57) has either one or three real solutions, thus we have a figure like Fig. 5d; this is how a gap is formed along the main characteristic of periodic orbits.

If the perturbation ε_1 becomes large, the instability regions and the gaps of Figs. 5c,d become large [20]. Higher order bifurcations and gaps can be derived in the same way.

The above theory cannot give the universal character of the successive pitchfork bifurcations. For that purpose, a renormalization analysis around the critical value $\varepsilon = \varepsilon_\infty$ is necessary, and this has been done by FEIGENBAUM [15] and COULLET and TRESSER [16] for dissipative systems and by WIDOM and KADANOFF [26] for conservative systems.

6. Three-Dimensional Systems

We consider now systems of three degrees of freedom, like three-dimensional galaxies. The Hamiltonian in such a case can be written in the form

$$H = \frac{1}{2} (\dot{x}^2 + \dot{y}^2 + \dot{z}^2 + \omega_1^2 x^2 + \omega_2^2 y^2 + \omega_3^2 z^2) + H_1 \tag{58}$$

where H_1 contains the perturbation terms. If there are two more integrals of motion

$$\Phi_1 = \Phi_1(x,y,z,\dot{x},\dot{y},\dot{z}) \ , \tag{59}$$

and

$$\Phi_2 = \Phi_2(x,y,z,\dot{x},\dot{y},\dot{z}) \ , \tag{60}$$

we can eliminate 2 variables between Eqs. (58-60), say $\dot{y}$ and $\dot{z}$, and find an integral surface of the form

$$\Phi = \Phi(x,y,z,\dot{x}) \ . \tag{61}$$

This equation represents the projection of a 3-dimensional surface on the 4-dimensional space $(x,y,z,\dot{x})$.

If we now use a "surface of section", say $z = 0$, we find an "invariant surface"

$$\Phi = \Phi(x,y,0,\dot{x}) \tag{62}$$

in the 3-D space $(x,y,\dot{x})$. Such a surface can be visualized by a stereoscopic picture (Fig. 6).

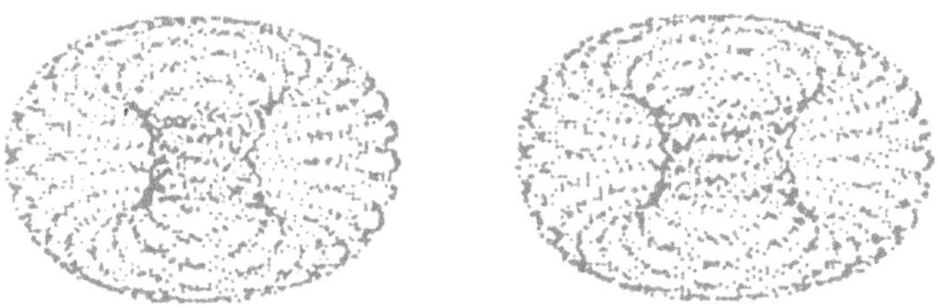

Fig. 6. A stereoscopic view of a 2-dimensional invariant surface, in the 3-dimensional space $(x,y,\dot{x})$.

The existence of 3-dimensional "integral surfaces" in the 6-dimensional phase space (whose projections are given by Eq. (61)) is secured by the KAM theorem for small enough perturbations. However, such surfaces do not form a continuous set (if the system is non-integrable), and they leave infinite resonant regions where the motion is stochastic. These regions are 5-dimensional [they are on a "surface" of constant energy (58)], therefore they are not restricted by the 3-dimensional integral surfaces (notice that Eq. (61) gives only a projection on a 4-dimensional space), and they may communicate with each other. Thus, a stochastic orbit may diffuse along various connected stochastic regions, and go very far from the initial trajectory. This phenomenon is called "Arnold diffusion" [27], and it exists only in systems of three or more degrees of freedom. In systems of two degrees of freedom, the integral surfaces are 2-dimensional on the 3-dimensional energy surface, and normally they are tori. Thus, if an orbit is inside such a torus it cannot go outside it, even if it is stochastic, because it cannot cross the toroidal integral surface (in fact an orbit that has a point on an integral surface remains for ever on it). Therefore, a linearly stable periodic orbit in a system of two degrees of freedom in general is stable for all times. On the contrary, in systems of three or more degrees of freedom, a linearly stable orbit is in general unstable because of Arnold diffusion. Arnold expressed this fact as a paradox. If we apply the KAM theorem to the solar system, we can prove that this system is stable for the majority of initial conditions (provided that the masses, eccentricities and inclinations of the planets are sufficiently small). Nevertheless because of Arnold's diffusion the solar system is unstable. A small random force will bring it into a stochastic region, therefore in the end the planets will escape, or collide with the sun.

The question that remains open is the time scale of this effect. The diffusion rate is extremely small in general. An estimate was given by NEKHOROSHEV [28]. However, such estimates are based on statistical arguments, and they have to be verified by numerical experiments. In a particular case that we studied a few years ago [29], we found two stochastic regions that appeared to be separated for very long times, although they are located close to each other (Fig. 7). The stochasticity in these regions was established by calculating the maximal Liapunov number along several orbits, i.e. the limit of the quantity

$$\chi = \frac{\ln|d|}{t}, \tag{63}$$

where d is the deviation of two nearby orbits after time t. As t increases, the quantity χ tends to a limit, which is different from zero (positive) for stochastic orbits (Fig. 7). The empirical χ seems to converge to constant values for large enough t, but in the two stochastic regions, the limits are different for more than 10^5 periods. Therefore, the Arnold diffusion cannot bring the two regions into contact for 10^5 periods at least.

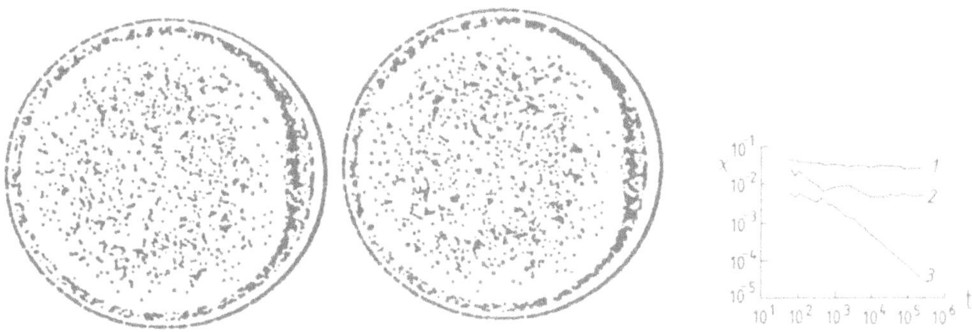

Fig. 7. Stereoscopic view of the consequents of two stochastic orbits. The regions filled by the two orbits do not communicate for very long times. The corresponding asymptotic values of χ are different (1 and 2). In orbit 3 (ordered motion) the asymptotic value of χ is probably zero.

In galactic dynamics 10^5 periods is much larger than the age of the Universe, which is about $1\text{-}2 \times 10^{10}$ years, i.e. about 100 periods, therefore Arnold diffusion can be safely ignored. Only strong interactions between resonances can produce a large degree of stochasticity, as in the cases of two degrees of freedom.

However, in Plasma Physics 10^5 periods is only a small fraction of a second, therefore in this case, the weak effects of Arnold diffusion may be important. Further work has to be done to establish the time scales of this effect in various systems.

Besides Arnold diffusion there are two more new effects, which do not appear in systems of two degrees of freedom but only in systems of three or more degrees of freedom. These are

(1) complex instability, and
(2) collisions of bifurcations.

Both of them refer to periodic orbits.

The eigenvalues of periodic orbits in systems of three degrees of freedom are given by an equation of the form

$$\lambda^4 + \alpha\lambda^3 + \beta\lambda^2 + \alpha\lambda + 1 = 0 \tag{64}$$

(there are also two trivial eigenvalues equal to 1).

Equation (64) can be separated into two factors

$$(\lambda^2 + b_1\lambda + 1)\,(\lambda^2 + b_2\lambda + 1) = 0 \, , \tag{65}$$

where

$$b_{\substack{1\\2}} = \frac{1}{2}\,(\alpha \pm \Delta^{1/2}) \tag{66}$$

and

$$\Delta = \alpha^2 - 4(\beta - 2) \, . \tag{67}$$

If $\Delta>0$ we have either $|b_1|<2$, $|b_2|<2$ (stability), or one $|b|$ larger than 2 and the other smaller than 2 (simple instability) or both $|b|$ larger than 2 (double instability). Complex instability appears if $\Delta<0$, when both b_i are complex, and the corresponding λ's are also complex but not on the unit circle, as in the stable case.

Complex instability does not appear in systems of 2 degrees of freedom. In systems of three or more degrees of freedom a stable orbit can become complex unstable if two real values b_1 and b_2 become equal and then complex conjugate. At this point, we do not have the bifurcation of another family of periodic orbits (Fig. 8c) as we have in many other transitions to instability, when one $|b|$ becomes equal to 2 (Figs. 8a,b). Therefore, when a complex instability is introduced in a conservative system, we do not have a sequence of infinite bifurcations as in conservative systems of two degrees of freedom. This means that in such systems, stochasticity is introduced abruptly and not after an infinite sequence of period doubling bifurcations.

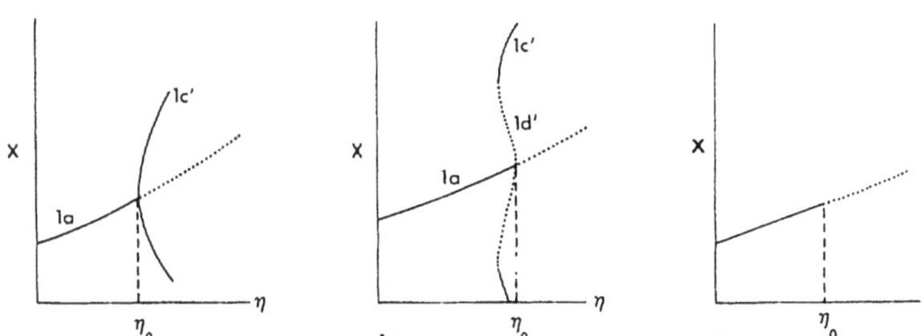

Fig. 8. The stability of a family of periodic orbits can be transmitted to another family (case a, direct bifurcation), or be lost (case b, inverse bifurcation and case c, complex instability). In case b a stable family (1c') joins the bifurcating family (1d') at a minimum η. A comparison of Figs. a and b shows the transition from direct to inverse bifurcations.

There is one more mechanism besides complex instability that forbids the transmission of the stability of a family to the next at a period doubling bifurcation. This is inverse bifurcation. If a family becomes unstable as a parameter increases beyond a critical value, say $\eta=\eta_0$, then if another family is produced by bifurcation for $\eta>\eta_0$, it is stable (Fig. 8a; bifurcation). If, however, the bifurcating family extends backwards (for $\eta<\eta_0$) it is unstable (Fig. 8b; inverse bifurcation). In the latter case, the stability of the original family is lost and it is not transmitted to the bifurcating family. If this phenomenon appears along a sequence of bifurcations, the sequence stops and no limiting value of the bifurcation ratio can be calculated. It seems that the termination of bifurcation sequences is a general phenomenon in systems of three degrees of freedom [21].

Finally, we will mention briefly the new phenomenon of "collision of bifurcations", that appears only in systems of three or more degrees of freedom. Suppose that we have two control parameters, ε and η, and for a fixed value of η we find the stability parameters b_1 and b_2 of various families of periodic orbits. At bifurcation either $|b_1|$ or $|b_2|$ is equal to 2. But the values of ε for which $|b_1| = 2$ are different from those for which $|b_2| = 2$. Only if we also change η can we find a value $\eta = \eta_{crit}$, where $|b_1| = |b_2| = 2$ for the same $\varepsilon = \varepsilon_{crit}$. This we call a "proper collision" of the two bifurcations. As η goes beyond the critical value, the bifurcations are separated again. But then we observe in general that the bifurcating families have different interconnections for $\eta > \eta_{crit}$ from those they had for $\eta < \eta_{crit}$. Thus, the bifurcation trees change and the various branches (bifurcating families) are connected in new ways. As a consequence, the regions where we have trapping or stochasticity of nonperiodic orbits change their topology.

The above examples show that systems of three (or more) degrees of freedom show interesting new phenomena. But much more work is needed in order to reach the degree of understanding that we now possess for systems of two degrees of freedom.

References

1. C.C. Lin and F.H. Shu: Astrophys. J. 140, 646 (1964); see also C.C. Lin and Y.Y. Lau: Stud. Appl. Math. 60, 97 (1979).
2. A.J. Kalnajs: Astrophys. J. 166, 275 (1971).
3. A. Toomre, in S.M. Fall and D. Lynden-Bell (eds.): The structure and evolution of normal galaxies (Cambridge Univ. Press, Cambridge 1981), p. 113.
4. G. Contopoulos: Astrophys. J. 201, 566 (1979).
5. G. Contopoulos: Astron. Astrophys. 64, 323 (1978).
6. G. Contopoulos: Z. f. Astrophysik 49, 27 (1960).
7. G. Contopoulos and M. Moutsoulas: Astron. J. 70, 817 (1965).
8. F.G. Gustavson: Astron. J. 71, 670 (1966).
9. M.N. Rosenbluth, R.A. Sagdeev, J.B Taylor and M. Zaslavskii: Nucl. Fusion 6, 297 (1966).
10. G. Contopoulos, in F. Nahon and M. Hénon (eds.): Les nouvelles méthodes de la dynamique stellaire (CNRS, Paris, 1966) ≡ Bull. Astron., Ser. 3, 2, Fasc. 1, 223 (1967).
11. B.V. Chirikov: Phys. Rep. 52, 263 (1969); see also M. Zaslavskii and B.V. Chirikov: Sov. Phys. Uspekhi 14, 549 (1972).
12. J.M. Greene: J. Math. Phys. 20, 1183 (1979).
13. R.S. MacKay: Ph.D. Thesis, Princeton University (1982).
14. S.J. Shenker and L.P. Kadanoff: J. Stat. Phys. 27, 631 (1982).
15. M.J. Feigenbaum: J. Stat. Phys. 19, 25 (1978).
16. P. Coullet and C. Tresser: J. Phys. C 5, 25 (1978).
17. G. Benettin, C. Cercignani, L. Galgani and A. Giorgilli: Lett. Nuovo Cim. 28, 1 (1980).
18. J.M. Greene, R.S MacKay, F. Vivaldi and M.J. Feigenbaum: Physica D 3, 468 (1981).
19. T. Bountis: Physica D 3, 577 (1981).
20. G. Contopoulos: Physica D 8, 142 (1983).
21. G. Contopoulos: Lett. Nuovo Cim. 38, 257 (1983).
22. G. Contopoulos: Astron. J. 75, 96; 108 (1970).
23. G. Contopoulos and M. Zikides: Astron. Astrophys. 90, 198 (1980).
24. D.S. Heggie: Celes. Mech. 29, 207 (1983).
25. R.M. May: Nature 261, 257 (1976).
26. M. Widom and L.P. Kadanoff: Physica D 5, 287 (1982).
27. V.I. Arnold and A. Avez: Ergodic problems of classical mechanics (Benjamin, N. York 1968).

28. N.N. Nekhoroshev: Russian Math. Surveys 32, 6, 1 (1977).
29. G. Contopoulos, L. Galgani and A. Giorgilli: Phys. Rev. A 18, 1183 (1978).

The above list is quite incomplete. Important books containing also many references are:

V.I. Arnold: 1979, Mathematical methods of classical mechanics, Springer Verlag, New York.
R.H.G. Helleman (ed.): 1980, Nonlinear dynamics, Ann. N. York Acad. Sci. 357, New York.
S. Jorna (ed.): 1978, Topics in nonlinear dynamics, Amer. Inst. Phys. 46, New York.
A.J. Lichtenberg and M.A. Lieberman: 1983, Regular and stochastic motion, Springer Verlag, New York.

Solitons in General Relativity and Supergravity

G.W. Gibbons
D.A.M.T.P., Silver Street, Cambridge CB3 9EW, United Kingdom

Lecture 1 Black Hole Solitons in Ungauged Extended Supergravity

1) The soliton concept for gravity.

In flat space physics we expect of a soliton or in Coleman's famous phrase a "classical lump" that it be a

a) time independent (static or stationary)
b) finite (positive) energy
c) non-singular
d) classically and semi-classically stable
solution of the classical equations of motion which frequently
e) spatially interpolates between different vacua.

The "vacua", since they contain nothing, must be space-time homogeneous solutions of the classical equations which may or may not be stable. If the soliton spatially interpolates between one vacuum in the interior and one at infinity the vacuum at infinity should be stable. Energy is, of course, measured relative to the vacuum at infinity. Since the soliton is an excitation above that vacuum or ground state, it should have positive energy and in order to have predictable dynamics it should be non-singular. If the soliton is to persist in time and possess particle-like properties it should itself be stable both classically and quantum mechanically.

Solitons frequently possess conserved, sometimes topological, charges and their stability arises because they minimize the energy for a fixed value of the charge. The fact that they are stationary usually implies that they extremize the energy. In a supersymmetric theory, the charge is often a central charge and the configurations extremizing it a partially supersymmetric one. The solitons then fit into supermultiplets. Because of the partial supersymmetry, these multiplets are "shortened". They are smaller than the usual massive multiplets. The archetypal example is that in 2-dimensions discussed by Witten and Olive [1]. The most interesting 4-dimensional example are the monopoles which are reviewed in [2].

If we apply these ideas to a gravity theory it is natural to demand of a soliton that it be a complete and non-singular stationary spacetime which is asymptotically flat and solves the relevant field equations. Actually it is too strong a restriction to demand non-singularity. Because of the well-known singularity theorems of Penrose and Hawking, almost all interesting solutions will contain singularities. However, provided these singularities lie inside event horizons, the dynamics of the spacetime outside the horizon will be predictable. Thus, we will be forced to assume the truth of the Cosmic Censorship Hypothesis [3].

An important role in our considerations will be played by the Positive Mass Theorem [4, 5], which states that for an asymptotically flat spacetime with a complete Cauchy surface and whose stress tensor satisfies the dominant energy condition, that the mass **M** is non-negative and vanishes if and only if the spacetime is flat. This result can be extended to spacetimes with singularities, provided they contain a Cauchy surface which is complete outside an inner boundary which is an apparent horizon [6]. An apparent horizon on a spacelike 3-surface is compact 2-surface whose outward null l^α normals have everywhere vanishing convergence, $l^{\alpha}{}_{;\alpha} = 0$.

We shall see in the next section that there are, at least in the ungauged supergravity models, no solitons without horizons. The existence of horizons raises the question of the quantum mechanical stability of our solutions, since the work of Hawking [7] shows that horizons will, in general, emit thermally because of pair creation by the gravitational field. This will, in general, render unstable quantum mechanically solutions which are stable classically. Thus the requirement of quantum mechanical stability is additional to the purely classical one. An analogy might be to the dyons in Yang-Mills theory, which may lose charge by pair creation.

An interesting discussion of the soliton concept in gravity theory in which these issues are discussed is [7]

2) Absence of solitons without horizons.

The ungauged extended supergravity theories all possess an important symmetry, the classical equations are invariant under a spacetime independent dilation. That is the metric $g_{\alpha\beta} \to \lambda^2 g_{\alpha\beta}$, the spin 1 field strengths $F^i_{\alpha\beta} \to \lambda F^i_{\alpha\beta}$ and the scalar fields $\phi^A \to \phi^A$. The classical Lagrangian density $L \to \lambda^2 L$. This symmetry arises because the particles are massless: it is broken if one gauges the theory, i.e. the field strengths become non-Abelian, since the unit of charge introduces a scale. (Remember, classically the coupling constant has a dimension $\sqrt{\hbar}$.) Such a rescaling takes a classical configuration of mass **M** to one of mass λ**M**, since by suitable rescaling the coordinates, one can obtain a new asymptotically flat metric which tends to the flat metric $\eta_{\alpha\beta}$ at infinity. Now any stationary solution extremizes the total energy amongst all possible initial data. This energy amongst all possible initial data. This is a general statement proved by Schutz and Sorkin [8]. Thus the deformation produced by rescaling the metric either $M = \infty$, which is not interesting, or $M = 0$. One may readily check that all ungauges extended supergravity bosonic matter stress tensors satisfy the dominant energy condition. Therefore, by the Positive mass theorem, the solution must be flat space – the trivial vacuum. We summarize this statement as "No solitons without Horizons". For pure gravity, or Einstein Maxwell theory, the result may be proven directly from the stationary field equations [9]. Note that in principle the argument should work in any spacetime dimension.

It is clear from this General Relativistic version of Derrick's Theorem [10] that in order to get an interesting solution of the equations we must break the scale invariance. This may be done by means of boundary conditions which break scale invariance. This is, in effect, how the Schwarzschild solution

$$ds^2 = (1 - 2\frac{GM}{r})dt^2 - (1 - 2\frac{GM}{r})^{-1}dr^2 - r^2(d\theta^2 + \sin^2\theta d\phi^2)$$

escapes this theorem. This is static and finite, positive mass. Clearly, the mass is not extremized since given one solution there exists another with mass $M + \delta M$ say. However, such a perturbation will change the area $A = 16\pi G^2M^2$ of the event horizon. The mass of the black hole data set will only extremize the mass M if one holds fixed the area A of the outermost apparent horizon. Formally, one finds that when one tries to prove that the mass is extremized, one encounters boundary terms which spoil the argument. Consider, for simplicity, the time symmetric vacuum case. The initial data is a 3-metric h_{ij} subject to the constraint

$$h^{ij}\ {}^3R_{ij} = 0,$$

the integration being over a spacelike surface with inner boundary the apparent horizon (a minimal 2-sphere in this case). The mass M is read off from the asymptotic metric $h_{ij} \approx \delta_{ij}(1 + \frac{2M}{r}) + 0(\frac{1}{r^2})$. Varying M subject to the constraint yields if integration by parts were permissible

$$\int V h^{ij}\ {}^3R_{ij} \sqrt{h}\ d^3x$$

$${}^3R = 0$$

$${}^3R_{ij} = \nabla_i \nabla_j V$$

when ∇_i is the covariant derivative with respect to the metric h_{ij}. But these are just the static field equations for the 4-metric:

$$ds^2 = V^2dt^2 - h_{ij}dx^idx^j$$

Thus, as stated above, static solutions extremize the mass provided boundary terms may be discarded. This will be possible if we keep the apparent area of the horizon fixed. In fact, there are reasons for believing that an Isoperimetric Inequality holds for black holes and their apparent horizons, i.e.

$$M \geqslant \sqrt{\left(\frac{A}{16\pi G^2}\right)}$$

with equality only in the Schwarzschild case. The evidence for this Isoperimetric Inequality is reviewed in [11]. In any event, by Israel's Theorem [12], the Schwarzschild metric is the unique static metric, regular outside a single event horizon.

One might ask whether black holes should be regarded as classical solitons. They are classically stable, however the mass may grow with time as matter or gravitational radiation falls into the horizon. Indeed classically, a theorem of Hawking states that the area of the event horizon (which lies outside the apparent horizon) cannot decrease. Similar remarks apply to the angular momentum of rotating Kerr black holes. An initial data set consisting of gravitational waves can disperse to infinity and collapse to form a black hole. One might ask whether it can do anything else? In particular, are there solutions of the classical equations which are periodic in time? Bound states of gravitational radiation or black holes in periodic orbits perhaps? The answer to this question is no. Because any such system must inevitably radiate gravitational radiation and hence lose mass to infinity periodic motions, other than those with an infinite period are not allowed. this result has recently been shown for pure gravity [13] and it almost certainly holds in the ungauged extended supergravity models. The import of this result is that there are no solutions analogous to the "breathers" of Sine Gordon theory.

In my view, it is questionable whether one regards Schwarzschild black holes as classical solitons, but it is clear that at the quantum level they are not, as will be explained in the next section.

3) Black Holes and the breakdown of supersymmetry at finite temperature.

In what follows, I shall speak only of the Schwarzschild black hole. Similar remarks apply to the rotating Kerr black holes, but the formulae are significantly more complicated without any significantly new qualitative features arising. As is well-known by now, a black hole of mass M acts like a hot body of temperature

$$T = \frac{1}{8\pi GM}$$

This means that it will emit massless particles at a rate inversely proportional to the square of the mass. This will lower the mass, and hence raise the temperature leading to an even larger emission rate. Presumably, although this is controversial, it will ultimately disappear altogether.

The temperature is most easily calculated by noting that if one introduces Kruskal coordinates

$$T = (1 - \frac{2GM}{r})^{1/2} \exp(\frac{r}{4GM}) \sinh(\frac{t}{4GM})$$

$$X = (1 - \frac{2GM}{r})^{1/2} \exp(\frac{r}{4GM}) \cosh(\frac{t}{4GM})$$

to remove the coordinate singularities at $r = 2GM$ the metric is periodic in the imaginary time coordinate $t = i\tau$ with period

$$\beta = \frac{1}{T} = 8\pi GM$$

Thus, a functional integral around this background is one at finite temperature. In fact, any boson field which is regular on the horizon must be periodic with β while any fermion field must be anti-periodic with period β [14].

This suggests that, like a heat bath in flat spacetime, [15] a black hole should violate supersymmetry, since the constant parameter of global supersymmetry, which should apply at infinity, cannot depend upon imaginary time. To make this idea more precise, recall that in $N = 1$ the supersymmetry transformations are

$$\delta e^a_\mu = -iK\bar{\epsilon}\gamma^o\psi_\mu$$

$$\delta\psi_\rho = \frac{1}{K}\nabla_\rho\epsilon$$

where e^a_μ are the vierbein fields and ψ_μ the Rarita-Schwinger field. A purely bosonic background possesses a supersymmetry if there exists a Killing spinor field ϵ such that

$$\delta_\epsilon\psi_\rho = 0$$

i.e.

$$\nabla_\rho\epsilon = 0$$

There are no such Killing spinors in the Schwarzschild background, since the only asymptotically flat spacetime with Killing spinors is flat space.

Thus, the Schwarzschild solution has no exact residual supersymmetry. However, what is of interest here are global supergauge transformations. We want ψ_μ's which fall off at infinity. Thus, the ϵ's should at most tend to constants at infinity. We want to factor the space of all such supergauge transformations ($\epsilon \to \epsilon_o$ at ∞) by those which tend to zero at infinity ($\epsilon \to 0$ at ∞). The resulting equivalence class of supergauge transformations is the global supergauge group, and it should be parameterized by giving a constant spinor at infinity. To effect this factorization, it is convenient to impose a gauge condition – e.g.

$$\gamma^\alpha \psi_\alpha = 0$$

This fixes ϵ up to solutions of the massless Dirac equation.

$$\gamma^\alpha \nabla_\alpha \epsilon = 0$$

If $\gamma^\alpha \phi_\alpha = 0$ is a good gauge condition, there should be no regular solutions of the Dirac equation which tend to zero at infinity and 4 real Majorana solutions which tend to constants at infinity. These 4 real constant solutions – or 2 complex – allow us to make global gauge transformations of our background; if the background is static the solutions should be time-independent. The gauged transformation solutions should also be static. These gauged transformed superpartners will fit into supermultiplets.

In the quantum theory for example one would associate with these solutions a spin $\frac{3}{2}$ solutions

$$\psi^i_\mu = \nabla_\mu \epsilon^i$$

where

$$\gamma^\alpha \nabla_\alpha \epsilon^i = 0$$

$$\epsilon^i \to \epsilon^i_o$$

with the $4\psi^i_\mu$ one would have 4 creation operators a^+_i and the original state $|s\rangle$ would be generate with the 16 states:

$a^+_i \; |s\rangle$, $a^+_i a^+_j \; |s\rangle$ $a^+_i a^+_j a^+_k \; |s\rangle$

$a^+_l a^+_i a^+_j a^+_k \; |s\rangle$

If ϵ^i were covariantly constant, as it is in flat space, where $|s\rangle$ represents the flat vacuum the 15 extra states vanish, i.e.

$$a^+_i \; |s\rangle = 0$$

For the Schwarzschild solution, one might have anticipated 4 zero modes and 15 superpartners of the original black holes (cf. [16]). However, this is wrong. The time-independent solutions of the Dirac equation may be found in isotropic coordinates defined by

$$r = \rho(1 + \frac{GM}{2\rho})^2$$

so that

$$ds^2 = \frac{(1 - \frac{GM}{2\rho})^2}{(1 + \frac{GM}{2\rho})^2}dt^2 - (1 + \frac{GM}{2\rho})^4 (d\rho^2 + \rho^2(d\theta^2 + \sin^2 d\phi^2))$$

the solutions are

$$\epsilon = \frac{1}{(1 - \frac{2G^2M^2}{4\rho^2})}\epsilon_o$$

Clearly, these solutions are singular on the horizon $r = 2GM$ or $\rho = GM/2$.

Thus, it is not possible to perform a supergauge transformation of the black hole in this way.

Cordero and Teitleboim [17] have suggested that

$$\epsilon = \frac{s}{16\pi GM}$$

is a suitable spinor with which to perform on supergauge transformation, where s is a constant spinor in the obvious spinor frame. But this spinor frame is singular on the horizon, since the tetrad is. One must pass to a non-singular tetrad, related to the present one by a boost, which becomes infinite as the horizon is approached. Alternatively, note that

$$\bar{\epsilon}\, \gamma_\mu \epsilon = \left[\frac{1 - \frac{GM}{2\rho}}{1 + \frac{GM}{2\rho}}\right]^{-1} \delta^o_\mu$$

if we choose $\gamma^o\, \epsilon = \epsilon$, but the right hand side is singular as $\rho \to GM/2$.

Thus, it seems that it is just not possible to perform a time-independent global supersymmetry transformation of a black hole. Of course nothing prevents one performing time-dependent supergauge transformations.

A similar thing happens at finite temperature. The finite temperature state is not supersymmetric - for example the euclidean metric or $\mathbb{R}^3 \times S^1$ admits no spinors which are both antiperiodic in imaginary time and covariantly constant. This is not surprising, since the thermal distribution of fermions and bosons is different. In fact, it would take an <u>infinite</u> amount of energy to perform a supersymmetry transformation of a heat bath, since the total energy of bosons differs from that of the fermions by an infinite amount.

In the quantum theory, black holes cannot correspond to stationary states, since they are unstable. If one prevents them from evaporating by sending in a thermal flux of particles from infinity they can be kept time independent. One can then use the classical Schwarzschild solution to approximate this mixed thermal state. However, this is different from using solitons to approximate the corresponding pure one-soliton state in the quantum theory. There just does not seem to be a pure state corresponding to a Schwarzschild black hole.

One can try to relate the absence of solitons in pure gravity to the existence of vacua. There are 3 homogeneous solutions of the vacuum Einstein Equations [18].

A) Flat space, invariant under the Poincare group.
B) A special case of the pp-wave metrics invariant under a 6-dimensional isometry group.
C) A solution due to Petrov invariant under a 4-dimensional isometry group.

Flat space has maximal supersymmetry, (i.e. 4 Majorana covariantly constant spinors). All the pp-waves have half the maximal supersymmetry and this may be taken as a characterization of them. Thus, the special case B) has half the maximal supersymmetry. Case C) has no supersymmetry. There appear to be no time-independent solutions interpolating between flat space and vacuum B) or C). We shall see that the situation is different in N=2 supergravity.

4) Extended supersymmetry: central charges and multiplet shortening.

The extended supersymmetry algebras allow the possibility of a qualitatively new feature - the existence of central charges [19]. These are central in that they commute with every other generator in the algebra. In particular, they commute with time translations and Lorentz rotations. Thus they are conserved invariant charges. They have the dimension of mass and are thus only non-zero in a massless theory if the dilation invariance is broken in the presence of solitons. Since the massless fields can carry off energy but not the central charges, the configurations with least energy for fixed central charge are likely to be stable. These least energy configurations are partially supersymmetric [1].

Specifically the basic anticommunicator becomes, in a Weyl basis

$$\{Q^i_A , \bar{Q}^j_{A'}\} = \sqrt{2}\, P_{AA'}\delta^{ij}$$

$$\{Q^i_A , Q^j_A\} = \epsilon_{AB} Z^{ij}$$

where ϵ_{AB} is the SL(2,C) invariant skew form and $Z^{ij} = -Z^{ji}$ and $1 = 1,2,--N$ where N is the number of supercharges.

If we move to the rest frame (assuming $P_\alpha \rho^\alpha \neq 0$) we have

$$\{Q_1^i \ , \ Q_1^{j\dagger}\} = \delta^{ij} M$$

$$\{Q_2^i \ , \ Q_2^{j\dagger}\} = \delta^{ij} M$$

where $Q_1^{i\dagger}$ is, say, spin up and $Q_2^{i\dagger}$ is spin down.

while

$$\{Q_1^i \ , \ Q_2^j\} = Z^{ij}$$

Consider the N = 2 case for simplicity. The last anticommunication relation becomes

$$\{Q_1^1 \ , \ Q_2^2\} = U + iV$$

$$\{Q_1^2 \ , \ Q_2^1\} = -U - iV$$

If $U + iV = |\lambda| e^{i\theta}$ we can multiply Q_A^i by $e^{+i\theta/2}$ to make the last two equations become

$$\{Q_1^1 \ , \ Q_1^2\} = |\lambda|$$

$$\{Q_1^2 \ , \ Q_2^1\} = -|\lambda|$$

Thus

$$\{Q_1^1 \pm Q_1^{2\dagger}, Q_1^{1\dagger} \pm Q_1^2\} = 2(M \mp |\lambda|)$$

$$\{Q_1^1 \pm Q_1^{2+} \ , \ Q_1^1 \mp Q_1^{2\dagger}\} = 0$$

These are the anticommulation relations of a pair of creation and annihilation operators creating states of spin ½. There is a similar set with lower index 1 changed to 2 which create states of spin -½. They yield immediately the inequality

$$M \geqslant |\lambda|$$

Further, if $M = |\lambda|$ we get multiplets of dimension 4 rather than 16 since the operators $Q_1^1 - Q_1^2$ and $Q_2^2 - Q_2^2$ act trivially on the states. We shall refer to the inequality as the Bogomolny inequality, though the supersymmetric version is due to Olive and Witten [1]. Ono can carry out a similar analysis for higher N. One finds that the skew matrix Z^{ij} has $\left[\frac{N}{2}\right]$ skew eigen-values λ_n and the mass cannot be less than the modulus of any λ_n , $M \geqslant |\lambda_n|$ for $n = 1,2,...\left[\frac{N}{2}\right]$

5) Black Holes in N = 2 ungauged extended Supergravity: Charge without Charge.

The physical fields of N = 2 ungauged supergravity [20] are a graviton, 2 majorana gravitini and a photon. They are all massless and all electrically neutral with respect to the U(1) charge of the theory. Nevertheless, classical solutions exist for which this charge is non-zero. In fact, the theory possesses a global "dual-chiral" invariance if the gravitino is chirally rotated $\psi^i_\mu \to \exp i\,\theta\frac{\gamma_5}{2}\,\psi^i_\mu$ and the Maxwell field $F_{\alpha\beta}$ dually rotated $F_{\alpha\beta} \to \exp i\theta{*}\,F_{\alpha\beta}$ the equations of motion are unaltered. This corresponds to the invariance of the algebra under $Q_A \to e^{i\frac{\theta}{2}}Q_A$, $Z_{ij} \to e^{i\theta}\,Z_{ij}$. Using the invariance, we can perform a dual transformation to make the charge of a soliton purely electric, which we shall assume, for simplicity of exposition, has been done from now on. However, one should bear in mind that both electric and magnetic charges are possible. These aro the Reissner-Nordstrom black holes or their rotating versions the Kerr-Newman solutions. As in section 3 we shall discuss only the non-rotating holes for simplicity. The metric is

$$ds^2 = (1 - \frac{2GM}{r} + G\frac{Q^2}{4\pi r^2})dt^2$$

$$- (1 - \frac{2GM}{r} + G\frac{Q^2}{4\pi r^2})^{-1}dr^2 + r^2(d\theta^2 + \sin^2\theta d\phi^2)$$

and in order that the singularity at $r = 0$ be hidden by an event horizon g_{oo} must change sign between $r = 0$ and $r = \infty$ which leads to the inequality

$$M \geqslant \frac{|Q|}{\kappa}$$

The temperature of a Reissner-Nordstrom black hole is easily worked out using the method described in section 3 and one obtains:

$$T = \frac{1}{4\pi G}\;\frac{\sqrt{(M^2 - Q^2/\kappa^2)}}{(M + \sqrt{(M^2 - Q^2/\kappa^2)})^2}$$

we sec that for the limiting or extreme value for the charge the temperature vanishes.

Consider now the quantum evolution of a black holes in this theory. Its mass will be reduced by the thermal emission of gravitons, photons and gravitini. However, its charge Q cannot change, sinco these particles do not carry this charge. The result is that the holes will evolve towards the state with least mass for fixed Q, i.e. the extreme zero-temperature

Reissner-Nordstrom black hole. It will, in fact, take an infinite amount of time to do so, since the specific heat (at constant charge) changes sign for $Q/\kappa > \frac{3}{4}M$ and the evolution slows down as $Q/\kappa \to M$. Thus the stable end points of gravitational collapse in this theory are zero-temperature black holes, rather than flat space as in the $N = 1$ theory.

Thus, the geometry of the 3-surface metric is if two asymptotically flat sheets are connected by an Einstein-Rosen Bridge or "throat". The electric flux lines pass from one asymptotically flat region to the other without ending on any sources. This is the phenomenon of "charge without charge". (Note, however, that the full spacetime contains singularities at $r = 0$ which do not lie on these t = constant surfaces.)

It is interesting to note that the metrics for $\frac{Q}{\kappa} < M$ are qualitatively different from those with $\frac{Q}{K} = M$. This is most easily seen from the Penrose diagrams. Another way is to look at the t = constant surfaces, Introducing isotropic coordinates the Reissner- Nordstrom metric restricted to the surface t = constant becomes

$$-\left[(1 + \frac{GM}{2\rho})^2 - (\frac{GQ}{8\pi\rho})^2\right]^2 \left\{d\rho^2 + \rho^2(d\theta^2 + \sin^2\theta d\phi^2)\right\}$$

where $0 < \rho < \infty$. The area of a sphere ρ = constant is

$$4\pi\left[\rho + GM + \frac{1}{4}G^2(M^2 - Q^2/\kappa^2)\frac{1}{\rho}\right]^2$$

If $M^2 > Q^2/\kappa^2$ this expression is symmetrical under the transformation:

$$\rho \to \frac{1}{\rho}(M^2 - \frac{Q^2}{\kappa^2})\frac{1}{4}$$

of the 11-dimensional supergravity theory. It describes a spontaneous compactification from 4 dimensions to 2 dimensions similar to that studied in [22].

On the other hand, if $M^2 = Q^2/K^2$ there is no other asymptotically flat region. In fact, as $M \to |Q|/K$ the throat becomes infinitely long. The horizon may then be thought of as an internal infinity.

The geometry of this throat approximates that of the Robinson-Bertotti metric. This is the product metric of the form $(AdS)_2 \times S^2$ where the radii of curvature of the two factors are equal in magnitude to $|Q|/\kappa$ [24]. This is of considerable interest, because the Robinson-Bertotti metric may be thought of as the analogue of the Freund-Rubin solution.

In fact, apart from certain plane waves the only homogeneous solution of the Einstein Maxwell equations, apart from flat spacetime, is the Robinson-Bertotti metric [2]. Thus, from the point of view of homogeneity we might claim that there are 2 vacua in Einstein Maxwell theory, and the extreme Reissner-Nordstrom black holes spatially interpolate between them. We shall see in the next section that this interpretation is in harmony with the supersymmetric interpretation.

6) The Bogomolny Inequality.

The striking similarity between the Bogolmolny inequality of supersymmetry and the inequality from black hole physics that $M \geq |Q|/K$ suggests that the positive energy argument may be extended to cover this case. This is in fact true [24, 6]. The relevant supersymmetry transformations for $N = 2$ is

$$\delta\psi_{\mu}^{i} = \frac{1}{\kappa}(\nabla_{\mu}\epsilon^{i} - \epsilon^{ij}\kappa F_{\alpha\beta}\gamma^{\alpha}\gamma^{\beta}\gamma_{\beta})\epsilon^{i}$$

$$= \frac{1}{\kappa}\hat{\nabla}^{ij}\epsilon^{j}$$

where the second equation maybe taken as a definition of the "hatted derivative". If one uses the hatted derivative in the Witten positive energy argument, one arrives at the statement that

$$M \geq \sqrt{\left(\frac{Q^2}{\kappa^2} + \frac{P^2}{\kappa^2}\right)}$$

with equality if, and only if, the spacetime admits solutions of the Killing spinor condition

$$\hat{\nabla}_{\mu}^{ij}\epsilon^{j} = 0$$

One can show [23, 24] that the only such spacetimes other than pp waves are the Israel-Wilson metrics [25] well known to classical relativists. It was shown by Hartle and Hawking [26] that the only regular black hole solutions in this class are the Papapetrou-Majumdar metrics [27] which describe an arbitrary number of extreme Reissner-Nordstrom black holes in neutral equilibrium, the gravitational attraction being just balanced by the electrostatic repulsion. This balance was named (perhaps misleadingly) antigravity by Scherk [28] who pointed out that matter multiplets with maximum central charge coupled to supergravity exhibit this phenomenon.

The Israel-Wilson metrics admit, in general, 4 Majorana or 2 Dirac Killing spinors - i.e. ½ the maximum number of supersymmetries for $N = 2$. However, in one special case - the Robinson-Bertotti metric - they admit 8 Majorana Killing spinors - i.e. the maximum allowed by supersymmetry. As I mentioned in section 5 the Robinson-Bertotti metrics are homogeneous product metrics with a non-vanishing Maxwell field. In the purely magnetic case the Maxwell field is proportional to the alternating tensor on the S^2 factor. This is precisely analogous to the Freund-Rubin soluton. Thus, the Robinson-Bertotti metric qualifies as an alternative, maximally supersymmetric, vacuum state to the usual flat vacuum. The Extreme Reissner-Nordstrom metrics spatially interpolate between them as one expects of a soliton.

7) <u>The zero modes.</u>

As explained in section 3 we are interested in pure gauge solutions of the spin $\frac{3}{2}$ equations. We may again imposed the gauge condition

$$\gamma^{\mu}\psi^{i}_{\mu} = 0$$

and find that if

$$\psi_{\mu} = \frac{1}{\kappa}\hat{\nabla}_{\mu}\epsilon$$

then ϵ satisfies the massless Dirac equation. Now, the following lemma is useful: If the metric takes the form:

$$ds^2 = A^2dt^2 - B^2d\underline{x}^2$$

where A and B depend only on $\underline{x}$ then the time-independent solutions of the Dirac equation have the form

$$\epsilon = \frac{\epsilon_0}{AB^2}$$

where ϵ_0 is a time-independent solution of the massless Dirac equation in flat space. Note that any spherically symmetric static metric can be cast in this "isotropic" from. It is clear that if $A - 0$, $B \neq \infty$, then the spinor field is singular on the horizon. This is true for the entire Reissner-Nordstrom family, except in the limiting case $M = |Q|/\kappa$ in which case

$$AB = 1$$

Thus the spinor field ϵ vanishes at the horizon rather than blowing up. If we take for ϵ_0 a constant spinor, we will get 4 spinor zero modes which tend to constants at infinity. Of these, 2 are in fact Killing spinors. They give a vanishing spin $\frac{3}{2}$ field

$$\psi_{\mu} = \frac{1}{\kappa}\hat{\nabla}_{\mu}\epsilon = 0$$

The remaining 2 zero modes can be used to perform supersymmetry transformations on the purely bosonic background. Thus, the extreme Reissner- Nordstrom holes fit into a supermultiplet of 4 states. The multiplet is shortened exactly as we would expect for a system saturating the Bogomolny bound.

8) Conclusion.

We have seen that the extreme Reissner-Nordstrom Black Holes behave much like solitons. They are stable against Hawking radiation. They have zero temperature. They have the least mass consistent with their conserved central charge and are (partially) supersymmetric. They also fit into shortened multiplets. The geometry spatially interpolates between two maximally supersymmetric vacua of the ungauged $N = 2$ theory.

However, there is no way known to me of fixing their mass. Because of the scale invariance of the theory and the lack of a coupling constant in the theory, the mass, or equivalently the charge, is arbitrary. Usually one expects the mass of a soliton to be a fixed quantity. In Yang-Mills theory the monopole mass is given in terms of the expectation value of the Higgs field. In the next lecture I will show how this defect can be remedied by using the Kaluza-Klein idea.

References.

[1] E. Witten and D. Olive, Phys. Lett., 99B 229 (1981).

[2] Monopoles in Quantum Field Theory. ed. N. Craigie, P. Goddard and W. Nahm, publ. World Scientific, (1982).

[3] R. Penrose, Ann. N. Y. Acad. Sci., 224 125 (1973).

[4] S. T. Yau and R. Schoen, Commun. Math. Phys., 65 45 (1979).

Commun. Math. Phys., 79 231 (1981).

[5] E. Witten, Commun. Math. Phys., 80 381 (1981).

[6] G. W. Gibbons, S. W. Hawking, G. Horowitz and M. J. Perry, Comm. Math. Phys., 88, 295, (1983).

[7] P. Hajicek, Nucl. Phys., B185 254 (1981).

[8] B. F. Schutz and R. Sorkin, Ann. Phys., 107 1 (1977).

[9] R. Serini, Acad. Na. Lincei. Mem. Cl. Sci. Mat. Nal., 27 235 (1918).

A. Einstein and W. Pauli, Ann. Math., 44 131 (1943).

A. Lichnerowicz., C. R. Acad. Sci. Paris, 222 432 (1975).

[10] C.H. Derrick, J.Math.Phys., 5 1252 (1962).

[11] G.W. Gibbons in "Global Riemannian Geometry", ed. T. Wilmore and N.J. Hitchin: Ellis Horwood, 1984.

[12] W. Israel, Phys.Rev., 164 1776 (1967).

[13] G.W. Gibbons and J.M. Stewart in "Classical General Relativity", ed. W.B. Bonnor and J. Islam, C.U.P., 1984.

[14] G.W. Gibbons and M.J. Perry, Phys.Rev.Letts., 36 985 (1976).

[15] L. Girardello, M.T. Grisaru and P. Salomonson, Proc.Roy.Soc., A358 467 (1978).

Nucl.Phys., B178 331 (1981).

[16] T. Yoneya, Phys.Rev., D17 2567 (1978)

[17] P. Cordero and C. Teitleboim, Phys.Letts., 78B 80 (1978).

[18] D. Kramer, H. Stephani, M. MacCallum and E. Herlt, "Exact Solution of Einstein's Field Equations", C.U.P. and V.E.B., Deutsches Verlag der Wissenschaften.

[19] R. Haag, J.T. Lopouszanshi and M.F. Sohnius, Nucl.Phys., B88 257 (1975).

[20] S. Ferrara and P. van Nieuwenhuizen, Phys.Rev.Letts., 37 1669 (1976)

[21] B. Carter, "Black Holes", eds. C. DeWitt and B.S. DeWitt, Gordon and Breach, New York (1973).

[22] D.Z. Freedman and G.W. Gibbons, Nucl.Phys., B233 24 (1984).

[23] P. Tod, Phys.Letts., B121 241 (1983).

[24] G.W. Gibbons and C.M. Hull, Phys.Letts., 109B 190 (1982).

[25] W. Israel and G.A. Wilson, J.Math.Phys., 13 865 (1972).

[26] J.B. Hartle and S.W. Hawking, Commun.Math.Phys., 26 87 (1982).

[27] A. Papapetrou, Proc.Roy.Irish Acad., A51 191 (1947).

S.D. Majumdas, Phys.Rev., 72 390 (1947).

[28] J. Scherk, Phys.Letts., 88B 265 (1979).

Lecture 2 Supersymmetric Monopoles in Kaluza-Klein Theory

1) Kaluza-Klein Theory.

In this lecture we shall deal with the simplest Kaluza-Klein theory - pure gravity in 5-dimensions and its (N = 8) supersymmetric extension. The main reason for this is that, as discussed in lecture 1, more sophisticated spontaneous compactifications entail dealing with an AdS_4 background in which the concept of a soliton is, as yet, rather obscure. As is well-known, if we assume that the 5-dimensional spacetime is independent of one coordinate - X^5 we may parameterize the metric as:

$$ds^2 = e^{\frac{4\sigma\kappa}{\sqrt{3}}}(dx^5 + 2\kappa A_\alpha dx^\alpha)^2 + e^{\frac{-2\sigma\kappa}{\sqrt{3}}} g_{\alpha\beta}dx^\alpha dx^\beta$$

where the scalar field σ, the vector field A_α and metric $g_{\alpha\beta}$ are independent of X^5. The 5-dimensional Einstein lagrangian becomes, up to a total derivative,

$$\sqrt{g}(\frac{R}{4\kappa^2} - \frac{1}{4}e^{2\sigma\kappa\sqrt{3}}F_{\alpha\beta}F^{\alpha\beta} + \tfrac{1}{2}g^{\alpha\beta}\partial_\alpha\sigma\partial_\beta\sigma$$

where R is the Ricci scalar of $g_{\alpha\beta}$ and $F_{\alpha\beta} = \partial_\alpha A_\beta - \partial_\beta A_\alpha$. The unusual coupling of scalars to the vectors means that in general an isolated, charged, system will possess a scalar charge Σ, which we may define by the requirement that

$$\sigma \to \frac{1}{4\pi}\frac{\Sigma}{r}$$

at infinity. This is because if $F_{\alpha\beta}F^{\alpha\beta} \neq 0$, it is inconsistent to set σ = constant. Two scalar sources then experience in the Born approximation an attractive force $\frac{\Sigma_1 \Sigma_2}{r^2}$. One point should be made about regularity requirements. We have parameterized the metric in this way so as to put the action in standard Einstein conformal gauge. Starting from this 4-dimensional form of the theory, it is natural to demand that the metric $g_{\alpha\beta}$

be complete and non-singular, the scalar field σ be everywhere bounded and the Maxwell tensor $F_{\alpha\beta}$ be regular. Those 4-dimensional regularity conditions, which guarantee a regular 5-metric, are too strong as far as the 5-dimensional metric is concerned, since there are regular 5-metrics which violate them as we shall see later. In particular, since a 5-metric may be regular even though $g_{55} = 0$ regular 5-metrics may give rise to scalar fields with logarithmic singularities. In what follows we shall take the liberal view that what matters is regularity in 5-dimensions rather than 4.

The vacuum of Kaluza-Klein theory is the flat metric on $S^1 \times$ Minkowski space. Killing spinors will exist, and hence the vacuum be supersymmetric and thus stable if we demand that spinor fields be periodic in X^5 rather than antiperiodic. This will also rule out the bounce solution of Witten [1] which would (in the antiperiodic case) render the vacuum unstable. The perturbative excitations around the vacuum consist of the massless, electrically neutral sector containing the graviton, $g_{\alpha\beta}$, graviphoton A_α, and dilaton σ. The dilaton maybe thought of as the goldstone particle arising from the broken dilation invariance of the vacuum. In addition, we obtain the tower or ladder of charged massive states with charge e and mass m given by

$$\frac{e^2}{K^2} = 4\frac{m^2}{K}$$

$$m = \frac{n}{R_K}$$

where $n = \pm 1, \pm 2, \ldots, \quad 0 \leqslant X^5 \leqslant R_K$. The lowest rung of the ladder $n^2 = 1$, are absolutely stable particles. They cannot decay into massless states, because that would not violate the conservation of the Noetherian charge e arising from the U(1) invariance of the vacuum [2]. Of course these particles, which following [2] we call "pyrgons", can annihilate with their antiparticles.

2) Gross-Perry-Sorkin Monopoles.

We now seek solutions of the 5-dimensional vacuum Einstein Equations which are independent of X^5 and also a time coordinate t. One class of such solutions has the ultrastatic form

$$ds^2 = dt^2 - h_{\alpha\beta}dx^\alpha dx^\beta$$

when $h_{\alpha\beta}$ is a complete riemannian 4-metric which has an additional U(1) isometry. That is, $h_{\alpha\beta}$ is the metric of a gravitational instanton with a

Killing vector. To be asymptotic to the Kaluza-Klein vacuum, $h_{\alpha\beta}$ must be asymptotically locally flat (ALF) i.e. tend to a twisted product of S^1 with $S \times \mathbb{R}$. If the bundle is twisted, the resulting objects will have a magnetic charge, in the language of [3] the NUT charge is now reinterpreted as magnetic charge. The self-dual Taub-NUT metrics provide an example [4, 5, 6]. The Rieman tensor $R_{\alpha\beta\gamma\delta}$ of $h_{\alpha\beta}$ satisfies

$$R_{\alpha\beta\gamma\delta} = \pm \tfrac{1}{2} \epsilon_{\alpha\beta}{}^{\mu\nu} R_{\mu\nu\gamma\delta}$$

where $\epsilon^{\mu\nu}_{\alpha\beta}$ is the alternating tensor. Explicitly, we have:

$$h_{\alpha\beta} dx^{\alpha} dx^{\beta} = W^2 d\underline{x}^2 - \frac{1}{W^2}(dx^5 + 2\kappa \underline{A}.d\underline{x})^2$$

$$\mathrm{grad}W = -\frac{\kappa}{W}\mathrm{curl}A$$

$$W = 1 + 4GM\Sigma_{p=1}^{p=n} \frac{1}{|\underline{X} - \underline{X}_p|}$$

The coordinate singularities at $\underline{X} = X_p$ can be removed [3] provided X^5 is identified

$$0 \leqslant X^5 \leqslant 4\kappa^2 M$$

One readily reads off the magnetic moment of each monopole as

$$\frac{P}{\kappa} = 2M$$

Now, since the identification of X^5 in Kaluza-Klein gives the unit of electric charge:

$$0 \leqslant X^5 \leqslant \frac{4\pi\kappa}{e}$$

we deduce that the Dirac quantization condition holds:

$$eP = 2\pi$$

Since $P = 2\frac{\pi}{e}$ and the ADM mass is M we have

$$M = \frac{\pi}{\kappa e}$$

so indeed we have obtained a quantization rate for the mass of our solitons. Notice, as alluded to in section 1, the points $\underline{X} = \underline{X}_p$ are 5-dimensional coordinate singularities. From the conservative point of view of 4-dimensions they appear as singularities. One may read off the scalar charge Σ. it is

$$\Sigma = \sqrt{3}GM$$

Thus the antigravity condition

$$M^2 = \Sigma^2 / G - \frac{P^2}{\kappa^2} - Q^2 / \kappa^2 = 0$$

is satisfied.

3) N = 8 Supergravity in 5-dimensions.

We shall now consider the supersymmetry algebra in 5-dimensions and then the supersymmetric extension of Einstein's theory in 5-dimensions given by Cremmer[7]. Remarks on the 5 dimensional algebra and central charges may also be found in [8]. The present analysis is essentially that given in [9]. We consider the N = 8 algebra in 4 dimensions for which, in 2 component form:

$$\{Q^i_A, \bar{Q}^j_{B'}\} = \sqrt{2}P_{AB'}\ \delta^{ij}$$

$$\{Q^i_A, Q^j_B\} = Z^{ij}E_{AB}$$

where $Z^{ij} = U^{ij} + i\ V^{ij}$

are the central charges. U^{ij} being P.T. even is magnetic and V^{ij} being PT even is odd is electric.

In the most symmetric case

$$V^{ij} = V\Omega^{ij}\ , \qquad U^{ij} = U\Omega^{ij}$$

where $\Omega^{ij}\ \Omega^{jk} = -\delta^{ik}$

Thus the algebra is Usp(8) invariant, Ω^{ij} being the symplectic form. 5-dimensional gamma matrices Γ^A, A = 0, 1, 2, 3, 5 may be chosen as

$$\Gamma^A = (\gamma^\alpha,\quad i\gamma^5)$$

The 5-dimensional charge conjugation matrix C_5 is given by

$$C_5 = C\gamma^5$$

where C is the 4-dimensional charge conjugation matrix and satisfies

$$\Gamma^t_A = C_5\Gamma_A\ C_5^{-1}$$

One cannot have Majorana 4-spinors in 5-dimensions, however one can have symplectic spinors.

These satisfy

$$(\bar{S}^i)_a = \Omega^{ij}(S^{jt}C_5)_a$$

In terms of the Weyl spinors we find that

$$S^{ia} = \begin{bmatrix} Q^{iA} \\ \Omega^{ij}\bar{Q}^j A^i \end{bmatrix}$$

One may now rewrite the algebra in the form

$$\{S^{ia}, \bar{S}^b_b\} = -U\delta^{ij}\ \delta^a_i + \delta^{ij}(\gamma^1 P_i - i\gamma^5 V)^a$$

This looks more 5-dimensional if we identify the electric central charge, V, with the 5th component of momentum

$$P^5 = V$$

$$\{S^{ia}, \bar{S}^j{}_b\} = -U\delta^{ij}\delta^a_b + \delta^{ij}(\Gamma^A P_A)^a_b$$

Thus, the magnetic charge is identified with a central charge in 5-dimensions.

The Bogomolny inequality now becomes

$$M \geqslant \surd(U^2 + (P^5)^2)$$

but, according to the usual Kaluza-Klein interpretation, the 5th component of momentum, P^5, and electric charge, Q, are related by

$$P^5 = \frac{Q}{2K}$$

whence

$$M \geqslant \surd(U^2 + \frac{Q^2}{4K^2})$$

Note the factor of 4 in the denominator of the last term inside the square-root sign. This differs from the inequality in Einstein-Maxwell theory given in lecture 2.

Now consider 5-dimensional, N = 8, supergravity. This was constructed by Cremmer [*]. The lagrangian has an $E_{6(+6)}$ global and Usp(8) local symmetry with fields as follows:

graviton	g_{AB}	1	$\underline{1}$
gravitino	ψ^i_A	$\underline{1}$	$\underline{8}$
vector	$A^{\alpha\beta}_A$	$\underline{27}$	$\underline{1}$
spin½	χ^{ijk}	$\underline{1}$	$\underline{48}$
scalar	$V^{ij}_{\alpha\beta}$	$\underline{27}$	$\underline{27}$

A, B = 0, 1, 2, 3, 5 are spacetime indices

i, j = 1, 2, ...8 are Usp(8) indices

α, β = 1, 2, ...8 are $E_{6(+6)}$ indices

The spinors are symplectic

$$\bar{\psi}^i = \Omega^{ij}\, \psi^{it}\, C_5$$

$$\bar{\chi}^{ijk} = \Omega^{il}\, \Omega^{jm}\, \Omega^{kn}\, \chi^{lmn\, t}\, C_5$$

$$\bar{\chi}^{ijk}\, \Omega_{ij} = 0$$

The matrix-valued scalars $V^{ij}_{\alpha\beta}$ provide a harmonic map into $E_{6(+6)}/Usp(8)$. The bosonic action is

$$-\frac{1}{4\kappa^2}R - \frac{1}{8}G_{\alpha\beta,\gamma\delta}\,F^{\alpha\beta}_{AB}\,F^{AB\gamma\delta}$$

$$+ \frac{1}{24\kappa^2}\,P_{Aijkl}\,P^{Aijkl} - \frac{\kappa}{12}\eta^{ABCDE}\,F^{\alpha\beta}_{AB}\,F_{CD\beta\gamma}A^{\gamma}_{E\alpha}$$

where

$$G_{\alpha\beta,\gamma\delta} = V^{ij}_{\alpha\beta}\,\Omega_{ik}\,\Omega_{jl}\,V^{kl}_{\gamma\delta}$$

is the metric on $E_{6(+6)}$ and

$$P^{ij}_{A\ kl} = \left[(V^{-1})_{kl}{}^{\alpha\beta}\,\partial_A\,V^{ij}{}_{\alpha\beta}\right]_{\perp}$$

For backgrounds for which the scalars are constant and the vector fields vanish, the supersymmetry transformations reduce to

$$\delta\,\psi^{i}_{A} = \frac{1}{\kappa}\,\nabla_A\,\epsilon^{i}$$

4: Supersymmetries and Zero-Modes.

The vacuum of Kaluza-Klein theory, $S^1 \times R^3 \times$ Time, has 32 real or 16 complex Killing spinors constructed from the 4 covariantly constant Dirac spinors. These are naturally independent of time. The solitons possess 2 covariantly constant time-independent Dirac fields which are eigenstates of γ^0 with eigenvalue +1. This is because the spatial curvature is self dual (or anti-self dual), and chiral eigenstates with negative (respectively positive) chirality don't feel the curvature. Now what one usually thinks of γ_5 is γ^0 (i.e. " γ^{extra} "). If one looks for solutions of the Dirac equation (which would be zero modes) one finds that those which tend to constants at infinity have the form

$$\tfrac{1}{2}(1 + \gamma^0)\xi$$

or

$$\frac{1}{2W}(1 - \gamma^0)\xi$$

where ξ is a constant spinor. In the first case, we get Killing spinors and in the second case regular zero modes. Thus, we get half the maximum number of supersymmetries and the maximum amount of multiplet shortening allowed in N = 8.

One may go through the Witten identity to derive the Bogomolny inequality. This also serves to fix the relation between the central charge U and the magnetic moment P. One obtains

$$U = \frac{P}{2\kappa}$$

whence

$$M \geqslant \sqrt{(\frac{P^2}{4\kappa^2} + \frac{Q^2}{4\kappa^2})}$$

5: Electric-Magnetic Duality.

The standard Kaluza-Kloin theory has a discrete, electric magnetic duality which is a symmetry of the equation of motion but not the lagrangian. The equations of motion for the vectors are

$$\nabla_\alpha (e^{2\sqrt{3}\kappa\sigma} F^{\alpha\beta} = 0$$

The Bianchi identity is

$$\nabla_\alpha ({}^*F^{\alpha\beta}) = 0$$

The substitution

$$e^{2\sqrt{\kappa}\sigma} F^{\alpha\beta} \to {}^*F^{\alpha\beta}$$

$$\sigma \to -\sigma$$

leaves both and the σ equation of motion:

$$-\nabla^2\sigma + \frac{\sqrt{3}\kappa}{2} F_{\alpha\beta}F^{\alpha\beta} e^{2\sqrt{3}\kappa\sigma}$$

unchanged.

For static metrics one can introduce electric and magnetic potentials ψ and χ. If the metric takes the form

$$ds^2 = e^{-2\sqrt{3}\kappa\sigma}(e^{2\eta}\, dt^2 - e^{-2\eta}\, \gamma_{ij} dx^i dx^j$$
$$-e^{4\sqrt{3}\kappa\sigma}(dx^5 - 2\kappa\psi\, dt - 2\kappa\, A_i dx^i)^2$$

where

$$\mathrm{grad}\chi = -e^{2(\eta - \kappa\sigma)} \quad \mathrm{curl}A$$

the static equations of motion can be derived from the effective 3-action:

$$\frac{3R}{\kappa^2} - 2\frac{\gamma^{ij}}{\kappa^2}\partial_i\eta\partial_j\eta - 2\partial_i\sigma\partial_j\sigma\gamma^{ij}$$
$$+ 2e^{-2\eta}\, \gamma^{ij}\, (\partial_i\psi\partial_j\psi e^{2\sqrt{3}\sigma} + \partial_i\chi\partial_j\chi e^{-2\sqrt{3}\sigma})$$

The duality:

$$\eta \to \eta$$
$$\gamma_{ij} \to \gamma_{ij}$$
$$\psi \to \chi$$
$$\chi \to \psi$$
$$\sigma \to -\sigma$$

taking electric fields into magnetic fields is now manifest.

One may use the duality to show that the monopoles are dual to some singular limits of black holes which antigravitate because they move at the speed of light in 5-dimensions.

The 4-dimensional black hole solutions were discusse in [17]. If one insists on regularity in 4-dimensions (c.f. section 1), one finds that the spherically symmetric holes are parameterized by their mass M and magnetic and electric charges Q and P. The scalar charge Σ is a function of these, not an independent parameter. Regularity, i.e. Cosmic Censorship, then requires

$$M^2 + \frac{4\pi\Sigma^2}{\kappa^2} \geqslant \frac{Q^2}{\kappa^2} + \frac{P^2}{\kappa^2}$$

In the case of equality we further require

$$QP \neq 0$$

Note that the inequalities are not, in this case, identical with the Bogomolny inequality.

Clearly, the duality interchanges the electric and magnetic charges and reverses the sign of Σ. The only supersymmetric objects are the monopoles:

$$Q = 0$$
$$P = 2\kappa M$$
$$\Sigma = 3\sqrt{3}GM$$

or their dual partners with

$$Q = \sqrt{2}\kappa M$$
$$P = 0$$
$$\Sigma = \sqrt{3}GM$$

In fact, the regular black holes with zero magnetic charge may all be obtained by boosting a Schwarzschild black hole in the 5-th direction. As the velocity approaches that of light, one must take a suitable limit and one gets precisely the dual partner of the monopoles. The singular limiting geometry is singular even in the 5-dimensional sense. Corresponding to the multi-monopole solutions, one also has multi-object solutions with purely electric charge. They antigravitate (since $M^2 + \frac{4\pi}{\kappa^2}\Sigma - \frac{Q^2}{\kappa^2} = 0$) because they are moving at the speed of light in 5 dimensions. The multi-object-metrics have the form

$$ds^2 = -W'(dx^5)^2 + 2dt dx^5 - d\underline{x}^2$$

where $W' = 1 + \Sigma_{p=1}^{p=n} \frac{4GM_p}{|\underline{x} - x_p|}$.

The "masses" M_p are in general arbitrary (though those obtained by duality are of course all equal). The metric is in fact a pp wave in 5- dimensions and so is naturally supersymmetric. The pp - waves represent classical gravitational waves in 5-dimensions. Thus, we see that the duality takes gravitational waves into monopoles. In the next section we shall discuss this duality in more detail.

6) Monopole-Pygron Duality.

In the last section we wish to extend the concept of duality discussed in section 5). There we saw that the monopoles are dual to gravitational waves in 5-dimensions. In 5-dimensions these gravitational waves fall into massless supermultiplets . Viewed in 4 dimensions these "waves" are massive multiplets which have 256 members. This is the same multiplet as that of the "pygrons". This suggests that the following duality may hold.

monopole	<->	pygron
massless sector	< >	sector sector
antimonopole	<->	anti-pyrgon

The noetherian central charge would then be dual to the topological NUT charge. The monopoles and the pyrgons would then be stable for essentially the same reason.

More specifically, I am suggesting that there exist operators creating monopoles and antimonopoles, and that these together with the operators creating the massless states satisfy an effective field theory which is identical to Cremmer's 5-dimensional supergravity.

Suppose this is true. Then the β -function for the electrically charged pyrgons and that for the magnetically charged monopoles must have the same form.

$$\mu\frac{dP}{d\mu} = \beta(P)$$

$$\mu\frac{de}{d\mu} = \beta(e)$$

to maintain the Dirac relation

$$eP = 2\pi$$

we must have

$$\beta(e) = \beta(P) = 0$$

Now to one-loop this is trivially true since Kaluza-Klein is one loop finite in 5-dimensions. What is being suggested here is that this is true to all orders. In fact $\beta(e)$ vanishes to one loop not only when the entire tower of massive states is taken into account but "floor by floor" by spin sum rules. This may indicate that the divergences of the 5-dimensional theory are less severe than appears at first sight.

The duality being conjectured here would have other consequences which could be checked. For example, the pyrgons all have magnetic dipole moments depending upon their spin. The states in the monopole multiplets with non-vanishing spin should have electric dipole moments. These are calculable, and the gyroelectric ratios should equal the gyromagnetic ratios.

The situation with respect to monopoles in N = 8 Kaluza-Klein theory appears to be very analogous to the Yang-Mills monopoles in N = 4 Yang-Mills theory [11]. The present lecture is just a beginning of what promises to be an interesting subject.

References.

[1] E. Witten, Nucl. Phys., B195 481 (1982).

[2] E.W. Kolb and R. Slansky, Phys. Lett., 135B 378 (1984).

[3] G.W. Gibbons and S.W. Hawking, Commun. Math. Phys., 66 291 (1979).

[4] S.W. Hawking, Phys. Letts., 60A 81 (1977).

[5] D. Gross and M.J. Perry, Nucl. Phys., B226 29 (1983).

[6] R. Sorkin, Phys. Rev. Lett., 51 87 (1983).

[7] E. Cremmer, " Supergravities in 5 Dimensions" in 'Superspace and Supergravity' ed. S.W. Hawking and M. Rocek, C.U.P., (1981).

[8] J. Lukierski and L. Rytel, Phys. Rev. D., 27 2354 (1983).

[9] G.W. Gibbons and M.J. Perry, "Soliton-Supermultiplets and Kaluza-Klein Theory", Princton Preprint (1984).

[10] G.W. Gibbons, Nucl. Phys., B207 337 (1982).

[11] C. Montonen and D. Olive, Phys. Lett., 72B 117 (1977)

D. Olive in "Monopoles in Quantum Field Theory", eds. N. Craigie, P. Goddard and W. Nahm, World Scientific (1981).

The Dynamics of Spacetime Curvature: Nonlinear Aspects*

Kip S. Thorne
California Institute of Technology, Pasadena, CA 91125, USA

1. Introduction

Although general relativity is among the oldest and most studied of all nonlinear theories in physics, it has played only a modest role in the exciting soliton and chaos developments of recent years. There are three causes of this, I think: (*i*) The absence of any experimental data for strong gravity situations, which has protected relativists from having their noses rubbed in chaos; (*ii*) the extreme difficulty of solving Einstein's equations, which has induced relativists to concentrate on situations of weak gravity or high symmetry where chaos is usually absent and soliton-like structures were recognized as such only recently; and (*iii*) inadequate communication between relativists and people working on solitons and chaos.

Each of these three factors is changing rapidly or may change in the coming decade: (*i*) Gravity experimenters may well succeed in detecting and studying the gravitational waves emitted by black-hole collisions and other highly dynamical, strong-gravity systems [1], and those waves may carry relic imprints of unexpected, non-linear dynamics. (*ii*) Numerical relativists are now solving, on supercomputers, the fully nonlinear Einstein equations for systems with just one degree of symmetry [2] and soon will embark on the completely asymmetric case where chaotic behavior is known to occur when gravity gets sufficiently strong [3]. (*iii*) A number of relativists are learning the mathematics of solitons and chaos and are beginning to apply them, with some success, to problems in general relativity. Most notable are the recent use of soliton-theoretic techniques to derive algorithms for generating the most general stationary, axisymmetric, asymptotically flat solution of the vacuum Einstein equations [4]; and the recent use of chaos-theoretic techniques to elucidate the dynamical evolution of spacetime curvature in "Mixmaster" cosmological models [5].

It is likely that over the coming years there will be much cross-fertilization between general relativity theory and other areas of nonlinear physics. The purpose of these lectures is to aid, a bit, in that cross fertilization by presenting a brief overview of some of the nonlinear phenomena that occur in general relativity theory. This overview is intended not for the relativity expert, but rather for physicists in other fields. However, so as to make these lecture notes reasonably short, I shall assume that the reader is already familiar with the basic concepts of general relativity as expounded, e.g., in chapter 1 of MTW [6]: the description of gravity as a curvature of spacetime; the metric $g_{\mu\nu}$ and squared interval $ds^2 = g_{\mu\nu}dx^\mu dx^\nu$ as descriptors of the geometry of spacetime; local inertial frames (≡ local Lorentz frames); the fact that all the laws of physics have the same universal form in every local inertial frame everywhere and everywhen in the universe ("Einstein equivalence principle"); and the fact that the Einstein field equations

$$G_{\mu\nu} = 8\pi T_{\mu\nu} \tag{1}$$

(with $G_{\mu\nu}$ the Einstein curvature tensor and $T_{\mu\nu}$ the stress-energy tensor) describe the manner by which nongravitational energy, momentum, and stress generate spacetime curvature. (Here and throughout I set Newton's gravitation constant G and the speed of light c to unity.)

It is a remarkable fact that, because of their nonlinearity, the Einstein field equations (1) permit spacetime to be curved even in the complete absence of any nongravitational energy,

*Supported in part by the U.S. National Science Foundation [AST82-14126] at Caltech and by the Cerro Tololo Interamerican Observatory [operated by the Association of Universities for Research in Astronomy, Inc., under contract with the National Science Foundation].

momentum, or stress, i.e. they permit curvature to generate curvature; gravity to generate gravity. This can be seen quite explicitly by writing down the vacuum Einstein equations $G_{\mu\nu} = 0$ as explicit differential equations for the "metric density" $\mathfrak{g}^{\mu\nu} \equiv (-g)^{1/2} g^{\mu\nu}$, where $g \equiv \det \|g_{\alpha\beta}\|$ [7]:

$$
\begin{aligned}
\mathfrak{g}^{\mu\nu}\mathfrak{g}^{\alpha\beta}{}_{,\mu\nu} = {} & \mathfrak{g}^{\alpha\nu}{}_{,\mu}\mathfrak{g}^{\beta\mu}{}_{,\nu} + \frac{1}{2}\mathfrak{g}^{\alpha\beta}\mathfrak{g}_{\lambda\mu}\mathfrak{g}^{\lambda\nu}{}_{,\rho}\mathfrak{g}^{\rho\mu}{}_{,\nu} \\
& + \mathfrak{g}_{\lambda\mu}\mathfrak{g}^{\nu\rho}\mathfrak{g}^{\alpha\lambda}{}_{,\nu}\mathfrak{g}^{\beta\mu}{}_{,\rho} - \mathfrak{g}^{\alpha\lambda}\mathfrak{g}_{\mu\nu}\mathfrak{g}^{\beta\nu}{}_{,\rho}\mathfrak{g}^{\mu\rho}{}_{,\lambda} - \mathfrak{g}^{\beta\lambda}\mathfrak{g}_{\mu\nu}\mathfrak{g}^{\alpha\nu}{}_{,\rho}\mathfrak{g}^{\mu\rho}{}_{,\lambda} \\
& + \frac{1}{8}(2\mathfrak{g}^{\alpha\lambda}\mathfrak{g}^{\beta\mu} - \mathfrak{g}^{\alpha\beta}\mathfrak{g}^{\lambda\mu})(2\mathfrak{g}_{\nu\rho}\mathfrak{g}_{\sigma\tau} - \mathfrak{g}_{\rho\sigma}\mathfrak{g}_{\nu\tau})\mathfrak{g}^{\nu\tau}{}_{,\lambda}\mathfrak{g}^{\rho\sigma}{}_{,\mu} \ .
\end{aligned}
\tag{2}
$$

Here $\mathfrak{g}_{\mu\nu}$ is the inverse of $\mathfrak{g}^{\mu\nu}$ (which means it is a highly nonlinear algebraic function of $\mathfrak{g}^{\mu\nu}$); repeated indices, one up and one down, are to be summed; commas denote partial derivatives

$$
\mathfrak{g}^{\alpha\beta}{}_{,\mu} \equiv \partial\mathfrak{g}^{\alpha\beta}/\partial x^{\mu}, \qquad \mathfrak{g}^{\alpha\beta}{}_{,\mu\nu} \equiv \partial^2\mathfrak{g}^{\alpha\beta}/\partial x^{\mu}\partial x^{\nu} \ ,
$$

and the coordinate system has been specialized so the metric is in "deDonder gauge", $\mathfrak{g}^{\alpha\beta}{}_{,\beta} = 0$. The key points about these horrendous looking equations (ten equations for the ten independent $\mathfrak{g}^{\alpha\beta}$) are these: (*i*) they are highly nonlinear; (*ii*) the left-hand side is a sort-of curved-spacetime wave operator acting on $\mathfrak{g}^{\alpha\beta}$, "$\Box$" $\mathfrak{g}^{\alpha\beta}$; (*iii*)the right-hand side is a sort of "stress-energy pseudotensor" for the gravitational field -- it is quadratic in the first derivatives of $\mathfrak{g}^{\mu\nu}$, and it acts as the source for "$\Box$" $\mathfrak{g}^{\alpha\beta}$. These are just the type of equations one would expect to go along with the descriptive phrase that "gravity generates gravity". And considering the complexity of these equations by comparison, e.g., with the Korteweg-deVries equation or the Navier-Stokes equation, it is obvious that our present understanding of the dynamics of vacuum spacetime curvature must be rudimentary indeed!

As rudimentary as that understanding may be, it nevertheless includes some remarkable phenomena -- e.g., gravitational waves, geons, black holes, white holes, wormholes, and spacetime singularities. These phenomena will be the focus of my lectures: I will try to convey, through diagrams, words and a few equations, some insight into their nonlinear dynamics.

2. Gravitational Waves

Gravitational waves are ripples in the curvature of spacetime with characteristic wavelength λ very short compared to the lengthscale R of the universe through which they propagate, $\lambda \ll R$. ISAACSON [8] has introduced a beautiful two-lengthscale approximation technique for solving the Einstein equations (2) for a vacuum region of spacetime containing gravitational waves: He writes

$$
\mathfrak{g}^{\alpha\beta} \equiv \mathfrak{g}^{\alpha\beta}_{(B)} - \bar{h}^{\alpha\beta}
\tag{3}
$$

where $\mathfrak{g}^{\alpha\beta}_{(B)} \equiv \langle \mathfrak{g}^{\alpha\beta} \rangle$ is the average of $\mathfrak{g}^{\alpha\beta}$over a few wavelengths λ and thus varies on the long lengthscale R, and $\bar{h}^{\alpha\beta}$ varies on the short lengthscale λ. $\mathfrak{g}^{\alpha\beta}_{(B)}$, which has magnitude unity, is the metric density of the "background spacetime" through which the waves propagate; and $\bar{h}^{\alpha\beta}$, which has magnitude h, is the gravitational-wave function. By expanding the Einstein equations (2) in powers of λ/R, and then splitting them into smooth parts $\langle G_{\mu\nu} \rangle = 0$ and rippled parts $G_{\mu\nu} - \langle G_{\mu\nu} \rangle = 0$, one obtains:

$$
\Box\bar{h}^{\mu\nu} \equiv \bar{h}^{\mu\nu|\alpha}{}_{|\alpha} = 0 + \text{(higher-order corrections)} \ ,
\tag{4a}
$$

$$
G^{(B)}_{\mu\nu} = 8\pi T^{(W)}_{\mu\nu} + \text{(higher-order corrections)} \ ,
\tag{4b}
$$

$$
T^{(W)}_{\mu\nu} \equiv \frac{1}{32\pi}\langle \bar{h}^{\alpha\beta}{}_{|\mu}\bar{h}^{\lambda\sigma}{}_{|\nu} \rangle g^{(B)}_{\alpha\lambda} g^{(B)}_{\beta\sigma} \ .
\tag{4c}
$$

Here the vertical bar denotes a covariant derivative in the background metric.

Equation (4a), with the higher-order corrections neglected, is a linear wave equation for the propagation of the gravitational waves in the background spacetime. It should be compared with the corresponding wave equation for the vector potential of short-wavelength electromagnetic waves, $\Box A^{\mu} \equiv A^{\mu|\alpha}{}_{|\alpha} = 0$. From the comparison it is obvious that gravitational waves will exhibit all the same propagation phenomena as electromagnetic waves: gravitational and cosmological redshifts, gravitational deflection when passing the sun or other sources of background gravity, gravitational focusing, diffraction at the focal point of a gravitational lens, ... [9].

Equation (4b), with the higher-order corrections neglected, describes the creation of background curvature by the stress-energy tensor $T^{(W)}_{\mu\nu}$ of the waves; and (4c) shows how to compute that stress-energy tensor from the spatial average of the square of the gradient of $\bar{h}^{\alpha\beta}$. Notice that because the Einstein curvature tensor of the background spacetime gives us a lengthscale $(G^{(B)}_{\mu\nu})^{-1/2}$ which must be $\gtrsim R$, and because the waves' stress-energy tensor (4c) has magnitude $(h/\lambda)^2$, the background Einstein equations (4b) say in order of magnitude that $1/R^2 \gtrsim (h/\lambda)^2$, i.e.

$$h \lesssim \lambda/R \ll 1 . \tag{5}$$

Thus, although Isaacson's analysis does not assume small amplitudes for the gravitational waves, his $\lambda \ll R$ assumption together with the vacuum Einstein equations demands small amplitudes, $h \ll 1$.

The higher-order corrections to the wave equation (4a) are of particular interest because of their nonlinearity. They have the form

$$\Box\bar{h}^{\mu\nu} = (\text{terms like } \bar{h}^{\alpha\beta|\mu}\bar{h}_{\alpha\beta}{}^{|\nu} - \langle\bar{h}^{\alpha\beta|\mu}\bar{h}_{\alpha\beta}{}^{|\nu}\rangle) ; \tag{6}$$

and because of this quadratic form they produce wave-wave coupling and the generation of harmonics -- phenomena familiar, e.g., in nonlinear optics and in the theory of weak plasma turbulence. In order of magnitude the right-hand side consists of a piece of $O(h^2/\lambda^2)$ which generates pure-gauge changes of $\bar{h}^{\mu\nu}$ that are of no physical interest (§ 2.6.2 of [9]), and a piece of $O(h^2/\lambda R)$ which generates first-harmonic waves and beats in resonance with them as they propagate, causing them to grow after a distance R to size $\sim h^2 \ll h$. Thus, *the very definition of a gravitational wave as a short-wavelength, $\lambda \ll R$, ripple in the curvature, because it enforces $h \ll 1$ via the Einstein equations, guarantees that nonlinear propagation effects such as wave-wave coupling and second-harmonic generation are of negligible magnitude!* On the other hand, as we shall see below, the background curvature produced by the waves' stress-energy can be of great importance.

Of course, Nature is not constrained to produce only short-wave, small-amplitude excitations of spacetime curvature. Long-wave, large-amplitude excitations are also possible; though our understanding of them is rather meager. Our principal source of understanding comes from a family of exact, analytic solutions of the vacuum Einstein equations which carry the name "plane-wave solutions" even though they can violate the usual $\lambda \ll R$ definition of a "gravitational wave." Written in a form due to ROSEN [10] one family of plane-wave solutions takes the form

$$\begin{aligned} &ds^2 = -dt^2 + dz^2 + L^2(e^{2\beta}dx^2 + e^{-2\beta}dy^2) , \\ &\quad \beta = \beta(t - z) \quad \text{is arbitrary,} \\ &\quad L = L(t - z) \quad \text{satisfies } L'' = -(\beta')^2 L \quad \text{where } ' = \partial/\partial(t - z) ; \end{aligned} \tag{7}$$

see Fig. 1a. The metric function $\beta(t - z)$ is the analog of the gravitational-wave amplitude $\bar{h}^{\alpha\beta}$ of the Isaacson formalism; the spatial average of L, $\langle L\rangle$, is the analog of $\tilde{g}^{\alpha\beta}_{(B)}$; and $L - \langle L\rangle$ is the

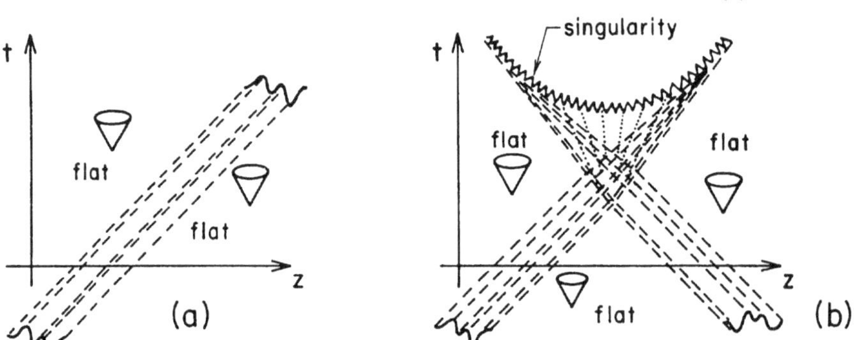

Fig. 1. (a) Spacetime diagram of the exact plane-wave solution (7) of the vacuum Einstein field equations. (b) Spacetime diagram for the collision of two plane waves. The waves focus each other gravitationally, producing a singularity of spacetime in the future. The little cones are light cones ($dz = \pm\, dt$); and in the regions labeled "flat" the curvature of spacetime vanishes.

analog of the higher-order corrections to the waves. Note that this exact solution of Einstein's equations is a "solitary wave" analogous to solutions of the Sine-Gordon and Korteweg-deVries equations [11]: although it is fully nonlinear, it preserves its form exactly as it propagates.

When two solitary Sine-Gordon or Korteweg-deVries waves collide, they interact for a short time and then emerge from the collision unscathed except for a phase shift [11]. Not so for exact gravitational plane waves. Numerous specific examples [12] and a general theorem [13] show that, if the spacetime before and after each wave is flat as in Fig. 1, then the waves gravitationally focus each other so strongly that they create a singularity of spacetime in their future! (Singularities will be discussed below.) The fact that this is true even when the two waves are arbitrarily weak indicates that, at least in part, the singularity is an artifact of the waves' infinite extent in the transverse x and y directions. Presumably for transversely bounded waves there is a threshold amplitude below which collisions produce no singularity and above which a singularity occurs.

3. Geons

John Archibald Wheeler, in the 1950s, realized that because gravitational waves carry energy and that energy gravitates, cf. (4b), a suitably organized bundle of gravitational waves should be able to hold itself together gravitationally. Wheeler called such an object a "gravitational geon" [14]; and his students BRILL and HARTLE [15] developed a mathematical theory of the structure of such geons using a two-lengthscale formalism analogous to (and predating) that of Isaacson.

Consider, as an example, a geon which at an initial moment of time $t = 0$ has vanishing momentum density $T^{(GW)}_{\hat{0}\hat{j}} = 0$ and spherically symmetric energy density $\rho(r) \equiv T^{(GW)}_{\hat{0}\hat{0}}$ as measured by a fiducial observer at rest inside the geon. (I use hats over indices to denote physical components as measured by such an observer.) Then the metric of the 3-dimensional background space, inside and outside the geon at $t = 0$, will have the standard spherical form [16]

$$ds^2 = \frac{dr^2}{1 - 2m/r} + r^2(d\theta^2 + \sin^2\theta\, d\varphi^2)\ ; \tag{8a}$$

and the Einstein equations at $t = 0$ ("initial-value equations") tell us that [16]

$$m(r) = \int_0^r 4\pi r^2 \rho\, dr\ . \tag{8b}$$

These are identically the same equations as occur in the analysis of a momentarily static, spherically symmetric star [16] with the sole exception that (i) the star's density of mass-energy ρ is replaced by the gravitational energy density $T^{(GW)}_{\hat{0}\hat{0}}$, and ($ii$) the star's full spatial metric (8) becomes the geon's background metric, to which one must add the gravitational-wave ripples in order to get the full geon metric.

The spatial geometry (8) can be depicted using an "embedding diagram" (§ 23.8 of MTW [6]): One imagines taking a two-dimensional, equatorial slice ($\theta = \pi/2$) through the star, removing it and embedding it in a flat, 3-dimensional Euclidean space without changing its intrinsic geometry. In the embedding space its shape must be

$$\bar{z}(\bar{r}) = \pm \int_0^{\bar{r}} \left[\frac{\bar{r}}{2m(\bar{r})} - 1 \right]^{-\frac{1}{2}} d\bar{r} \tag{9}$$

in order that the metric $ds^2 = d\bar{z}^2 + d\bar{r}^2 + \bar{r}^2 d\bar{\varphi}^2$ on it have the form (8a). This shape, augmented by the gravitational-wave ripples, is depicted in Fig. 2 for the case of geons with spatially constant internal energy density $\rho = \rho_0$ and various volumes [16]. Note that the background geometry (geometry with ripples averaged out) is spherical inside the geon, with radius of curvature $a = (3/8\pi\rho_0)^{\frac{1}{2}} \approx (10^8 \text{ light years})(\rho_0/1 \text{ erg cm}^{-3})^{-\frac{1}{2}}$.

A geon of roughly the volume shown in Fig. 2(a) can hold itself together for much longer than the gravitational-wave travel time around its circumference, providing the waves initially are traveling primarily circumferentially rather than radially, and providing the radial distribution of energy $\rho(r)$ is adjusted appropriately [14], [15]. On the other hand, there necessarily is some leakage of waves out of the geon; and ultimately the geon will be destroyed either by this

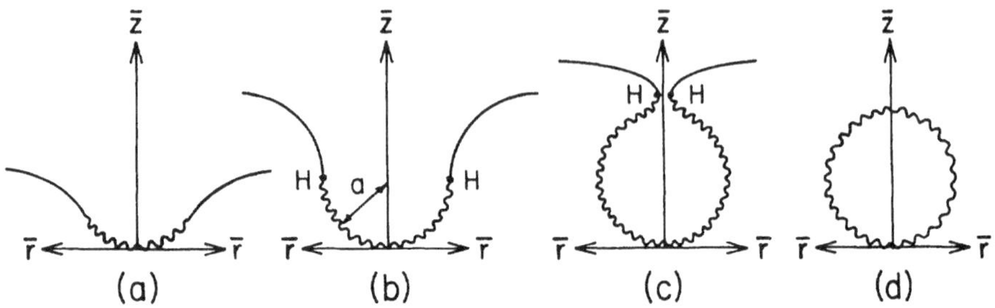

Fig. 2. Embedding diagrams for the equatorial, spatial 2-geometry of momentarily static geons with constant gravitational-wave energy density ρ_0. Each embedding diagram is shown in profile; it must be rotated about the $\bar{z}$ axis to form the full embedding diagram.

leakage or by the e-folding of a collective-mode instability [14], [15]. If the geon has much less volume than in Fig. 2(a), its background gravity will be so weak that the geon will disrupt itself very quickly and explode. If the geon's volume is somewhat greater than in Fig. 2(a), its background gravity will be so strong that it will pull the geon into dynamical gravitational collapse, producing a black hole. The geons in Figs. 2(b) and 2(c), for example, are so massive that, even though their background geometries are momentarily static, there is a black-hole horizon at the "throats" labeled "H" and the geon interiors will never be able to communicate with the external universe. The geon in Fig. 2(c) has almost enough gravitational-wave-occupied volume to close itself off from the rest of the universe; and that in Fig. 2(d) does close itself off: it is a closed "Friedman-like" universe in which the mass-energy required for closure is entirely due to gravitational waves. This closed universe, like the geons (b) and (c), is destined to collapse to a singularity in a time of order $\pi a/c \simeq 3 \times 10^6$ years $(\rho_0/ 1\ \mathrm{erg\ cm^{-3}})^{-1/2}$.

It seems very unlikely that any natural processes in the real universe ever conspire to produce geons. Nevertheless, geons are of considerable pedagogical interest for the insight they give us about the nonlinearities of vacuum, general relativistic gravity.

4. Black Holes

The gravitational collapse of a geon such as that in Fig. 2(a) is qualitatively similar to that of a star made of gas. The only significant difference is that some of the geon's gravitational waves will leak out during the collapse--but in the case of a realistic star some of the star's outer, gaseous layers may similarly be ejected. Some important details of the collapse are depicted in Fig. 3 (cf. Fig. 32.1 and Box 32.1 of MTW [6])

Figure 3(a) depicts the dynamics of the collapse by means of a spacetime diagram of the geon's equatorial plane, $\theta = \pi/2$. The collapse produces a spacetime singularity surrounded by a horizon, i.e. a "black hole." The interior of the horizon cannot send signals to the external universe because all signals must travel inside or on the light cones, and the interior light cones are all tilted inward toward the singularity. One can deduce the tilt of the light cones mathematically from the exact solution for the spacetime metric at late times t, after all escaping gravitational waves have leaked away and after all collapsing waves have hit the singularity:

$$ds^2 = -\left(1 - \frac{r_g}{r}\right)dt^2 + \frac{2r_g}{r}dtdr + \left(1 + \frac{r_g}{r}\right)dr^2 + r^2(d\theta^2 + \sin^2\theta d\varphi^2) \tag{10}$$

(equation (2) of Box 31.2 of MTW [6] with $\tilde{V} = t + r$ and $2M = r_g$). Here $r_g \sim a \sim (10^6$ light years$)(\rho_0/ 1\ \mathrm{erg\ cm^{-3}})^{-1/2}$ is the "gravitational radius" of the black hole; and the light cones of Fig. 3(a) are computed by setting $ds^2 = 0$. The spatial geometries of the collapsing geon and resulting black hole are depicted in Fig. 3(b) by means of embedding diagrams. The geon's embedding diagram is the same as Fig. 2(a), but with the required rotation about the $\bar{z}$ axis carried out. As the collapse proceeds, the gravitational wave region shrinks toward a point, leaving behind a cusp in the spatial geometry -- the black hole's central singularity.

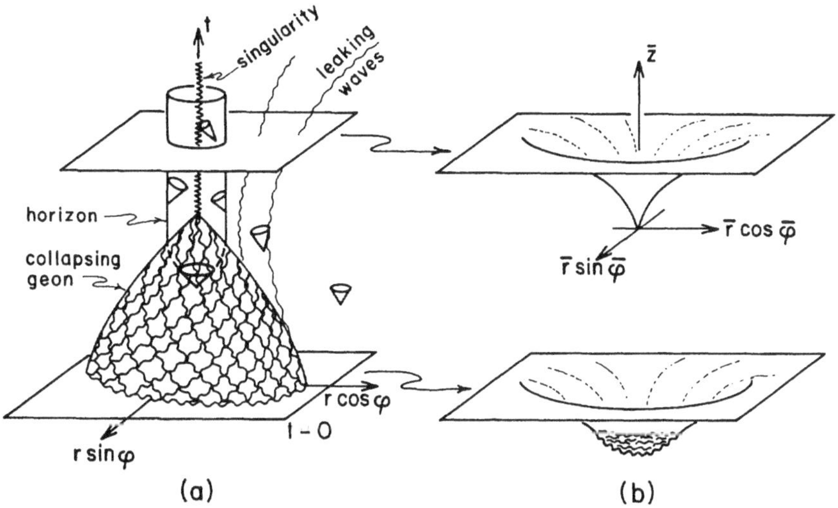

Fig. 3. (a) Spacetime diagram for the gravitational collapse of a geon [e.g. that of Fig. 2(a)] to form a black hole. (b) Embedding diagrams showing the spatial geometry of the geon at the initial moment of time and the spatial geometry of the black hole created by the geon's collapse.

Black holes are remarkable, soliton-like objects: As shown by Israel, Carter, Hawking, and Robinson [17], once a black hole has settled down into an equilibrium state (and this happens quickly, in a time $\tau_* \sim 10 r_g / c$), its spacetime geometry is determined completely and uniquely by just three parameters: its mass M, angular momentum J, and electric charge Q. Equation (10) is the special case for $J = Q = 0$; in it $r_g = 2GM/c^2$. From a hole's final equilibrium geometry it is impossible to deduce whether the hole was formed by the collapse of a geon or the collapse of a star made of matter or the collapse of a star made of antimatter. Moreover (providing the conjecture of cosmic censorship is true; see below), once formed a black hole can never be destroyed by any classical process. And although quantum field theory predicts quantum mechanical evaporation of the hole by the emission of "Hawking radiation" [18], the timescale for that evaporation is so long, $\hat{\tau} \simeq (2\times10^{66}\,\text{years})(M/M_\odot)^3$, as to be of no astrophysical interest for holes of stellar mass.

While geons almost certainly do not exist in the real universe, black holes probably do. Astronomers see large numbers of massive stars that are likely to collapse to form black holes when they die; and they also see strong but indirect evidence for the presence of black holes in some compact X-ray sources and in the nuclei of quasars and some galaxies [19].

5. White Holes

Because the Einstein field equations are invariant under time reversal, one can obtain a new solution by time reversing any old solution. When the collapse-to-a-black-hole solution of Fig. 3(a) is time reversed, the result is the "white-hole" [20] solution of Fig. 4(a)

In the bottom half of Fig. 4(a) the spacetime geometry of this white hole is described by the time reversal of the black-hole metric (10):

$$ds^2 = -\left(1 - \frac{r_g}{r}\right)dt^2 - \frac{2r_g}{r}dt\,dr + \left(1 + \frac{r_g}{r}\right)dr^2 + r^2(d\theta^2 + \sin^2\theta\, d\varphi^2)\,. \tag{11}$$

From this formula one can deduce that the singularity at $r = 0$ is surrounded by a "past horizon" at $r = r_g$. The interior of this past horizon can never receive signals from the exterior, but it can send signals to the exterior; hence the name "white hole." At some moment of time, t, the singularity (which can be regarded as an unexploded remnant of the big-bang singularity) transforms itself into a bundle of short-wavelength gravitational waves, i.e. a geon, which explodes outward through the past horizon and into the external universe. As the geon explodes, the strength of gravity near its center lessens and the past horizon shrinks and disappears.

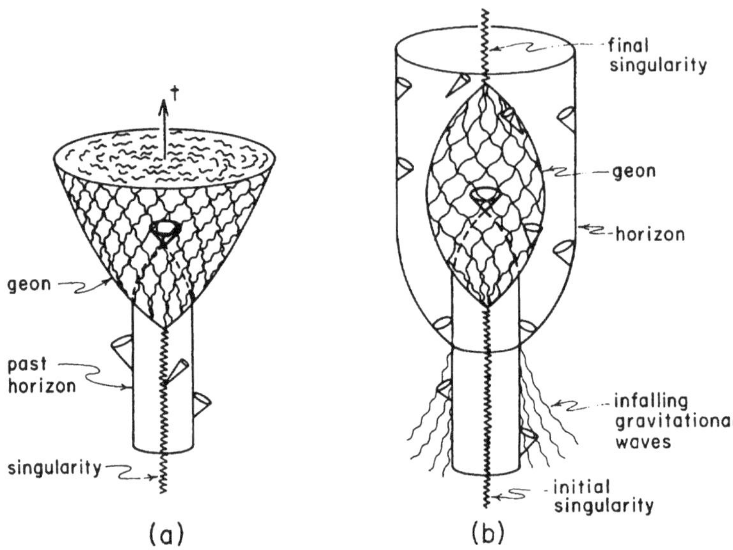

Fig. 4. (a) White-hole solution to the vacuum Einstein field equations: An initial singularity surrounded by a "past horizon" explodes to form a bundle of gravitational radiation (geon). (b) The conversion of a white hole into a black hole due to infalling gravitational waves triggering an instability.

In the classic papers on white holes [20] the exploding object was a hot, gaseous fireball which, it was suggested, might be the power source of a quasar. However, the equations of general relativity permit one to postulate, equally well, that the singularity produces an exploding geon.

White-hole models for quasars were abandoned in the early 1970s when (*i*) ZEL'DOVICH, NOVIKOV, and STAROBINSKY [21] showed that, because of quantum effects, the explosion of the white-hole singularity cannot be delayed from the big-bang era until the era of quasar formation, billions of years later; and when (*ii*) EARDLEY [22] showed that a white hole is highly unstable against transforming itself into a black hole. Since Eardley's instability is a nonlinear, classical effect rather than a quantum effect, it is of interest for these lectures. The instability is triggered by the infall of gravitational waves (or electromagnetic waves or matter), as shown in Fig. 4(b). The white hole, like a black hole, gravitationally attracts all forms of energy, including gravitational waves. An infalling packet of gravitational waves, no matter how weak initially, will become arbitrarily energetic as it approaches the past horizon due to the gravitational blue shift [(locally measured energy) $\propto (1 - r_g/r)^{-1/2} \propto \exp(t/2r_g)$]. The resulting sheet of energy on the past horizon gravitates so strongly that it creates a black-hole horizon around the past horizon; and when the singularity later explodes, its resulting geon is trapped inside the black-hole horizon and is forced to recollapse and form a new singularity.

6. Wormholes

A wormhole is another nonlinear gravitational object that probably does not exist in the real universe, but nevertheless is of pedagogical interest. The classic wormhole solution of the vacuum Einstein field equations is that of SCHWARZSCHILD and FLAMM [23], which has spatial geometry

$$ds^2 = \frac{dr^2}{1 - r_g/r} + r^2(d\theta^2 + \sin^2\theta d\varphi^2) , \tag{12}$$

with radius r double valued as shown in the embedding formula (9)

$$\bar{z} = \pm [4r_g(r - r_g)]^{1/2} , \quad ds^2 = d\bar{z}^2 + d\bar{r}^2 + \bar{r}^2 d\bar{\varphi}^2 , \tag{13}$$

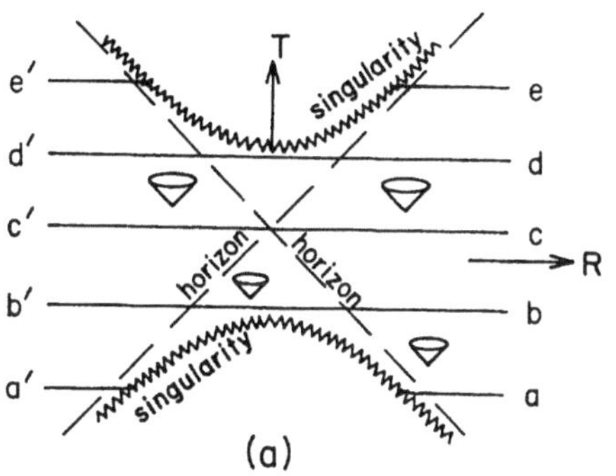

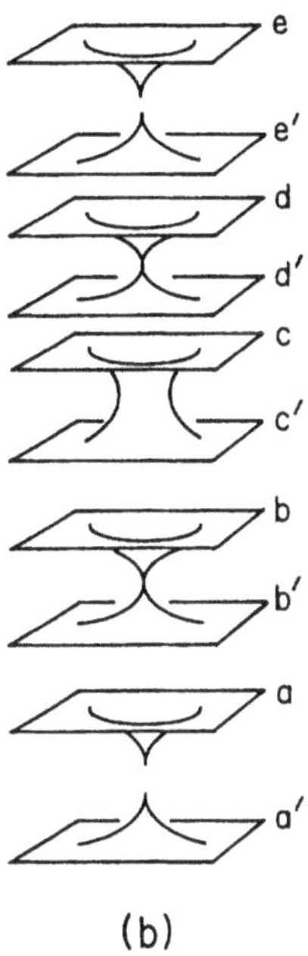

Fig. 5. (a) Spacetime diagram in Kruskal-Szekeres coordinates for the evolution of a Schwarzschild-Flamm wormhole. (b) Embedding diagrams depicting the evolution.

and in the corresponding embedding diagram cc′ of Fig. 5 (b). As shown by the embedding, this solution describes a "bridge" or "wormhole" that begins in one asymptotically flat universe and extends down through the embedding space into another asymptotically flat universe. We can think of the upper universe as our own and the lower as an alien universe that we might wish to visit.

Unfortunately, the dynamical evolution of the wormhole prevents such a visit from being made. The full dynamical evolution is described most clearly by writing Schwarzschild's solution of the vacuum Einstein field equations in a form due to KRUSKAL and SZEKERES [24]

$$ds^2 = \frac{4r_g^3}{r} e^{-r/r_g}(-dT^2 + dR^2) + r^2(d\theta^2 + \sin^2\theta d\varphi^2) , \tag{14a}$$

where $r(T,R)$ is given by

$$\left(\frac{r}{r_g} - 1\right) e^{r/r_g} = R^2 - T^2 , \tag{14b}$$

and r_g is the gravitational radius of the wormhole (equal to $1/2\pi$ times the circumference of the wormhole in Fig. 5(b) c-c′). A $T-R$ spacetime diagram for this solution is depicted in Fig. 5(a). Note that the radial light cones in this coordinate system are everywhere inclined at 45 degrees to the vertical, $dT = \pm dR$. The two regions $r = 0$ (i.e. $T = \pm \sqrt{1 + R^2}$) are physical singularities, and the regions $r = r_g$ (i.e. $T = \pm R$) are horizons. The evolution of the wormhole can be understood by making embedding diagrams for the equatorial ($\theta = \pi/2$) slices of constant Kruskal-Szekeres time T; Fig. 5 (b).[1] At early times (slice a-a′) the two universes are totally disconnected, and each possesses a singularity (cusp in the embedding geometry). At some moment of time T the singularities join onto each other, forming a nonsingular wormhole between the universes (slice b-b′). The throat of the wormhole then expands until it reaches a maximum circumference of $C = 2\pi r_g$ [slice c-c′; equations (12) and (13)]; and it then recontracts and pinches off (slice d-d′, leaving the two universes with disconnected singularities (slice e-e′). From the spacetime diagram of Fig. 5(a) it should be obvious that the wormhole expansion,

[1]Actually, to obtain the embedding diagrams a-a′ and e-e′ one must choose a slice that deviates slightly from T = constant as it nears the singularity: It must hit the singularity tangentially rather than at a finite angle in the $T-R$ coordinate plane; for an explanation of why see Fig. 31.6 of MTW [6]. The freedom to choose one's 3-dimensional slices of constant "time" in a variety of ways (the "many-fingered" nature of time in general relativity as described in § 21.8 of MTW [6]) causes the precise shapes of the embedding diagrams to be somewhat arbitrary.

recontraction, and pinchoff proceed so rapidly that no person, object, or signal can pass through the wormhole safely from our universe to the alien universe. On the other hand, a person can begin in our universe and fall through the horizon into a region near the final singularity, where he can receive signals from the alien universe. But he must pay the ultimate price for the knowledge so gained: he inevitably will be destroyed in the future singularity and he will never be able to send signals back out into our universe to tell us what he has learned.

If the wormhole is threaded by an electric or magnetic field (Reissner-Nordstrom solution of Einstein equations [25]) or is rotating (Kerr solution [26]), and is not perturbed in any way, its evolution can permit travel from our universe into some other universe. However, first-order perturbation theory suggests that the smallest of perturbations will trigger an instability which converts the route to the other universe into an impassable physical singularity [27].

7. Singularities

Time and again in the above discussion we have encountered singularities in the evolution of fully nonlinear solutions of Einstein's equations. Prior to 1964 it was widely believed that such singularities are artifacts of the high degree of spatial symmetry possessed by these solutions. However, PENROSE and HAWKING [28] have proved, using powerful techniques from differential topology, that singularities are endemic in nonlinear solutions of Einstein's equations: no amount of asymmetry can prevent the creation of a singularity once sufficient energy (material, electromagnetic, or gravity-wave) has become sufficiently concentrated.

It seems likely, but nobody has yet proved it true, that the singularities created by the gravitational collapse of a star or geon or wormhole are always hidden from external view by a black-hole horizon. The assumption that this is so is called the "Conjecture of Cosmic Censorship" [29]. All efforts to prove Cosmic censorship or to find a convincing counterexample have met with failure.

Because classical general relativity is unable to predict the future evolution of a singularity (e.g., in Fig. 5(a) it cannot predict whether the white hole will explode to form a geon or a gaseous fireball), Cosmic Censorship is a crucial assumption underlying several key theorems about the evolution of spacetime curvature. For example, the second law of black-hole mechanics (that the surface area of a black hole can never decrease) relies on Cosmic Censorship. Without Cosmic Censorship a naked singularity might influence the horizon's evolution in some unpredictable way that violates the second law [30].

It is also widely believed, but has not yet been proved (§34.6 of MTW [6]), that the generic singularities predicted by the Penrose-Hawking theorems have the mathematical and physical structure discovered and explored by BELINSKY, KHALATNIKOV, and LIFSHITZ [3]. This structure can be described in terms of the "tidal gravitational forces" felt by an observer who falls into the singularity.

In Newtonian gravity theory tidal forces are produced by the gradient of the gravitational acceleration or, equivalently, by the double gradient of the Newtonian gravitational potential, $\partial^2\Phi/\partial x^j \partial x^k \equiv E_{jk}$. (Here x^j and x^k are spatial Cartesian coordinates.) Two particles with vector separation ζ^k experience a relative gravitational acceleration

$$d^2\zeta_j/dt^2 = -E_{jk}\zeta^k \ . \tag{15}$$

It is this relative acceleration, due to the potentials Φ of the moon and the sun, that produces tides on the earth's oceans.

In general relativity tidal forces are produced by spacetime curvature: An observer falling toward a singularity feels, in his local inertial frame, tidal accelerations with the form (15), where $E_{jk} = R_{j0k0}$ = (space-time-space-time part of the Riemann curvature tensor); cf. Sec. 1.6 of MTW [6].

The Belinsky-Khalatnikov-Lifshitz (BKL) generic singularity is spatially inhomogeneous [3]: different observers falling into spatially different parts of the singularity see different, uncorrelated tidal fields E_{jk}. However, the pattern of evolution is the same at all spatial locations--and it is the same as in the spatially homogeneous "Mixmaster" solution of the vacuum Einstein field equations [31]: In the local inertial frame of an infalling observer there is a preferred choice of Cartesian coordinate axes (x, y, z) which diagonalizes the tidal field E_{jk}. In terms of these "principal axes" the component E_{xx} produces a tidal compression $\ddot{\zeta}_x/\zeta_x = -E_{xx}$ along the x direction (or, if E_{xx} is negative, it produces a tidal stretching). Similarly E_{yy} produces a tidal compression $\ddot{\zeta}_y/\zeta_y = -E_{yy}$ along the y direction, and E_{zz} produces compression along the z direction. A typical example of the evolution of these tidal accelerations near a BKL generic singularity is depicted in Fig. 6.

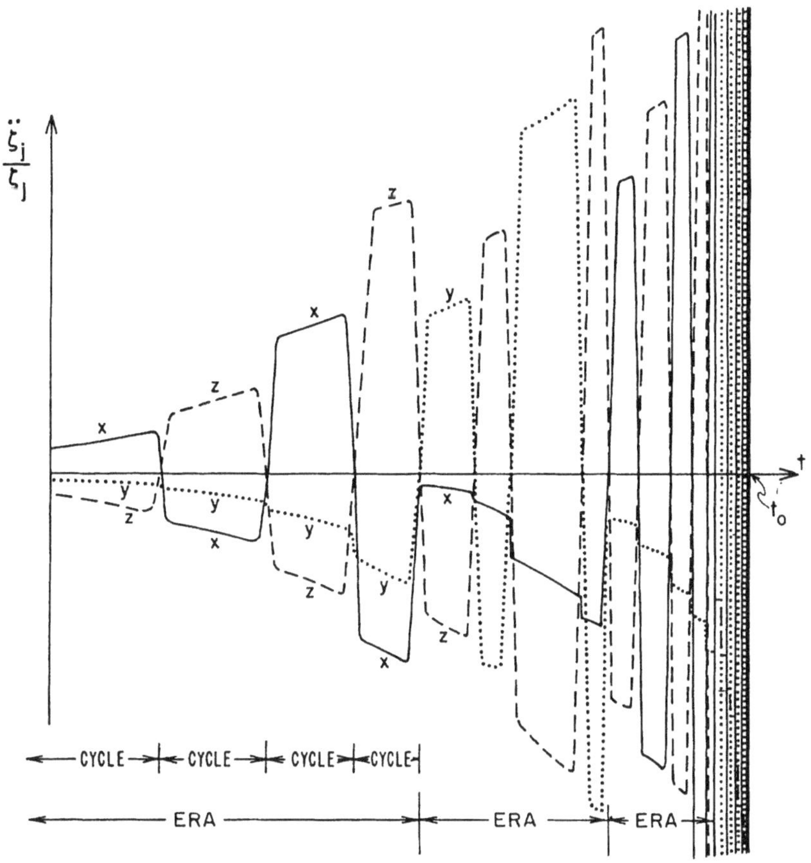

Fig. 6. Tidal gravitational accelerations along the principal axes of a freely falling observer as the observer falls into a generic BKL singularity.

The evolution is characterized by an infinite sequence of cycles which are grouped together into eras. During each era one of the principal axes experiences a continual tidal compression, while the other two axes oscillate between compression and stretch. At a change of era there is a change in the roles of the axes; e.g., at the first change of era in Fig. 6 the continual compression axis switches from y to x, and the oscillating axes switch from x and z to y and z. The tidal accelerations during the n'th cycle are expressible analytically in terms of a parameter U_n and the time (as measured by the free-falling observer) until the singularity will be reached, $t_0 - t$:

$$\text{Continual compression axis:} \quad \frac{\ddot{\zeta}}{\zeta} = -\frac{U_n(1+U_n)}{(1+U_n+U_n^2)^2}\frac{1}{(t_0-t)^2}, \tag{16a}$$

$$\text{Temporary compression axis:} \quad \frac{\ddot{\zeta}}{\zeta} = -\frac{U_n^2(1+U_n)}{(1+U_n+U_n^2)^2}\frac{1}{(t_0-t)^2}, \tag{16b}$$

Temporary expansion axis: $$\frac{\ddot{\zeta}}{\zeta} = +\frac{U_n(1 + U_n)^2}{(1 + U_n + U_n^2)^2}\frac{1}{(t_0 - t)^2}\,. \qquad (16c)$$

The parameter U_n is governed by the same "Gauss map" as appears in the expression for a number in terms of continued fractions:

$$\begin{aligned} U_{n+1} &= U_n - 1 \text{ if } U_n > 1 \quad \text{(change of cycle; fixed era)} \\ &= 1/U_n \text{ if } 0 < U_n < 1 \quad \text{(change of era).} \end{aligned} \qquad (17)$$

For example, one possible sequence is 4.321..., 3.321..., 2.321..., 1.321..., 0.321..., 1/0.321... = 3.12... (new era), 2.12..., 1.12..., 0.12..., 1/0.12... = 8... (new era), 7..., etc. Obviously, the map (17) and corresponding tidal accelerations (16) are unstable in the sense that two sequences with slightly different initial values of U_n become arbitrarily different at large n. The change of axis roles between eras is also governed by an unstable map, as is the time duration of each cycle. In order of magnitude the duration of a cycle at time t is $\Delta t \sim (t_0 - t)\ln[1/(t_0 - t)]$, and the total number of cycles and eras is infinite. (In actuality, when $t_0 - t$ reaches $(\hbar G/c^5)^{1/2} \simeq 10^{-43}$ seconds, classical general relativity breaks down; and the as-yet-unknown laws of quantum gravity take over and modify the evolution [32].)

The classical evolution of Fig. 6 and equations (16), (17) is "chaotic" and can be analyzed by techniques such as those that FEIGENBAUM [33] presents in this volume. KHALATNIKOV, LIFSHITZ, and colleagues [5] have carried out such an analysis.

8. Techniques of Future Research: Soliton Theory and Numerical Relativity

Our present understanding of nonlinear phenomena in vacuum gravity, as sketched in these lectures, is based primarily on exact solutions of the Einstein field equations, with some crucial additional information coming from perturbation theory, from two-lengthscale expansion methods, and from techniques of differential topology.

I expect that in the next few years significant new insights will come from two directions: (*i*) studies of the physical interpretations of the plethora of new exact solutions that are now being generated by soliton-theoretic techniques [4], [34]; and (*ii*) numerical solutions of the Einstein field equations [2]. These lines of research might change dramatically our view of nonlinear gravity.

References

1 See, e.g., the experimental lectures in N. Dereulle and T. Piran, eds: *Gravitational Radiation* (North Holland, Amsterdam, 1983); also K. S. Thorne: Rev. Mod. Phys. 52, 285 (1980).

2 See, e.g., the lectures of T. Piran in *Gravitational Radiation*, ref. [1]; also L. Smarr, ed.: *Sources of Gravitational Radiation* (Cambridge University Press, Cambridge, 1979).

3 V. A. Belinsky and I. M. Khalatnikov: Sov. Phys.--JETP 32, 169 (1970); I. M. Khalatnikov and E. M. Lifshitz: Phys. Rev. Lett. 24, 76 (1970); V. A. Belinsky, I. M. Khalatnikov, and E. M. Lifshitz: Adv. Phys. 31, 639 (1982).

4 C. Hoenselaers, W. Kinnersley, and B. C. Xanthopoulos: J. Math. Phys. 20, 2530 (1979); V. A. Belinsky and V. E. Zakharov: Sov. Phys.--JETP 50, 1 (1979); I. Hauser and F. J. Ernst: Phys. Rev. D 20, 362 (1979); C. M. Cosgrove: J. Math. Phys. 23, 615 (1982).

5 I. M. Khalatnikov, E. M. Lifshitz, K. M. Khanin, L. N. Shchur, and Ya. G. Sinai: in *General Relativity and Gravitation*, Proceedings of the Tenth International Conference on General Relativity and Gravitation, eds. B. Bertotti, F. deFelice, and A. Pascolini (Dordrecht, Reidel), 1984), p. 343.

6 C. W. Misner, K. S. Thorne, and J. A. Wheeler: *Gravitation* (Freeman, San Francisco, 1973).

7 L. D. Landau and E. M. Lifshitz: *The Classical Theory of Fields* (Addison-Wesley, Reading, Mass. 1962) §100; also ref [6] §20.3.

8 R. A. Isaacson: Phys. Rev. 166, 1263 and 1272 (1968); also ref. [6] §§35.13-35.15.

9 See, e.g., the lectures of K. S. Thorne: in *Gravitational Radiation*, ref. [1].

10 N. Rosen: Phys. Z. Sowjeto 12, 366 (1937); ref[6] §§35.9-35.11.

11 See, e.g., G. B. Whitham: *Linear and Nonlinear Waves* (Wiley, New York, 1974).

12 K. Kahn and R. Penrose: Nature 229, 185 (1971); P. Szekeres: J. Math. Phys. 13, 286 (1972); Y. Nutku and M. Halil: Phys. Rev. Lett. 39, 1379 (1977).

13 F. J. Tipler: Phys. Rev. D 22, 2929 (1980).

14 J. A. Wheeler: *Geometrodynamics* (Academic Press, 1962); J. A. Wheeler: in *Relativity, Groups, and Topology*, ed. C. and B. DeWitt (Gordon and Breach, 1964), esp. pp. 409-438.

15 D. R. Brill and J. B. Hartle: Phys. Rev. 135, B271 (1964).

16 B. K. Harrison, K. S. Thorne, M. Wakano, and J. A. Wheeler: *Gravitation Theory and Gravitational Collapse* (University of Chicago Press, 1965), esp. chapters 2 and 8.

17 See, e.g., B. Carter: in *General Relativity, an Einstein Centenary Survey*, eds. S. W. Hawking and W. Israel (Cambridge University Press, 1979), chapter 6.

18 S. W. Hawking: Commun. Math. Phys. 43, 199 (1975).

19 See, e.g., R. D. Blandford and K. S. Thorne: in *General Relativity, an Einstein Centenary Survey*, eds. S. W. Hawking and W. Israel (Cambridge University Press, 1979), chapter 8.

20 I. D. Novikov: Sov. Astron--A. J. 8, 857 (1964); Y. Ne'eman: Astrophys. J. 141, 1303 (1965)

21 Ya. B. Zel'dovich, I. D. Novikov, and A. A. Starobinsky: Sov. Phys.--JETP 39, 933 (1975).

22 D. M. Eardley: Phys. Rev. Lett. 33, 442 (1974).

23 K. Schwarzschild: Sitzber, Deut. Akad. Wiss. Berlin, Kl. Math.-Phys. Tech. 1916, 189 (1916); L. Flamm: Physik Z. 17, 448 (1916); Fig. 31.5 of MTW [6].

24 M. D. Kruskal: Phys. Rev. 116, 1743 (1960); G. Szekeres: Publ. Mat. Debrecen 7, 285 (1960); chapter 31 of MTW [6].

25 H. Reissner: Ann. Phys. (Germany) 50, 106 (1916); G. Nordstrom: Proc. Kon. Ned. Akad. Wet. 20, 1238 (1918); J. C. Graves and D. R. Brill: Phys. Rev. 120, 1507 (1960).

26 R. P. Kerr: Phys. Rev. Lett. 11, 237 (1963); R. H. Boyer and R. W. Lindquist: J. Math. Phys. 8, 265 (1967); B. Carter: Phys. Rev. 174, 1559 (1968).

27 M. Simpson and R. Penrose: Int. J. Theor. Phys. 7, 183 (1973); J. M. Macnamara: Proc. Roy. Soc. London A358, 499 and A364, 121 (1978); Y. Gürsel, I. D. Novikov, V. D. Sandberg, and A. A. Starobinsky: Phys. Rev. D 20, 1260 (1979); R. A. Matzner, N. Zamorano, and V. D. Sandberg: Phys. Rev. D 19, 2821 (1979).

28 R. Penrose: Phys. Rev. Lett. 14, 57 (1965); S. W. Hawking and R. Penrose: Proc. Roy. Soc. London A314, 529 (1969); S. W. Hawking and G. F. R. Ellis: *The Large Scale Structure of Space-Time* (Cambridge University Press, 1973).

29 R. Penrose: Rev. Nuovo Cimento 1, special number, 252 (1969); V. Moncrief and D. M. Eardley: Gen. Rel. Grav. 13, 887 (1981).

30 S. W. Hawking: Phys. Rev. Lett. 26, 1344 (1971); §34.5 of MTW [6].

31 C. W. Misner: Phys. Rev. Lett. 22, 1071 (1969); V. A. Belinsky, I. M. Khalatnikov, and E. M. Lifshitz: Usp. Fiz. Nauk 102, 463 (1970); Box 30.1 of MTW [6].

32 J. A. Wheeler: in *Relativity, Groups, and Topology*, eds. C. DeWitt and B. S. DeWitt (Gordon and Breach, 1964), p. 317; chapter 44 of MTW [6]; J. A. Wheeler: in *Problems in the Foundations of Physics*, ed. G. T. diFrancia (North Holland, 1979), p. 395.

33 M. Feigenbaum: lecture notes in this volume.

34 V. A. Belinsky and V. E. Zakharov: Sov. Phys.--JETP 48, 985 (1978); V. A. Belinsky and M. Francaviglia: Gen. Rel. Grav. 14, 213 (1982).

Part V **High Energy Physics**

Unification, Supersymmetry and Cosmology

John Ellis
European Organization for Nuclear Research (C.E.R.N.), CH-1211 Geneva 23, Switzerland

1. The Standard Model

1.1. How to make a model and win a prize

The following step-by-step guide serves as a model for grand unified theories as well as reminding us of the structure of the Standard Model.

(a) Choose a group: The Standard Model choice pioneered by Glashow is SU(2)xU(1).

(b) Choose representations for the fermions: In the Standard Model left-handed fermions are taken to be doublets of SU(2), while right-handed fermions are taken to be singlets and the U(1) hypercharges are chosen to give the observed electromagnetic charges: $Q_{em} = I_3 + Y$.

(c) Choose Higgs representations (or find an alternative to elementary Higgs fields): The Standard Model just contains a Higgs doublet (ϕ^+, ϕ^0), as needed to give masses to fermions $m_f \bar{f}_L f_R$.

(d) Specify pattern of symmetry breaking. The Standard Model has $SU(2)_L xU(1)_Y$ broken down to $U(1)_{em}$. In general this step requires a suitable fixing of the Higgs potential

(e) Publish and await experimental confirmation: If you are successful at this step, go to (f). If you are not successful, go to (a).

(f) Go to Stockholm: If you are ambitious, return to step (a) and start again!

1.2. The Standard Model Lagrangian

Let us now remind ourselves of the form of the Standard Model Lagrangian and what it does. There is a gauge boson piece

$$L_a = -\frac{1}{4}(G_{\mu\nu a} G_a^{\mu\nu}) - \frac{1}{4}(F_{\mu\nu}F^{\mu\nu}) \tag{1.1}$$

where the field strength

$$G_{\mu\nu a} \equiv \partial_\mu W_{\nu a} - \partial_\nu W_{\mu a} + g\varepsilon_{abc} W_{\mu b} W_{\nu c} \quad [SU(2)] \quad \text{and} \tag{1.2a}$$

$$F_{\mu\nu} = \partial_\mu B_\nu - \partial_\nu B_\mu \quad [U(1)]. \tag{1.2b}$$

There are fermion kinetic terms

$$L_F = -\sum_f [\bar{f}_L \gamma^\mu (\partial_\mu + ig\frac{\tau_a}{2} W_{\mu a} + ig' Y_L B_\mu) f_L + \bar{f}_R \gamma^\mu (\partial_\mu + ig' Y_R B_\mu) f_R] \tag{1.3}$$

where my notation is $f_{L,R} \equiv \frac{1}{2}(1 \mp \gamma_5)f$. Note that in eq. (1.3) the SU(2) doublet left-handed fermion fields have $W_{\mu a}$ terms in their covariant derivatives, whereas the SU(2) singlet right-handed fermions do not. There is a kinetic term for the Higgs doublet ϕ:

$$L_\phi = -|(\partial_\mu + ig \frac{\tau_a}{2} W_{\mu a} + \frac{ig'}{2} B_\mu) f_R|^2. \tag{1.4}$$

Note that up to now the normalization of the U(1) gauge coupling g' and of the corresponding hypercharges had been arbitrary: eq. (1.4) fixes the normalization so that $Y(\phi) = 1/2$. There are also fermion-Higgs Yukawa couplings

$$L_Y = -\sum_{ff'} \{H_{ff'}(\bar{f}_L\phi) f_R + \text{h.c.}\} \tag{1.5}$$

where $H_{ff'}$ is an undiagonalized coupling matrix. Finally there is the Higgs potential

$$L_V = -V(\phi) \; : \; V = -\mu^2(\phi^+\phi) + \frac{\lambda}{2} (\phi^+\phi)^2 \; . \tag{1.6}$$

Thanks to this potential the Higgs field develops a non-zero vacuum expectation value:

$$< 0|\phi|0 > \; = \frac{v}{\sqrt{2}} \begin{pmatrix} 0 \\ 1 \end{pmatrix} \quad : \quad v = \sqrt{\frac{2\mu^2}{\lambda}} \; . \tag{1.7}$$

This symmetry breaking generates masses for the $W^\pm$ and Z^0 bosons:

$$m_W = \frac{gv}{2} \, , \; m_{Z^0} = \sqrt{g^2 + g'^2} \; \frac{v}{2} \; : \; Z^0 \; \equiv \; \frac{gW^0 - g'B}{\sqrt{g^2 + g'^2}} \tag{1.8}$$

while the photon $\gamma = (g'W^0 + gB)/\sqrt{g^2 + g'^2}$ remains massless because $U(1)_{em}$ is not broken by having $< 0|\phi^0|0 > \neq 0$. This vacuum expectation value also gives a mass matrix $m_{ff'} = H_{ff'}v$ to the quarks which must be diagonalized.

The exchange of $W^\pm$ bosons gives rise to an effective 4-fermion interaction at low energies:

$$\frac{1}{4} L_{eff} = \frac{G_F}{\sqrt{2}} \left[J^+_\mu \; J^\mu_- \right] \tag{1.9}$$

with the normalization

$$\frac{G_F}{\sqrt{2}} = \frac{g^2}{8m_W^2} \quad : \quad v = (\sqrt{2} \; G_F)^{-\frac{1}{2}} \tag{1.10}$$

which will find analogues in our subsequent grand unified theories. The photon couples to the combination $I_3 + Y$ of $SU(2) \times U(1)_Y$ generators, which we must identify with Q_{em}. The values of the hypercharges Y must be chosen with incredible precision to ensure that (for example) $|Q_e/Q_n| = 1+\mathcal{O}(10^{-20})$. The Standard Model provides no explanation for this: we must look further for an explanation. The coupling constant for the photon is

$$e \;\equiv\; \frac{gg'}{\sqrt{g^2 + g'^2}} \; . \tag{1.11}$$

Instead of working with g and g' it is frequently useful to work with θ_W

$$\cos\theta_W = \frac{g}{\sqrt{g^2 + g'^2}} \, , \quad \sin\theta_W = \frac{g'}{\sqrt{g^2 + g'^2}} \, ,$$

$$e = g\sin\theta_W, \quad m_{Z^0} = \frac{m_{W^\pm}}{\cos\theta_W} \tag{1.12}$$

in terms of which the neutral boson Z^0 couples to the neutral current

$$J^0_\mu \equiv J^3_\mu - \sin^2\theta_W J^{em}_\mu \tag{1.13}$$

with a coupling $\sqrt{g^2 + g'^2} = g/\cos\theta_W$. In this case Z^0 exchange gives the same strength of effective interaction as does $W^\pm$ exchange:

$$\frac{1}{4} L_{eff} = \frac{G_F}{\sqrt{2}} \left(J^+_\mu J^\mu_- + \rho J^0_\mu J^\mu_0 \right) \tag{1.14}$$

where

$$\rho \equiv \frac{m_W^2}{m_Z^2 \cos^2\theta_W} \tag{1.15}$$

is one for any number of Higgs doublets, and

$$\rho = \frac{\sum_\phi (I^2 - I_3^2 + I) <0|\phi_{I,I_3}|0>^2}{2\sum_\phi <0|\phi_{I,I_3}|0>^2 I_3^2} \tag{1.16}$$

for an arbitrary collection of Higgs multiplets.

1.3 Predictions

The $SU(2)_L x U(1)_Y$ neutral currents are characterized by two parameters $\sin^2\theta_W$(1.12) and ρ (1.15) which give a good [1] fit to the available low energy data with the values

$$\sin^2\theta_W = 0.215 \pm 0.012, \qquad \rho^2 = 1.00 \pm 0.05 \tag{1.17}$$

after including radiative corrections [2]. After diagonalization of the quark mass matrix $m_{ff'} = H_{ff'}v$ the charged currents are characterized by the unitary Kobayashi-Maskawa matrix [3]

$$U = \begin{pmatrix} U_{ud} & U_{us} & U_{ub} \\ U_{cd} & U_{cs} & U_{cb} \\ U_{td} & U_{ts} & U_{tb} \end{pmatrix} \tag{1.18}$$

which can be parametrized by 3 angles θ_i and 1 CP-violating phase δ:

$$U = \begin{pmatrix} C_1 & -S_1C_3 & -S_1S_3 \\ S_1C_2 & C_1C_2C_3 - S_2S_3e^{i\delta} & C_1C_2S_3 + S_2C_3e^{i\delta} \\ S_1S_2 & C_1S_2C_3 + C_2S_3e^{i\delta} & C_1S_2S_3 - C_2C_3e^{i\delta} \end{pmatrix} \tag{1.19}$$

where $C_i \equiv \cos\theta_i$, $S_i \equiv \sin\theta_i (i = 1, 2, 3)$. This parametrization fits the data with [4]

$$|U_{ud}| = 0.9737 \pm 0.0025, \quad |U_{us}| = 0.231 \pm 0.003$$
$$|U_{cd}| \sim 0.2, \quad |U_{cs}| \geq \frac{1}{2}, \quad |U_{cb}| \approx 0.05 \pm 0.01, \; |U_{ub}| \ll |U_{cb}| \tag{1.20}$$

but no one yet understands the origin of these values. Encouraged by the success of the Standard Model at low energies, one can predict the properties of the vector bosons. Their masses are expected to be

$$m_{W^\pm} = \left(\sqrt{\frac{\pi\alpha}{\sqrt{2}G_F}} \Big/ \sin\theta_W = \frac{37.4}{\sin\theta_W}\right) + \begin{array}{c}\text{radiative}\\ \text{corrections}\end{array} \approx 83\text{GeV}$$

$$m_{Z^0} = \left(\sqrt{\frac{\pi\alpha}{\sqrt{2}G_F}} \Big/ \sin\theta_W\cos\theta_W = \frac{m_W}{\cos\theta_W}\right) + \begin{array}{c}\text{radiative}\\ \text{corrections}\end{array} \approx 93\text{GeV} \qquad (1.21)$$

We expect the $W \to e\nu$ decay width to be

$$\Gamma(W \to e\nu) = \frac{G_F m_W^3}{6\pi\sqrt{2}} \qquad (1.22)$$

with other prominent decay modes in the ratios

$$\Gamma(W \to e\nu) : \Gamma(W \to \mu\nu) : \Gamma(W \to \tau\nu) : \Gamma(W \to \bar{q}q')$$

$$= 1 : 1 : 1 : 3|U_{qq'}|^2. \qquad (1.23)$$

Thus we expect a leptonic branching ratio

$$B(W \to e\nu \text{ or } \mu\nu) = \frac{1}{4N_G} \sim 8\,\% \text{ if } N_G = 3 \text{ generations} \qquad (1.24)$$

while the total decay width $\Gamma(W \to \text{all}) \sim N_G\text{GeV} \sim 3\text{GeV}$.

The Z^0 has one partial decay width independent of $\sin^2\theta_W$:

$$\Gamma(Z^0 \to \nu\bar{\nu}) = N_\nu \frac{G_F m_Z^2}{12\pi\sqrt{2}} \qquad (1.25)$$

while

$$\Gamma(Z^0 \to l^+l^-) : \Gamma(Z^0 \to \nu_e\bar{\nu}_e) : \Gamma\left(Z^0 \to q\left(\tfrac{2}{3}\right)\bar{q}\left(-\tfrac{2}{3}\right)\right) : \Gamma\left(Z^0 \to q\left(-\tfrac{1}{3}\right)\bar{q}\left(+\tfrac{1}{3}\right)\right)$$

$$= 1 + (1-4\sin^2\theta_W)^2 : 2 : 3\left(1 + \left(1 - \tfrac{8}{3}\sin^2\theta_W\right)^2\right) : 3\left(1 + \left(1 - \tfrac{4}{3}\sin^2\theta_W\right)^2\right) \qquad (1.26)$$

giving a leptonic branching ratio

$$B(Z^0 \to e^+e^- \text{ or } \mu^+\mu^-) \sim \frac{1}{11N_G} \sim 3\% \qquad (1.27)$$

and a total decay width $\Gamma(Z^0 \to \text{all}) \sim N_G\text{GeV} \sim 3\text{GeV}$.

Undiscovered until recently was the t quark. It was known from e^+e^- experiments [5] to have a mass above 23GeV, and there were weak arguments [6] based on weak decays that it should have a mass below about 40GeV, but no clear theoretical prediction for its mass. The t quark should decay predominantly via the charged current into a b quark, with

$$B(t \to be\nu) \sim B(t \to b\mu\nu) \sim 10\% \qquad (1.28)$$

unless there were a charged Higgs particle with mass less than $m_t - m_b$, in which case $t \to bH^+$ would dominate t decays.

The minimal Standard Model with one Higgs doublet has no charged Higgs boson but does contain one physical neutral Higgs boson of unknown mass:

$$m_{H^0} = \sqrt{2}\,\mu \; : \; \mu^2 = \lambda(2\sqrt{2}G_F)^{-1} \tag{1.29}$$

which should decay dominantly into the heaviest available fermions:

$$\begin{array}{lllll} \Gamma(H^0 \to \bar{b}b) & : \Gamma(H^0 \to c\bar{c}) & : \Gamma(H^0 \to \tau^+\tau^-) & : \Gamma(H^0 \to \mu^+\mu^-) & : \Gamma(H^0 \to e^+e^-) \\ \approx 3m_b^2 & : 3m_c^2 & : m_\tau^2 & : m_\mu^2 & : m_e^2 \\ \approx 75 & : 7 & : 3 & : 10^{-2} & : \frac{1}{4} \times 10^{-6} \end{array} \tag{1.30}$$

or into W^+W^- and Z^0Z^0 if $m_{H^0} > 2m_{W^\pm}, 2m_{Z^0}$. The total decay width of the Higgs approaches 1TeV as its mass is increased to 1TeV, indicating that a heavy Higgs is strongly interacting.

1.4. Recent Discoveries

The oldest recent discovery was that [7] of the $W^\pm$:

$$m_{W^\pm} = \begin{cases} 80.9 \pm 1.5 \pm 2.4 \text{ GeV (UA1)} \\ \\ 83.1 \pm 1.9 \pm 1.3 \text{ GeV (UA2)} \end{cases} \tag{1.31}$$

cf. a theoretical value of $83.0^{+2.9}_{-2.7}$ GeV for $\sin^2\theta_W = 0.210 \pm 0.014$. More recently there has been the Z^0 with [8]

$$m_{Z^0} = \begin{cases} 95.6 \pm 1.5 \pm 2.9 \text{ GeV (UA1)} \\ \\ 92.7 \pm 1.7 \pm 1.4 \text{ GeV (UA2)} \end{cases} \tag{1.32}$$

cf.a theoretical value of $93.8^{+2.4}_{-2.2}$ GeV. The UA1 collaboration has observed $4Z^0 \to e^+e^-(\gamma)$ and $5Z^0 \to \mu^+\mu^-(\gamma)$, while UA2 has observed $8Z^0 \to e^+e^-(\gamma)$. These enable one to establish upper limits on the number of light neutrino species. From the direct upper limits on the Breit-Wigner width of the Z^0 one deduces

$$\begin{aligned} \Gamma_{Z^0} &< 5\text{GeV} \Rightarrow N_\nu < 18 \quad \text{(UA1)} \\ \Gamma_{Z^0} &< 6.1\text{GeV} \Rightarrow N_\nu < 24 \quad \text{(UA2)}. \end{aligned} \tag{1.33}$$

On the basis of a comparison of the relative production rates $\sigma_{Z^0} \times B(Z^0 \to e^+e^-)/\sigma_{W^\pm} \times B(W^\pm \to e^\pm\nu)$ the UA2 collaboration [9] infers indirectly that

$$\begin{aligned} \Gamma_{Z^0} < 2.7\text{GeV} \Rightarrow N_\nu &\lesssim 3 \text{ (90\% confidence)} \\ &< 5 \text{ (95\% confidence)} \end{aligned} \tag{1.34}$$

which approaches the precision of cosmological limits [10].

Most recently, UA1 has reported evidence for the t quark from $W \to t\bar{b}$ decay and/or gg, $\bar{q}q$ or $Z^0 \to \bar{t}t$ decay, followed by $t \to b(e \text{ or } \mu)\nu$ [11]. One expects from $W \to t\bar{b}$ final states containing 2 hadronic jets, an e or a μ, and missing energy carried off by a neutrino. UA1 sees 3 candidates for $t \to be\nu$ and 3 for $t \to b\mu\nu$, giving $m(l\nu j_1 j_2) \sim$ 80 GeV and $m(l\nu j_2) \sim$ (30 to 50) GeV. All these events are compatible with

$$m_t \approx (30 \text{ to } 50) \text{ GeV} \tag{1.35}$$

but the existence of the t quark is not yet proven unambiguously by experiment: it is however the most conservative interpretation of the data.

1.5. Still to discover

The existence of the ν_τ has been inferred from τ decay and the agreement between experimental and theoretical estimates of the τ lifetime, but the ν_τ has not yet been observed directly. Its interactions could perhaps be seen in beam dump experiments. Crucial unconfirmed aspects of non-Abelian gauge theories are the 3-boson vertices, which could perhaps be sought in $\bar{p}p \to (W + \gamma) + X$ (the $W^+W^-\gamma$ vertex) and in $e^+e^- \to W^+W^-$ (both the $W^+W^-\gamma$ and $W^+W^-Z^0$ vertices are essential for the gauge theory cancellations so that $\sigma(W^+W^-) \sim 1/E_{cm}^2$ at high energies). Most intriguing of all is the neutral Higgs H^0 of unknown mass but known couplings $gm_f/2m_W$ in the Standard Model. It could be sought in toponium decays:

$$\frac{\Gamma(\theta \to H^0 + \gamma)}{\Gamma(\theta \to \gamma^* \to e^+e^-)} = \frac{G_F m_\theta^2}{4\sqrt{2}\,\pi\alpha}\left[1 - \frac{m_{H^0}^2}{m_\theta^2}\right] \tag{1.36}$$

which is above 10% for $m_\theta > 40\text{GeV}$, or in Z^0 decays:

$$B(Z^0 \to H^0 + l^+l^- \text{ or } H^0 + \gamma) \geqslant 10^{-6} \quad \text{for } m_{H^0} < 50\text{GeV} \tag{1.37}$$

or in the reaction $e^+e^- \to Z^{0*} \to Z^0 + H^0$ which has an observable cross-section for $m_{H^0} < E_{cm} - m_{Z^0} - 10\text{GeV}$, so that LEP II ($E_{cm} \sim 200\text{GeV}$) should be able to see H^0 up to 100GeV. The Crystal Ball Collaboration at DESY has recently reported seeing a state $\zeta(8.3\text{GeV})$ in $\Upsilon \to \gamma$ + (multihadrons $\approx c\bar{c}$? few particles $\approx \tau^+\tau^-$?) [12], but its branching ratio is 2 orders of magnitude too high for a conventional Higgs and the ζ was not seen in $\Upsilon \to \gamma + \zeta$ decay, as a Higgs would have been. Another false Messiah to reject alongside last year's [13] ξ (2.2)?

2. From the Standard Model to Grand Unification

2.1. Defects of the Standard Model

These abound in every sector. For example, in the gauge sector there are 3 independent gauge groups - $SU(3)_C \times SU(2)_L \times U(1)_Y$ - with 3 distinct gauge couplings - g_3, g_2 and g' (recall that $g'/g = \tan\theta_W$ is an arbitrary parameter in the Weinberg-Salam model). The fermions sit in very untidy representations:

$$\begin{array}{cccccccc} \begin{pmatrix} u \\ d \end{pmatrix}_L & \begin{pmatrix} c \\ s \end{pmatrix}_L & \begin{pmatrix} t \\ b \end{pmatrix}_L & (+\ldots?) & \begin{pmatrix} \nu_e \\ e^- \end{pmatrix}_L & \begin{pmatrix} \nu_\mu \\ \mu^- \end{pmatrix}_L & \begin{pmatrix} \nu_\tau \\ \tau^- \end{pmatrix}_L & (+\ldots?) \\ u_R & c_R & t_R & & e_R & \mu_R & \tau_R & \\ d_R & s_R & b_R & & & (+\nu_R?) & & \end{array} \tag{2.1}$$

requiring at least 5 irreducible representations per generation, and an unkown number of generations in toto. We have certain phenomenological restrictions on the number of generations [9, 10]:

$$\text{experiment: } N_G \geqslant 3, \quad \Gamma_{Z^0}: N_G < 5?, \quad \text{cosmology: } N_G < 4? \tag{2.2}$$

but no fundamental understanding of their number. Moreover, we have the mystery of charge quantization: why are the electron and proton charges equal and opposite to 1 in 10^{20}? In the Standard Model, the hypercharge assignments, and hence the electromagnetic charges are completely arbitrary. Another comment about fermions concerns the anomalous triangle diagrams, which are a source of current non-conservation and thereby responsible for $\pi^0 \to \gamma\gamma$ decay. It would be disastrous if one of the currents coupled to an SU(2)xU(1) gauge boson were anomalous, since the renormalizability of the gauge theory requires their conservation. Some groups are automatically "safe" in that they only have real fermion representations, so that the loops of left-and right-handed fermions automatically cancel. An example of such a safe group is SU(2): $\sum T_3^2 = 0$ for any representation. This means that the only potentially anomalous triangle diagram in the SU(2)xU(1) model is that with two SU(2) currents and one U(1) current: $\sum T_3^2 Y = 0$. This is equivalent, by the safety of SU(2) and $Y = Q_{em} - T_3$, to $\sum T_3^2 Q_{em} = 0$. Since all fermions with $T_3 \neq 0$ have $T_3 = \pm\frac{1}{2}$, the anomaly is cancelled if

$$\sum Q_{em} = 0. \tag{2.3}$$

This condition is satisfied by the Standard Model fermions

$$3Q_{em}(u) + 3Q_{em}(d) + Q_{em}(e) = 3 \times \frac{2}{3} + 3 \times \left(\frac{-1}{3}\right) - 1 = 0 \tag{2.4}$$

if one assumes the mysterious charge quantization condition.

This is a convenient place to make a technical comment about fermions. Gauge boson vertices couple fermions of the same helicity, e.g. $\bar{f}_L \gamma^\mu f_L W_\mu$ and $\bar{f}_{R,L} \gamma^\mu f_{R,L} B_\mu$, and in principle any pair of fermions of the same helicity could be coupled by a gauge boson. Moreover, the antiparticle of every L(R) fermion is R(L)-for example the antineutrino $\bar{\nu}$ is right-handed. It is therefore both possible and desirable to work in a basis where all the fermions are regarded as left-handed. This means replacing

$$q_R \to q_L^c, \quad e_R \to e_L^c, \quad \bar{f}_R \gamma^\mu f_R \to -\bar{f}_L^c \gamma^\mu f_L^c . \tag{2.5}$$

In this convention, the SU(3)xSU(2) content of a single generation of left-handed fermions is

$$(u, d)_L + (u_L^c) + (d_L^c) + (\nu_e, e^-)_L + (e_L^c)$$
$$= (3,2) + (\bar{3},1) + (\bar{3},1) + (1,2) + (1,1) \tag{2.6}$$

which should be kept in mind for our subsequent construction of GUTs.

The Higgses are responsible for many defects of the Standard Model. While many regard them as an ugly appendix to the theory, they are the only known way to give m_W, m_Z and $m_f \neq 0$ while retaining renormalizability. For a while, technicolour theories in which Higgses were composites of new strongly interacting fermions were popular [14]. However, these are now out of favour, one reason being the continuing absence of any charged Higgs-like particle expected in most theories to have a mass less than $\mathcal{O}$(15 to 40)GeV. The lower end of this range has been excluded by e^+e^- experiments [9] while almost the whole range is excluded by the apparently canonical $t \to bl\nu$ decay modes seen recently [11], which would have been overwhelmed by $t \to bH^+$ if $m_{H^+} < m_t - m_f$. Elementary Higgses pose severe problems of naturalness, since radiative corrections to their masses are quadratically divergent and it is different to maintain the hierarchy of mass scales expected in GUTs. We return to these problems in Section 5. Finally, Higgs-fermion couplings introduce many parameters which are not constrained by the gauge principle. The matrix couplings $H_{ff'}$ of Section 1 yield $2N_G$ quark masses, N_G charged lepton masses and $(N_G-1)^2$ Kobayashi-Maskawa mixing angles and phases, for a total of

$$N_G^2 + N_G + 1 \geqslant 13 \quad \text{parameters for } N_G \geqslant 3 . \tag{2.7}$$

All in all, even if one neglects the arbitrariness in the choice of fermion and Higgs representations, U(1) hypercharges, etc., the Standard Model contains many parameters: 3 gauge couplings g_3, g_2 and g', 2 nonperturbative vacuum angles

θ_2 and θ_3, 9 fermion masses, 4 mixing angles and phases, and 2 parameters in the Higgs potential for a total of 20. No matter what your religion, probably you believe in fewer than 20 parameters, perhaps 1,2, or 3. Thus you appreciate the need to simplify the Standard Model, either by simplifying the gauge structure, or the flavor structure, or both. GUTs aim primarily at simplifying the gauge structure, as we now see.

2.2. GUT philosophy

We will seek [15, 16] a (semi) simple gauge group G, which has a single gauge coupling g_G. We suppose that this is spontaneously broken down to the Standard Model through elementary Higgses:

$$G \rightarrow SU(3)_c \times SU(2)_L \times U(1)_Y \rightarrow SU(3)_c \times U(1)_{em} . \tag{2.8}$$

How is this possible, seeing that present experiments show that

$$g_3 \gg g_2 \neq g' \quad ? \tag{2.9}$$

The answer is provided by the renormalization group which tells us that couplings are not constant, but vary with the energy scale Q at which they are measured. The archetype is the asymptotic freedom of QCD [4]:

$$\alpha_3(Q) \approx \frac{12\pi}{(33-2N_q)\ln Q^2/\Lambda^2} \quad : \quad \Lambda = \mathcal{O}(100)\text{MeV} . \tag{2.10}$$

Equation (2.10) manifests the famous decrease of $\alpha_3(Q)$ as Q increases. Since our only problem (2.9) was that $\alpha_3 \gg \alpha_2$, α, at present energies, it gives us hope that grand unification may be possible at a sufficiently high energy scale.

This grand unification scenario involves the assumed existence of a "desert" in which no new physics appears. Notice that α_2, and α' vary with Q in much the same way as α_3 (2.10), but more slowly in the Standard Model. It is trivial for any two of the couplings to meet at some scale: the non-trivial aspect of the GUT hypothesis is that α_3, α_2 and α' should all become equal at the same scale m_X where G is broken (2.8). Above this scale, all three couplings stay equal to α_G: whether α_G increases or decreases for $Q > m_X$ depends how complicated the GUT is. In order for this philosophy of gauge unification to work without needing to worry about quantum gravity, it is necessary that m_X be less than the Planck mass $m_P \approx 10^{19}$ GeV at which quantum gravity effects become important. In order to estimate m_X we must make an enormous extrapolation and assume no new physics intervenes to alter the rate of approach of the couplings α_i. Since GUT interactions in general violate baryon number, the observed baryon stability requires $m_X > \mathcal{O}(10^{14})$ GeV. As we will see in more detail shortly, the slow evolution (2.10) of the gauge couplings means that one estimates

$$m_X = 1\text{GeV} \times \exp\left(\frac{\mathcal{O}(1)}{\alpha_{em}}\right) . \tag{2.11}$$

For this to be between 10^{14}GeV (baryon stability) and 10^{19}GeV (neglect of quantum gravity) we find that [17]

$$\frac{1}{120} > \alpha_{em} > \frac{1}{170} . \tag{2.12}$$

Since these values bracket the observed value of α_{em}, and since there is no other good argument why α_{em} should have the value it does, we are encouraged to look at GUTs more quantitatively.

2.3. Variation of gauge couplings

It is convenient to rewrite the equation (2.10) describing the energy variation of $\alpha_3(Q)$, and similarly of its comrade $\alpha_2(Q)$, in the unified form:

$$\frac{1}{\alpha_i}(Q) = \frac{1}{\alpha_G(m_X)} + \frac{1}{12\pi}[11i - 4N_G]\ln\left(m_X^2 / Q^2\right): i = 2, 3 . \qquad (2.13)$$

Notice that the fermion contributions are expressed in terms of the number of fermion generations, and that they are equal. Notice also that the integration constant Λ in equation (2.10) has been replaced by a combination of the GUT coupling strength at m_X and a $\ln(m_X^2/Q^2)$ factor which guarantees equality as $Q \to m_X$. The analogous equation for the U(1) coupling is

$$\frac{1}{\alpha'(Q)} = \frac{1}{\alpha'(m_X)} + [-\frac{20}{3} N_G]\ln\left(m_X^2 / Q^2\right) \qquad (2.14)$$

Notice that we have yet to determine the appropiate GUT normalization of $\alpha' \equiv g'^2/4\pi$ which ensures equality to α_G at $Q \geqslant m_X$. Combining the SU(2) and SU(3) equations (2.13) we get

$$\frac{1}{\alpha_3(Q)} - \frac{1}{\alpha_2(Q)} = \frac{11}{12\pi}\ln\left(Q^2/m_X^2\right) \qquad (2.15)$$

which yields m_X^2/Q^2

$$m_X^2/Q^2 = \exp\left[\frac{12\pi}{11}\left(\frac{1}{\alpha_{em}(Q)\sin^2\theta_W(Q)} - \frac{1}{\alpha_3(Q)}\right)\right] \qquad (2.16)$$

when exponentiated. Evaluating this expression with $Q = \mathcal{O}(1)$GeV, where $\alpha_3(Q) = \mathcal{O}(1)$, we find [18]

$$m_X = 1\text{GeV} \times \exp\left(\frac{6\pi}{11\sin^2\theta_W}\frac{1}{\alpha_{em}} + \ldots\right) \qquad (2.17)$$

as advertised previously in (2.11).

Now we are ready to tackle the normalization of the U(1) gauge factor which was previously unspecified because it was an Abelian group. Recall the universality (2.13) of the fermion contributions to the variations of $\alpha_3(Q)$ and $\alpha_2(Q)$: this would be explicable if the known fermions of one generation formed a complete GUT representation, or at least included all the fermions in a GUT representation which had non-trivial $SU(3)_C$ and $SU(2)_L$ transformation properties. This hypothesis is also consistent with the anomaly condition $\sum Q_{em} = 0$ (2.3) which necessarily holds if Q_{em} is a generator of a (semi) simple GUT group, and a single generation contains all the fermions in a GUT representation which have $Q_{em} \neq 0$. Let us suppose that the known fermions do indeed fill a GUT representation and that Q_{em} is the following combination of normalized GUT generators:

$$Q_{em} = T_3 + cT_0 . \qquad (2.18)$$

Comparing with the usual hypercharge Y,

$$Y = cT_0, \qquad g' = \frac{1}{c}g_1 \qquad (2.19)$$

where g_1 is the gauge coupling associated with the normalized I = 0 generator T_0 (2.18). We can compute the coefficient c (2.18, 2.19) by noting that

$$\sum_{\text{generation}} Q_{em}^2 = (1 + c^2)\sum_{\text{generation}} T_3^2 \qquad (2.20)$$

since $\sum T_3 T_0 = 0$ and $\sum T_0^2 = \sum T_3^2$ if a generation is a complete GUT representation. The left-hand side of (2.20) is 16/3, while the right-hand side is 2, yielding

$$c^2 = 5/3, \qquad g' = \sqrt{\frac{3}{5}}\, g_1 \tag{2.21}$$

for the correct normalization of the U(1) generator. Substituting (2.21) into (2.14) we find the universal

$$\frac{1}{\alpha_1(Q)} = \frac{1}{\alpha_G(m_X)} + \frac{1}{12\pi}\left[-4N_G\right] \ln\left(m_X^2/Q^2\right) \tag{2.22}$$

as expected if a fermion generation contains all the fermions with non-zero U(1) charge in some GUT representation.

2.4. Calculation of $\sin^2 Q_W$

We are now in a position to calculate $\sin^2 Q_W$ independently of any specific GUT model. Recall that

$$\sin^2\theta_W = \frac{g'^2}{g_2^2 + g'^2} = \frac{3g_1^2/5}{g_2^2 + 3g_1^2/5} \tag{2.23}$$

using the normalization relation (2.21). Hence in the symmetry limit when $Q \geqslant m_X$ we have $\alpha_1 = \alpha_2$ and so

$$\sin^2\theta_W = 3/8 . \tag{2.24}$$

Using previous Eqs. (2.13, 2.22) for the evolution of $\alpha_2(Q)$, $\alpha_1(Q)$ one easily computes the renormalized value

$$\sin^2\theta_W(Q) = \frac{3}{8}\left[1 - \frac{\alpha_{em}}{4\pi}\,\frac{110}{9}\ln m_X^2/Q^2\right] + \ldots \tag{2.25}$$

A simple-minded evaluation of this formula at $Q = \mathcal{O}(m_W)$ yields

$$\sin^2\theta_W \sim 0.2 \tag{2.26}$$

which agrees qualitatively with experiment. This successful result was independent of the details of any GUT, but just a consequence of the general GUT philosophy. It motivates us to look in more detail at specific models.

3. Simple GUT Models

3.1. How to make (another) model and win (another) prize

Here we proceed using the rules mentioned in Section 1.

(a) Choose a group. There are some constraints: it must have rank 4 (i.e., $\geqslant 4$ simultaneously diagonalizable generators) in order to contain $SU(3)_C \times SU(2)_L \times U(1)_Y$ (SU(n) has rank (n - 1), so SU(3) has rank 2 and SU(2) has rank 1, as does U(1)). The group should be (semi) simple so that it has just one coupling or perhaps be the product of identical (semi) simple factors with a discrete permutation symmetry imposed to make all the g_i equal. This latter possibility is not very elegant, but was important historically. It is easy then to catalogue the available groups of the minimal rank 4:

$$[SU(2)]^4,\ [O(5)]^2,\ [SU(3)]^2,\ [G_2]^2,\ O(8),\ O(9),\ Sp(8),\ F_4,\ SU(5) . \tag{3.1}$$

It is easy to see why most of these are unsuitable choices for GUT groups. For a start, $[SU(2)]^4$ and $[O(5)]^2$ contain no SU(3) factor for QCD. Of course, $[SU(3)]^2$ does-in fact it also contains a possible weak SU(3) group. However, this would necessarily contain Q_{em} as a generator, and both the quarks and the leptons would have to have $\sum Q_{em} = 0$ separately, which seems to be untrue experimentally. Next, the candidates $[G_2]^2$, O(8), O(9), Sp(8) and F_4 only have real fermion representations, while the known fermions of each generation must sit in a complex representation of SU(3) x SU(2):

$$(3,2) + 2(\overline{3}.1) + (1,2) + (1,1) \tag{3.2}$$

as we saw in Section 2. Therefore the only remaining candidate GUT group of rank 4 is SU(5), which naturally accommodates SU(3) and SU(2) [16]:

$$\left(\begin{array}{ccc|cc} & SU(3) & & & \\ & & & & \\ & & & & \\ \hline & & & & \\ & & & & SU(2) \end{array}\right) \begin{array}{c} 1 \\ 2 \\ 3 \\ 4 \\ 5 \end{array} \tag{3.3}$$

with U(1) as the commuting traceless diagonal generator.

(b) Choose fermion representations. As seen in (3.2), each generation contains 15 fermion helicity states, and so useful SU(5) representations must have dimension ≤15. There are 3 representations available with the following SU(3) x SU(2) decompositions:

$$\underline{5}: \ \psi_\alpha = (3,1) + (1,2); \quad \underline{10}: \ \chi_{[\alpha\beta]} = (3,2) + (\overline{3},1) + (1,1);$$
$$\underline{15}: \ \chi_{\{\alpha\beta\}} = (6,1) + (3,2) + (1,3). \tag{3.4}$$

Of these, the 15 is not useful for fermions because it contains a colour sextet and a weak triplet: it may come in useful later for Higgses. However, the conjugate $\overline{5}$ and the 10 are perfect for a conventional generation of fermions:

$$\overline{F}^\alpha = \begin{pmatrix} d_R^c \\ d_Y^c \\ d_B^c \\ \hline e^- \\ -\nu_e \end{pmatrix}_L \qquad T_{\alpha\beta} = \frac{1}{\sqrt{2}} \left(\begin{array}{ccc|cc} 0 & u_B^c & -u_Y^c & -u_R & -d_R \\ -u_B^c & 0 & u_R^c & -u_Y & -d_Y \\ u_Y^c & -u_R^c & 0 & -u_B & -d_B \\ \hline u_R & u_Y & u_B & 0 & -e^+ \\ d_R & d_Y & d_B & e^+ & 0 \end{array}\right)_L \tag{3.5}$$

$$= (\overline{3},1) + (1,2) \qquad\qquad = (3,2) + (\overline{3},1) + (1,1) \text{ of } SU(3) \times SU(2)$$

and similarly for heavier generations. There is no need for a Procrustes to fit the fundamental fermions into SU(5).

Note that the electromagnetic charges are now fixed, with $\sum Q_{em} = 0$ as in any GUT. This condition strongly implies that the electromagnetic charges are simple rational numbers, as seen from the $\overline{F}$ in (3.5)

$$(\bar{e}, \nu_e) \Rightarrow \sum Q_{em} = -1, \quad \sum_{\overline{F}} Q_{em} = 0 \Rightarrow Q_d = -\frac{1}{3}\ . \tag{3.6}$$

Charge quantization is no longer a mystery. The triangle anomalies should cancel for any GUT. In the case of SU(5) this is trivial because we have the same fundamental fermions as in the Standard Model, which is guaranteed to have no anomalies as soon as $\sum Q_{em} = 0$ as is the case here.

(c) Choice of Higgs representations. These have two jobs, to break SU(5) and to give fermion masses. There are two stages of SU(5) symmetry breaking:

$$SU(5) \to SU(3)_C \times SU(2)_L \times U(1)_Y \to SU(3)_C \times U(1)_{em} . \tag{3.7}$$

The rank 4 is conserved at the first stage which severely restricts the choice of possible representations. The simplest possibility is the adjoint $\underline{24}$ ϕ^α_β with a vacuum expectation value

$$<0|\phi|0> = V \begin{bmatrix} 1 & 0 & 0 & \vdots & 0 & 0 \\ 0 & 1 & 0 & \vdots & 0 & 0 \\ 0 & 0 & 1 & \vdots & 0 & 0 \\ \cdots & \cdots & \cdots & + & \cdots & \cdots \\ 0 & 0 & 0 & \vdots & -3/2 & 0 \\ 0 & 0 & 0 & \vdots & 0 & -3/2 \end{bmatrix} : V = \mathcal{O}(10^{15})\text{GeV}. \tag{3.8}$$

At the second stage we need a representation containing an SU(2) doublet, with the simplest choice being a $\underline{5}$

$$H_\alpha : <0|H|0> = \frac{v}{\sqrt{2}}(0\ 0\ 0 \vdots\ 0\ 1) : v = \mathcal{O}(10^2)\text{GeV}. \tag{3.9}$$

This can indeed give fermion masses via the couplings

$$\lambda H^{*\alpha} T_{\alpha\beta} \bar{F}^\beta : <0|H_5|0> \neq 0 \Rightarrow \lambda v T_{5\beta} \bar{F}^\beta = \lambda v(e^c e + d^c d) \tag{3.10a}$$

and

$$\mu \varepsilon^{\alpha\beta\gamma\delta\varepsilon} H_\alpha T_{\beta\gamma} T_{\delta\varepsilon} : <0|H_5|0> \neq 0 \Rightarrow \lambda_\mu (T_{12} T_{34} + \text{etc.}) = \mu v(u^c u) . \tag{3.10b}$$

Thus the first term (3.10a) gives identical masses to the charge -1 leptons and to the charge -1/3 quarks, while the second term (3.10b) gives masses to the charge +2/3 quarks.

The remaining steps (d), (e), (f) and (g) of Section 1 will be deferred. Let us now

3.2. Try again

In the previous section we found the unique model of rank 4: what possible GUTs exist with rank 5?

(a) Choose a group[19,20]. The available $[(\text{semi})\text{simple}]^n$ groups are

$$[SU(2)]^5, \quad SO(11), \quad Sp(10), \quad SU(6) \text{ and } SO(10) \tag{3.11}$$

Of these, the first has no SU(3) factor and SO(11) and Sp(10) have only real fermion representations. Therefore the only possible choices are SU(6) and SO(10).

(b) Choose fermion representations. We want 15 helicity states per generation, and SU(6) has a representation

$$\underline{15}: \chi_{[\alpha\beta]} = (3,2) + (3,1) + (\bar{3},1) + (1,2) + (1,1) \text{ of } SU(3) \times SU(2). \tag{3.12}$$

Unfortunately, this decomposition does not fit the desired pattern (3.2). One can construct SU(6) with more helicity states, but by the time one has cancelled their anomalies, they look much like "trivial" extensions of SU(5). The SO(10) group has, however, an interesting spinorial representation:

$$\underline{16} = \underline{\bar{5}} + \underline{10} + \underline{1} \quad \text{of SU(5)}. \tag{3.13}$$

The first two pieces are those we already met in our analysis of SU(5), while the extra singlet is a candidate for a right-handed neutrino. Therefore we continue with SO(10) [19,20].

(c) Choose Higgs representations. These have the twin jobs of breaking SO(10) and of giving fermion masses. At the first stage we have a wider choice of Higgs representations because we are reducing the rank from 5 to 4. Our choice of Higgses also depends on the intermediate energy symmetry group we wish to retain. One possibility is

$$SO(10) \to SU(5) \to SU(3)_C \times SU(2)_L \times U(1)_Y \to SU(3)_C \times U(1)_{em}. \tag{3.14}$$

In this case the last two stages of symmetry breaking can be achieved with extensions of the corresponding SU(5) representations: the adjoint $\underline{24}$ of SU(5) becomes an adjoint $\underline{45}$ of SO(10), and the fundamenteal $\underline{5}$ of SU(5) becomes the vector $\underline{10}$ of SO(10). To realize the first stage of symmetry breaking in (3.14) we need another Higgs representation, and the minimal choice is a spinorial $\underline{16} = \underline{\bar{5}} + \underline{10} + \underline{1}$ of SU(5). A vacuum expectation value for the SU(5) singlet breaks SO(10) down to SU(5). Other intermediate symmetries are possible, such as

$$SO(10) \to SU(4) \times SU(2)_L \times SU(2)_R \to SU(3)_C \times SU(2)_L \times U(1)_Y. \tag{3.15}$$

The fermion $\underline{16}$ of SO(10) has the following $SU(4) \times SU(2)_L \times SU(2)_R$ decomposition:

$$\begin{aligned}\underline{16} &= (4,\, 2,\, 1) + (\bar{4},\, 1,\, 2)\\ &= \begin{pmatrix} u_{R,Y,B} \; ; \; \nu_e \\ d_{R,Y,B} \; ; \; e^- \end{pmatrix}_L + \begin{pmatrix} u^c_{R,Y,B} \; ; \; \nu^c_e \\ d^c_{R,Y,B} \; ; \; e^c \end{pmatrix}_L .\end{aligned} \tag{3.16}$$

The SU(4) contains lepton number as a fourth colour, and SU(2) gauges the left-handed conjugate fermions (i.e., the right-handed fermions). Many larger Higgs representations have been discussed in connection with alternative patterns of SO(10) breaking, such as, $\underline{54}$, $\underline{120}$ and $\underline{126}$ representations, but I will not discuss them in detail here.

3.3. Predictions of Minimal SU(5)

$\sin^2\theta_W$: In the previous lecture we made a general analysis suggesting that $\sin^2\theta_W \sim 0.2$ in a wide range of GUTs. In the specific context of SU(5) we can estimate $\sin^2\theta_W$ more precisely by including the effects of the light Higgs doublets (which modifies the variations of $\alpha_2(Q)$ and $\alpha_1(Q)$) [21], including two-loop effects on the variation of the couplings between m_W and m_X, treating carefully the symmetry breaking thresholds at m_W and m_X (which needs some assumption of minimality of the GUT), and assuming some input value of Λ_{QCD}. This must be specified by a two-loop renormalization prescription for some definite number of flavours. Most importantly, one must include the variation of α_{em} between $Q^2 = 0$ (where it is measured, for example, using Josephson junctions) and $Q^2 = \mathcal{O}(m_W^2)$ where electroweak "unification" takes place. Fermionic vacuum polarization diagrams cause [22]

$$\frac{1}{\alpha_{em}(Q^2 = 0)} = 137 \to \frac{1}{\alpha_{em}(Q^2 = m_W^2)} = 128 \pm \frac{1}{2}. \tag{3.17}$$

One can now compute a more accurate value of m_X (including $\mathcal{O}(1)\ln\alpha_{em}$ and $\mathcal{O}(1)$ terms in the exponential (2.17)):

$$m_X = \Lambda_{\overline{MS}}\text{ (4 flavours)} \times (1\tfrac{1}{2} \pm \tfrac{1}{2}) \times 10^{15}. \tag{3.18}$$

Putting in a best guess for $\Lambda_{\overline{MS}}$:

$$\Lambda_{\overline{MS}}\text{ (4 flavours)} = (100\text{ to }200)\text{ MeV} \tag{3.19}$$

one gets

$$m_X = (1\text{ to }4) \times 10^{14}\text{ GeV} \tag{3.20}$$

and the corresponding value of $\sin^2\theta_W$ (using the $\overline{MS}$ prescription at m_W) is:

$$\sin^2\hat{\theta}_W(m_W) = 0.208 + 0.004(N_H - 1) + 0.006\ln\left(\frac{400\text{MeV}}{\Lambda_{\overline{MS}}}\right) \tag{3.21}$$

$$= 0.214 \pm 0.002 \tag{3.21}$$

if one takes $N_H = 1$ light Higgs doublet and uses the range (3.19) of $\Lambda_{\overline{MS}}$. The agreement with experiment is perfect.

<u>fermion masses</u> [20, 23]: These are related to Yukawa couplings: $m_f = \lambda_f v$ which are subject to renormalization in the same way as gauge couplings. A complete calculation of the renormalization of m_f should start from the renormalization of λ_f. However, the one-loop renormalization by SU(3) gluons alone is equivalent to renormalization of the running mass in the quark propagator $S_q(Q)$:

$$S_q^{-1}(Q) = \not{Q} - m_q(Q) \tag{3.22}$$

where $m_q(Q)$ is defined as a coefficient in a Dirac matrix expansion of the propagator evaluated at momentum Q. The physical mass is naively the solution of $Q^2 = m_q^2(Q^2)$. The mass operator has a calculable anomalous dimension which yields

$$\left[\frac{m_f(Q)}{m_\tau(Q)}\right] = \left[\frac{\alpha_3(Q)}{\alpha_3(m_X)}\right]^{\frac{4}{11 - 4N_G/3}} + \text{(higher order effects)} \tag{3.23}$$

if one starts from the naive SU(5) symmetry prediction that $m_b = m_\tau$. A numerical evaluation including higher order effects yields [24]

$$m_b = \left(5\text{ to }5\tfrac{1}{2}\right)\text{ GeV}, \tag{3.24}$$

another success. The corresponding prediction for m_s is qualitatively successful:

$$m_s = m_\mu \Rightarrow m_s \sim 0.5\text{GeV} \tag{3.25}$$

while there is a failure for the lightest generation:

$$m_\mu/m_e \sim 200 \Rightarrow m_s/m_d \sim 200 \tag{3.26}$$

whereas the generally accepted value of the ratio is O(20).

Despite this failure, SU(5) is qualitatively successful at interelating quark and lepton masses, and the predictions (3.24, 3.25) provide the first evidence that the quarks and leptons conventionally assigned to the same generation actually have anything to do with each other. Some people would claim that the prediction (3.25) for m_s is a factor of 3 too high, but my view is that the evidence for $m_s \sim 150$MeV is not overwhelming [4]. In any case the naive SU(5) predictions are only "wrong" for the lightest fermions-perhaps there is an additional "small" source of mass beyond the minimal SU(5) we have discussed so far. One possibility is to complicate the theory by adding a $\underline{45}$ of Higgs, in which case one can arrange [25] to get

$$m_s \approx \frac{1}{3} m_\mu, \quad m_d \approx 3m_e. \tag{3.27}$$

Another possibility is to include quartic (Higgs)^2-fermion- fermion couplings with small coefficients $\mathcal{O}(1/m_P)$ [26] : for example

$$\frac{1}{m_P} H^\alpha \phi_\alpha^{\ \beta} T_{\beta\gamma} \bar{F}^\gamma \Rightarrow \delta m_{q,l} = \mathcal{O}\left(\frac{m_X m_W}{m_P}\right) = \mathcal{O}(10)\mathrm{MeV} \tag{3.28}$$

which could correct the m_d/m_e ratio, and might even be the origin of all the first generation fermion masses.

3.4. Parameters in SU(5)

Disregarding these fermion mass problems, how many parameters do we have in minimal SU(5)? We have 1 gauge coupling, 1 nonperturbative vacuum angle and 6 independent fermion masses, all improvements on the Standard Model. However, we now have 6 generalized Kobayashi-Maskawa mixing angles and phases and the Higgs potential requires at least 7 parameters as we will see later. Thus the Higgs sector is worse than in the Standard Model and we have a total of 21 parameters. There is still room for progress!

4. New Interactions in GUTs

4.1. Baryon number violation in SU(5)

Let us briefly recall the forms (3.5) of the fermion representations in SU(5), and of the adjoint representation containing the gauge bosons:

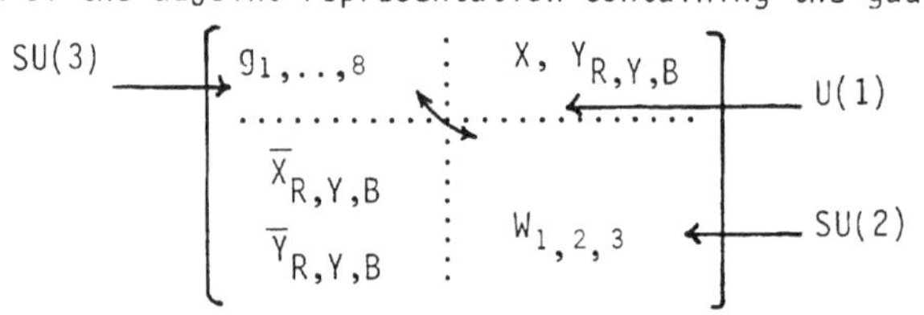

(4.1)

It is apparent from (4.1) that the new gauge bosons X and Y couple the first 3 indices of SU(5) to the last 2. It is also apparent from (3.5) that the X and Y bosons therefore couple quarks to leptons in both the $\underline{\bar{5}}$ and the $\underline{10}$ of SU(5)-and quarks to antiquarks-in the $\underline{10}$ of SU(5). The baryon number violating couplings of the X and Y bosons are [23]

$$\frac{g}{\sqrt{2}} X_{i\mu}\left(-\bar{d}^i_R\gamma^\mu e^+_R + \epsilon_{ijk}\bar{u}^c_{k_L}\gamma^\mu u_{j_L} + \bar{d}^i_L\gamma^\mu e^+_L\right) + \begin{pmatrix}\text{herm.}\\ \text{conj.}\end{pmatrix}$$

$$\frac{g}{\sqrt{2}} Y_{i\mu}\left(\bar{d}^i_R\gamma^\mu \nu^c_{e_R} + \epsilon_{ijk}\bar{u}^c_{k_L}\gamma^\mu d_{j_L} - \bar{u}^i_L\gamma^\mu e^+_L\right) + \begin{pmatrix}\text{herm.}\\ \text{conj.}\end{pmatrix} \tag{4.2}$$

where the first terms come from the $\underline{\bar{5}}$ and the last two terms come from the $\underline{10}$. Of course, these new gauge bosons acquire masses when SU(5) is broken to $SU(3)_C \times SU(2)_L \times U(1)$, which is why we called the grand unification scale m_X. Therefore X and Y exchange gives rise to an effective 4-fermion interaction at low energies, which takes the form [23]

$$\frac{1}{4} L_{GU} = \frac{1}{\sqrt{2}} G_{GU}\left[\left(\epsilon_{ijk}\bar{u}^c_{R_L}\gamma_\mu u_{jL}\right)\left(2\bar{e}^+_L\gamma^\mu d_{i_L} + \bar{e}^+_R\gamma^\mu d_{i_R}\right)\right.$$

$$-\left(\epsilon_{ijk}\bar{u}^c_{k_L}\gamma_\mu d_{jL}\right)\left(\bar{\nu}^c_e\gamma^\mu d_{i_R}\right) + \left(\begin{matrix}\text{herm.}\\ \text{conj.}\end{matrix}\right)\Bigg]. \tag{4.3}$$

This is reminiscent of the SU(2) x U(1) effective interaction of Section 1, and has the coupling strength

$$\frac{G_{GU}}{\sqrt{2}} = \frac{g^2}{8m_X^2} = \frac{g^2}{8m_Y^2} \quad \text{cf.} \quad \frac{G_F}{\sqrt{2}} = \frac{G^2}{8m_W^2} . \tag{4.4}$$

Notable features of L_{GU} (4.3) are that its strength is proportional to $1/m_X^2$-which follows from dimensional analysis and heavy boson exchange-that it conserves (B-L) we-will see shortly whether this is a general feature-and that it only involves the first generation. Later we will explore effects of generalized Cabibbo-Kobayashi-Maskawa mixing.

4.2. Generality of (B-L) Conservation

Although not absolutely conserved in most GUT models, (B-L) is approximately conserved at low energies in all GUT models, a fact guaranteed by $SU_C(3) \times SU(2)_L \times U(1)_Y$ invariance [27, 28], as we now see using a mathematical trick invented by Weinberg [28]. He introduced a quantum number called F parity with the following assignment

$$F = +1: \quad q, L; \quad F = -1: \bar{q}, \bar{L}, \text{ gauge and Higgs bosons, derivatives } \partial_\mu . \tag{4.5}$$

Multiplicative conservation of F is guaranteed by a combination of Lorentz invariance and $SU(2)_L$. Lorentz invariance decrees that particles sit in representations (a, b) where a and b are half integers: for example $f_L \in (\frac{1}{2}, 0)$, $f_R \in (0, \frac{1}{2})$, etc. Moreover, Lorentz invariance guarantees that $(-1)^{2a}$ and $(-1)^{2b}$ are separately conserved. Since SU(2) invariance implies that $(-1)^{2I}$ is conserved, it follows that $F \equiv (-1)^{2a+2I}$ is automatically conserved. It is easy to check that with this definition of F one arrives at the assignments (4.5). For example

$$\left.\begin{matrix} f_L: a = I = \frac{1}{2}, \; b = 0 \\ f_R: a = I = 0, \; b = \frac{1}{2} \end{matrix}\right\} F = +1 \qquad \left.\begin{matrix} W, Z: a = b = \frac{1}{2} \\ g, \gamma, \partial_\mu \; I = 0, 1 \end{matrix}\right\} F = -1 . \tag{4.6}$$

Let us now classify all low dimensional interactions with F = +1. For d = 6 we have

$$\frac{1}{M^2} qqqL : F = +1, \text{ allowed}; \; \frac{1}{M^2} qqq\bar{L}: F = -1, \text{ disallowed}. \tag{4.7}$$

Thus we see that (B-L) conservation is automatic for d = 6 interactions. Therefore even in SO(10) and other GUTs we expect

$$p,n \to (L^+, \bar{\nu}) + \chi, \quad p,n \not\to (L,\nu) + \chi . \tag{4.8}$$

In general, however, we will encounter different forms of (B-L) conserving interaction from those found in minimal SU(5). For d = 7 we have

$$\frac{1}{M^3} qqqLB : F = -1, \text{ disallowed}; \; \frac{1}{M^3} qqq\bar{L}B : F = +1, \text{ allowed}. \tag{4.9}$$

Therefore (B-L) is violated and (B+L) is conserved by d=7 interactions, but the extra factor of 1/M in their coefficients (4.9) means that d = 7 decay amplitudes are suppresed relative to $d=6$ by $\mathcal{O}(m_W \text{ or } \Lambda_{QCD}/m_X) = O(10^{-12})$. If we want baryons to decay to 3 antileptons, we must go to $d = 9$:

$$\frac{1}{M^5}\, qqqLLL: \; F = +1, \text{ allowed} \tag{4.10}$$

in which case we need $M \lesssim 10^5$GeV to get observably rapid baryon decay. Finally to get $N \to LLL$ decay, we need $d = 10$:

$$\frac{1}{M^6}\, qqq\overline{LLLB}: \; F = +1, \text{ allowed} \tag{4.11}$$

which requires $M \lesssim 10^4$ GeV to be observable. A far cry from our $m_X \sim 10^{15}$ GeV!

Accepting the dominance by (B-L) conserving interactions, we should now ask about the Cabibbo mixing between generations, a question answerable in minimal SU(5) which has the following effective interaction for the first two generations:

$$\begin{aligned}\frac{1}{2}\,L_{GU} = e^{i\phi}\,\frac{1}{\sqrt{2}}\,G_{GU} \;\; &\{(\varepsilon_{ijk} u^C_{k_L} \gamma^\mu u_{j_L})[(1+\cos^2\theta_C)\bar{e}^+_L + \sin\theta_C\cos\theta_C \bar{\mu}^+_L]\gamma_\mu d_{i_L} \\ &+ [(1+\sin^2\theta_C)\bar{\mu}^+_L + \sin\theta_C\cos\theta_C \bar{e}^+_L]\gamma_\mu s_{i_L} + \bar{e}^+_R \gamma_\mu d_{i_R} + \bar{\mu}^+_R \gamma_\mu s_{i_R}] \\ &- [\varepsilon_{ijk}\bar{u}^C_{k_L}\gamma_\mu (d_{j_L}\cos\theta_C + s_{j_L}\sin\theta_C)][\bar{\nu}^C_{e_R}\gamma^\mu d_{i_R} + \bar{\nu}^C_{\mu_R}\gamma^\mu s_{i_R}]\} \\ &+ (\text{herm. conj.}).\end{aligned} \tag{4.12}$$

From this we can deduce an expected pattern of Cabibbo favored and suppressed decay modes [29]:

$$\begin{aligned}\frac{\Gamma(N\to\mu^+ + \text{non-strange})}{\Gamma(N\to e^+ + \text{non-strange})} &= \frac{\sin^2\theta_C\cos^2\theta_C}{(1+\cos^2\theta_C)^2+1} \\ \frac{\Gamma(N\to e^+ + \text{strange})}{\Gamma(N\to\mu^+ + \text{strange})} &= \frac{\sin^2\theta_C\cos^2\theta_C}{(1+\sin^2\theta_C)^2+1}\,.\end{aligned} \tag{4.13}$$

Thus we expect dominance by the final states

$$p,\, n \to e^+(\text{or}\,\bar{\nu}_e) + \text{pions}; \qquad p,\, n \to \mu^+(\text{or}\,\bar{\nu}_\mu) + K + \text{pions}.$$

If observed, these would be the ultimate confirmation of our GUT ideas about the association of quarks and leptons in generations.

4.3. Calculation of the baryon decay rate

We have seen that the baryon decay amplitude A in minimal GUTs $\alpha\ 1/m_X^2$. Hence the decay rate $\Gamma \;\alpha\; |A|^2 \;\alpha\; 1/m_X^4$ and the baryon lifetime

$$\tau = c(m_X^4/m_N^5) \tag{4.15}$$

where we have extracted the factor of $1/M_N^5$ motivated by dimensional analysis, leaving us with a dimensionless numerical coefficient c which must be determined. There are two key ingredients in the calculation [23] of c. First is the short distance enhancement factor which is analogous to that computed in nonleptonic decays and whose QCD part determined by the anomalous dimension of the qqq operator:

$$(L_{GU})_\mu = \left[\frac{\alpha_3(\mu)}{\alpha_G(m_X)}\right]^{\frac{2}{11-4N_G/3}} L_{GU}\,. \tag{4.16}$$

Second is the estimation of matrix elements $<\bar{L}$, mesons$|\ (L_{GU})_\mu|N>$ of the baryon decay operator renormalized at some large distance: $\mu = \mathcal{O}(1)$GeV. Most calculations use a non-relativistc SU(6) model or a bag model for the nucleon and mesons [30]. There is a considerable spread in the absolute values of c:

$$\tau_N = (0.25 \text{ to } 10) \times 10^{30} \text{ years} \times \left(\frac{m_X}{4 \times 10^{14}\text{GeV}}\right)^4 \tag{4.17}$$

but consensus on the dominant decay modes to be expected:

$$B(p \to e^+ \pi^\circ) \sim 30\%; \quad B(n \to e^+ \pi^-) \sim 70\% . \tag{4.18}$$

An alternative way of estimating the rate for $N \to \bar{L}$ + pseudoscalar is to use current algebra and PCAC [31]:

$$< \bar{L}, P|(L_{GU})_\mu|N> \approx \frac{1}{f_p} < \bar{L}|\,(L'_{GU})_\mu|N> + \quad \text{(pole term)} \tag{4.19}$$

where

$$L'_{GU} \equiv [\,Q_5^P, L_{GU}] . \tag{4.20}$$

This procedure is questionable because $E_p \approx m_N/2$, but model calculations confirm the expected ratio between the two terms in (4.18) [32]

$$1\!:\ g_A \text{ becomes } 1\!:\ \mathcal{O}(1) . \tag{4.21}$$

Therefore this approach may give estimates of the decay amplitudes which are approximately correct. Both contributions in (4.19) are proportional to the 3 quark overlap

$$\alpha \equiv < 0|(\varepsilon_{ijk} d_{iR} u_{jR}) u_{k_L}|N>. \tag{4.22}$$

There are two approaches to estimating α: one uses QCD sum rules and the other tries to relate α to knowledge about the baryon wave function at short distances gleaned from measurements of baryon form factors at large Q^2. This latter method yields

$$\alpha \approx 0.03\text{GeV}^2 \tag{4.23}$$

which gives [33]

$$\tau(p \to e^+ \pi^\circ) \approx 1.3 \times 10^{29} \text{ years} \times \left(\frac{m_X}{4 \times 10^{14}\text{GeV}}\right)^4 \tag{4.24}$$

a result close to the lower end of the range (4.18).

4.4. The Great Baryon Stability Disaster

Our previous estimate was that m_X = (1 to 4) x 10^{14}GeV in minimal SU(5), in which case our results (4.18 and 4.24) yield

$$\tau(p \to e^+ \pi^\circ) \leqslant 10^{29} \text{ to } 10^{31} \text{ years}. \tag{2.25}$$

This range conflicts with the experimental limit [34]

$$\tau(p \to e^+ \pi^\circ) \geqslant 2 \times 10^{32} \text{ years} \tag{4.26}$$

which seems to exclude the minimal SU(5) GUT, but not more complicated GUTs where

m_X can be larger. It is worth recalling that $\sin^2\theta_W$ depends on $\mathrm{Ln}(m_X^2/m_W^2)$ while τ_N depends on m_X^4. Thus the nucleon lifetime is much more sensitive to the precise value of m_X, and it is easy to modify one's GUT so that τ_N is acceptably long, while the change in $\sin^2\theta_W$ is small, and within the experimental errors. As we will see in later Sections, supersymmetric GUTs are examples of such theories. Perhaps we should not yet despair of observing nucleon decay: all the experiments now running have candidates for $p \to \mu^+ K^\circ$, consistent with a lifetime an order of magnitude below the bound (4.29), and $p \to \bar{\nu} K^+$, $n \to \bar{\nu} K^\circ$ decays are also not excluded at this level.

4.5. Neutrino Masses

There are very stringent upper limits on neutrino masses

$$m_{\nu_e}/m_e \lesssim \mathcal{O}(10^{-4}), \quad m_{\nu_\mu}/m_\mu \lesssim \frac{1}{200}, \quad m_{\nu_\tau}/m_\tau \lesssim \frac{1}{10} \tag{4.27}$$

and even a claim [35] that $m_{\nu_e} \simeq 30\mathrm{eV}$. There is no fundamental reason why the neutrino masses should be strictly zero, because there is no exact (gauge) symmetry principle to forbid them. If neutrino masses are zero, why? If they are massive, why are their masses so small (4.27)? We recall that the source of masses for conventional fermions q, L is a Higgs coupling $\bar{f}_R$ and f_L:

$$g_{H\bar{f}f}\, H_{\Delta I = \frac{1}{2}}\, \bar{f}_R f_L \to m_f^D \equiv g_{H\bar{f}f} < 0|H_{\Delta I = \frac{1}{2}}|0>. \tag{4.28}$$

Therefore one can rephrase our previous questions. If a ν_R exists, why is the Dirac mass m_ν^D so small (if it is)? If no ν_R exists, can m_ν be non-zero?

The answer to the second question is yes: neutrinos can have Majorana masses. Recall that the antiparticle of a left (right)-handed fermion is a right (left)-handed fermion. Hence $\bar{f}_R$ can be regarded as a left-handed conjugate fermion f_L^c. Thus the mass term (4.28) is due to the coupling of a pair of different left-handed fermions. The coupling of a pair of identical left-handed fermions to form a mass term is only possible for fields with no colour and no electromagnetic charge, i.e. the ν_L, not q_L or e_L. Such a Majorana mass term has the structure

$$m_\nu^M \nu_L^T C \nu_L = m_\nu^M \bar{\nu}_L^c \nu_L \equiv m_\nu^M \text{"}\nu_L \nu_L\text{"} \tag{4.29}$$

where C is the charge conjugation matrix. To achieve such a mass term (4.29) we need to violate lepton number: $\Delta L = 2$, e.g.,

$$\tilde{g}_{H\bar{\nu}\nu}\, \tilde{H}_{\substack{\Delta I = 1 \\ \Delta L = 2}}\, \nu_L \nu_L \Rightarrow m_\nu^M = \tilde{g}_{H\bar{\nu}\nu} < 0|\tilde{H}_{\substack{\Delta I = 1 \\ \Delta L = 2}}|0>. \tag{4.30}$$

Such a Higgs is not included in the Standard Model or in minimal SU(5) but it could be added. To do so would require a $\underline{15}$ of Higgs:

$$\nu_L \nu_L \varepsilon \underline{5} \times \underline{5} = \underline{\overline{10}} + \underline{\overline{15}} \tag{4.31}$$

since the $\underline{\overline{10}}$ contains no $Q_{em} = 0$ state, but the $\underline{\overline{15}}$ contains an $I = 1$, $Q_{em} = 0$ state. Alternatively, one could postulate [36] a nonrenormalizable coupling of a pair of $\underline{5}$ Higgses:

$$\frac{1}{M}\left(H_{\Delta I = \frac{1}{2}}\nu_L\right)\left(H_{\Delta I = \frac{1}{2}}\nu_L\right) \Rightarrow m_\nu^M = \frac{<0|H_{\Delta I = \frac{1}{2}}|0>^2}{M} = 0\left(\frac{m_W^2}{M}\right). \tag{4.32}$$

Here M must be a very heavy mass of order m_X or m_P, so the Majorana mass (4.32) must be very small. What could be the origin of M? A possible (most likely?) source is a heavy fermion. But this fermion would have the same couplings to $H\nu_L$ as the mythical ν_R. But then why does the ν not have a large Dirac mass? The answer is that it does, but this is harmless because the ν_R is heavy, as we see from looking at the full neutrino mass matrix [37]:

$$(\nu_L, \bar{\nu}_R) \begin{pmatrix} m_\nu^M & m_\nu^D \\ m_\nu^D & M_\nu^M \end{pmatrix} \begin{pmatrix} \nu_L \\ \bar{\nu}_R \end{pmatrix} . \tag{4.33}$$

The bottom right-hand entry M_ν^M is naturally large $\mathcal{O}(m_X$ or $m_P)$ because it is SU(5) invariant if the ν_R is an SU(5) singlet. The off-diagonal m_ν^D are $\mathcal{O}(m_q, m_L)$ while in many models m_ν^M is naturally very small $\mathcal{O}(m_W^2/m_X)$. An example of a model with this structure is SO(10), where the spinor <u>16</u> of fermions contains an SU(5) singlet candidate for $\bar{\nu}_R$, the <u>10</u> of Higgs gives m_ν^D, and a $\overline{\underline{126}}$ of Higgs contains an SU(5) singlet to give a large M_ν^M and an SU(5) <u>15</u> to give m_ν^M. Diagonalizing (4.33) we find

$$\begin{aligned} "\nu_L" &= \nu_L + \mathcal{O}\left(\frac{m_W}{m_X}\right)\bar{\nu}_R \quad : \quad m_{\nu_L} = \mathcal{O}\left(\frac{m_W^2}{m_X}\right) \\ "\nu_R" &= \nu_R + \mathcal{O}\left(\frac{m_W}{m_X}\right)\bar{\nu}_L \quad : \quad m_{\nu_R} = \mathcal{O}(m_X). \end{aligned} \tag{4.34}$$

Thus it is natural to expect $m_\nu \neq 0$ in most GUTs, and natural to expect $m_\nu \ll m_{q,L}$, with a typical range being $\mathcal{O}(10^{-5}\text{eV}$ to $10^2\text{eV})$. In many models there is a hierarchy: $m_\nu \alpha (m_q$ or $m_L)^{1 \text{ or } 2}$, but this is not inevitable. In general the familiar weak interaction eigenstates ν_e, ν_μ, ν_τ are not mass eigenstates, which leads us to expect neutrino oscillations. These are now being sought by experiment, but none have yet been definitely observed.

5. The Hierarchy Problem

5.1. Introduction

GUTs contain at least 2 very different mass scales, e.g.

$$SU(5) \underset{m_X}{\longrightarrow} SU(3)_C \times SU(2)_L \times U(1)_Y \underset{m_W}{\longrightarrow} SU(3)_C \times U(1)_{em} \tag{5.1}$$

which are provided by elementary Higgses: $m_H = \mathcal{O}(m_X$ or $m_W)$, at our present level of understanding. The hierarchy problem [38] is that of reconciling the existence of these two very different mass scales. The problem can be divided into 2 parts. One is to understand the origin of the hierarchy: why is $m_W \ll m_X$, m_P? Can one generate m_W dynamically? The second aspect is to maintain the hierarchy: how can it be stabilized against radiative corrections? Related to this second aspect is the problem of naturalness: how can one avoid the generic difficulty with elementary Higgs fields that radiative corrections to the Higgs mass squared $\delta m_H^2 \gg m_H^2$ itself?

5.2. Higgses in SU(5)

To appreciate the seriousness of the hierarchy problem, let us study the SU(5) GUT as a prototype [23]. The required pattern of symmetry breaking Higgs vacuum expectation values is:

$$\underline{24}\ :\ \phi^\alpha_\beta : \langle 0|\phi|0\rangle = V\begin{pmatrix} 1 & 0 & 0 & \vdots & 0 & 0\\ 0 & 1 & 0 & \vdots & 0 & 0\\ 0 & 0 & 1 & \vdots & 0 & 0\\ \cdots & \cdots & \cdots & \vdots & \cdots & \cdots\\ 0 & 0 & 0 & \vdots & -3/2 & 0\\ 0 & 0 & 0 & \vdots & 0 & -3/2 \end{pmatrix} : m_X^2 = m_Y^2 = \frac{25}{8} g_5^2 V^2 \quad (5.2)$$

$$\underline{5}:\ H_\alpha : \langle 0|H|0\rangle = \frac{v}{\sqrt{2}}(000\vdots 01) : m_W^2 = \frac{1}{4} g_2^2 v^2. \quad (5.3)$$

To achieve this, step one is to set up a Higgs potential for ϕ:

$$V(\phi) = -\mu^2(\mathrm{Tr}\phi^2) + \frac{1}{4}a(\mathrm{Tr}\phi^2)^2 + \frac{b}{2}(\mathrm{Tr}\phi^4) \quad (5.4)$$

which has the desired minimum (5.2) if

$$\mu^2,\ b > 0;\quad a > \frac{-7}{15}b \quad (5.5)$$

in which case V is given by:

$$\mu^2 = \left(\frac{15}{2}a + \frac{7}{2}b\right)V^2. \quad (5.6)$$

No difficulties have arisen so far. Step two is to set up a Higgs potential for M:

$$V(H) = -\frac{1}{2}\nu^2|H|^2 + \frac{1}{4}\lambda(|H|^2)^2 \quad (5.7)$$

which has the desired minimum (5.3) if

$$\nu^2,\ \lambda > 0 \Rightarrow \nu^2 = \frac{\lambda}{2}v^2. \quad (5.8)$$

This also seems easy: what is the problem?

In fact there are two problems. One is that there is at this stage a massless triplet of coloured Higgses left uneaten by gauge bosons:

$$\hat{H}_i = H_i + \frac{\sqrt{2}v}{5V}\phi_{5_i} \quad : i = 1, 2, 3. \quad (5.9)$$

These Higgses must be heavy because they mediate baryon decay:

$$\frac{\Gamma(p, n \underset{H}{\to} \bar{L} + X)}{\Gamma(p, n \underset{X}{\to} \bar{L} + X)} = \mathcal{O}\left(\frac{m^4_{q,L}}{m^4_W}\ \frac{m^4_X}{m^4_{\hat{H}}}\right) \quad (5.10)$$

and should therefore have a mass $> \mathcal{O}(10^{11})$ GeV to be compatible with the observed baryon stability [33]. A second problem is that radiative corrections couple the ϕ and H Higgses to each other, and for the theory to be renormalizable we must include $\phi^2|H|^2$ interactions:

$$V(\phi, H) = \alpha|H|^2(\mathrm{Tr}\phi^2) + \beta\bar{H}\phi^2 H \quad (5.11)$$

which actually solve our first problem, since now,

$$m^2_{\hat{H}} \;=\; -\frac{5}{2}\beta V^2 \;=\; \mathcal{O}(m^2_X). \tag{5.12}$$

However, the new terms (5.11) modify our previous equations for V:

$$\mu^2 \;=\; \left(\frac{15}{2}a + \frac{7}{2}b\right) V^2 \;+\; \left(\alpha + \frac{9}{30}\beta\right)v^2 \tag{5.13}$$

a small and manageable change, and

$$\nu^2 \;=\; \frac{\lambda}{2}v^2 + \left(15\alpha + \frac{9}{2}\beta\right)V^2 \tag{5.14}$$

which is a big and worrisome change.

Previously (5.8) we would have chosen $v^2 = \mathcal{O}(m^2_W)$ - small and possibly surprising, but certainly possible. Now we must choose $v^2 = \mathcal{O}(m^2_X)$ in such a way that it will cancel with

$$\delta m^2_H = \left(15\alpha + \frac{9}{2}\beta\right)V^2 \tag{5.15}$$

which is 24 orders of magnitude larger than m^2_W. We need

$$\nu^2 - \left(15\alpha + \frac{9}{2}\beta\right)V^2 \;=\; \mathcal{O}(v^2) \;=\; \mathcal{O}(10^{-24})V^2. \tag{5.16}$$

Even if this miraculous cancellation occurs at the tree level, it is destroyed by radiative corrections.

5.3. Radiative Corrections and Naturalness

If we calculate loop corrections to the $\phi^2|H|^2$ couplings (5.11) we find

$$\delta(\alpha,\beta) \;=\; \mathcal{O}\left(\frac{g^2}{4\pi^2}\right) \times (\alpha,\ \beta) \tag{5.17}$$

and hence

$$\delta m^2_H \;=\; \mathcal{O}\left(\frac{g^2}{4\pi^2}\right) \times (\alpha,\ \beta) \times V^2 \tag{5.18}$$

which is still perhaps 22 orders of magnitude larger than m^2_W. To ensure $\delta m^2_H \lesssim \mathcal{O}(m^2_W)$ we must cancel the radiative corrections through $\mathcal{O}(12)$ loops. This sounds like a difficult miracle, but it is not the end of our problems with radiative corrections.

Elementary scalar masses are quadratically divergent in perturbation theory.

$$\delta m^2_H \;=\; \frac{g^2}{(2\pi)^4}\int^{\Lambda} d^4k\,\frac{1}{k^2} \sim \mathcal{O}\left(\frac{g^2}{4\pi^2}\right)\Lambda^2 \tag{5.19}$$

where Λ is some cut-off introduced in an ad hoc manner. It seems "unnatural" if $\delta m^2_H > m^2_H$, so we would like a cut-off

$$\Lambda \;\lesssim\; \mathcal{O}\left(m_H/\sqrt{\alpha_{em}}\right) \;=\; \mathcal{O}(1\text{TeV}). \tag{5.20}$$

But in GUTs as we have discussed them so far there is no cut-off below $\mathcal{O}(10^{15})$GeV. Does one need to worry about this seeming lack of naturalness? Counterarguments are that even these quadratic divergences are in fact renormalizable, and indeed absent in some regularization schemes such as dimensional regularization. However, they are present in more "physical" renormalization schemes such as momentum space renormalization or Pauli-Villars. Moreover, these quadratic divergences appear to be symptomatic of underlying diseases, such as the GUT hierarchy problems mentioned earlier, and the tendency [39] in quantum gravity for elementary scalars to acquire $\delta m_H^2 = \mathcal{O}(m_P^2)$ from propagation through space-time foam. Therefore we worry about quadratic divergences.

5.4. A Symmetry to Protect m_H?

Our basic problem is that we have no symmetry principle which protects elementary scalar masses and keeps them small, in the same way as chiral symmetry protects elementary fermion masses:

$$\delta m_f = \mathcal{O}\left(\frac{g^2}{4\pi^2}\right) m_f, \quad m_f = 0 \Rightarrow \delta m_f = 0, \quad \delta m_f \ll m_f \tag{5.21}$$

Can we find in a more complicated theory some symmetry to protect m_H?

The technicolour approach [14] is to use chiral symmetry to protect the Higgs scalar in the same way that in QCD the chiral symmetry associated with massless quarks guarantees the masslessness of the pions. Thus inspired, one invents new interactions which are strong on a scale of $\mathcal{O}(1)$TeV and bind "techniquarks" Q_T to form bound state Higgses $H = (Q_T\bar{Q}_T)$. Then $m_{Q_T} = 0$ guarantees $m_H = 0$. A possible way to think of this idea is that the technicolour analogue of Λ_{QCD} introduces an effective cut-off of order 1TeV into the loop integrals (5.19) that worried us previously, corresponding to breaking up the previously elementary Higgs. Unfortunately, it is difficult [40] to give fermions masses in technicolour theories, and existing models have problems with the absence of flavour-changing neutral currents as well as the absence of light charged "Higgses" mentioned earlier.

An alternative idea is based on the observation that the quadratically divergent boson and fermion loop contributions to δm_H^2 have <u>different signs</u>. Can one cancel the boson divergences against the fermion divergences? To do this one obviously needs a direct correspondence between boson and fermion degrees of freedom, one needs similar masses:

$$\delta m_H^2 = \mathcal{O}\left(\frac{g^2}{4\pi^2}\right) \left|m_B^2 - m_F^2\right| \tag{5.22}$$

so that to get $\delta m_H^2 < \mathcal{O}(m_W^2)$ one needs

$$\left|m_B^2 - m_F^2\right| = \mathcal{O}(\Lambda^2) = \mathcal{O}(1\text{TeV}^2) \tag{5.23}$$

and one must have identical couplings for the bosons and fermions. The unique theory with these properties is supersymmetry (susy) [41].

5.5. What is supersymmetry?

It is a novel kind of symmetry which interrelates bosons and fermions:

$$Q|F\rangle = |B\rangle, \qquad Q|B\rangle = |F\rangle \tag{5.24}$$

where Q_α is a spin $\frac{1}{2}$ charge. The unique possibility for their algebra is

$$\{Q_\alpha^i, Q_j^{+\alpha}\} = (\gamma^\mu)_\alpha^\alpha P_\mu \delta_j^i + \begin{pmatrix}\text{central}\\ \text{charges}\end{pmatrix} \tag{5.25}$$

where i,j = 1, 2,...,N and central charges only appear when $N > 1$. The case N = 1 is called simple sypersymmetry, while $N > 1$ is called extended supersymmetry. In gauge theories, particle helicities are restricted to the range $|h| \lesssim 1$ and at most N = 4 supersymmetric charges can be accommodated:

$$h = +1, \underset{Q}{\longrightarrow} +\frac{1}{2}, \underset{Q}{\longrightarrow} 0, \underset{Q}{\longrightarrow} -\frac{1}{2}, \underset{Q}{\longrightarrow} -1 . \tag{5.26}$$

If one extends gravity to supergravity with local susy, one can have $|h| \lesssim 2$ and hence $N \lesssim 8$:

$$h = +2, \underset{Q}{\longrightarrow} +\frac{3}{2}, \underset{Q}{\longrightarrow} +1, \underset{Q}{\longrightarrow} +\frac{1}{2}, \underset{Q}{\longrightarrow} 0, \underset{Q}{\longrightarrow} -\frac{1}{2}, \underset{Q}{\longrightarrow} -1, \underset{Q}{\longrightarrow} -\frac{3}{2}, \underset{Q}{\longrightarrow} -2 . \tag{5.27}$$

However, in the phenomenological applications considered in subsequent sections we will be restricted to N = 1. In this case the building blocks are the following simple supermultiplets:

$$\text{gauge:} \quad h = \begin{pmatrix} 1 \\ 1/2 \end{pmatrix} \quad ; \text{chiral:} \quad h = \begin{pmatrix} 1/2 \\ 0 \end{pmatrix}. \tag{5.28}$$

We will see in the next sections how these can be used to construct realistic theories.

6. Supersymmetric GUTs

6.1. Structure of supersymmetric theories

We start this section with a description of some of the properties of supersymmetric theories which are relevant for the construction of models. Recall (5.28) the supermultiplet building blocks for N = 1 supersymmetry . For N = 2 the corresponding supermultiplets are:

$$\text{gauge:} \quad \begin{pmatrix} & 1 & \\ 1/2 & & 1/2 \\ & 0 & \end{pmatrix} \qquad \text{matter:} \quad \begin{pmatrix} & 1/2 & \\ 0, & & 0 \\ & -\frac{1}{2} & \end{pmatrix}. \tag{6.1}$$

Notice that in each case there are 2 states of intermediate helicity, corresponding to the two susy charges. Note also that the gauge fermions sit in an adjoint representation of the gauge group, while the $\pm\frac{1}{2}$ helicity fermions in the matter supermultiplet sit in identical representations of the gauge group. This means that the N = 2 fermions must fill out a self-conjugate, real representation, which is in conflict with phenomenological observation. Thus N = 2 is not convenient for realistic model-building, and the same is also true for $N \geqslant 3$. For reference, two important extended supermultiplets are the N = 4 gauge supermultiplet:

helicity	+1	$+\frac{1}{2}$	0	$-\frac{1}{2}$	-1
multiplicity	1	4	6	4	1

(6.2)

and the N = 8 graviton supermultiplet

helicity	+2	$+\frac{3}{2}$	+1	$+\frac{1}{2}$	0	$-\frac{1}{2}$	-1	$-\frac{3}{2}$	-2
multiplicity	1	8	28	56	70	56	28	8	1

. (6.3)

To construct realistic theories we must just embed the known elementary particles in gauge and chiral simple supermultiplets, which gives us the following spectrum:

Particle	Spin	Sparticle		spin
quark q	$\frac{1}{2}$	squark	$\tilde{q}$	0
lepton L	$\frac{1}{2}$	slepton	$\tilde{L}$	0
photon γ	1	photino	$\tilde{\gamma}$	1/2
gluon g	1	gluino	$\tilde{g}$	1/2
W	1	wino	$\tilde{W}$	1/2
Z	1	zino	$\tilde{Z}$	1/2
Higgs H	0	shiggs	$\tilde{H}$	1/2

From the fact that no charged sparticle has been seen [5] in e^+e^- collisions we deduce

$$m_{\tilde{q}},\ m_{\tilde{L}},\ m_{\tilde{W}^\pm},\ m_{\tilde{H}^\pm} \gtrsim \mathcal{O}(20)\,\text{GeV} \tag{6.4}$$

while the absence of hadronic sparticles in hadron-hadron collisions [42] tells us that

$$m_{\tilde{g}} \gtrsim \mathcal{O}(3)\,\text{GeV}. \tag{6.5}$$

As we shall see later, cosmology suggests [43,44] that

$$m_{\tilde{\gamma}} \gtrsim \mathcal{O}(1)\,\text{GeV}. \tag{6.6}$$

Apart from these weak constraints and the general limit (5.23), we have no clear indications as to the masses of the sparticles.

On the other hand, sparticle couplings are completely fixed by supersymmetry: gauge boson couplings are related to gaugino couplings [45]:

$$g\bar{f}_L\gamma^\mu f_L G_\mu \rightarrow \sqrt{2}\, g\, (\bar{f}_L \tilde{G}_L)\tilde{f}_L + (\text{herm.conj.}) \tag{6.7}$$

and Yukawa couplings of known particles are related to Yukawa couplings of sparticles [45]:

$$\lambda(f_L f_L')H \rightarrow \lambda(f_L\tilde{H}_L)\tilde{f}_L' + \lambda\tilde{f}_L(f_L'\tilde{H}_L) \tag{6.8}$$

where in (6.7) and (6.8) we use the notation

$$(f_L f_L') \equiv \varepsilon^{\alpha\beta} f_\alpha f_\beta' = f_L^T C f_L'. \tag{6.9}$$

Also related to the couplings (6.7, 6.8) are quartic scalar couplings [45]:

$$\text{D terms} \quad : \frac{1}{2}\sum_\alpha g_\alpha^2 D_\alpha D_\alpha \equiv \frac{1}{2}\sum_\alpha g_\alpha^2\, |\tilde{f}^* \underline{T}^\alpha \tilde{f}|^2 \tag{6.10}$$

where α denotes the gauge group with coupling g_α and representation matrices $\underline{T}^\alpha$, and [45]

$$\text{F terms} \quad : \sum_i |F_i|^2 \equiv \sum_i |\partial P/\partial\phi_i|^2 \tag{6.11}$$

where $P(\phi)$ is the so-called superpotential which is a polynomial that can be at most cubic in the chiral fields if the theory is to be renormalizable:

$$P = \mu_{ij}\phi_i\phi_j + \lambda_{ijk}\phi_i\phi_j\phi_k + \ldots \tag{6.12}$$

In Eqs. (6.11) and (6.12) the ϕ_i are the scalar members of the chiral supermultiplets, also denoted by $\tilde{f}_i$. The Yukawa couplings (6.8) are also directly related [45] to the superpontential (6.12):

$$(f_i f_i)\left(\frac{\partial^2 P}{\partial\phi_i\partial\phi_j}\right) = \mu_{ij}(f_i f_j) + \lambda_{ijk}(f_i f_j)\phi_k + \ldots \tag{6.13}$$

Because of the structure (6.10, 6.11) of the effective scalar potential it is positive semidefinite:

$$V = \sum_i |F_i|^2 + \frac{1}{2}\sum_\alpha g_\alpha^2 D_\alpha D_\alpha \gtrsim 0. \tag{6.14}$$

The energy is strictly positive if and only if supersymmetry is spontaneously broken. To see this, consider the vaccum expectation value of the anticommutator (5.25):

$$< 0 \,|\{Q_\alpha, Q^{+\alpha}\}\, |0 > \; = \; < 0|\gamma^\mu P_\mu|0 > \tag{6.15}$$

Translation invariance guarantees $< 0|P_{1,2,3}|\,0 > = 0$, so the right-hand side is just the vacuum energy density E_0. But the left-hand side is non-zero if and only if there is a fermion, the Goldstino χ, coupled to the vacuum by the supersymmetry charge Q:

$$< 0|Q|\chi > = f_\chi^2 \neq 0 \Rightarrow E_0 = f_\chi^4. \tag{6.16}$$

Here the Goldstino constant f_χ is analogous to the pion decay constant f_π.

Supersymmetry must be broken, for if not scalar and fermion masses would be identical:

$$|m_\phi|^2{}_{ik} = \mu_{ij}\,\mu^\star_{jk} = |m_f|^2{}_{ik} \tag{6.17}$$

which conflicts with experiment. Susy must be broken either spontaneously as in equation (6.16), or explicitly, or both. We have two options for making the vacuum energy density (6.14) strictly positive. Either the D terms (6.10) are non-zero [46], or the F terms (6.11) [47], or a combination of the two. Both options have their disadvantages [48]. To make the D terms (6.10) positive we need a new U(1) gauge group (where is its gauge boson?) with many additional fields (to avoid a supersymmetric minimum and to avoid anomalies) and then the renormalization group tells us that the α_i tend to blow up at scales $\ll m_P$. To make the F terms (6.11) positive we need a very artificial superpotential P with many otherwise unmotivated chiral fields.

A preferred option [48] is the spontaneous breakdown of local susy (supergravity) in which case the super-Higgs effect causes the spin - $\frac{1}{2}$ Goldstino to be eaten by the gravitino, giving it a mass

$$m_{3/2} = \mathcal{O}(f_\chi^2/m_P) \tag{6.18}$$

in the same way that the Higgs effect causes a spin-0 Goldtone boson to be eaten by a gauge boson, giving it a mass

$$m_V = \mathcal{O}(< 0|H|0 > g_V)\,. \tag{6.19}$$

We will have more to say about local susy (supergravity) in the next lecture.

Suffice to say here that when spontaneously broken it gives an effective theory at low energies ($E \ll m_P$) of the form

$$\text{(global susy gauge theory) + (soft susy breaking terms)} \tag{6.20}$$

where the later include, for example, $\sum_i m_i^2|\phi_i|^2$ with $m_i = \mathcal{O}(m_{3/2})$.

6.2. How Susy is Useful

We have seen that a susy theory is characterized by gauge couplings g_α and a superpotential $P(\phi_i)$. What is the effect of radiative corrections? There are no quadratic divergences: susy (6.17) ensures that radiative corrections to scalar masses are no worse behaved than those to fermion masses. The gauge couplings g_α are renormalized in the normal way. There is no renormalization of superpotential parameters [49] apart from wave function renormalization $\phi_i \to Z_{ij}\phi_j$. Hence renormalizability does not require one to include all possible terms in P: if not present at the tree level, a coupling λ_{ijk} will not be generated by loops. Thus if one specifies the couplings in such a way that, for example, $m_H^2 \ll m_X^2$ at the tree level, then this relation is maintained to all orders of perturbation theory. Thus, the technical problem of maintaining the hierarchy despite the depradations of radiative corrections is solved. Of course, the more fundamental problem of the tree-level origin of the hierarchy is left unsolved.

6.3. Constructions of susy GUTs

We are now in a position to construct [50] simple susy GUTs based on SU(5). We replace the previous fermions, $\underline{\bar{5}}$ ($\bar{F}^\alpha$) and $\underline{10}$ ($T_{\alpha,\beta}$), and Higgses, $\underline{24}$ (ϕ) and $\underline{9}$ (H_α), by chiral supermultiplets. A minor complication is that we now need an even number of 5-dimensional Higgs supermultiplets $n\times(\underline{5} + \underline{\bar{5}})$, for two reasons. One is the nature of superpotential couplings which requires a $\underline{\bar{5}}$ ($\bar{H}$) to couple to $\bar{F}$ and T:

$$P \ni \lambda\, \bar{F}^\alpha T_{\alpha\beta} \bar{H}^\beta \Rightarrow m_d, m_e \tag{6.21a}$$

and a $\underline{5}$(H) to couple to two $\underline{10}$'s T:

$$P \ni \mu\varepsilon^{\alpha\beta\gamma\delta\varepsilon} H_\alpha T_{\beta\gamma} T_{\delta\varepsilon} \Rightarrow m_u \ . \tag{6.21b}$$

Also, one needs both a $\underline{5}$ and $\underline{\bar{5}}$ so that their gauge anomalies due to shiggs fermion loops cancel out. The Higgs superpotential couplings in the simplest SU(5) GUT take the form

$$P(\phi) = \lambda_1 \left[\frac{1}{3}(\mathrm{Tr}\phi^3) + \frac{m}{2}(\mathrm{Tr}\phi^2)\right] \ : \ m = \mathcal{O}(m_X) \tag{6.22}$$

which vanishes when

$$\langle 0|\phi|0\rangle = 2m \begin{pmatrix} 1 & 0 & 0 & . & 0 & 0 \\ 0 & 1 & 0 & . & 0 & 0 \\ 0 & 0 & 1 & . & 0 & 0 \\ \cdots & \cdots & \cdots & \cdots & \cdots & \cdots \\ 0 & 0 & 0 & . & -3/2 & 0 \\ 0 & 0 & 0 & . & 0 & -3/2 \end{pmatrix} \tag{6.23}$$

and

$$P(\phi,H) = \lambda_2 \overline{H}^{\alpha}(\phi_\alpha^\beta + 3m'\delta_\alpha^\beta)H_\beta . \tag{6.24}$$

After the spontaneous breakdown (6.23) of $SU(5) \to SU(3)_C \times SU(2)_L \times U(1)_Y$, the effective low energy superpotential for the Higgses H and $\overline{H}$ becomes

$$\widetilde{P}(H) = \lambda_2 \overline{H}^{4,5} H_{4,5} (-3m + 3m') + \lambda_2 \overline{H}^{1,2,3} H_{1,2,3} [2m + 3m']. \tag{6.25}$$

We can cancel the first term in (6.25) so that the Higgs doublet is massless by choosing [50]

$$m = m' = \mathcal{O}(m_X) \Rightarrow m_{H_D} = 0 \tag{6.26}$$

in which case the mass of the Higgs triplets given by the second term in (6.25) becomes

$$m_{H_T} = 5m = \mathcal{O}(m_X). \tag{6.27}$$

Thanks to the miraculous no-renormalization theorems [49], the hierarchy (6.26, 6.27) between the large and small mass-scales is maintained in all orders of perturbations theory.

A slightly more elegant alternative to the SU(5) theory outlined above is one [51] with a so-called "sliding singlet" in which $P(\phi,H)$ (6.24) is replaced by

$$P(\phi,H,X) = \overline{H}^{\alpha}(\lambda_2\phi_\alpha^\beta + \lambda_3 X\delta_\alpha^\beta)H_\beta \tag{6.28}$$

where X is an SU(5) singlet. The corresponding F terms in the potential are

$$V \ni |(\lambda_2\phi_\alpha^\beta + \lambda_3 X\delta_\alpha^\beta)H_\beta|^2 + |\overline{H}^{\alpha}(\lambda_2\phi_\alpha^\beta + \lambda_3 X\delta_\alpha^\beta)|^2. \tag{6.29}$$

When $\langle 0|H_5|0\rangle, \langle 0|\overline{H}^5|0\rangle \neq 0$, to minimize V $\langle 0|X|0\rangle$ adjusts itself so that

$$\langle 0|(\lambda_2\phi_{55} + \lambda_3 X)|0\rangle = 0 \tag{6.30}$$

and them $m_{H_D} = 0$ automatically. Although this seems simple in theory, arranging it in practice in a realistic theory is not so simple.

6.4. Susy GUT phenomenology

How are our previous calculations of m_X, $\sin^2\theta_W$, m_b/m_τ and the baryon lifetime modified in susy GUTs? Due to the presence of additional low-mass sparticles such as gauginos, squarks, sleptons and shiggses, the evolution of the low energy couplings is different [52] from conventional GUTs:

$$\frac{1}{\alpha_3(Q)} = \frac{1}{\alpha_G(m_X)} + \frac{1}{12\pi}[27 - 6N_G]\ln Q^2/m_X^2 \quad \text{and} \tag{6.31}$$

$$\frac{1}{\alpha_2(Q)} = \frac{1}{\alpha_G(m_X)} + \frac{1}{12\pi}\left[18 - 6N_G - \frac{3}{2}N_H\right]\ln Q^2/m_X^2 \tag{6.32}$$

where the number of light Higgs doublets is now even: $N_H = 2, 4, \ldots$ The couplings α_3 (6.31) and α_2 (6.32) approach each other at a different rate from minimal GUTs:

$$\frac{1}{\alpha_3(Q)} - \frac{1}{\alpha_2(Q)} = \frac{1}{12\pi}\left\{9 + \frac{3}{2}N_H\right\}\ln Q^2/m_X^2 \tag{6.33}$$

leading to a modified estimate for m_X:

$$m_X = (6 \pm 3) \times 10^{16} \times \Lambda_{\overline{MS}} \tag{6.34}$$

corresponding to $m_X \approx 10^{16}$ GeV if $\Lambda_{\overline{MS}} \approx 150$ MeV. The correspondingly modified value of $\sin^2\theta_W$ is

$$\sin^2\theta_W = 0.236 \pm 0.003 \tag{6.35}$$

for $\Lambda_{\overline{MS}}$ = (100 to 200)MeV, still less than 2 standard deviations away from the best experimental value. Finally, m_b/m_τ has essentially the same value in susy GUTs as in conventional minimal GUTs.

Naively, one might have expected $L_{GU} \propto 1/m_X^2$ and hence $\tau_N \propto m_X^4$, so that the estimate (6.34) of m_X would give a much longer lifetime for the nucleon in susy GUTs. However, this is not the case, for there are additional diagrams [53] in susy GUTs involving superheavy shiggs triplet exchange which yield

$$\bar{L}_{GU} \propto \frac{1}{m_{\tilde{H}_T}} \left[(qq)\,\tilde{q}\,\tilde{L} + (qL)\widetilde{qq} \right]. \tag{6.36}$$

To get from $\tilde{L}$ (6.36) to baryon decay we must transform the squarks and sleptons into quarks and leptons by gaugino exchange. The order of magnitude of the resulting one-loop effective interaction is such that [54]

$$\tilde{G}_{GU} = \frac{\mathcal{O}(10^{-12})}{m_{\tilde{H}_T}\, m_{\tilde{V}}} \tag{6.37}$$

where the small numerator contains small Yukawa couplings, $1/16\pi^2$ loop factors, etc. Taking $m_{\tilde{V}} = \mathcal{O}(m_W)$ we finally estimate

$$\tau_N(\text{susy GUT}) \approx \tau_N\,(\text{minimal GUT}) \tag{6.38}$$

but with considerably larger uncertainties. However, the expected baryon decay modes are completely different from conventional GUTs. The presence of Higgs Yukawa couplings and the necessity to antisymmetrize L_{GU} (6.36) in flavour space (a consequence of combining SU(3) colour and susy) means that second generation fermions appear in the decay. The charge $Q_{em}(H_T) = +1/3$ of the exchanged Higgs tries to yield $\bar{\nu}_\mu \bar{s}$ or $\mu^+ \bar{c}$ final states, but the latter are excluded by phase space. Therefore [54, 55]

$$p \to \bar{\nu}K^+, \quad n \to \bar{\nu}K^0 \tag{6.39}$$

are expected to be the dominant nucleon decay modes in minimal susy GUTs, though other possibilities are open [56] in more complicated susy GUTs.

7. Supersymmetry at low energies

7.1. Structure of N = 1 supergravity

Recall from the previous lecture that global susy is broken spontaneously if and only if the vacuum energy density $E > 0$. This suggests that problems will arise when we consider gravity, because the vacuum energy will become a positive effective cosmological constant. However, this is a premature conclusion: since the problem arises in connection with gravity, we should first write down a consistent theory of supersymmetry and gravity, i.e. supergravity [57], and this we now do.

The particle content of supergravity is the same set of gauge and chiral supermultiplets that we had before, plus the graviton/gravitino supermultiplet. The supergravity action contains renormalizable gauge and Yukawa interactions supplemented by non-renormalizable interactions with fields of dimension $d > 4$, divided by the appropriate powers of the Planck mass: $1/m_P^{d-4}$. The structure of the Yukawa couplings of the chiral supermultiplets is fixed by the Kähler potential G which replaces and extends the superpotential P of the previous Section [58]. The Kähler potential is a real function of the complex chiral fields ϕ^i. Examples of its rôle include the following:

$$G \ni |\phi^i|^2 \Rightarrow \text{canonical kinetic terms} \tag{7.1a}$$

$$G \ni \ln P(\phi) + (\text{herm. conj.}) \Rightarrow \text{superpotential } P(\phi). \tag{7.1b}$$

However, more complicated forms of G are possible, in which the basic pieces (7.1) are supplemented by extra terms which give additional $\mathcal{O}(1/m_P^n)$ interactions. Henceforth, unless otherwise stated, we use natural units in which $m_P/\sqrt{8\pi} = 1$.

For forms of the Kähler potential more general than the simple examples (7.1), the kinetic terms of the spin-zero fields are [58]

$$G_i^{\ j}\partial_\mu\phi^i\,\partial^\mu\phi_j^* \quad \text{where} \tag{7.2}$$

$$G_i \equiv \frac{\partial G}{\partial\phi^i}, \quad G_i^{\ j} = \frac{\partial^2 G}{\partial\phi^i\partial\phi_j^*}, \quad \text{etc.} \tag{7.3}$$

The Goldstino is

$$\chi \propto G_i f^i \tag{7.4}$$

and the scalar potential

$$V = e^G\left[G_i(G''^{-1})^i_{\ j}G^j - 3\right] + (\text{D terms}) \tag{7.5}$$

where $(G''^{-1})^i_{\ j}$ is the inverse of $G_i^{\ j}$. If we consider the basic (7.1) example

$$G = \sum_i |\phi^i|^2 + [\ln P(\phi) + \text{herm. conj.}] \quad \text{then} \tag{7.6}$$

$$e^G = \exp\left(\sum_i |\phi^i|^2\right) \times |P(\phi)|^2 \quad \text{while} \tag{7.7}$$

$$G_i = \phi_i^* + \frac{1}{P}\,\partial P/\partial\phi^i \tag{7.8}$$

$$G_i^{\ j} = \delta_i^{\ j}. \tag{7.9}$$

In this case the potential (7.5) can be written as follows (dropping D terms):

$$V = \exp\left(\sum_i|\phi^i|^2\right)\left[\sum_i\left|\frac{\partial P}{\partial\phi^i} + \phi_i^*\,P(\phi)\right|^2 - 3\,|P(\phi)|^2\right]. \tag{7.10}$$

Very importantly, when the super-Higgs mechanism operates

$$m_{3/2} = e^{G/2} . \qquad (7.11)$$

In addition to these rôles, the Kähler potential also determines chiral fermion couplings, but these are not important for the moment.

Recall that in the conventional Higgs mechanism, a massless Goldstone boson of a spontaneously broken global symmetry is eaten by a gauge boson (which only has helicity ±1 states) and becomes its helicity 0 state, thereby allowing it to acquire a mass, e.g.

$$m_W = \frac{g}{\sqrt{2}} < 0|H|0> = \frac{g}{2} v . \qquad (7.12)$$

In parallel to this story, the massless spin - $\frac{1}{2}$ Goldstino fermion of spontaneously broken global supersymmetry is eaten by the gravitino (which only has helicity ±3/2 states) and becomes its helicity $\pm\frac{1}{2}$ states, thereby allowing it to acquire a mass (7.11). Because of the all-important minus sign in equation (7.5) (see also the simple example (7.10)) one can simultaneously break local susy spontaneously ($< 0|P|0> \neq 0$ in the simple example (7.6) substituted into the general formula (7.11) fixes $m_{3/2} \neq 0$), break global susy spontaneously ($< 0|\partial P/\partial\phi^i|0 > \neq 0$ as in Section 6) and nevertheless keep <u>zero vacuum energy</u>. Thus, if we suppose that $< 0|\phi^i|0 > \ll 1$, then (7.7) implies

$$e^G = |P(\phi)|^2 \qquad (7.13)$$

and therefore (using (7.11)):

$$m_{3/2} = \mathcal{O}\left(\frac{1}{m_P^2}\right) \times < 0|P|0> . \qquad (7.14)$$

The requirement of zero vacuum energy is

$$\sum_i \left|\frac{\partial P}{\partial\phi^i} + \phi_i^\star P(\phi)\right|^2 = 3|P(\phi)|^2 \qquad (7.15)$$

which becomes (if we assume again that $< 0|\phi^i|0>\ll 1$)

$$\sum_i \left|\frac{\partial P}{\partial\phi^i}\right|^2 \approx 3|P|^2 . \qquad (7.16)$$

Hence equation (7.14) also implies

$$m^2_{3/2} = \mathcal{O}\left(\frac{1}{m_P^2}\right) \times \sum_i \left|\frac{\partial P}{\partial\phi^i}\right|^2 \qquad (7.17)$$

as advertized in Section 6.

In order to study the relevant effective low-energy theory, we consider the limit

$$m_P \to \infty , \qquad m_{3/2} \text{ fixed} . \qquad (7.18)$$

The simple model (7.6, 7.10) then yields [59]

$$V \approx \sum_i \left|\frac{\partial P}{\partial\phi^i}\right|^2 + m^2_{3/2} \sum_i |\phi^i|^2 + m_{3/2}\left[\sum_i \phi^i \frac{\partial P}{\partial\phi^i} + \text{herm. conj.}\right] \qquad (7.19)$$

in which we discern scalar masses equal to $m_{3/2}$, and other soft global susy-breaking terms. In general, one can expect

$$V = \sum_i \left|\frac{\partial P}{\partial \phi^i}\right|^2 + m_0^2 \sum_i |\phi^i|^2 + [AP_3(\phi) + BP_2(\phi) + \text{herm. conj.}] \qquad (7.20)$$

where m_0^2 is a model-dependent scalar $(\text{mass})^2$ parameter of order $m_{3/2}^2$, and A and B are model-dependent numerical parameters of order 1, where $P_3(\phi)$ and $P_2(\phi)$ are the trilinear and bilinear terms respectively in the global superpotential $P(\phi)$. It is now apparent why we choose to keep $m_{3/2}$ fixed and $\ll m_P$. We would like the Higgs and other scalar masses to be $\mathcal{O}(m_W)$. Equations (7.19) and (7.20) say that this occurs naturally, if $m_{3/2} = \mathcal{O}(m_W)$. To set this value of m_W, we need [48] a "hidden" sector of chiral fields. This is because all the known "observable" chiral fields have $< 0|\phi^i|0 > = 0$ or $\ll m_P$, and therefore cannot generate $< 0|P|0 >$ or $< 0|\sum_i|\partial P/\partial\phi^i|^2|0 >$ of large enough order of magnitude. Therefore we postulate the existence of additional "hidden" gauge singlet chiral superfields Z which bear the burden of generating the gravitino mass:

$$m_{3/2} = e^{G/2} \approx e^{\hat{G}(Z)/2}. \qquad (7.21)$$

The "hidden" fields Z are very weakly coupled to ordinary matter, which is disappointing for experimentalists and troublesome for cosmology (see section 10). By incorporating non-minimal kinetic terms for the gauge fields, it is also possible to generate susy breaking gaugino masses $m_{1/2}$ of order $m_{3/2}$ (7.21).

7.2. Weak Gauge Symmetry Breaking

There are two ways of realizing this in the type of effective low energy theory (7.20) derived from supergravity that we have developed. One is to put in $SU(2)_L \times U(1)_Y$ breaking at the tree level [60]. However, this requires the addition of an extra light gauge singlet chiral field, which raises problems of naturalness since radiative corrections tend to take its mass scale off to m_X in a GUT. Moreover, to arrange the potential one must introduce a new mass parameter of $\mathcal{O}(m_W)$ to avoid a true vacuum with $< 0|\tilde{e}|0 > \neq 0$ which violates electromagnetic charge conservation [61]. Alternatively, one can generate $SU(2)_L \times U(1)_Y$ breaking by radiative corrections [62]. The idea is that the hidden sector in supergravity provides input values for the soft susy-breaking parameters in (7.20) at a renormalization scale $\mu = \mathcal{O}(m_P)$, and these then evolve through multiplicative (as long as $\mu \gg m$) renormalization group equations [63] of the general form

$$\mu \frac{\partial m^2}{\partial \mu} = [\,\mathcal{O}(\alpha) \text{ matrix}\,]\, m^2(\mu) \qquad (7.22)$$

as one decreases the scale μ. The plan then is that at some renormalization scale μ_0 the Higgs $(\text{mass})^2$ becomes negative thereby triggering weak gauge symmetry breaking, while other $(\text{mass})^2$ parameters such as $m_{\tilde{q}}^2$ and $m_{\tilde{L}}^2$ remain positive. Since the equations (7.22) governing the evolution of m_H^2 are multiplicative for $\mu \gg m$, the scale μ_0 at which m_H^2 becomes negative is fixed by dimensional transmutation [64].

$$\mu_0 = m_P \exp\left[\frac{-\mathcal{O}(1)}{\alpha}\right] \; : \; m_W = \frac{g}{2} < 0|H|0 > = \mathcal{O}(\mu_0). \qquad (7.23)$$

How does one arrange $m_H^2 < 0$ while $m_{\tilde{q}}^2, m_{\tilde{L}}^2 > 0$?

This can be done with a heavy t quark [63, 65], which contributes to (7.22) in the following way:

$$\mu \frac{\partial m_H^2}{\partial \mu} = (+)\,\mathcal{O}\left(\frac{g^2_{Ht\bar{t}}}{4\pi}\right) m_{\tilde{t}}^2 \ (-)\,\mathcal{O}\left(\frac{g^2}{4\pi}\right) m_{1/2}^2 . \qquad (7.24)$$

If we now suppose that the t quark coupling $g_{Ht\bar{t}}$ overwhelms the gauge coupling g, then (7.24) tells us that m_H^2 increases as μ increases, or conversely as μ is decreased from m_P, m_H^2 decreases and may become negative at $\mu_0 = \mathcal{O}(m_W)$ (7.23). How heavy does this "heavy" t quark have to be? The mechanism works for some value of A, m_0 and $m_1/2$ whenever $m_t \gtrsim m_b$, and it is certainly possible to get

$$m_t \approx 40\text{GeV} \qquad (7.25)$$

as suggested by experiment[11]. If so, in most such models [64]

$$m_{\tilde{q}}\,,\ m_{\tilde{L}} < m_W \qquad (7.26)$$

although this is not a general property of susy models. Unsettlingly, it is also possible that our present vacuum is unstable [64, 66], and that we will eventually tunnel into another with electromagnetic charge and colour conservation violated, but with a tunnelling lifetime much longer than the present age of the Universe.

7.3. Introduction to sparticle phenomenology

In most models, there is a multiplicatively-conserved quantum number R = +1 for particles, -1 for sparticles: See for example the couplings $\bar{f}\,\tilde{f}\,\tilde{G}$, $\tilde{f}^*\tilde{f}\ G$, $f\ f\ \tilde{H}$ and $(|\tilde{f}|^2)^2$ of Section 6. This means that sparticles must be pair-produced, e.g.

$$e^+e^- \to \tilde{e}^+\tilde{e}^- \quad \text{or} \quad \tilde{\mu}^+\tilde{\mu}^-, \quad \bar{p}p \to \tilde{\bar{q}}\,\tilde{q} + X,\ \tilde{g}\,\tilde{q} + X \qquad (7.27)$$

that heavier sparticles decay into lighter sparticles, e.g.

$$\tilde{q} \to q\,\tilde{\gamma} \quad \text{or} \quad q\tilde{g}\,, \quad g \to q\,\bar{q}\tilde{\gamma} \qquad (7.28)$$

and that the lightest sparticle is absolutely stable. In this case, cosmology tells us that the lightest sparticle is probably electromagnetically neutral and not strongly interacting. Otherwise, the relic stable sparticles from the Big Bang (see Sections 9 and 10) would tend to condense into stars, planets and us along with ordinary matter, and in a concentration

$$n(\tilde{X})/n(B) \approx \mathcal{O}(10^{-10} \text{ to } 10^{-6}). \qquad (7.29)$$

Experimental searches for heavy stable particles establish upper limits way below the estimate (7.29), leaving as candidates for the lightest sparticle the $\tilde{\gamma}$, a neutral shiggs $\tilde{H}^0$ (but more detailed cosmological studies suggest $m_{\tilde{H}^0} > m_{\tilde{\gamma}}$), a $\tilde{\nu}$ (but this is often heavier than the $\tilde{\gamma}$ in models), or the gravitino (even if lighter than the $\tilde{\gamma}$, which is disfavoured in models, it would not be produced copiously in particle collisions, and the $\tilde{\gamma}$ would have a very long lifetime for decay into it). Therefore the most likely signature for supersymmetry is the production of photinos at the ends of susy decay chains, which then escape from the apparatus much as do neutrinos, because they also are weakly interacting. The escaping photinos therefore provide a missing energy-momentum signature, e.g.

$$e^+e^- \to \tilde{e}^+\tilde{e}^-\,; \quad \tilde{e}^+ \to e^+\tilde{\gamma}, \quad \tilde{e}^- \to e^-\tilde{\gamma} \qquad (7.30)$$

in which 50% of the centre-of-mass energy is lost, or

$$p\bar{p} \to \begin{cases} \tilde{q}\,\tilde{q} + X, & \tilde{q} \to q\tilde{\gamma}\ ,\ \bar{\tilde{q}} \to \bar{q}\tilde{\gamma} \\ \tilde{g}\,\tilde{g} & \tilde{g} \to q\bar{q}\tilde{\gamma} \end{cases} \tag{7.31}$$

in which there is a missing transverse momentum of order $m_{\tilde{q}}$ or $m_{\tilde{g}}$. Searches for signatures like (7.30) have yielded no result, from which we infer [5]

$$m_{\tilde{e},\ \tilde{\mu},\ \tilde{\tau},\ \tilde{q}\ \ \tilde{W}^{\pm},\ \tilde{H}^{\pm}} \gtrsim \mathcal{O}(20)\,\mathrm{GeV}. \tag{7.32}$$

Note that there are two spin-0 sparticles $\tilde{q}_{L,R}$ or $\tilde{L}_{L,R}$ for every quark or charged lepton flavour, which are the spartners of the $q_{L,R}$ or $L_{L,R}$ respectively. These can mix through the soft susy terms in the potential (7.20). In the next Section we will see what $p\bar{p}$ experiments have to say about the signatures (7.31).

8. Supersymmetry and CERN Collider data

8.1. Introduction

Several categories of "funny" events have recently been reported from the CERN collider:

$$Z^0 \to L^+L^-\gamma,\quad Lj\not{p}_T,\quad jj,\quad \gamma\not{p}_T,\quad j\not{p}_T \tag{8.1}$$

where j denotes a hadronic jet, and $\not{p}_T$ missing p_T. Are any of these new physics, and do any of them have anything to do with supersymmetry? For each category (8.1) we will first recall characteristics of the events, then collect estimates of conventional physics backgrounds, (see the Table) and then review exotic explanations such as susy. The $\not{p}_T$ events (8.1) are the best candidates for susy: if that is what they are, then as we will see, other events much as $e^+e^- \to \gamma\not{p}_T$, $e^+e^- \to \tilde{L}\,\bar{\tilde{L}}$, $e^+e^- \to \tilde{t}\bar{\tilde{t}}$ and J/ψ or $\Upsilon \to H^0 + \gamma$ may be on the verge of discovery [67].

Table 1: Categories of "funny" events.

Type	Observed by	Number	Background
$Z^0 \to L^+L^-\gamma$	UA1, UA2	3	QED: 2.9σ effect
$Lj\not{p}_T$	UA2	3 - 4	QCD: 1 to 3 events
jj bump	UA2	69 ± 17	none: ≈ 4σ effect
$\gamma\not{p}_T$	UA1	1	γZ^0: 4 x 10^{-3} events
$j\not{p}_T$	UA1	6	gZ^0: 0.2 events b,c: 0.1 events

8.2. $Z^0 \to L^+L^-\gamma$ events

Parameters of these 3 events are published elsewhere [9]. Here we note the following comments on them. The muons are not very well measured in the $\mu^+\mu^-\gamma$ event, and it could be $(Z^0 \to \mu^+\mu^-) + \gamma$. The (L^+L^-) invariant masses are not equal within the errors, and neither are the high nor low $(L\gamma)$ invariant masses. The 3 events are all close to the edge of the Z^0 decay Dalitz plot, but not actually at the edge, since the $\theta_{L\gamma}$ openning angles are non-zero.

The conventional physics background is from QED bremsstrahlung, which gives a small rate [68] for events with large E_γ and large $\theta_{L\gamma}$:

$$R \equiv \frac{P(e^+e^-\gamma:\ E_\gamma/m_Z > \varepsilon,\ \theta_{L\gamma} > \delta)}{P(Z^0 \to e^+e^- \text{ or } e^+e^-\gamma)} \approx \frac{\alpha}{\pi}\left[(4\ln 2\varepsilon + 3)\ln\delta/2 + \frac{\pi^2}{3} - \frac{7}{4}\right] \qquad (8.2)$$

which is 0.02(3×10^{-3}) for $\delta=10^0$, $\varepsilon = 0.1(0.25)$. However, there is a much larger rate if one integrates over all configurations less probable than the ones observed,

$$R \approx 1.5\% \text{ per event} . \qquad (8.3)$$

With this estimate, the probability of seeing 3 such events among the 16 observed $Z^0 \to L^+ L^-(\gamma)$ decays is 0.2%, making it a 2.9σ effect [69]. While unlikely, this is not impossible, and one should wait for more data before overinterpreting the observed events. Nevertheless, some "exotic" explanations have been proposed.

Susy has no explanation, since it wants events with missing energy. Among composite model suggestions have been $Z^0 \to L\ (L^* \to L\gamma)$. For this to work, one must take the high $m(L\gamma)$ combination, since PEP and PETRA have excluded any L^* in the range of the low $m(L\gamma)$ combinations. However, the high $m(L\gamma)$ combinations are <u>different</u> in the two well measured $e^+e^-\gamma$ events. Moreover, this explanation does not fit the Dalitz plot because it does not explain why the γ from the L^* is almost parallel to the other lepton. Another composite suggestion has been $Z^0 \to \gamma(X \to L^+L^-)$. Here one needs at least two new bosons X, since the two $m(e^+e^-)$ are different, and these are actually expected in some composite models with chiral symmetry. However, these models do not explain why one of the $L^\pm$ from the X emerges almost parallel to the γ. My advice to the composite modellers is: wait and see!

8.3. $Lj\not p_T$ events

UA2 has reported [70] 3 of these events where the $\not p_T$ vector is opposite to an e in azimuthal angle, and one where the $\not p_T$ and e have similar azimuthal angles. The latter could well be t decay, and we will not discuss it further here. Among the features of the three "opposite" events we notice that while the $(L\not p_T)$ pair is always compatible with $W \to e\nu$ decay, $|\not p_T|$ is always larger than $|p_T^e|$. If they are interpreted as $(W \to e\nu)$ + jets, the invariant masses peak around 150GeV, but this is where one would expect a background peak, given the experimental cuts. UA1 has no $ej\not p_T$ events, but does have [71] a $\mu j\not p_T$ which is marginally compatible with $(W \to \mu\nu)$ + jet, but has the striking feature that the μ is almost parallel to the jet.

The dominant conventional physics background is QCD bremsstrahlung: W + g etc. This has been estimated in two ways, experimentally [70] by scaling from hadronic (jj) + jet events, and a theoretical calculation [72] of g bremsstrahlung which gives similar results:

background	event A	B	C
experimental	0.4	0.012	0.007
theoretical	0.35	0.02	0.39 .

(8.4)

For the last event, the experimental estimate takes account of the fact that there are two jets with $|\not p_T| \ll E_T$, while the theoretical estimate does not. Since the background is not negligible-especially for event A - again we should wait for more data.

However, composite model people are not patient, and propose [73] that we are seeing an excited quark $q^* \to q + W$. Susy people want $\not p_T$, and suggest [74] that these events could be $q\bar{q} \to \tilde{W}^{\pm}\tilde{g}$ or $qg \to \tilde{W}^{\pm}\tilde{q}$ followed by $\tilde{W} \to e\tilde{\nu}$ or $\nu(\tilde{e} \to e\tilde{\gamma})$ and $\tilde{g} \to q\bar{q}\tilde{\gamma}$ or $\tilde{q} \to q\tilde{\gamma}$. In this case the kinematics need not be compatible with W decay, and one could understand why $|\not{p}_T| < |p_T^e|$ in all the observed events, since the $\not{p}_T$ is being carried by more than one missing neutral particle. However, a problem with this explanation seems to be getting a large enough event rate. Wait and see!

8.4. jj Bump

Since the statistical significance of this bump is only 3 or 4σ, and it occurs near the tail of a steeply falling spectrum, I prefer to wait and see on this one too. Composite modellers would like it to be $q^* \to q + g$, but susy people do not want it because it has no $\not{p}_T$.

8.5. "γ" + Missing p_T

UA1 reports [71] two events of this type, but one of them could be a $W \to e\nu$ decay, since the detected electromagnetic shower is behind a hole in the central detector where a charged track could have been missed. Therefore I only worry greatly about one "γ"$\not{p}_T$ event, which could be a collimated neutral jet from one or more π^0 or η. The "γ"$\not{p}_T$ combination is compatible with a Z^0 decay.

If one does not allow the "γ" to be a hadronic jet, the only conventional physics background is $q\bar{q} \to Z^0 + \gamma$, but this is only expected at the level of 4×10^{-3} of an event. If only there were more of them!

A possible composite model explanation is $Z^0 \to \bar{\nu}(\nu^* \to \nu\gamma)$, but this would require a configuration as bizarre as the $L^+L^-\gamma$ events, but sitting close to a different boundary of the Dalitz plot, with the ν and $\bar{\nu}$ almost parallel. There is no explanation in conventional susy phenomenology for events with a single final state γ, so one would like the "γ" to be an (exceptional) neutral hadronic jet. However, there are unconventional susy models [76] in which a lone γ could appear, e.g. if $m_{\tilde{\gamma}} > m_{\tilde{H}^0}$, $\tilde{\gamma} \to \gamma + \tilde{H}^0$, $\tilde{H}^0$ escapes invisibly, or if $< 0|\tilde{\nu}_\tau|0 > \neq 0$, $\tilde{\gamma} \to \gamma + \nu_\tau$. My preference here is for the (exceptional) neutral hadronic jet explanation. More events of this type are eagerly awaited.

8.6. j$\not{p}_T$ events

UA1 reports [71] 6 monojet + $\not{p}_T$ events and 1 trijet + $\not{p}_T$ event. One of the monojet events is kinematically compatible with W decay, and is believed to be $W \to \tau\bar{\nu}$, $\tau \to$ hadrons + ν, while the other events have too large values of $\not{p}_T$. UA2 is not yet in a position to pick out such events above background, since they do not have a hermetic detector.

UA1 claims that conventional QCD jet fluctuations are a negligible background in their apparatus, much less than one event for $|\not{p}_T| > 30$GeV. The background from $q\bar{q} \to g + (Z^0 \to \bar{W})$ is also small:

$\lvert\not{p}_T\rvert >$	30	40	50	60 GeV
# events seen	6	5	2	1
# events expected	0.5	0.2	0.1	0.03

(8.5)

The background from heavy quark semileptonic decay: (b or c) $\to$ (e or μ) + ν + jet, where the charged lepton is not seen, is also expected to be small-less than 0.1 event with $|\not{p}_T|$ above 40GeV.

Composite modellers [73] would like to explain these events as $q^* \to q + (Z^0 \to \nu\,\bar{\nu})$, and good luck to them.

Qualitatively, these events are just what the susy people wanted, events with $\not{p}_T$ and one or more jets. Quantitatively, the known sparticle couplings enable one to calculate the expected number of events reliably as a function of the sparticle masses. One can then use the small events with large $\not{p}_T$ conservatively , to give lower bounds on $m_{\tilde{q}}$ or $m_{\tilde{g}}$. More daringly, one can try to interpret the observed events in terms of sparticle production. We concentrate [77] on strongly interacting $\tilde{g}$ and $\tilde{q}$ since they have the largest cross-sections. In addition to calculating the cross-sections for $p\bar{p} \to \tilde{g}\,\tilde{g} + X$ (assuming $m_{\tilde{q}} \gg m_{\tilde{g}}$) and for $p\bar{p} \to \tilde{q}\,\bar{\tilde{q}} + X$ (assuming $m_{\tilde{g}} \gg m_{\tilde{q}}$), we also implement experimental cuts modelled on those used by UA1.

One might have expected 4 jet events from $(\tilde{g} \to q\bar{q}\tilde{\gamma})^2$, but the experimental cuts leave one with mainly monojets for $m_{\tilde{g}} < 40\text{GeV}$. From the relatively small number of events with $|\not{p}_T| < 30\text{GeV}$ one can infer [77, 78] the lower limit

$$m_{\tilde{g}} > \mathcal{O}(40)\text{GeV} \tag{8.6}$$

to be compared with the previous limit of about 3GeV. More excitingly, $m_{\tilde{g}} = \mathcal{O}(40)\text{GeV}$ is compatible [79] with the observed number of monojet and trijet events with $\not{p}_T$. As for squarks, if $m_{\tilde{q}} < 40\text{GeV}$ then the bound (8.6) forbids $\tilde{q} \to q\,\tilde{g}$ decay, $\tilde{q} \to q\tilde{\gamma}$ decays would be expected to dominate, and hence two jet events from $(\tilde{q} \to q\,\tilde{\gamma})^2$. However, once again the experimental cuts yield mainly monojets for $M_{\tilde{q}} < 40\text{GeV}$. Also, from the relatively small number of events with $|\not{p}_T| < 30\text{GeV}$ one can deduce [79]

$$m_{\tilde{q}} > \mathcal{O}(40)\text{GeV} \tag{8.7}$$

compared with the previous limit of about 20GeV (7.32). Moreover, $m_{\tilde{g}} = \mathcal{O}(40)\text{GeV}$ is also compatible [79] with the observed number of monojets and multijets.

There are possibly two reasons [79,67] to prefer squarks as the origin of the observed $j\not{p}_T$ events. One is that the two-body $\tilde{q} \to q\tilde{\gamma}$ decay kinematics provides a harder $\not{p}_T$ spectrum which may be favoured by the data. The other reason is that $\tilde{g} \to q\,\bar{q}\tilde{\gamma}$ gives fatter monojets ($\langle m_j \rangle \approx 10\text{GeV}$) than does $\tilde{q} \to q\tilde{\gamma}$ decay, while the observed jets have $m_j \lesssim \mathcal{O}(5)\text{GeV}$. Gluon bremsstrahlung gives a small non-zero m_j in $\tilde{q} \to q\,\tilde{\gamma}$ decay [67]:

$$\langle m_j \rangle \approx 0.05\, m_{\tilde{q}} \approx 2\text{GeV} \tag{8.8}$$

for $m_{\tilde{q}} \approx 40\text{GeV}$, which is compatible with the observed monojets.

Clearly the conservative attitude is to wait for more monojet data before taking susy seriously. However, if one is foolhardy, one can ask the following question.

8.7. What if...?

Suppose [67] UA1 has detected squarks with masses around 40GeV, and $m_{\tilde{g}} \gtrsim 40\text{GeV}$. What are the implications for other sparticles and sexperiments? Much of the sparticle spectrum is fixed by three parameters, the gravitino mass $m_{3/2}$, the gaugino mass $m_{1/2}$ and the ratio $\bar{v}/v$ of Higgs vacuum expectation values. In the minimal GUT model of Sections 6 and 7 with $\bar{v} = v$ one has [67]

$$m^2_{\tilde{q}} \approx m^2_{3/2} + 5.4\, m^2_{1/2}$$

$$m^2_{\tilde{e}_L} = m^2_{\tilde{\nu}} = m^2_{3/2} + 0.5\, m^2_{1/2} \qquad (8.9)$$

$$m^2_{\tilde{e}_R} = m^2_{3/2} + 0.15\, m^2_{1/2}$$

$$m_{\tilde{g}} = 2.6 m_{1/2},\quad m_{\tilde{\gamma}} = 0.48 m_{1/2}$$

after incorporating renormalization as discussed in Section 7. One combination of the parameters is fixed by taking $m_{\tilde{q}} \simeq 40\mathrm{GeV}$, and other important constraints are

$$m_{\tilde{g}} \geqslant 40\mathrm{GeV}, \quad m_{\tilde{e}} \geqslant 20\mathrm{GeV} \quad \text{and} \quad m^2_{\tilde{\nu}} \geqslant 0 \qquad (8.10)$$

[equation (8.6)] [equation (7.32)] [ensures $<0|\tilde{\nu}|0> = 0$].

These constraints are barely compatible: a consistent solution has

$$\bar{v}/v \approx 1,\quad m_{\tilde{\gamma}} \approx 7\mathrm{GeV},\quad m_{\tilde{g}} \approx 40\mathrm{GeV},\quad m_{\tilde{e}_R} \approx 20\mathrm{GeV}. \qquad (8.11)$$

These results predict excitement soon for many experiments. In addition to the many flavours of squark with mass $\mathcal{O}(40)$GeV, there should also be gluinos with mass $\mathcal{O}(40)$GeV, and many sleptons with masses between $\mathcal{O}(20)$ and $\mathcal{O}(30)$GeV. The rate for $e^+e^- \to \gamma\, \not{p}_T$ due to $e^+e^- \to \tilde{\gamma}\,\tilde{\gamma}\,\gamma$ mediated by $\tilde{e}$ exchange [80] should be very close to the present experimental upper limit established by the MAC collaboration [81] at PEP.

One can go further and speculate that weak gauge symmetry breaking is driven by the 40GeV t quark of UA1, as discussed in section 7.2. In that case consistency would prefer

$$m_{\tilde{q}},\, m_{\tilde{g}} \simeq 45 \text{ to } 50\mathrm{GeV}$$

$$m_{\tilde{\gamma}} \approx 8 \text{ to } 10\mathrm{GeV}. \qquad (8.12)$$

The A parameter of equation (7.20) is now determined to be about 3, which does not greatly affect the squark mass spectrum, except for the t squarks, of which the lighter may well lie [82] in the range

$$m_t - m_{\tilde{\gamma}} < m_{\tilde{t}} < m_t. \qquad (8.13)$$

In this case one might discover $e^+e^- \to \tilde{t}\,\bar{\tilde{t}}$ before $e^+e^- \to t\,\bar{t}$, and toponium decays could be very unusual-$\Gamma(\tilde{t}\,\bar{\tilde{t}}) \gg \Gamma(ggg)$. Such models also have a light Higgs boson H^0 which weighs a few GeV and could appear in J/ψ or $\Upsilon \to H^0 + \gamma$ decays (cf. Section 1).

8.8. Caution

One must be careful not to overinterpret the CERN Collider data. My personal opinion is that the $\gamma\not{p}_T$ and $j\not{p}_T$ events seem to emerge the furthest above their backgrounds. These events are not incompatible with susy, but one must wait for confirmation. If these events are susy, confirmation will soon arrive.

9. GUTs and Cosmology

9.1. Big Bang Cosmology

The Universe is approximately homogeneous and isotropic on large scales, but Einstein soon discovered that there are no such solutions to the equations of general relativity which are stable. Einstein found an expanding solution which he did not like and so modified his equations by introducing a cosmological constant, something he regarded as his greatest error, but more of this later. Nowadays cosmologists regard the expanding Universe and the Big Bang as their Standard Model, for the following 3 major reasons. One is the Hubble expansion, seen in the recession of distant objects at a rate

$$H_0 = 100h_0 \text{ km/sec per Megaparsec separation} \tag{9.1}$$

where observers variously estimate $h_0 \sim \frac{1}{2}$ to 1. Second is the microwave background radiation. It is expected that at earlier times when the Universe was smaller it would have had a higher temperature. In particular, when its scale size was about 10^3 to 10^4 times smaller than today, it was expected that the temperature would have been high enough to ionize atoms, so that the Universe would have contained a "soup" of photons, electrons and nuclei. The two latter would have recombined at lower temperatures, while the photons would have escaped to be observed today. Redshifted by the subsequent expansion of the Universe, they should be present today as a thermal spectrum of microwave photons with a temperature T_0 of about 3°K - as was observed experimentally 20 years ago. The density of these microwave photons is given by thermodynamics:

$$n_\gamma \approx 400 \left(\frac{T_0}{2.7°K}\right)^3 \tag{9.2}$$

The third reason for believing in the Big Bang is primordial nucleosynthesis. The Universe is observed to contain $Y = \mathcal{O}(25 \text{ to } 30)\%$ by weight of ^4He, while stars are not believed capable to have produced more than $\mathcal{O}(5)\%$. The remaining 25% is believed to have been formed along with other light nuclei (D, ^3He, ^7Li) by nuclear fusion in the early Universe when its temperature was $\mathcal{O}(10^9 \text{ to } 10^{10}\,°K) = \mathcal{O}(0.1 \text{ to } 1)$MeV. The success of these primordial nucleosynthesis calculations seems to confirm that the scale size of the Universe was once 10^{-9} smaller than today. In order to get consistently correct abundances for ^4He, D + ^3He and ^7Li one cannot have too many neutrinos [10]:

$$N_\nu \lesssim 3 \text{ or } 4 \tag{9.3}$$

for $Y \leqslant 25\%$, $\Delta Y \simeq 2\%$ for each extra neutrino, and one needs a baryon-to-photon ratio

$$n_B/n_\gamma \simeq (3 \text{ to } 6) \times 10^{-10} \tag{9.4}$$

which is approximately consistent with observation in the current Universe, and corresponds to a mass density $\mathcal{O}(1/10)$ of that required to close the Universe.

Now that we believe in the Big Bang, we want to extrapolate to sizes $\mathcal{O}(10^{-18})$ smaller (and temperatures $\mathcal{O}(10^{18})$ higher) than during Big Bang nucleosynthesis. The crucial Einstein equation [83] we need is that for the Hubble expansion:

$$H^2 = \left(\frac{\dot{R}}{R}\right)^2 = \frac{8\pi G_N}{3}\rho - \frac{k}{R^2} \tag{9.5}$$

where ρ is the energy density and the curvature k is positive, zero or negative, depending whether the Universe will expand for ever or is closed. Nowadays the curvature term is not much larger than the energy density, which means that it was negligible at earlier times. The energy density ρ could include contributions from relativistic matter, for which photons serve as a prototype:

$$\rho \quad = \quad \frac{1}{2}\, g(T)\rho_\gamma(T) \quad : \quad \rho_\gamma(T) \quad = \quad a\, Tn_\gamma(T),\; n_\gamma(T) \quad = \quad cT^3 \tag{9.6}$$

where

$$g(T) \quad \equiv \quad \#(\text{boson helicity states}) + \frac{7}{8}\; \#\,(\text{fermion helicity states}) \tag{9.7}$$

and a, c are constants $\mathcal{O}(1)$ if the distribution is equilibrium. From (9.6) we find

$$\rho \;\alpha\; T^4 \;\Rightarrow\; t \;\alpha\; T^{-2},\;\; R^2. \tag{9.8}$$

For non-relativistic matter such as protons and neutrons

$$\rho \quad = \quad \sum_i m_i n_i(T) \quad : \quad n_i(T) \quad = \quad c_i T^3 \tag{9.9}$$

where the c_i are dynamically determined ($\mathcal{O}(10^{-9}$ to $10^{-10})$ for p, n). From (9.9) we find

$$\rho \;\alpha\; T^3 \;\Rightarrow\; t \;\alpha\; T^{-3/2},\;\; R^{3/2}. \tag{9.10}$$

Another interesting possibility is that ρ is dominated by a cosmological constant

$$\rho \quad = \quad \Lambda \;\Rightarrow\; R \;\alpha\; \exp\left(t\; \frac{8\pi G_N \Lambda}{3}\right) \tag{9.11}$$

as we will discuss in connection with the inflationary models of Section 10.

9.2. The Origin of Matter

The density of baryons visible in the Universe is

$$n_B \quad = \quad \mathcal{O}(0.1)/\text{metre}^3 \tag{9.12}$$

which when divided by the photon density (9.2) gives $n_B/n_\gamma = \mathcal{O}(10^{-9}$ to $10^{-10})$. It is generally believed that there can be no substantial concentrations of antimatter in our cluster of galaxies, otherwise we would have detected radiation from annihilation with conventional matter. Can one understand these facts in a baryon symmetric cosmology [84]: $n_B = n_{\overline{B}}$? In such a Universe $B - \overline{B}$ annihilation would have yielded

$$\frac{n_B}{n_\gamma} \quad = \quad \frac{n_{\overline{B}}}{n_\gamma} \quad = \quad \mathcal{O}(10^{-18}) \tag{9.13}$$

which is orders of magnitude too small for n_B/n_γ, and one does not see how the B and $\overline{B}$ could have been separated to distances larger than a cluster of galaxies. Causality means that any $B - \overline{B}$ separation mechanism could only separate them by distances up to the horizon

$$d \;\sim\; t \;\sim\; T^{-2} \tag{9.14}$$

in a radiation-dominated Universe, and

$$d(\text{galaxy}) \;\longleftrightarrow\; T \;\ll\; 1\text{MeV} \tag{9.15}$$

at which temperature no efficient separation mechanism is known.

There would also be statistical fluctuations in a baryon symmetric Universe which would yield

$$\frac{n_B - n_{\bar{B}}}{n_\gamma} = \frac{1}{\sqrt{n_\gamma}} = \mathcal{O}(10^{-39} \text{ to } 10^{-40}) \tag{9.16}$$

in the volume of a galaxy. Again too small!

So we must look at baryon asymmetric cosmology in which $n_B > n_{\bar{B}}$: was this an initial condition, or did this asymmetry evolve from an initially symmetric Universe? In 1967 Sakharov [85] set out the necessary conditions for generating $n_B > n_{\bar{B}}$, and hence a large enough n_B/n_γ, via particle interactions:

1) One needs interactions which do not conserve baryon number B. These are available in GUTs.

2) These interactions must be able to distinguish between particles and anti-particles, i.e., they must violate charge conjugation C because

$$C(B) = \bar{B} \tag{9.17a}$$

and its combination with parity P, because

$$CP(B \text{ moving left} + B \text{ moving right}) = (\bar{B} \text{ moving right} + \bar{B} \text{ moving left}) \tag{9.17b}$$

These symmetries are automatically violated by the B violating interaction GUTs.

3) One needs an arrow of time which corresponds to a breakdown of thermal equilibrium. Particle interactions are always invariant under the combination of transformations CPT where T here means time reversal. Hence if T is a good symmetry, so also is CP, and $n_B = n_{\bar{B}}$ (9.17b). In other words, number densities are specified in a state of thermal equilibrium, and there is no possibility to evolve from $n_B = n_{\bar{B}}$ to $n_B > n_{\bar{B}}$. This arrow of time is provided by the expansion of the Universe.

GUT interactions with $\Delta B \neq 0$ are important and in equilibrium only when $T \geqslant m_X \approx 10^{15}$ GeV $\approx 10^{28}$°K. They drop out of equilibrium as required by condition (3) above only when $T < m_X$. The preferred scenario [86, 87] for baryosynthesis in GUTs is the out-of-equilibrium decay of a heavy particle, eg. a vector boson X or a Higgs boson H. Initially, equilibrium ensures $n_X = n_{\bar{X}}$. Let us suppose that as occurs in minimal SU(5), the dominant decays of the X are qq and $\overline{Lq}$ ($X \to \overline{qq}$ and Lq). Invariance under CPT ensures $\Gamma(X \to \text{all}) = \Gamma(\bar{X} \to \text{all})$ ($\tau_X = \tau_{\bar{X}}$), but C and CP violation permit unequal partial decay widths:

$$p \equiv \frac{\Gamma(X \to \overline{Lq})}{\Gamma(X \to \text{all})} \neq \bar{p} \equiv \frac{\Gamma(\bar{X} \to Lq)}{\Gamma(X \to \text{all})} \tag{9.18}$$

which produces a final state asymmetry

$$\frac{n_q - n_{\bar{q}}}{n_q + n_{\bar{q}}} = \frac{[2(1-p)-p] - [2-3\bar{p}]}{[2(1-p)+p] + [2-\bar{p}]} \approx \frac{3(\bar{p}-p)}{4-p-\bar{p}} = \mathcal{O}(\bar{p}-p) \tag{9.19}$$

which is non-zero if $\bar{p} \neq p$. Subsequently, essentially all the $\bar{q}$ are annihilated by q to give γ, and we are left with a small surplus of q which combine: $qqq \to B$ to yield finally

$$n_B/n_\gamma = \mathcal{O}(\bar{p}-p). \tag{9.20}$$

What is the mechanism for getting an asymmetry (9.18) in X decay? At the tree level, $p = \bar{p}$, but one can already get $\bar{p} \neq p$ from one-loop diagrams:

$$(\bar{p}-p) = \mathcal{O}\left(\frac{\alpha}{\pi}\right) \times \delta \tag{9.21}$$

where α is a gauge (or Yukawa in the case of a Higgs H) coupling squared, and δ is a CP violating phase. In minimal SU(5) one needs [87] a 3-loop diagram in H decay:

$$(\bar{p}-p) = \mathcal{O}\left(\frac{\alpha}{\pi}\right)^3 \left(m_q/m_W\right)^6 \delta' \tag{9.22}$$

where δ' is a CP-violating phase which is distinct from that of Kobayashi-Maskawa (see Section 4). The result (9.22) gives a baryon asymmetry which is much too small: $n_B/n_\gamma \lesssim \mathcal{O}(10^{-15})$. To get a realistic number, one must complicate the theory, for example by including 2 or more <u>5</u> of Higgses H.

There is no direct relation between n_B/n_γ and the CP-violating observable ε of the K°-$\bar{K}^\circ$ system in minimal SU(5), since δ' (9.22) is distinct from the Kobayashi-Maskawa phase δ. <u>A fortiori</u> there is no such relation in non-minimal SU(5) with more Higgses. However, there is a qualitative observation [88] that theories which are sufficiently complicated to get n_B/n_γ large enough also contain contributions to the QCD vacuum angle θ_{QCD} which may well be large enough to make the neutron electric dipole moment observable in the near future.

The need to go to a GUT beyond minimal SU(5) introduces undetermined parameters which prevent a quantitative calculation of n_B/n_γ. Moreover, the value of n_B/n_γ (9.19) may be diluted, for example by $2 \leftrightarrow 2$ $\Delta B \neq 0$ interactions mediated by X boson exchange at $T \sim m_X$ [89]. The efficiency for generating n_B/n_γ from a given value of $(\bar{p}-p)$ depends on the values of m_X and α_X. It drops unacceptably low for $m_X < 10^{15}$ GeV if $\alpha_X \approx 1/40$ as for minimal SU(5) gauge bosons, but if $\alpha_H \sim 10^{-3}$ for Higgs bosons one can go down to $m_H \sim 10^{13}$GeV. There could also be dilution of n_B/n_γ by extra entropy generated after $T \sim m_X$, perhaps during the Weingberg-Salam $SU(2)_L \times U(1)_Y \to U(1)_{em}$ phase transition, or during the QCD quark-hadron phase-transition.

Successful baryosynthesis impose constraints on the nature of GUT CP violation. "Hard" CP violation due to coupling parameters in the Lagrangian, as occurs in Kobayashi-Maskawa SU(5), is perfectly acceptable. There are however problems for spontaneous CP violation through random choices in the phases of Higgs vacuum expectation values (or of quark condensates as in technicolour). One must arrange for the same choice of CP violating phase to be made in a domain large enough to include our cluster of galaxies. Moreover, this must be done at $T \gtrsim m_X$, when causally connected domains are much too small (9.15), unless we inflate them as in Section 10.

GUT baryosynthesis strongly suggests the nature of the perturbations needed for the formation of galaxies, clusters, etc. <u>A priori</u> these could either be <u>isothermal</u>, in which case n_B varies while $n_\gamma \propto T^3$ is fixed -which is naively favoured by the smoothness of the microwave background radiation, or they could be <u>adiabatic</u>, in which n_B/n_γ is fixed while $n_\gamma \propto T^3$ varies. The latter is favoured by GUTs, since the microphysics parameters specify a value of n_B/n_γ which is independent of the details of the environment. In this case, there is a question of consistency with the microwave background observations, a problem to which we return in Section 10.

9.3. Magnetic Monopoles

Their existence is assured in GUTs, which is a severe embarrassment for cosmology. Why are they so confidently expected? They appear [90] whenever $U(1)_{em}$ is embedded in a spontaneously broken (semi) simple non-Abelian group G such as $SU(5) \to SU(3)_C \times SU(2)_L \times U(1)_Y \to SU(3)_C \times U(1)_{em}$. The topological group $\pi_3(G) = \mathbb{Z}$ which means that magnetic charge g is quantized:

$$eg = \frac{n}{2} . \tag{9.23}$$

Suppose that the minimum of the Higgs potential occurs for a value $|H| = v$ of the Higgs field. In general there is a rotational symmetry and in a monopole configuration the Higgs field points in different internal space directions in different external space directions, looking like a hedgehog. At the centre $|H|=0$, and $|H| \to v$ at infinite distances. The overall energy of the system is the monopole mass

$$m_M = m_X \times \mathcal{O}\left(\frac{1}{\alpha_X}\right). \tag{9.24}$$

More generally, it is believed that any theory predicting electromagnetic charge quantization contains states with magnetic charges. Another example is provided by Kaluza-Klein theories [91] which contain monopoles of mass $\mathcal{O}(10^{19})$GeV.

Monopoles are embarrassing to GUT cosmologists [92] because of the severe upper limits on their abundance. The success of conventional Big Bang nucleosynthesis implies

$$n_M/n_\gamma \lesssim \mathcal{O}(10^{-19}) \tag{9.25}$$

while the upper limit on the present mass density of the Universe implies

$$n_M/n_\gamma \leqslant \mathcal{O}(10^{-25}) \tag{9.26}$$

with more speculative upper limits coming from galactic magnetic fields [93] and from neutron stars [94]. In contrast to (9.25), (9.26) it is impossible [92] to annihilate a large initial abundance below

$$\frac{n_M}{n_\gamma} \approx \mathcal{O}(10^{-10}) \tag{9.27}$$

and one expects many more than (9.25, 9.26) to have been produced in conventional cosmology. At high Temperature finite temperature corrections to the Higgs potential stabilize $|H|=0$ and it is only when T falls below $\mathcal{O}(m_X)$ that the global minimum at $|H|\neq 0$ develops. As the Universe cools, one expects domains no larger than the horizon size at that epoch to form, throughout which the Higgs field points in a correlated direction. One expects misfits in the relative orientations of these domains which produce $\mathcal{O}(1/10)$ of a monopole per causal domain. Whatever the value of m_X is the expected range $\gtrsim 10^{14}$GeV, one expects far too many independent domains and hence far too many monopoles [92] by comparison with the bounds (9.25, 9.26).

10. Inflation and Supersymmetric Cosmology

10.1 Inflation

This a possible solution [95] to the monopole and many other cosmological problems. Why is the Universe so old? The only scale in the Hubble expansion (9.5) is m_P: why is the Universe so much older than $t_P \approx 10^{-43}$ sec? Why is the density of the Universe so close to the critical density required to close it: $\rho \geqslant \mathcal{O}(1/10)\rho_c$?

This means that the k/R^2 curvature term in (9.5) is not much larger than the first term: since $\rho \propto T^4 \propto R^{-4}$ for relativistic matter, and $\rho \propto T^3 \propto R^{-3}$ for non-relativistic matter, it is surprising that the curvature has not yet taken over. Why is the Universe homogeneous and isotropic on scales much larger than those of domains causally connected at earlier stages in the expansion? For example, the microwave background radiation is isotropic to 1 part in 10^4, although radiation coming from opposite directions was $\mathcal{O}(100)$ horizon lengths apart at recombination. Where are the monopoles, gravitinos, etc, expected in conventional cosmology?

The inflationary idea [95] is that the early Universe underwent a period of exponential expansion driven by Higgs vacuum energy:

$$\left(\frac{\dot{R}}{R}\right)^2 = \frac{8\pi}{3m_P^2}\rho : \rho = \Lambda \Rightarrow R = R_0 \exp\left(\sqrt{\frac{8\pi\Lambda}{m_P^2}}\, t\right). \tag{10.1}$$

This exponential expansion suppresses the curvature term in the Hubble expansion term, so that it is small today, even through the Universe is "old" on the Planck scale. The exponential expansion means that domains of the Universe which are so far apart that they could not have talked to each other in the past history of a conventional Universe, actually were close enough to be causally connected in the early stages of an inflationary universe. This makes possible homogeneity and isotropy on a large scale. Exponential expansions of the broken SU(5) domains mean fewer misfits at the joints to form monopoles. The exponentially large number of particles produced at the end of the inflationary epoch when Higgs vacuum energy is converted into matter dilutes the initial density of unwanted particles such as gravitinos to acceptably low levels.

In order to use the Higgs vacuum energy $V_0 = \Lambda$ to inflate sufficiently, one must keep the Higgs field at a value with $V \approx V_0$ for a "long" time:

$$t_0 \geqslant 65 \Big/ \sqrt{\frac{8\pi\Lambda}{3m_P^2}} \tag{10.2}$$

so as to get an expansion factor

$$R/R_0 \geqslant e^{65} \approx 10^{28} \tag{10.3}$$

which solves the homogeneity and isotropy problems. Then the Higgs field makes a transition towards the true low-temperature vacuum, oscillating about it, generating particles and heating up the Universe to a reheating temperature T_R. This transition to the new vacuum should be second order, since a first order transition is never completed [96]: the regions between bubbles of the new vacuum continue to expand exponentially. Clearly, baryosynthesis must take place at or after the end of the inflationary epoch, or else the baryon number would be drowned in the entropy produced during the terminal oscillations.

In old inflationary models [95], the inflation took place while the Higgs field was trapped at the origin, perhaps by its low temperature potential, or perhaps by finite temperature effects. Problems included the incompletion [96] of the phase-transition, mentioned earlier, and monopoles. Since the choice of SU(5) direction was made after inflation, the domains would be small, and there would be many monopoles. A solution to these difficulties was proposed [97] in the form of new inflation, which takes place after the Higgs phase-transition has started (e.g. after the SU(5) breaking direction has been chosen). The bubbles of the old inflationary Universe themselves expand exponentially in the new inflationary scenario. To get a long roll-over period during which $V \sim V_0$ one needs a very flat potential. A prototype potential considered [97] was "Coleman-Weinberg" SU(5) with no mass term μ and the smallest possible $\mathcal{O}(g^4)$ quartic interactions:

$$V = A\phi^4 \left(\ln\phi^2/v^2 - \frac{1}{2}\right) + D\phi^2 \tag{10.3}$$

where

$$A = \frac{1}{64\pi^2 v^4} \left(\sum_B m_B^4 - \sum_F m_F^4 \right) \quad \text{and} \tag{10.4}$$

$$D = \frac{1}{2} \left(\mu^2 + \frac{75}{8} g^2 T^2 + bR - 3\lambda < \phi^2 > \right) . \tag{10.5}$$

Here the sum in (10.4) runs over all bosons and fermions getting their masses from $< 0|\phi|0 > \neq 0$, and contributions from finite temperatures, space time curvature and fluctuations in ϕ^2 (λ is an effective quartic coupling) have been included in the effective Higgs mass squared term (10.5). However, there were problems[98] with the model (10.3, 10.4, 10.5) in reconciling the need for a long roll-over time $\geqslant 65/\sqrt{8\pi V_0/3m_P^2}$ with a smooth start to the roll-over, in the sense of requiring the bounce action for a complete De Sitter volume to make a simultaneous transition over the barrier close to $\phi = 0$. These two conditions yield

$$\frac{4A^{3/2}v^4}{\sqrt{3m_P^2}} < D < \frac{4}{130} \pi \frac{Av^4}{m_P^2} \tag{10.6}$$

which are only consistent if

$$A < \mathcal{O}(10^{-3}), \quad D < \mathcal{O}(10^{-12}) v^2 \left(\frac{v}{10^{15}\mathrm{GeV}} \right)^2 . \tag{10.7}$$

For comparison, in minimal SU(5) the value of A(10.4) is $\geqslant 1/10$. Another fundamental difficulty is with the magnitude of perturbations [99]. For the first time these are now calculable on the basis of microphysics. A fluctuation inside the horizon is expanded exponentially and now reappears inside our visible horizon as a galaxy or cluster. The spectrum of perturbations is almost scale-independent because of the time-translation invariance during the inflationary expansion [99]. However, the "perturbations" are $\mathcal{O}(50)$ in the minimal SU(5) Coleman-Weinberg model (10.3) [99].

10.2 Supersymmetric Cosmology

Supersymmetry can cure many of the evils of the new inflation [98]. The fine-tuning of parameters (10.7) now becomes technically natural. In particular, one can now make the effective ϕ^4 coupling A arbitrarily small: clearly the formula (10.4) is zero in exact susy. An acceptable choice of superpotential coupling λ_Y is

$$\lambda_Y \lesssim \mathcal{O}(10^{-3}) \left(\frac{v}{m_P} \right) . \tag{10.8}$$

(Note that less fine-tuning is required for larger values of v). Perturbations can be adjusted downwards to be $\mathcal{O}(10^{-4})$ or less by making λ_Y small enough [100].

There is, unfortunately, no prediction for the order of magnitude of perturbations unless one can fix λ_Y in some other way. The great inflationary victory of approximately scale-independent perturbations is always retained, however.

What are the implications of supersymmetry for other aspects of cosmology? Baryosynthesis is slightly more difficult than before, since additional degrees of freedom, a larger value of α_X and new interactions make it easier to dilute the baryon number $2 \leftrightarrow 2$ interactions [101]. A novelty is the cosmological appearance of gravitinos, whose primordial abundance is presumably comparable with that of photons or neutrinos. This imposes our upper limit on the mass of a stable gravitino of about 1 KeV [102], compared with $\mathcal{O}(100)$eV for a neutrino. However, models tend to prefer heavier gravitinos weighing $\mathcal{O}(100)$GeV which are unstable, with a lifetime of $\mathcal{O}(10^7$ to $10^8)$ sec and a high likelihood for $3/2 \to \gamma + \tilde{\gamma}$ decays. Their relatively long lifetime means that they are present during and after nucleosynthesis, and there are strong upper limits on their abundance from considerations of the mass density of the Universe during these epochs. Moreover, the photons produced in $3/2 \to \gamma + \tilde{\gamma}$ decays tend to disrupt light nuclei, and possibly the microwave background radiation [103, 104]. All these considerations limit the gravitino abundance to be much smaller than that of photons. The primordial abundance of gravitinos could be diluted by inflation, but in contrast to monopoles the production of gravitinos subsequent to inflation is not negligible. They can be regenerated by particle collisions after inflation to give an abundance [104]

$$\frac{n_{3/2}}{n_\gamma} = \mathcal{O}(10^{-11}) \left(\frac{T_R}{10^9\,\mathrm{GeV}}\right) \tag{10.9}$$

where T_R is the reheating temperature after inflation. Based on this calculation, nucleosynthesis gives [104]

$$\left(\frac{m_{3/2}}{100\,\mathrm{GeV}}\right) \times \left(\frac{T_R}{10^9\,\mathrm{GeV}}\right) \lesssim \mathcal{O}(10^7) \tag{10.10}$$

while constraints on the entropy produced in 3/2 decay yield

$$\left(\frac{m_{3/2}}{100\,\mathrm{GeV}}\right)^{-1/2} \times \left(\frac{T_R}{10^9\,\mathrm{GeV}}\right) \lesssim \mathcal{O}(10^4)\ . \tag{10.11}$$

Limits on the dissociation of light nuclei give

$$\left(\frac{m_{3/2}}{100\,\mathrm{GeV}}\right) \times \left(\frac{T_R}{10^9\,\mathrm{GeV}}\right) \lesssim 2 \times 10^{10} \times \left(\frac{n_B}{n_\gamma}\right) \tag{10.12}$$

while the microwave background radiation will not be disrupted if

$$\left(\frac{m_{3/2}}{100\,\mathrm{GeV}}\right)^{1/2} \times \left(\frac{T_R}{10^9\,\mathrm{GeV}}\right) \lesssim \mathcal{O}(10^2). \tag{10.13}$$

Of all these constraints, that (10.12) from light nuclei seems to be the most severe, yielding [104]

$$T_R \lesssim \mathcal{O}(10^{10})\,\mathrm{GeV} \tag{10.14}$$

as an upper limit on the reheating temperature T_R after inflation [105]. This makes baryosynthesis more difficult, since it requires $m_X \lesssim \mathcal{O}(10^{11})$GeV, which is not compatible with the GUT philosophy, or $m_H \lesssim \mathcal{O}(10^{11})$GeV, which is barely compatible with a lower limit from baryon decay. It suggests that baryons are about to decay, with the main decay modes being [106]

$$p \to \mu^+ K^0,\ \ \bar{\nu} K^+; \qquad n \to \bar{\nu} K^0. \tag{10.15}$$

Alternatives to this possibility are that $m_{3/2} \gtrsim \mathcal{O}(10^4)$GeV, in which case they decay too rapidly to cause any problems, or $m_{3/2} \lesssim \mathcal{O}(1)$KeV in which case they can safely be stable. It is possible to construct models with such gravitino masses, in the context of previous attempts to calculate the weak interaction scale and completely "solve" the hierarchy problem [107]. In this framework, one can construct models with inflation, baryosynthesis, baryons which are sufficiently stable, and all in a "natural" way [108]. However, in these models the nucleon lifetime tends to be unobservably long.

Other supersymmetric particles may also be important relics from the Big Bang. As discussed in Lecture 7 these are most likely to be neutral and not strongly interacting. Of the available candidates, sneutrinos $\tilde{\nu}$ have a large enough annihilation rate for their cosmological mass density to be acceptably low, whatever the neutrino mass [109]. Neutral shiggses $\tilde{H}^0$ have a small annihilation rate $\alpha(m_f/m_W)^4$ for $\tilde{H}^0\tilde{H}^0 \to f\bar{f}$, and should either be heavier than the b or t quark so as to annihilate sufficiently, or else $m_{\tilde{H}^0} > m_{\tilde{\gamma}}$ so that all the $\tilde{H}^0$ decay [44]. The photinos $\tilde{\gamma}$ annihilate in the P-wave : $\sigma(\tilde{\gamma}\tilde{\gamma} \to f\bar{f}) \propto \left(1/m_{\tilde{f}}\right)^4$. Hence their present density increases as $m_{\tilde{f}}$ increases. In the most optimistic case [43, 44]

$$m_{\tilde{\gamma}} \gtrsim \mathcal{O}\left(\frac{1}{2}\right) \text{GeV} \quad \text{for } m_{\tilde{L}}, m_{\tilde{q}} \simeq 20\text{GeV}. \tag{10.16}$$

Both photinos and possibly light gravitinos are candidates for the dark matter, unseen by telescopes, which seems to fill the Universe with as much density as the visible matter, and may be sufficient to push its density up to the critical value required to close the Universe. Moreover, dark matter accelerates the growth of perturbations and may help adiabatic perturbations be compatible with upper limits on variations in the microwave background radiation. Other candidates for the dark matter include neutrinos weighing $\mathcal{O}(10)$eV, but these tend not to produce dark matter on the small scales where it has recently been detected, and they tend to form galaxies too late [110]. Both these problems may be avoided by gravitinos with masses $\mathcal{O}(1)$keV, and especially by photinos weighing $\mathcal{O}(1)$GeV.

10.3 Conclusions

The early Universe may be the only laboratory for testing GUTs, while GUTs and supersymmetry may be the only tools for tackling fundamental cosmological questions. Already they have provided cosmology with many interesting suggestions, for baryosynthesis, inflation, adiabatic perturbation with an almost scale-free spectrum, dark matter and galaxy formation. Conversely, cosmology has provided GUTs with constraints on the number of neutrino species, limits on particle masses, and evidence for more complicated GUTs as well as supersymmetry. It is to be expected that the present mutually beneficial symbiosis between the physics of the very large (cosmology) and of the very small (particle physics) will continue into the future.

References

1. W.J. Marciano: Proc. 4th Topical Workshop on $p\bar{p}$ collider physics, Bern 1984, ed. H. Hänni and J. Schacher, CERN 84-09, p. 165.
2. A. Sirlin and W.J. Marciano: Nucl. Phys. B189, 442 (1981); C.H. Lewellyn Smith and J.F. Wheater: Phys. Lett. 105B, 486 (1981); D. Bardin, P. Christova and O. Fedorenko: Nucl. Phys. B197, 1 (1982); and references contained in ref. 1.
3. M. Kobayashi and T. Maskawa: Prog. Theor. Phys. 49, 652 (1973)
4. E. de Rafael: Article in this volume
5. S. Yamada: Proc. 1983 Int. Symposium on Lepton and Photon Interactions at High Energies, ed. D.G. Cassel and D.L. Kreinick (Cornell, 1984), p. 525.
6. P. Ginsparg, S. Glashow and M.B. Wise: Phys. Rev. Lett. 50, 1415 (1983)
7. UA1 Collaboration, G. Arnison et al: Phys. Lett. 122B, 103 (1983); UA2 Collaboration, M. Banner et al: Phys. Lett. 122B, 476 (1983)

8. UA1 Collaboration, G. Arnison et: Phys. Lett. 126, 398 (1983) UA2 Collaboration R. Bagnaia et: Phys.Lett. 129B, 130 (1983)
9. UA2 Collaboration, P. Bagnaia et al: Zeit. für Physik C24, 1 (1984)
10. J. Yang, M.S. Turner, G. Steigman, D.N. Schramm and K.A. Olive: Ap. J. 281, 493 (1984); K.A. Olive, D.N. Schramm, G. Steigman, M.S. Turner and J. Yang: Ap. J. 246, 557 (1981)
11. UA1 Collaboration, G. Arnison: CERN preprint CERN-EP/84-134 (1984)
12. Crystal Ball Collaboration, SLAC/DESY preprint PUB-3380/84-064 (1984)
13. D. Hitlin: Proc. 1983 Int. Symposium on Lepton and Photon Interactions at High Energies, ed. D.G. Cassel and D.L. Kreinick (Cornell, 1984) p. 746
14. For reviews, see E. Farhi and L. Susskind: Phys. Rep. 74C, 277 (1981); J. Ellis: Proc. 1981 SLAC Summer Institute on Particle Physics, ed. A. Mosher, SLAC-245 (1982) p. 621
15. J.C. Pati and A. Salam: Phys. Rev. Lett. 31, 661 (1973); Phys. Rev. D8, 1240 (1973); Phys.Rev. D10, 275 (1974)
16. H. Georgi and S. Glashow: Phys. Rev. Lett. 32, 438 (1974)
17. J. Ellis and D.V. Nanopoulos: Nature 292, 436 (1981)
18. H. Georgi, H. Quinn and S. Weinberg: Phys. Rev. Lett. 33, 451 (1974)
19. H. Georgi: Particles and Fields-1974, ed. C.E. Carlson (A.I.P., N.Y., 1975); H. Fritzsch and P. Minkowski: Ann. Phys. (N.Y.) 93, 193 (1975)
20. M.S. Chanowitz, J. Ellis and M.K. Gaillard: Nucl. Phys. B128, 506 (1977)
21. W.J. Marciano and A. Sirlin: Phys. Rev. Lett. 46, 163 (1981); C.H. Lewellyn Smith, G.G. Ross and J.F. Wheater: Nucl. Phys. B177, 263 (1981)
22. T.J. Goldman and D.A. Ross: Phys. Lett. 84B, 208 (1979); W.J. Marciano: Phys. Rev. D20, 274 (1980)
23. A.J. Buras, J. Ellis, M.K. Gaillard and D.V. Nanopoulos: Nucl. Phys. B139, 66 (1978)
24. D.V. Nanopoulos and D.A. Ross: Nucl. Phys. B157, 273 (1979) and Phys. Lett. 108B, 351 (1982)
25. H. Georgi and C.Jarlskog: Phys. Lett. 86B, 297 (1979)
26. J. Ellis and M.K. Gaillard: Phys. Lett. 88B, 315 (1979)
27. F.A. Wilczek and A. Zee: Phys. Rev. Lett. 43, 157 (1979); H.A. Weldon and A. Zee: Nucl. Phys. B173, 269 (1980)
28. S. Weinberg: Phys. Rev. Lett. 43, 1566 (1979) and Phys. Rev. D22, 1694 (1980)
29. J. Ellis, M.K. Gaillard and D.V. Nanopoulos: Phys. Lett. 88B, 320 (1980)
30. W.J. Marciano: Proc. 1983 Int. Symposium on Lepton and Photon Interactions at High Energies, ed. D.G. Cassel and D.L. Kreinick (Cornell, 1984), p. 80
31. V.S. Berezinski, B.L. Ioffe and Ya. I. Kogan: Phys. Lett. 105B, 33 (1981); Y. Tomozawa: Phys. Rev. Lett. 46, 463 (1981)
32. N. Isgur and M.B. Wise: Phys. Lett. 117B, 179 (1982)
33. S.J. Brodsky, J. Ellis, J.S. Hagelin and C.T. Sachrajda: Nucl. Phys. B238, 56 (1984)
34. IMB Collaboration, R.M. Bionta et al: Phys. Rev. Lett. 91, 27 (1983)
35. V.A. Lyubimov et al: Yad. Fiz. 32, 30 (1980) and Phys. Lett. 94B, 266 (1980)
36. R. Barbieri, J. Ellis and M.K. Gaillard: Phys. Lett. 90B, 249 (1980)
37. M. Gell-Mann, P. Ramond and R. Slansky: unpublished (1979); T. Yamagida: Proc. Workshop on the Unified Theory and the Baryon Number in the Universe (KEK, Japan, 1979)
38. E. Gildener and S. Weinberg: Phys. Rev. D13, 3333 (1976); E. Gildener: Phys. Rev. D14, 1667 (1976)
39. S. Hawking, D.N. Page and C.N. Pope: Phys. Lett. 86B, 175 (1979) and Nucl. Phys. B170 (FS1), 283 (1980)
40. S. Dimopoulos and L. Susskind: Nucl. Phys. B155, 237 (1979); E. Eichten and K. Lane: Phys. Lett. 90B, 125 (1980)
41. Y.A. Gol'fand and E.P. Likhtman: Pis'ma Zh. Eksp. Teor. Fiz. 13, 323 (1971); D. Volkov and V.P.Akulov: Phys. Lett. 46B, 109 (1973); J. Wess and B. Zumino: Nucl. Phys. B70, 39 (1974)
42. F. Bergsma et al: Phys. Lett. 121B, 429 (1983); R.C. Ball et al: Univ. of Michigan preprint UMHE 83/13/UW Ex-83-234 (1983)
43. H. Goldberg: Phys. Rev. Lett. 50, 1419 (1983)
44. J. Ellis, J.S. Hagelin, D.V. Nanopoulos, K.A. Olive and M. Srednicki: Nucl. Phys. B238, 453 (1984)

45. P. Fayet and S. Ferrara: Phys. Rep. 32C, 251 (1977)
46. P. Fayet and J. Iliopoulos: Phys. Lett. 51B, 461 (1974)
47. L. O'Raifeartaigh: Phys. Lett. 56B, 41 (1975) and Nucl. Phys. B96, 331 (1975); P. Fayet: Phys. Lett. 58B, 67 (1975)
48. J. Ellis: Proc. 1983 Int. Symposium on Lepton and Photon Interactions at High Energies, ed. D.G. Cassel and D.L. Kreinick (Cornell, 1984), p. 439
49. J. Wess and B. Zumino: Phys. Lett. 49B, 52 (1974); J. Iliopoulos and B. Zumino: Nucl. Phys. B76, 310 (1974); S. Ferrara, J. Iliopoulos and B. Zumino: Nucl. Phys. B159, 420 (1979)
50. S. Dimopoulos and H. Georgi: Nucl. Phys. B193, 150 (1981); N. Sakai: Zeit. für Phys. C11, 153 (1982)
51. E. Witten: Phys. Lett. 105B, 267 (1981)
52. S. Dimopoulos, S. Raby and F. Wilczek: Phys. Rev. D24, 1681(1981); L.E. Ibáñez and G.G. Ross: Phys. Lett. 105B, 439 (1982); M.B. Einhorn and D.R.T. Jones: Nucl. Phys. B196, 475 (1982)
53. S. Weinberg:Phys. Rev. D26, 287 (1982); N. Sakai and T. Yanagida: Nucl. Phys. B197, 533 (1982)
54. J. Ellis, D.V. Nanopoulos and S. Rudaz: Nucl. Phys. B202, 43 (1982)
55. S. Dimopoulos, S. Raby and F. Wilczek: Phys. Lett. 112B, 133 (1982)
56. D.V. Nanopoulos and K.A. Tamvakis:Phys. Lett. 113B, 151 (1982) and 114B, 235 (1982); A. Masiero, D.V. Nanopoulos, K.A. Tamvakis and T. Yanagida: Phys. Lett. 115B, 298 (1982) and Zeit. für Phys. C17, 33 (1983); Y. Igarashi, J. Kubo and S. Sakakibara: Phys. Lett. 116B, 349 (1982)
57. P. Van Nieuwenhuizen: Phys. Rep. 68C, 189 (1981)
58. E. Cremmer et al: Phys. Lett. 79B, 23 (1978) and Nucl. Phys. B147, 105 (1979), E. Cremmer, S. Ferrara, L. Girardello and A. Van Proeyen: Phys. Lett. 116B, 231 (1982) and Nucl. Phys. B212, 413 (1983)
59. H.P. Nilles, M. Srednicki and D. Wyler: Phys. Lett. 120B, 324 (1982); L.J. Hall, J. Lykken and S. Weinberg: Phys. Rev. D27, 2359 (1983)
60. A. Chamseddine, R. Arnowitt and P. Nath: Phys. Rev. Lett. 49, 970 (1982); R. Barbieri, S. Ferrara and C.A. Savoy: Phys. Lett. 119B, 343 (1982)
61. J.M. Frère, D.R.T. Jones and S. Raby: Nucl. Phys. B222, 11 (1983)
62. J. Ellis,D.V. Nanopoulos and K.A. Tamvakis : Phys. Lett. 121B, 123 (1983); L.E. Ibáñez: Phys. Lett. 118B, 73 (1982) and Nucl. Phys. B218, 514 (1983)
63. K. Inoue, A. Kakuto, H. Komatsu andd S. Takeshita : Prog. Theor. Phys. 68, 927 (1982)
64. J.Ellis, J.S. Hagelin, D.V. Nanopoulos and K.A. Tamvakis : Phys. Lett. 125B, 275 (1983)
65. L.E. Ibáñez and G.G. Ross: Phys. Lett. 110B, 215 (1982); J. Ellis, L.E. Ibáñez and G.G. Ross: Phys. Lett. 113B, 283 (1982) and Nucl. Phys. B221, 29 (1983)
66. M. Claudson, L. J. Hall and I. Hinchliffe: Nucl. Phys. B228, 501 (1983)
67. J. Ellis and M. Sher: CERN preprint TH-3968 (1984)
68. F.A. Berends, R. Kleiss, J.P. Revol and J.P. Vialle: CERN preprint EP/84-117 (1984) and references therein.
69. T. Himel: seminar at SLAC, July 1984
70. UA2 Collaboration, P. Bagnaia et al: Phys. Lett. 139B, 105 (1984)
71. UA1 Collaboration, G. Arnison et al: Phys. Lett. 139B, 115 (1984)
72. G. Altarelli, R.K. Ellis, M. Greco and G. Martinelli: CERN preprint TH-3851 (1984) and references therein.
73. A.de Rújula, L. Maiani and R. Petronzio: Phys. Lett. 140B, 253 (1984)
74. H.E. Haber and G.L. Kane: Phys. Lett. 142B, 212 (1984); R.M. Barnett, H.E. Haber and G.L. Kane: Private communications (1984)
75. J.R. Hansen: Proc. 4th Topical Workshop on $p\bar{p}$ Collider Physics, Bern 1984, ed. H. Hänni and J. Schacher, CERN 84/09, p. 20
76. J. Ellis, G. Gelmini, C.Jarlskog, G.G. Ross and J.W. F. Valle: Rutherford Appleton Laboratory preprint RAL-84-089 (1984)
77. J. Ellis and H. Kowalski : Phys. Lett. 142B, 441 (1984)
78. E. Reya and D.P. Roy: Phys. Lett. 141B, 442 (1984) and Phys. Rev. Lett 53, 881 (1984)
79. J. Ellis and H. Kowalski: DESY preprint 84-045 (1984)
80. P. Fayet: Phys. Lett. 117B, 460 (1982); J. Ellis and J.S. Hagelin: Phys. Lett. 122B, 303 (1983)

81. MAC Collaboration, R. Prepost - private communication (1984).
82. J. Ellis and S. Rudaz: Phys. Lett. 128B, 248 (1983)
83. S. Weinberg: Gravitation and Cosmology (J.S. Wiley, 1972)
84. M.S. Turner: Gauge Theories in High Energy Physics, ed. M.K. Gaillard and R. Stora (North-Holland, 1983), p. 387
85. A.D. Sakharov: Pis'ma Zh. Eksp. Teor. Fiz. 5, 32 (1967)
86. M. Yoshimura : Phys. Rev. Lett. 41, 381 (1978) (E42, 746 (1979)) and Phys. Lett. 88B, 294 (1979) A. Yu. Ignatiev, N.V. Krasnikov, V.A. Kuzmin and A.N. Tavkhelidze: Phys. Lett 76B, 436 (1978); S. Dimopoulos and L. Susskind: Phys. Rev. D18, 4500 (1978) and Phys. Lett. 81B, 416 (1979), D. Toussaint, S.B. Treiman, F. Wilczek and A. Zee: Phys. Rev. D19, 1036 (1979); S. Weinberg: Phys. Rev. Lett. 42, 850 (1979); D.V. Nanopoulos and S. Weinberg: Phys. Rev. D20, 2484 (1979)
87. J. Ellis, M.K. Gaillard and D.V. Nanopoulos: Phys. Lett. 80B, 360 (1979) (E 82B, 464 (1979))
88. J. Ellis, M.K. Gaillard, D.V. Nanopoulos and S. Rudaz: Phys. Lett. 99B, 101 (1981)
89. E.W. Kolb and S. Wolfram: Phys. Lett. 91B, 217 (1980) and Nucl. Phys. B172, 224 (1980); J. Fry, K.A. Olive and M.S. Turner: Phys. Rev. D22, 2953, 2977 (1980)
90. G.'t Hooft: Nucl. Phys. B79, 276 (1974); A.M. Polyakov: Zh. Eksp. Teor. Fiz. Pis'ma Red. 20, 194 (1974)
91. G. Gibbons: Lectures at ELAF'84 (Santiago, 1984)
92. J. Preskill: Phys. Rev. Lett. 43, 1365 (1979)
93. M.S. Turner, E.N. Parker and T.J. Bogdan: Phys. Rev. D26, 1296 (1982); see however E.E.Salpeter, S.L. Shapiro and I. Wasserman: Phys. Rev. Lett. 49, 1114 (1982); J. Arons and R. Blandford: Phys. Rev. Lett. 50, 544 (1983)
94. E.W. Kolb, S.A. Colgate and J.A. Harvey: Phys. Rev. Lett. 49, 1373 (1982); S. Dimopoulos, J. Preskill and F.A.Wilczek: Phys. Lett. 119B, 320 (1982); F.A. Bais, J. Ellis, D.V. Nanopoulos and K.A. Olive: Nucl. Phys. B219, 189 (1983)
95. A. Guth: Phys. Rev. D23, 347 (1981); D. Kazanas: Ap. J. 241, L59 (1980)
96. A. Guth and E. Weinberg: Phys. Rev. D23, 826 (1981)
97. A. D. Linde: Phys. Lett. 108B,389 (1982); A. Albrecht and P. Steinhardt: Phys. Rev. Lett. 48, 1220 (1982)
98. J. Ellis, D.V. Nanopoulos, K.A. Olive and K. Tamvakis: Phys. Lett. 118B, 335 (1982) and Nucl. Phys. B221, 524 (1983)
99. A. Guth and S.Y. Pi: Phys. Rev. Lett. 49, 1110 (1982); S. Hawking: Phys. Lett. 115B, 295 (1982); J. Bardeen, P. Steinhardt and M.S. Turner: Phys. Rev. D28, 679 (1983)
100. J. Ellis, D.V. Nanopoulos, K.A. Olive and K. Tamvakis: Phys. Lett. 120B, 331 (1983)
101. J.N. Fry and M.S. Turner: Phys. Lett. 125B, 379 (1983)
102. H. Pagels and J. Primack: Phys. Rev. Lett. 48, 223 (1982)
103. S. Weinberg: Phys. Rev. Lett. 48, 1303 (1982)
104. J. Ellis, J.E. Kim and D.V. Nanopoulos: Phys. Lett. 145B, 181 (1984)
105. J. Ellis, A.D. Linde and D.V. Nanopoulos: Phys. Lett 118B, 58 (1982)
106. D.V. Nanopoulos and K.A. Tamvakis: Ref. 56
107. J. Ellis, A.B. Lahanas, D.V. Nanopoulos and K. Tamvakis: Phys. Lett. 134B, 429 (1984); J. Ellis, C. Kounnas and D.V. Nanopoulos: Nucl. Phys. B241, 406 (1984)
108. J. Ellis et al: CERN preprint TH-3902 (1984)
109. J.S. Hagelin, G.L. Kane and S. Raby: Nucl. Phys. B241, 638 (1984); L.E. Ibáñez: Phys. Lett. 137B, 160 (1984)
110. J. Silk: Proc. 1st. ESO-CERN Symposium, Nov. 1983, ed. G. Setti and L. Van Hove (ESO-CERN, 1984), p. 225

Lectures on Quark Flavor Mixing in the Standard Model

Eduardo de Rafael

Centre de Physique Théorique, Section II*, CNRS-Luminy-Case 907
F-13288 Marseille Cedex 9, France, and

Max-Planck-Institut für Physik und Astrophysik, Werner-Heisenberg-Institut für Physik, Föhringer Ring 6, D-8000 München 40, Fed. Rep. of Germany

The present status of the theory of elementary particle physics, and the standard model in particular, reminds me of the following lines of a poem by the great Chilean poet Pablo Neruda which he wrote in an "Oda a Federico Garcia Lorca", another great poet of the Spanish language:

Sobre todo de noche
De noche hay muchas estrellas,
Todas dentro de un rio
Como una cinta junto a las ventanas
De las casas llenas de pobres gentes.

I find there is a striking analogy between *muchas estrellas* and the many parameters of the standard model; between *las pobres gentes* and the poor physicists who are trying desperately to put them *todas dentro de un rio*; i.e., all in a super grand unified theory.

I. OVERVIEW OF THE STANDARD MODEL

1.1 General Introduction

The standard model is a quantum field theory of gauge fields based on the group SU(3) x SU(2) x U(1) which describes the interactions of the fundamental constituents of matter, the spin-1/2 leptons and quarks. Quarks carry a color degree of freedom which is assigned to the fundamental representation of an SU(3)-group called $SU(3)_{color}$. Leptons are singlets of $SU(3)_{color}$. Both leptons and quarks come in various flavors:

LEPTONS				QUARKS			
electron	(e)	e-neutrino	(ν_e)	down	(d)	up	(u)
muon	(μ)	μ-neutrino	(ν_μ)	strange	(s)	charm	(c)
tau	(τ)	τ-neutrino	(ν_τ)	bottom	(b)	top	(t)
	...				...		

The left-handed components of each generation (i.e., each line above) of matter fields are assigned into doublets of the fundamental representation of a SU(2)-group called $SU(2)_{left}$ as follows:

---

*) Permanent address

$$L_e(x) \equiv \frac{1}{2}(1-\gamma_5)\begin{pmatrix}\nu_e(x)\\ e(x)\end{pmatrix} \quad ; \quad L_u(x) \equiv \frac{1}{2}(1-\gamma_5)\begin{pmatrix}u(x)\\ d(x)\end{pmatrix} \tag{1.1a}$$

$$L_\mu(x) \equiv \frac{1}{2}(1-\gamma_5)\begin{pmatrix}\nu_\mu(x)\\ \mu(x)\end{pmatrix} \quad ; \quad L_c(x) \equiv \frac{1}{2}(1-\gamma_5)\begin{pmatrix}c(x)\\ s(x)\end{pmatrix} \tag{1.1b}$$

$$L_\tau(x) \equiv \frac{1}{2}(1-\gamma_5)\begin{pmatrix}\nu_\tau(x)\\ \tau(x)\end{pmatrix} \quad ; \quad L_t(x) \equiv \frac{1}{2}(1-\gamma_5)\begin{pmatrix}t(x)\\ b(x)\end{pmatrix} \tag{1.1c}$$

$$\cdots \qquad\qquad \cdots \quad .$$

The corresponding right-handed components

$$e_R(x) \equiv \frac{1}{2}(1+\gamma_5)e(x) \; ; \; u_R(x) \equiv \frac{1}{2}(1+\gamma_5)u(x) ; \; d_R(x) \equiv \frac{1}{2}(1+\gamma_5)d(x) \tag{1.2a,b,c}$$

$$\cdots \qquad \cdots \qquad \cdots$$

are singlets of $SU(2)_L$. Neutrinos are assumed to be massless; and therefore they only appear as components of left-handed doublets. It is, however, possible to accommodate massive neutrinos by introducing their appropriate right-handed components in $SU(2)_L$ singlets.

Both left-handed doublets and right-handed singlets carry weak hypercharge associated to an abelian U(1) group called $U(1)_y$ with particular assignments which are made so as to reproduce the desired electric charges for leptons Q = -1 and Q = 0; and quarks Q = 2/3 and Q = -1/3. The electric charge, the third component of weak isospin and the weak hypercharge are constrained by the relation:

$$Q = I_3^W + \frac{Y_W}{2} . \tag{1.3}$$

Therefore, the required assignments are:

$$Y_W L_e(x) = (-1) L_e(x) \tag{1.4a}$$

$$Y_W e_R(x) = (-2) e_R(x) \tag{1.4b}$$

$$\cdots$$

$$Y_W L_u(x) = \frac{1}{3} L_u(x) \tag{1.5a}$$

$$Y_W u_R(x) = \frac{4}{3} u_R(x) \tag{1.5b}$$

$$Y_W d_R(x) = -\frac{2}{3} d_R(x) \tag{1.5c}$$

$$\cdots$$

Associated to the 3-generators of weak isospin $\vec{I}_W$ there are 3-gauge vector boson fields

$$\vec{I}_W \Longrightarrow W_\mu^{(1)}(x) , \; W_\mu^{(2)}(x) , \; W_\mu^{(3)}(x) . \tag{1.6}$$

Associated to the generator of weak hypercharge Y_W there is one neutral gauge vector boson field

$$Y_W \Longrightarrow B_\mu(x) \quad , \tag{1.7}$$

and associated to the 8-generators of $SU(3)_{color}$ rotations there are 8-gauge vector boson fields, the so-called gluons:

$$T^{(A)}_{color} \Longrightarrow G^{(A)}_\mu(x) \qquad A = 1,2,\dots 8 \ . \tag{1.8}$$

There are three independent coupling constants in the standard model associated to each of the factor groups which describe the various types of couplings:

$$g_s \Longleftrightarrow SU(3)_c \quad : \tag{1.9a}$$

Gluons ····· Gluons
(Quarks)$_{color}$

$$g \Longleftrightarrow SU(2)_L \quad : \tag{1.9b}$$

W ···· W
Quarks Leptons

$$g' \Longleftrightarrow U(1)_{Y_W} \quad : \tag{1.9c}$$

B ··· Quarks
B ··· Leptons .

At the massless level, the Lagrangian of the standard model is a Yang-Mills type Lagrangian [1] with leptons and quarks coupled to the gauge vector bosons via appropriate covariant derivatives:

$$\begin{aligned} \mathcal{L}(x)_{SU(3)_c \otimes SU(2)_L \otimes U(1)_{Y_W}} = \\ -\frac{1}{4}\sum_A G^{(A)}_{\mu\nu}(x)\, G^{(A)\mu\nu}(x) - \frac{1}{4}\vec{F}_{\mu\nu}(x)\,\vec{F}^{\mu\nu}(x) - \frac{1}{4} B_{\mu\nu}(x)\, B^{\mu\nu}(x) \\ + i\sum_{\ell=e,\mu,\tau} \bar{L}_\ell(x)\, \not{D}\, L_\ell(x) + i\sum_\ell \bar{R}_\ell(x)\, \not{D}\, R_\ell(x) \\ + i\sum_{q_\uparrow = u,c,t} \bar{L}_{q_\uparrow}(x)\, \not{D}\, L_{q_\uparrow}(x) + i\sum_{q_\uparrow} \bar{R}_{q_\uparrow}(x)\, \not{D}\, R_{q_\uparrow}(x) + i\sum_{q_\downarrow = d,s,b} \bar{R}_{q_\downarrow}(x)\, \not{D}\, R_{q_\downarrow} \ . \end{aligned} \tag{1.10}$$

Here $G^{(A)}_{\mu\nu}(x)$ and $\vec{F}_{\mu\nu}(x)$ are the Yang-Mills strength tensors associated to the gluon fields $G^{(A)}_\mu(x)$ and the $\vec{W}_\mu(x)$ fields:

$$G^{(A)}_{\mu\nu}(x) = \partial_\mu G^{(A)}_\nu(x) - \partial_\nu G^{(A)}_\mu(x) + g_s f_{ABC}\, G^{(B)}_\mu(x)\, G^{(C)}_\nu(x) \ , \tag{1.11}$$

with f_{ABC} the structure constants of the $SU(3)_c$ Lie algebra; and

$$\vec{F}_{\mu\nu}(x) = \partial_\mu \vec{W}_\nu(x) - \partial_\nu \vec{W}_\mu(x) + g\, \vec{W}_\mu(x) \times \vec{W}_\nu(x) \ . \tag{1.12}$$

The tensor field $B_{\mu\nu}(x)$ is abelian:

$$B_{\mu\nu}(x) = \partial_\mu B_\nu(x) - \partial_\nu B_\mu(x) . \tag{1.13}$$

The specific form of the covariant derivatives $\not{D} \equiv \gamma^\mu D_\mu$ is fixed by the requirement of gauge invariance of the Lagrangian under local gauge transformations. When acting on lepton fields D_μ takes the following form:

$$D_\mu \equiv \partial_\mu + ig \frac{\vec{\tau}}{2} \cdot \vec{W}_\mu(x) - i \frac{1}{2} g' B_\mu(x) \ldots L_\ell(x); \tag{1.14a}$$

$$D_\mu \equiv \partial_\mu \qquad\qquad - i \; g' B_\mu(x) \ldots R_\ell(x); \tag{1.14b}$$

and, when acting on quark fields:

$$D_\mu \equiv \left(\partial_\mu + ig \frac{\vec{\tau}}{2} \cdot \vec{W}_\mu(x) + i \frac{1}{6} g' B_\mu(x)\right) \delta_{\alpha\beta} - ig_s \frac{\lambda^{(A)}_{\alpha\beta}}{2} G^{(A)}_\mu(x) \ldots L_{q_\uparrow}; \tag{1.15a}$$

$$D_\mu \equiv \left(\partial_\mu \qquad + i \frac{2}{3} g' B_\mu(x)\right) \delta_{\alpha\beta} - ig_s \frac{\lambda^{(A)}_{\alpha\beta}}{2} G^{(A)}_\mu(x) \ldots R_{q_\uparrow}; \tag{1.15b}$$

$$D_\mu \equiv \left(\partial_\mu \qquad - i \frac{1}{3} g' B_\mu(x)\right) \delta_{\alpha\beta} - ig_s \frac{\lambda^{(A)}_{\alpha\beta}}{2} G^{(A)}_\mu(x) \ldots R_{q_\downarrow}; \tag{1.15c}$$

where α and β are quark color indices; $\lambda^{(A)}_{\alpha\beta}$ are the eight SU(3) Gell-Mann matrices and summation over the index A = 1,2,...,8 is understood.

Low energy weak interactions phenomenology is well described by an effective four fermion interaction, the Fermi Lagrangian

$$\mathcal{L}(x)_{Fermi} = - \frac{G_F}{\sqrt{2}} j^\dagger_\mu(x) j^\mu(x) , \tag{1.16}$$

with $j^\dagger_\mu(x)$ the charged current combination ($\theta \equiv \theta_{\text{Cabibbo}}$),

$$\begin{aligned} j^\dagger_\mu(x) = {} & \cos\theta \, \bar{u}(x)\gamma_\mu(1-\gamma_5)d(x) + \sin\theta \, \bar{u}(x)\gamma_\mu(1-\gamma_5)s(x) \\ & + \bar{\nu}_e(x)\gamma_\mu(1-\gamma_5)e^-(x) + \bar{\nu}_\mu(x)\gamma_\mu(1-\gamma_5)\mu^-(x) . \end{aligned} \tag{1.17}$$

Search for a renormalizable Lagrangian field theory which in some limit would give the Fermi theory as an effective Lagrangian, has been the main motivation to explore and study gauge field theories. Early since the appearance of the Fermi theory it was recognized that an alternative formulation holds via the coupling of the weak charged current to a heavy vector boson field $W_\mu(x)$:

$$\mathcal{L}_{IVB}(x) = - \frac{g}{2\sqrt{2}} j^\dagger_\mu(x) W^\mu(x) + h.c. \tag{1.18}$$

At the tree level the two theories are equivalent when the momentum transfer in a given process is such that $q^2 \ll M_W^2$:

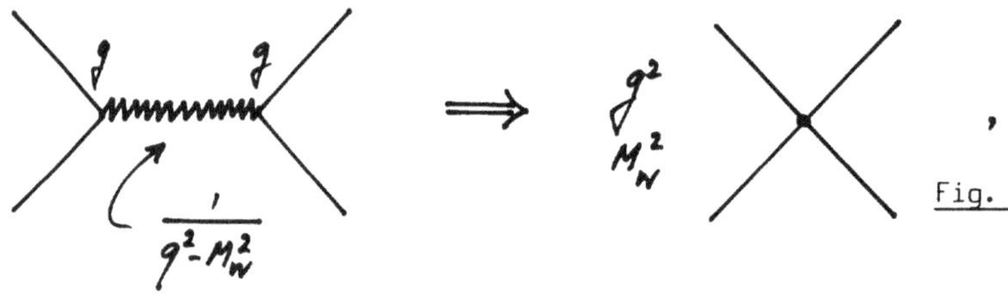

Fig. I.1

provided the identification

$$\frac{G_F}{\sqrt{2}} = \frac{g^2}{8M_W^2} \tag{1.19}$$

is made. In other words, the intermediate vector boson W must be a heavy object to explain the short-range character of the weak force. The problem of introducing masses in a gauge field theory without spoiling the gauge invariance properties, which are crucial for the renormalizability of the theory, was a stumbling block for quite a while. We now know one mechanism which is acceptable, the spontaneous breaking of local symmetry. In the standard model [2] this is accomplished in a minimal way by introducing one doublet of color singlet scalar fields,

$$\underline{\Phi}(x) \equiv (\phi^+(x), \phi^0(x)),$$

belonging to the fundamental representation of $SU(2)_L$ and adding to the previous Lagrangian in Eq.(1.10) the scalar couplings:

$$\mathcal{L}(x) \to \mathcal{L}(x) + D^\mu \underline{\Phi}^\dagger(x) D_\mu \underline{\Phi}(x) - \mu^2 \underline{\Phi}^\dagger(x) \underline{\Phi}(x) - \lambda(\underline{\Phi}^\dagger(x) \underline{\Phi}(x))^2, \tag{1.20}$$

with D_μ the covariant derivative

$$D_\mu \equiv \partial_\mu + ig \frac{\vec{\tau}}{2} \cdot \vec{W}_\mu(x) + ig' \frac{1}{2} B_\mu(x) . \tag{1.21}$$

When $\mu^2 < 0$, the minimum of the so-called Higgs potential

$$V(\underline{\Phi}^\dagger\underline{\Phi}) \equiv \mu^2 \underline{\Phi}^\dagger \underline{\Phi} + \lambda(\underline{\Phi}^\dagger\underline{\Phi})^2 \tag{1.22}$$

occurs at

$$|\underline{\Phi}| = \left(-\mu^2/_{2\lambda}\right)^{1/2} \equiv \frac{v}{\sqrt{2}} , \tag{1.23}$$

and not at $\underline{\Phi} = 0$, as in the usual $\mu^2 > 0$ case. The minimum value $v/\sqrt{2}$ corresponds to the vacuum expectation value of the Higgs field $\underline{\Phi}$ at the tree level. The symmetry is spontaneously broken [3] in the sense that the generators $\vec{I}_W$ and Y_W do not leave any longer the vacuum invariant. Only the combination in Eq.(1.3) corresponding to the electric charge generator Q and, of course, the eight color generators $T^{(A)}_{color}$ remain as symmetry operators. The gauge symmetry is spontaneously broken from

$$SU(3)_{color} \otimes SU(2)_L \otimes U(1)_{Y_W} \Rightarrow SU(3)_{color} \otimes U(1)_Q \; . \tag{1.24}$$

With

$$\phi^0(x) = \frac{\phi_1(x) + i\,\phi_2(x)}{\sqrt{2}} \quad ; \; \phi_1 \text{ and } \phi_2 \text{ hermitean} , \tag{1.25}$$

it is possible to choose, without loss of generality,

$$\langle 0|\phi_2|0\rangle = \langle 0|\phi^+|0\rangle = 0 \tag{1.26a}$$

and

$$\langle 0|\phi_1|0\rangle = v \quad \text{real} \; . \tag{1.26b}$$

If now we introduce a shifted field

$$\phi_1 = \chi + v \quad , \qquad \text{such that} \qquad \langle 0|\chi|0\rangle = 0 \; , \tag{1.27}$$

it can be readily seen from the Lagrangian that the fields

$$\phi^+(x) \; , \quad \phi_2(x) \; , \quad \phi^-(x) \tag{1.28}$$

remain massless - the would-be Goldstone fields associated to each of the generators which no longer leave the vacuum invariant - while $\chi(x)$, corresponding to the so-called Higgs particle, acquires a mass

$$M_H = v\sqrt{2\lambda} \; . \tag{1.29}$$

The massless scalar fields ϕ^+ , ϕ_2 , ϕ^- can be eliminated by choosing a specific gauge - the unitary gauge - where the charged vector boson fields

$$W_\mu^{(\pm)}(x) = \frac{W_\mu^{(1)}(x) \mp i\, W_\mu^{(2)}(x)}{\sqrt{2}} \; , \tag{1.30}$$

and the neutral vector boson field

$$Z_\mu^{(0)}(x) = \frac{g W_\mu^{(3)}(x) - g' B_\mu(x)}{\sqrt{g^2 + g'^2}} \tag{1.31}$$

associated to the (broken) generator orthogonal to Q appear with three polarization states corresponding to massive spin 1 particles, with masses

$$M_{W^\pm} = v\, g/2 \qquad \text{and} \qquad M_{Z^0} = \frac{v}{2}\sqrt{g^2+g'^2} \; . \tag{1.32a,b}$$

The massless photon field is

$$A_\mu(x) = \frac{g' W_\mu^{(3)}(x) + g B_\mu(x)}{\sqrt{g^2+g'^2}} , \tag{1.33}$$

and the identification of its coupling to the electromagnetic current with the electric charge leads to the relation

$$-|e| = \frac{-gg'}{\sqrt{g^2+g'^2}} . \tag{1.34}$$

Furthermore, the identification of $W^\pm(x)$ with the intermediate vector boson field in Eq.(1.18) leads to another relation:

$$\frac{G_F}{\sqrt{2}} = \frac{1}{8M_W^2} g^2 . \tag{1.35}$$

It is then convenient to introduce an angle θ_W[2a], the so-called Weinberg angle,

$$tg\,\theta_W = g'/g , \tag{1.36}$$

and write in terms of it the three independent relations:

$$e = g \sin\theta_W \quad ; \quad \frac{M_W^2}{M_Z^2} = \cos^2\theta_W \quad ; \quad \frac{G_F}{\sqrt{2}} = \frac{g^2}{8M_W^2} . \tag{1.37a,b,c}$$

Lepton masses in the standard model are generated by Yukawa type couplings of the Higgs doublet with fermions. The only isosinglet combination possible is

$$\mathcal{L}_{Yukawa}(x) = \sum_{\ell=e,\mu,\tau} -g_\ell \bar{R}_\ell(x) \Phi^\dagger(x) L_\ell(x) + h.c. . \tag{1.38}$$

The term in $\Phi^\dagger(x)$ from the v.e.v. $v/\sqrt{2}$ generates then a mass-like coupling

$$\sum_{\ell=e,\mu,\tau} -g_\ell \, v/\sqrt{2} \; \bar{\psi}_\ell(x) \psi_\ell(x) \tag{1.39}$$

when we identify

$$m_\ell = g_\ell \, v/\sqrt{2} , \quad \ell = e,\mu,\tau . \tag{1.40}$$

Lepton masses are therefore also generated by the mass dimension which originates in the v.e.v. of the Higgs field. However, the actual mass values are arbitrary since the Yukawa couplings g_ℓ are not fixed by any principle in the standard model.

Quark masses are also generated by Yukawa like couplings. Here, however, we have two possibilities to construct an isoscalar:

$$\sum_{ij} g_{ij}\, \bar{L}_{q_i^\uparrow}(x) \cdot \underline{\Phi}(x) R_{q_j^\downarrow}(x) + \tilde{g}_{ij}\, \bar{L}_{q_i^\uparrow}(x) \cdot \tilde{\underline{\Phi}}(x) R_{q_j^\uparrow}(x) + h.c. \quad ; \tag{1.41}$$

$$Y_W : (-1/3)\ (1)\ (-2/3) \qquad (-1/3)\ (-1)\ (4/3)$$

where $\tilde{\underline{\Phi}}(x)$ denotes the isospinor

$$\tilde{\underline{\Phi}}(x) = i\tau_2\, \underline{\Phi}^*(x) = \begin{pmatrix} \phi^0(x) \\ -\phi^+(x) \end{pmatrix}^* . \tag{1.42}$$

We get again mass-like couplings from the v.e.v. piece of the neutral scalar field $\phi_1(x)$:

$$\mathcal{L}_{Masses}(x) = \bar{q}_L^{\uparrow}\, \mathcal{M}(2/3)\, q_R^{\uparrow} + \bar{q}_L^{\downarrow}\, \mathcal{M}(-1/3)\, q_R^{\downarrow} + h.c., \tag{1.43}$$

where we have introduced the notation

$$q_{L,R}^{\uparrow}(x) \equiv \frac{1}{2}(1\mp\gamma_5)\begin{pmatrix} u(x) \\ c(x) \\ t(x) \end{pmatrix} \quad ; \quad q_{L,R}^{\downarrow}(x) \equiv \frac{1}{2}(1\mp\gamma_5)\begin{pmatrix} d(x) \\ s(x) \\ b(x) \end{pmatrix} \quad ; \tag{1.44a,b}$$

$$\mathcal{M}(2/3) \equiv \tilde{g}_{ij}\, v/\sqrt{2} \quad \text{and} \quad \mathcal{M}(-1/3) \equiv g_{ij}\, v/\sqrt{2} \quad ; \tag{1.45a,b}$$

$$i = u,c,t \; ; \; j = u,c,t \qquad i = d,s,b \; ; \; j = d,s,b \, ,$$

i.e., $\mathcal{M}(2/3)$ and $\mathcal{M}(-1/3)$ are 3 x 3 matrices. These matrices can be diagonalized using the following procedure. We search for unitary matrices

$$\mathcal{U}_L(Q) \quad \text{and} \quad \mathcal{U}_R(Q) \quad ; \quad Q = 2/3\,,\ -1/3 \tag{1.46a,b}$$

such that

$$\mathcal{U}_L(Q)\, \mathcal{M}(Q)\, \mathcal{U}_R^{+}(Q) = \hat{M}(Q) \text{ diagonal}, \; Q = 2/3, -1/3 \, . \tag{1.47a,b}$$

They can be found as follows. Form the quantity

$$\left(\mathcal{U}_L(Q)\mathcal{M}(Q)\mathcal{U}_R^{+}(Q)\right)^{+}\left(\mathcal{U}_L(Q)\mathcal{M}(Q)\mathcal{U}_R^{+}(Q)\right) = \left(\hat{M}(Q)\right)^2 ,$$

i.e.,

$$\mathcal{U}_R(Q)\, \mathcal{M}^{+}(Q)\, \mathcal{M}(Q)\, \mathcal{U}_R^{+}(Q) = \left(\hat{M}(Q)\right)^2 , \tag{1.48a,b}$$

and also

$$\mathcal{U}_L(Q)\mathcal{M}(Q)\mathcal{U}_R^{+}(Q)\left(\mathcal{U}_L(Q)\mathcal{M}(Q)\mathcal{U}_R^{+}(Q)\right)^{+} = \left(\hat{M}(Q)\right)^2 ,$$

i.e.,

$$\mathcal{U}_L(Q)\, \mathcal{M}(Q)\, \mathcal{M}^{+}(Q)\, \mathcal{U}_L^{+}(Q) = \left(\hat{M}(Q)\right)^2 . \tag{1.49a,b}$$

Since $\mathcal{M}^\dagger(Q)\mathcal{M}(Q)$ and $\mathcal{M}(Q)\mathcal{M}^\dagger(Q)$ are hermitian matrices, Eqs.(1.39) and (1.40) define $\mathcal{U}_R(Q)$ and $\mathcal{U}_L(Q)$ up to a diagonal unitary matrix. Therefore

$$\mathcal{L}_{MASSES}(x) = \bar{q}_L^{\uparrow}\mathcal{U}_L^\dagger(2/3)\mathcal{U}_L(2/3)\mathcal{M}(2/3)\mathcal{U}_R^\dagger(2/3)\mathcal{U}_R(2/3)q_R^{\uparrow}$$

$$+\bar{q}_L^{\downarrow}\mathcal{U}_L^\dagger(-1/3)\mathcal{U}_L(-1/3)\mathcal{M}(-1/3)\mathcal{U}_R^\dagger(-1/3)\mathcal{U}_R(-1/3)q_R^{\downarrow} \tag{1.50}$$

$$= \bar{\psi}_L^{\uparrow}(x)\hat{M}(2/3)\psi_R^{\uparrow}(x) + \bar{\psi}_L^{\downarrow}(x)\hat{M}(-1/3)\psi_R^{\downarrow}(x) ,$$

where

$$\psi_{L,R}^{\uparrow}(x) = \mathcal{U}_{L,R}(2/3)\,q_{L,R}^{\uparrow}(x) \; ; \; \psi_{L,R}^{\downarrow}(x) = \mathcal{U}_{L,R}(-1/3)\,q_{L,R}^{\downarrow}(x) \tag{1.51a,b}$$

are the new rotated quark fields which correspond to eigenstates of mass. Again, like in the case of leptons, the mass eigenvalues are arbitrary because the Yukawa couplings g_{ij} and $\tilde{g}_{ij}$ are neither fixed nor constrained by the standard model. The procedure of quark mass diagonalization introduces, however, flavor mixing in the charged current sector: the interaction term

$$\bar{q}_L^{\uparrow}(x)\gamma^\mu q_L^{\downarrow}(x)W_\mu^+(x) + h.c. , \tag{1.52a}$$

when rotated to quark fields which are mass eigenstates, becomes

$$\bar{\psi}_L^{\uparrow}(x)\gamma^\mu\,\mathcal{U}_L(2/3)\mathcal{U}^\dagger(-1/3)\,\psi_L^{\downarrow}(x)\;W_\mu^+(x) + h.c. \tag{1.52b}$$

The matrix

$$V \equiv \mathcal{U}_L(2/3)\mathcal{U}_L^\dagger(-1/3) , \tag{1.53}$$

which, by definition, is a unitary matrix acting on the flavor degrees of freedom of charge -1/3 quarks - the $\downarrow$ type quarks - is the generalization of the Cabibbo rotation in the old Fermi theory. It corresponds exactly to the mechanism proposed by Glashow, Iliopoulos and Maiani [4] to ensure the absence of flavor changing neutral currents in the case of two generations.

In general, for n-generations, V is now called the Cabibbo-Kobayashi-Maskawa matrix. It was Kobayashi and Maskawa [5] who first pointed out that in the case of 3-generations the most general form for V allows the presence of a CP-violating phase.

1.2 Predictive Power of the Standard Model and its Successes

The standard model $SU(3)_{color} \times SU(2)_{left} \times U(1)_y$ is rich in phenomenological predictions which have been spectacularly confirmed by experiment. Here we shall briefly review the salient features.

i) In the standard model, the interaction between the gauge bosons and the currents induced by leptons and quarks is described by three terms

$$\mathcal{L}_{int}^{(leptons,\,quarks)}(x) = +e A_\mu(x)\left(L_\gamma^\mu(x) - J_\gamma^\mu(x)\right)$$

$$-g/\sqrt{2}\left[W_\mu^{(+)}(x)\left(L_W^{\mu+}(x) + J_W^{\mu+}(x)\right) + h.c.\right] - \frac{1}{2}\left(g^2+g'^2\right)^{1/2} Z_\mu(x)\left(L_Z^\mu(x) + J_Z^\mu(x)\right). \quad (1.54)$$

The first term is the usual electromagnetic interaction

$$L_\gamma^\mu(x) = \sum_{\ell=e,\mu,\tau} \bar{\psi}_\ell(x)\gamma^\mu\psi_\ell(x) \quad , \quad (1.55a)$$

$$J_\gamma^\mu(x) = 2/3 \sum_{\uparrow=u,c,t} \bar{\psi}_\uparrow(x)\gamma^\mu\psi_\uparrow(x) - 1/3 \sum_{\downarrow=d,s,b} \bar{\psi}_\downarrow(x)\gamma^\mu\psi_\downarrow(x) \,. \quad (1.55b)$$

The successful predictions of quantum electrodynamics (QED) are thus incorporated in the standard model.

The second term describes the weak interactions of charged currents

$$L_W^{\mu+}(x) = \sum_{\ell=e,\mu,\tau} \bar{\nu}_\ell(x)\gamma^\mu\left(\frac{1-\gamma_5}{2}\right)\ell^-(x) \quad , \quad (1.56a)$$

$$J_W^{\mu+}(x) = \sum_{\substack{\downarrow=d,s,b\\ \uparrow=u,c,t}} \bar{\psi}_\uparrow(x)\gamma^\mu\left(\frac{1-\gamma_5}{2}\right) V_{\uparrow\downarrow}\,\psi_\downarrow(x) \,. \quad (1.56b)$$

The effective Hamiltonian $O(g^2)$ induced by these currents reproduces the successful low energy phenomenology predicted by the Fermi-Cabibbo theory.

The third term, specific to the standard model, predicts the existence of <u>neutral currents</u> both in the leptonic sector and in the flavor conserving hadronic sector

$$L_Z^\mu(x) = \sum_{\ell=e,\mu,\tau} \bar{\nu}_\ell(x)\gamma^\mu\left(\frac{1-\gamma_5}{2}\right)\nu_\ell(x) - \bar{\ell}(x)\gamma^\mu\left(\frac{1-\gamma_5}{2}\right)\ell(x) + 2\sin^2\theta_W\,\bar{\ell}(x)\gamma^\mu\ell(x),$$

$$J_Z^\mu(x) = \sum_{\uparrow=u,c,t} \bar{\psi}_\uparrow(x)\gamma^\mu\left(\frac{1-\gamma_5}{2}\right)\psi_\uparrow(x) - \sum_{\downarrow=d,s,b} \bar{\psi}_\downarrow(x)\gamma^\mu\left(\frac{1-\gamma_5}{2}\right)\psi_\downarrow(x) - 2\sin^2\theta_W\, J_\gamma^\mu(x). \quad (1.57a,b)$$

Both predictions are now quantitatively confirmed by a series of experiments (for recent reviews see Refs. [R1,2,3]).

ii) The pattern of two left-handed quark doublets

$$\frac{1}{2}(1-\gamma_5)\begin{pmatrix} u \\ d\cos\theta + s\sin\theta \end{pmatrix} \; ; \; \frac{1}{2}(1-\gamma_5)\begin{pmatrix} c \\ -d\sin\theta + s\cos\theta \end{pmatrix}, \quad (1.58a,b)$$

proposed by Glashow, Iliopoulos and Maiani, postulated the existence of a new charge 2/3 heavy quark c, called charm. A few years later new narrow states were observed [6] which were subsequently interpreted as bound states of charm and anticharm states. Charm particles have now an entry in the particle data group tables - PDG-tables [R.4] -, much in the same footing as strange particles.

iii) The pattern of three color left-handed quark doublets and one left-handed lepton doublet for each generation has the property [7] that it eliminates the anomaly in the Ward identities associated with graphs like

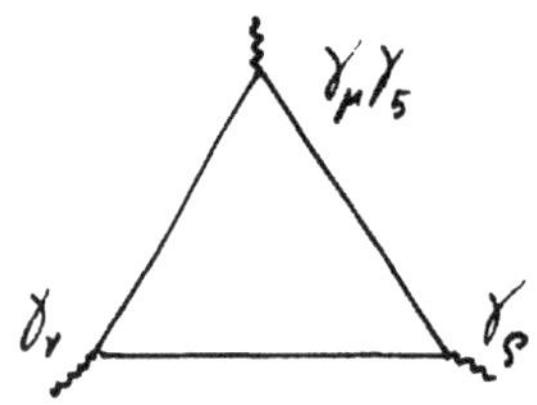

Fig. I.2

In the standard model, the condition for cancellation can be simply expressed as

$$\mathrm{Tr}\, Q = 0 \tag{1.59}$$

and it occurs separately for each generation:

$$\underset{\text{(leptons)}}{-1 + 0} \quad + \quad \underset{\text{color}}{(3)} \quad \times \quad \underset{\text{(quarks)}}{\left(\tfrac{2}{3} + -\tfrac{1}{3}\right)} = 0$$

It implies in particular that once a new lepton is observed, there must be two new flavors of quarks. The discovery of the tau lepton [8] was a hint that two new quark flavors should exist. The charge -1/3 bottom quark has been discovered in e^+e^- collisions [9]; and very recently, evidence for the existence of a corresponding charge 2/3 top quark has been presented by the UA2-collaboration working at the CERN-SPS collider [10].

iv) A crucial test that gauge theories are indeed the adequate framework to describe the weak interactions is the existence of the vector bosons $W^\pm$ and Z. In the standard model, and to lowest order in perturbation theory (no radiative corrections)

$$M_W = \left(\frac{\pi\alpha}{\sqrt{2}\, G_F}\right)^{\frac{1}{2}} \frac{1}{\sin\theta_W} \simeq \frac{37.3\ \mathrm{GeV}}{\sin\theta_W} \quad , \tag{1.60}$$

and

$$M_Z = \frac{1}{\cos\theta_W} M_W \quad . \tag{1.61}$$

In fact, the last relation holds even in the presence of several Higgs doublets and singlets. In general, for arbitrary scalar fields with weak isopsin $T_W^{(i)}$, third component $T_3^{(i)}$ and vacuum expectation value of the neutral member of the ith component v_i, one has [11]:

$$\rho \equiv \frac{M_W^2}{M_Z^2 \cos^2\vartheta_W} = \frac{\sum_i \left[T^{(i)}(T^{(i)}+1) - (T_3^{(i)})^2 \right] v_i^2}{2 \sum_i (T_3^{(i)})^2 v_i^2} . \qquad (1.62)$$

The parameters ρ and $\sin^2\vartheta_W$ have been determined from a wealth of neutral current experiments (for recent reviews see Refs. [R.1,2,3]) like: ν and $\bar{\nu}$ deep inelastic scattering from isoscalar targets and by protons; elastic $\nu_\mu p$ and $\bar{\nu}_\mu p$ scattering; exclusive and inclusive π-production in neutral current events; the asymmetry in electron-deuteron deep inelastic scattering; elastic νe scattering; and the forward-backward asymmetry in $e^+e^- \to \mu^+\mu^-$. At present, the best determination of ρ and $\sin^2\vartheta_W$ comes from data on ν deep inelastic scattering and the e - D experiment, with the results

$$\rho = 0.992 \pm 0.017 \pm (0.011) \quad , \quad \sin^2\vartheta_W = 0.224 \pm 0.015 \pm (0.012) , \qquad (1.63a,b)$$

where the errors in parenthesis are an estimate of theoretical uncertainties. If ρ is restricted to one,

$$\sin^2\vartheta_W = 0.229 \pm 0.009 \pm (0.005) . \qquad (1.64)$$

Inserting $\sin^2\vartheta_W \simeq 0.22$ in the lowest order expressions, one gets the theoretical estimates

$$M_W \simeq 80 \text{ GeV} \quad \text{and} \quad M_Z \simeq 90 \text{ GeV} . \qquad (1.65a,b)$$

The existence of the $W^\pm$ and the Z^0 has been confirmed by the UA1 and UA2 collaborations working at the CERN-SPS collider [12]. The mass values quoted by these groups (these are December 1983 values) are in GeV:

UA1 : $M_W = 80.9 \pm 1.5$ (stat.) ; $M_Z = 95.6 \pm 1.4$ (stat.), (1.66)

UA2 : $M_W = 81.0 \pm 2.5 \pm 1.3$; $M_Z = 91.9 \pm 1.3 \pm 1.4$. (1.67)

The comparison with the standard model predictions requires the inclusion of radiative correction effects which are rather large. These have been discussed by a number of authors (see, e.g. ref. [13] where a complete set of references can also be found). The possibility to perform radiative corrections to the desired order of accuracy is, of course, a byproduct of the renormalizability property of the standard model which has been proven in a series of brilliant papers initiated by 't Hooft papers [14]. (For a historical account where other relevant references can be found, see M. Veltman's talk at the Bonn conference [15]; and also refs. [R.5,6,7]).

v) The dynamics governed by the unbroken SU(3) color subgroup of the standard model is called quantum chromodynamics (QCD). The most spectacular successes of QCD apply to strong interaction phenomena governed by short-distance behavior where the asymptotic freedom property of QCD [16] allows a perturbative treatment in terms of a running coupling constant

$$\alpha_s(Q^2) \approx \left(\log Q^2/\Lambda^2\right)^{-1} \tag{1.68}$$

where Q^2 is some characteristic momentum transfer of the process conjugate to the short-distance variable, and Λ is an intrinsic QCD mass scale which is renormalization invariant but changes slightly from one renormalization scheme to another 80 MeV $\lesssim \Lambda <$ 300 MeV (see the review articles [R.12,13]).

The asymptotic freedom behaviour of QCD explains the phenomenological successes of the Feynman parton model to describe deep inelastic scattering processes and predicts quantitatively scaling violations to the parton picture (for recent reviews see [R.9,10,11,12,13]). QCD also provides a semiquantitative framework to the phenomenology of jet production in hadronic collisions and large p_T phenomena.

Another important feature of QCD is that it gives an explicit realization of the current algebra framework developed in the 60's. The currents of the algebra of currents are the Noether currents associated to global gauge transformations acting on the flavor degrees of freedom of the quarks. The successes of current algebra become thus predictions of QCD [R.13]. At the same time, the dynamical input provided by QCD leads to a large variety of so-called QCD sum rules with interesting new predictions and constraints on hadronic spectroscopy ([17], for recent reviews where references can be found see, e.g. refs. [18], [19]).

From another point of view the formulation of QCD on a lattice and the subsequent Monte Carlo simulations provide a promising framework to compute a wealth of hadronic phenomena governed by long distance behavior when the quark confinement dynamics eludes a perturbative treatment (see, e.g. ref. [R.15] for an excellent introductory textbook).

1.3 Untested Aspects of the Standard Model

a. A characteristic feature of the non-abelian gauge structure of the standard model is the appearance of three vector boson vertices like

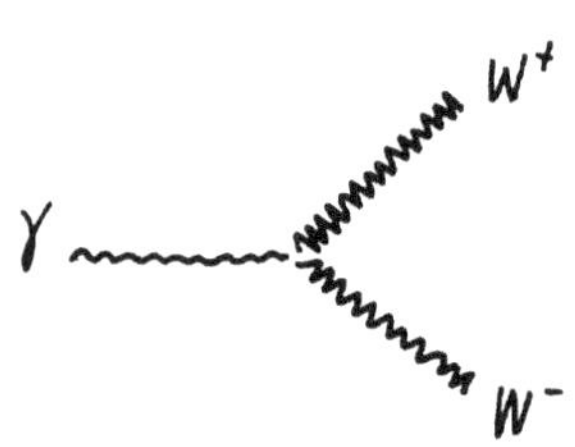

Fig. I.3

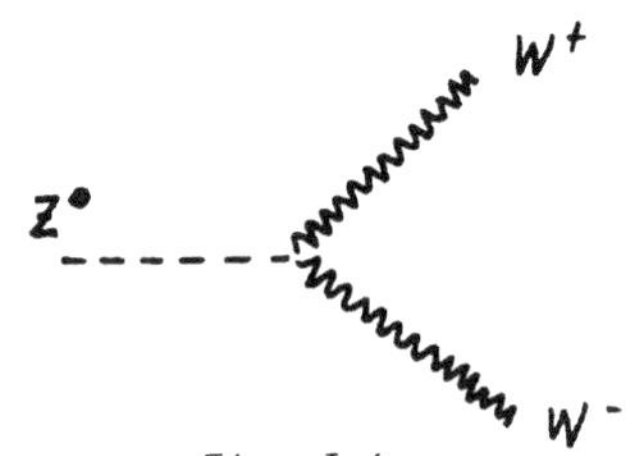

Fig. I.4

The first diagram in particular corresponds to an electromagnetic coupling of the charged W with anomalous magnetic moment $K_W = 1$. This coupling makes finite by itself the weak contribution to the anomalous magnetic moment of the muon coming from the graph

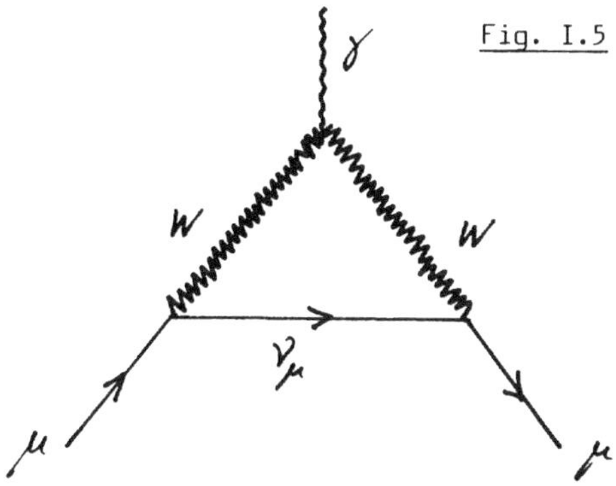

Fig. I.5

with the result [20], in the limit of small lepton mass,

$$a_\mu^W = \frac{G_F m_\mu^2}{8\pi^2\sqrt{2}} \; \frac{10}{3} \quad . \tag{1.69}$$

The other contributions at the one-loop level come from the diagrams

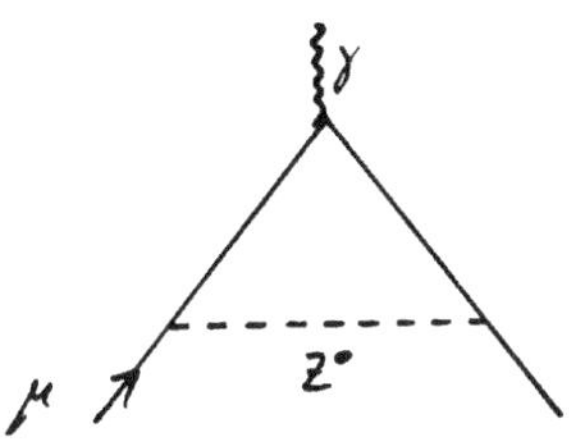

and

Fig. I.6

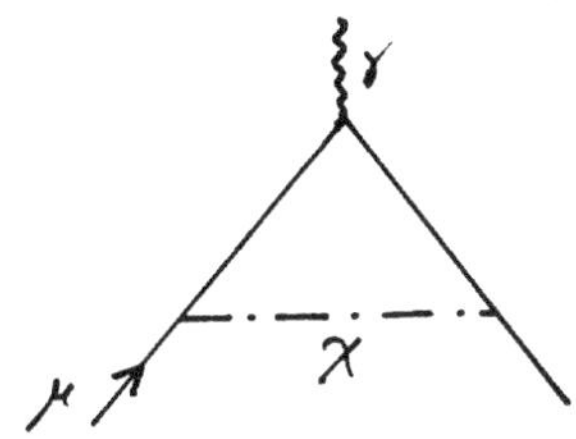

Fig. I.7

with the results [20] (for a review of other contributions to g-2 see ref. [R.15])

$$a_\mu^Z = \frac{G_F m_\mu^2}{8\pi^2\sqrt{2}} \; \frac{4}{3}\left(1-2\frac{M_W^2}{M_Z^2}\right) - \frac{\alpha}{\pi}\;\frac{1}{3}\;\frac{m_\mu^2}{M_Z^2} \quad , \quad \text{and} \tag{1.70}$$

$$a_\mu^\chi = \frac{G_F m_\mu^2}{8\pi^2\sqrt{2}} \int_0^1 dx \; \frac{2x^2(1-x)}{x^2+(1-x)\frac{M_\chi^2}{m_\mu^2}} \quad . \tag{1.71}$$

With the recent improvement on the calculation of other contributions to the muon anomaly [21], it may become possible to use a new generation of g-2 muon experiments as a test of the electromagnetic coupling of W's predicted by gauge theories.

The three vector boson vertex plays of course an important role in the W-pair production from e^+e^- annihilations:

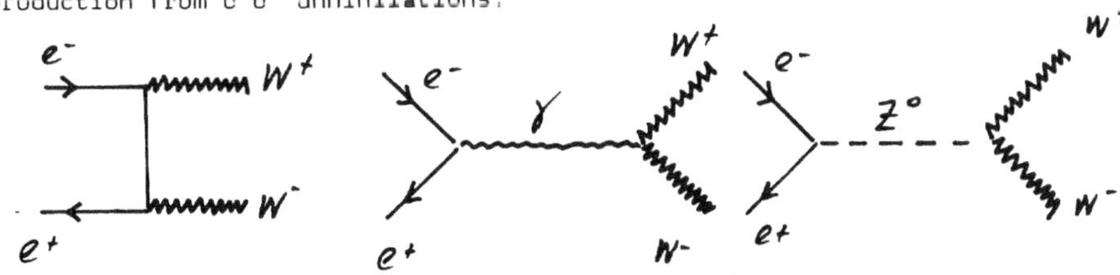

Fig. I.8

which hopefully will be seen at LEP in a few years' time.

b. The Higgs mechanism plays a crucial role in giving masses to the by now observed vector bosons W and Z and predicts the existence of a massive scalar particle (in the minimal one doublet version). Lower limits on the Higgs mass have been obtained from the stability requirement of the vacuum in the presence of radiative corrections. (For a very lucid description see the lectures of L. Maiani at Gif 79 School, ref. [22]). The bound obtained from the explicit calculation of the one-loop corrections to the effective potential, in the presence of one heavy fermion, with mass M_t < 80 GeV, leads to the result

$$M_\chi^2 \geqslant \frac{G_F}{8\pi^2\sqrt{2}} \left[3\left(2M_W^4 + M_Z^4\right) - 12\, M_t^4\right] \quad . \tag{1.72}$$

For M_t = 40 GeV, one gets

$$M_\chi \gtrsim 6.6 \quad \text{GeV} \quad . \tag{1.73}$$

Another crucial role of the Higgs coupling to gauge vector bosons is to assure the unitarity constraint of physical amplitudes at very high energies. As an example, let us consider again the $e^+e^- \rightarrow W^+W^-$ process illustrated by the diagram in Fig. I.8. As the total C.M. energy squared s grows, one obtains from the contribution of the electron axial current

$$\left.\frac{d\sigma}{dt}\right|_{|t|\sim s} \sim \frac{m_e^2 s}{M_Z^4} \frac{1}{s^2} \quad , \tag{1.74}$$

while unitarity requires

$$\left.\frac{d\sigma}{dt}\right|_{|t|\sim s} \sim \frac{1}{s^2} \quad . \tag{1.75}$$

The bad term proportional to m_e^2 is precisely cancelled by the Higgs annihilation diagram

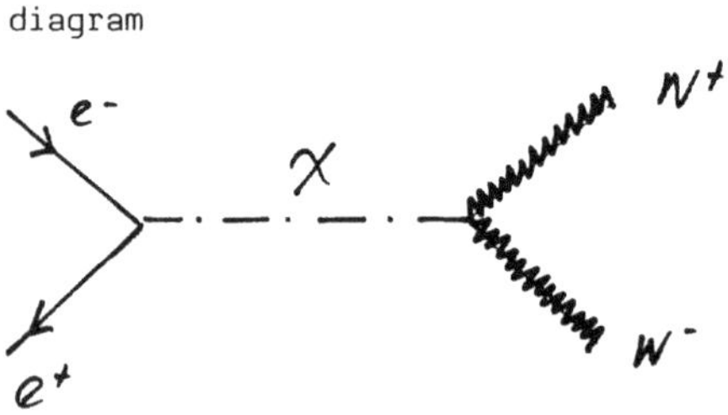

Fig. I.9

If, however, the Higgs particle has a big mass $M_H \gg M_{W,Z}$, the effective interactions among W's, Z's and χ will no longer be weak; and indeed for $M_\chi \gtrsim$ 1 TeV they become strong in the sense that the effective coupling constant - like the quartic coupling λ - in the Higgs potential is O(1). It is therefore expected that M_χ < 1 TeV; or, what is a tantamount statement, a breakdown of the standard model at energies of the order of or larger than the TeV region.

It is obvious that search for the hitherto unobserved Higgs particle(s) is an issue of fundamental priority in particle physics.

1.4 Limitations of the Standard Model

The standard model gives no answer to a series of fundamental questions:

i) Why the generations?

The old question "why the muon?" is now replaced by "why the generations?"

ii) The $SU(3)_c \times SU(2)_L \times U(1)_y$ theory incorporates successfully the observed phenomenology associated to parity violation, but it does not tell us much about the origin of parity violation.

iii) The same is true for the violation of CP-invariance. Furthermore, in this case, the standard mechanism which attributes to the phase in the Cabibbo-Kobayashi-Maskawa matrix V the source of CP-violation has not yet been fully tested; and, indeed, various experiments in progress may well reveal interesting surprises.

iv) Perhaps the ugliest feature of the standard model is the proliferation of uncalculable fundamental parameters: the $SU(3)_{color}$ coupling constant g is trade by the QCD-mass scale Λ which has to be determined from experiment. The $SU(2)_L$ coupling constant g, the weak hypercharge coupling constant g', and the vacuum expectation value v are trade by fixing three of the physical parameters α_{QED}; G_F; $\theta_{Weinberg}$; M_z and M_w from experiment. If we limit ourselves to three generations (why three ...?) we are still left with 6 arbitrary mass values for the leptons and 6 arbitrary mass values for the quarks. The lepton masses are directly accessible to experimental measurement, but the quark masses (especially the light quark masses) have to be extracted from hadronic spectroscopy which is not a very straightforward thing [R-14]. Furthermore, for three generations, there appear three Cabibbo like mixing angles and a CP-violating phase δ which again have to be determined phenomenologically.

v) Another important issue in the standard model is the question of the so-called strong CP-problem which originates with the presence of a term

$$\frac{g^2}{64\pi^2}\,\theta\,\varepsilon^{\mu\nu\rho\sigma}\sum_A G^{(A)}_{\mu\nu}(x)\,G^{(A)}_{\rho\sigma}(x)$$

in the QCD-Lagrangian and its interplay with the global U(1)-invariance. Several options with interesting phenomenological implications are still open. (For a recent review see the article of G. Veneziano in ref. [23].)

Our choice of selected topics in the standard model to be discussed in the following sections is very much inspired by the limitations mentioned above. The phenomenological determination of the Cabibbo-Kobayashi-Maskawa matrix elements which we shall discuss in the next section is an extremely important issue. Failure to reach consistency from different sources of phenomenological determination of these matrix elements would be most welcomed to pin down new directions to explore beyond the standard model.

The next topic we have chosen is the $\Delta I = 1/2$ rule for strangeness changing non-leptonic processes. This is because our present understanding of the dynamical origin of this selection rule is based on a very subtle interplay between QCD and weak vector boson exchanges. The same mechanism leads to a milliweak CP-violating transition which could be tested by the high precision experiments underway. This is why we devote the last topic to the question of violation of CP-invariance in the K^o-$\bar{K}^o$ system.

II PHENOMENOLOGICAL DETERMINATION OF THE CABIBBO-KOBAYASHI-MASKAWA MATRIX ELEMENTS

2.1 General Properties of the CKM Matrix

In the first section we have seen that in the standard model $SU(3)_c \times SU(2)_L \times U(1)_y$ the quark fields are in left-handed doublets

$$\begin{pmatrix} u \\ d' \end{pmatrix}_L , \quad \begin{pmatrix} c \\ s' \end{pmatrix}_L , \quad \begin{pmatrix} t \\ b' \end{pmatrix}_L , \quad \ldots \tag{2.1}$$

and right-handed singlets. The "primed" fields are not mass eigenstates, but they are related to the mass eigenstates by the matrix V

$$\underbrace{\mathcal{U}_L(2/3)\,\mathcal{U}_L^\dagger(-1/3)}_{\longleftarrow\ V\ \longrightarrow}\ \psi_L^{\downarrow} \tag{2.2}$$

acting on a column vector of left-handed fields of the down type (charge -1/3):

$$\begin{pmatrix} \times & \times & \times & \\ \times & \times & \times & \cdots \\ \times & \times & \times & \\ & \vdots & & \end{pmatrix} \begin{pmatrix} d_L(x) \\ s_L(x) \\ b_L(x) \\ \vdots \end{pmatrix} . \tag{2.3}$$

The nxn matrix V, which by construction is a unitary matrix

$$V V^\dagger = V^\dagger V = \mathbb{1} , \tag{2.4}$$

is the so-called Cabibbo-Kobayashi-Maskawa matrix. The unitary condition is equivalent to the set of equations

$$\sum_{k=1}^{n} \rho_{ik}\rho_{kj}\, e^{i(\vartheta_{ik}-\vartheta_{jk})} = \delta_{ij} , \quad i \leq j = 1,2,\ldots n$$

among the moduli ρ_{ij} and phases ϑ_{ij} of the matrix elements $V_{ij} = \rho_{ij} e^{i\vartheta_{ij}}$. This set of equations gives

$$1 + 2 + \dots n-1 = \frac{1}{2} n\,(n-1)$$

independent relations among the ϑ_{ij} phases, and

$$\frac{1}{2} n\,(n-1) + n = \frac{1}{2} n\,(n+1)$$

independent relations among the ρ_{ij} moduli. Therefore the unitarity condition reduces the number nxn of arbitrary complex numbers in V to

$$n^2 - \frac{1}{2} n\,(n-1) = \frac{1}{2} n\,(n-1) \quad \text{moduli} \quad \text{and}$$

$$n^2 - \frac{1}{2} n\,(n-1) = \frac{1}{2} n\,(n+1) \text{ phases .}$$

However, of these phases there are 2n-1 which are of no physical relevance since, without lost of generality, they can be rotated away by appropriately fixing the 2n-1 relative phases of the 2n quark fields. The general nxn CKM matrix has therefore

$$1/2\; n\,(n-1) \quad \text{moduli} \quad \text{and} \tag{2.5a}$$

$$1/2\; n\,(n+1) - (2n-1) = 1/2(n-1)(n-2) \text{ phases .} \tag{2.5b}$$

The number of moduli corresponds to the number n_ϑ of orthogonal rotations in an n-dimensional space. The rest of the physical parameters n_δ are observable phases:

For n = 2 $\quad n_\vartheta = 1$ the Cabibbo rotation (2.6a)

For n = 3 $\quad n_\vartheta = 3$ the 3-Euler angles of rotations in a 3-dimensional space

and $n_\delta = 1$ one phase (2.6b)

For n = 4 $\quad n_\vartheta = 6; \quad n_\delta = 3$. (2.6c)

2.2 Parametrization of the n=3 CKM Matrix

The conventional choice of rotation angles in the 3-generation parametrization of the CKM matrix is as follows: we first perform a rotation of angle ϑ_3 around the d-like axis; then a rotation of angle ϑ_1 around the new b-like axis, and finally a rotation of angle ϑ_2 around the new d-like axis. With the notation

$$\cos\vartheta_i \equiv c_i \quad ; \quad \sin\vartheta_i \equiv s_i \quad , \quad i = 1,2,3 \quad , \tag{2.7}$$

the product of the 3-rotations corresponds to the product of matrices

$$\begin{pmatrix} 1 & & \\ & c_2 & s_2 \\ & -s_2 & c_2 \end{pmatrix} \begin{pmatrix} c_1 & s_1 & \\ -s_1 & c_1 & \\ & & 1 \end{pmatrix} \begin{pmatrix} 1 & & \\ & c_3 & s_3 \\ & -s_3 & c_3 \end{pmatrix} .$$

The introduction of the observable phase δ is conventionally done by inserting the matrix

$$\begin{pmatrix} 1 & & \\ & 1 & \\ & & \pm e^{i\delta} \end{pmatrix}$$

between the 1st and 2nd Euler rotations. Most people now have adopted this convention with the $+e^{i\delta}$ sign. Several authors, however, choose $-e^{i\delta}$ in the last Euler rotation. In these lectures we shall adopt the convention corresponding to the product

$$\begin{pmatrix} 1 & & \\ & c_2 & s_2 \\ & -s_2 & c_2 \end{pmatrix}\begin{pmatrix} c_1 & s_1 & \\ -s_1 & c_1 & \\ & & 1 \end{pmatrix}\begin{pmatrix} 1 & & \\ & 1 & \\ & & e^{i\delta} \end{pmatrix}\begin{pmatrix} 1 & & \\ & c_3 & s_3 \\ & -s_3 & c_3 \end{pmatrix}$$

$$= \begin{pmatrix} c_1 & s_1 c_3 & s_1 s_3 \\ -s_1 c_2 & c_1 c_2 c_3 - s_2 s_3 e^{i\delta} & c_1 c_2 s_3 + s_2 c_3 e^{i\delta} \\ s_1 s_2 & -c_1 s_2 c_3 - c_2 s_3 e^{i\delta} & -c_1 s_2 s_3 + c_2 c_3 e^{i\delta} \end{pmatrix}, \tag{2.8}$$

the same as, e.g. in Okun's book quoted in the references. Then, in the limit where $\vartheta_2 = \vartheta_3 = 0$, ϑ_1 becomes the usual Cabibbo angle with the correct sign. The relative quark phases are adjusted so that the 3-angles ϑ_1, ϑ_2 and ϑ_3 are taken to lie in the first quadrant. Then the phase δ is free to lie within

$$-\pi \leq \delta \leq \pi \quad . \tag{2.9}$$

2.3 The Matrix Element V_{ud}

Traditionally, this matrix element has been obtained from a comparison of the $0^+ \to 0^+$ nuclear β-decays of ^{14}O and ^{26}Al with the Fermi rate for $\mu \to e\nu\bar{\nu}$ decay.

The most recent analyses of the determination of $|V_{ud}|$ [1] based on Wilkinson's reanalyses of superallowed Fermi transitions [2] gives the value

$$|V_{ud}| = 0.9735 \pm 0.0015 \quad . \tag{2.10}$$

This determination takes into account the radiative corrections calculated by Sirlin [3]. It assumes, of course, the absence of leptonic mixing. We then have

$$C_1 = 0.9735 \pm 0.0015 \quad , \tag{2.11}$$

and therefore

$$S_1 = 0.2287 \begin{smallmatrix} +\ 0.0063 \\ -\ 0.0065 \end{smallmatrix} \quad . \tag{2.12}$$

These values are those entered in Table I at the end of section II.

2.4 The Matrix Element V_{us}

The determination of $|V_{us}|$ comes from the analysis of strange particle decay data: K_{l3} and semileptonic hyperon decays. There is a recent analysis by Leutwyler and Roos [4] which gives

$$|V_{us}| = 0.221 \pm 0.002 \quad , \tag{2.13}$$

slightly higher than the previous determination ($|V_{us}| = 0.219 \pm 0.002$, see, e.g. ref. [5]). This value comes from a simultaneous fit to the K_{e3} data [PDG-(R.4)]:

$$\Gamma_{K^+_{e3}} = 2.565\ (1 \pm 0.011) \times 10^{-15}\ \text{MeV}, \tag{2.14a}$$

$$\Gamma_{K^0_{e3}} = 4.915\ (1 \pm 0.015) \times 10^{-15}\ \text{MeV}, \tag{2.14b}$$

and the recent results on $\Lambda \to pe\nu$, $\Sigma^- \to ne\nu$ and $\Xi^- \to \Lambda e\nu$ of the CERN WA2 experiment [1]. The value of $|V_{us}|$ which follows from the K_{e3} data only, with the recent theoretical analysis of Leutwyler and Roos [4] is

$$|V_{us}|_{K_{e3}} = 0.220 \pm 0.002 \quad ; \tag{2.15a}$$

whilst from the data of the WA2 collaboration, they obtain

$$|V_{us}|_{Y_{e3}} = 0.231 \pm 0.005 \quad . \tag{2.15b}$$

The traditional difference from the determination of $|V_{us}|$ from K_{l3} decays and Y_{l3} decays still persists but the simultaneous fit of the two sets of data shows that it is not statistically significant.

With

$$S_1C_3 = 0.221 \pm 0.002 \tag{2.16}$$

and the input value for S_1 from $|V_{ud}|$, we get

$$C_3 = 0.9663 \begin{array}{l} + 0.0337 \\ - 0.0342 \end{array} \quad \text{and} \tag{2.17a}$$

$$S_3 = 0.2571 \begin{array}{l} + 0.1054 \\ - 0.2571 \end{array} \quad . \tag{2.17b}$$

The value of S_1C_3 obtained in eq. (2.16) is the one we insert in Table I at the end of this section.

2.5 Unitarity Limit on the Matrix Element $|V_{ub}|$

Information on $|V_{ub}|$ can be obtained from the recent measurements of the lifetime of the B-meson and the experimental observation that b-quarks decay preferently into c-quarks. However, before we enter into the detailed discussion of this topic, it is interesting to point out that the unitarity constraint,

$$\sum_{i=d,s,b,\ldots} |V_{ui}|^2 = 1 \quad , \tag{2.18}$$

already restricts $|V_{ub}|$ to

$$|V_{ub}|_{UNITARITY} \lesssim 0.0852 \quad . \tag{2.19}$$

In the case of 3-generations, the same unitarity constraint gives

$$|V_{ub}| = 0.0592 \begin{array}{l} +\ 0.0260 \\ -\ 0.0592 \end{array} \quad . \tag{2.20}$$

2.6 The Matrix Elements V_{cd} and V_{cs}

The information we have on these matrix elements comes from opposite sign dimuon production in $\bar{\nu}$-reactions; and from the difference of dimuon production in ν and $\bar{\nu}$ reactions. The phenomenological analyses made by Kleinknecht and Renk [5] using the CHDS data [6] from the CERN experiment gives the results

$$|V_{cd}| = 0.24 \pm 0.03 \qquad \text{and} \tag{2.21}$$

$$|V_{cs}| \gtrsim 0.77 \quad . \tag{2.22}$$

It is important to keep in mind that the extraction of these numbers from the data is not a straightforward thing. Let us review the main steps in the derivation:

i) Opposite sign dimuons in $\bar{\nu}$-reactions are expected to be produced by the fundamental reactions

$$\bar{\nu}_\mu + \bar{d} \to \mu^+ \bar{c} \to \mu^+\mu^- X \qquad \sim |V_{cd}| \tag{2.23a}$$

and

$$\bar{\nu}_\mu + \bar{s} \to \mu^+ \bar{c} \to \mu^+\mu^- X \qquad \sim |V_{cs}| , \tag{2.23b}$$

i.e. anticharm production from the sea quarks in the nucleon target with subsequent semileptonic decay of the $\bar{c}$-quark. The first reaction proceeds with a rate $|V_{cd}|^2$, the second one with a rate $|V_{cs}|^2$.

ii) Assuming SU(3)-symmetry for the quark structure functions in the nucleons,

$$\xi(x) = \bar{u}(x) = \bar{d}(x) = s(x) \quad , \tag{2.24}$$

and Bjorken scaling, the physical $\bar{\nu}$ cross section on an isoscalar target will then be

$$\sigma(\bar{\nu}+N \to \mu^+\bar{c}\Gamma)_{ISOSCALAR} = \left(|V_{cs}|^2 + |V_{cd}|^2\right) \sigma(\text{rea}) \tag{2.25}$$

with

$$\sigma(\text{rea}) = \frac{2}{\pi} G_F^2 M_N E \int_0^{1-w} dx \int_{w/1-x}^{1} dy \left(x' - \frac{m_c^2}{2M_N E}\right) \xi(x') , \tag{2.26a}$$

$$x' = x + m_c^2/2M_N E_y \qquad \text{and} \qquad \omega = (M_D^2 + 2M_D M_N)/2M_N E \,. \tag{2.26b}$$

iii) The expected mechanism for $\mu^+\mu^-$ production implies that

$$\sigma(\bar{\nu}N \to \mu^+\mu^- X) = \sigma(\bar{\nu}N \to \mu^+\bar{c})\, BR(\bar{c} \to \mu^- X) \,. \tag{2.27}$$

Then, $|V_{cs}|^2 + |V_{cd}|^2$ can be extracted from the CDHS measurements of $\sigma(\bar{\nu}N \to \mu^+\mu^- X)$; the branching ratio

$$BR(\bar{c} \to \mu^- X) \simeq (9 \pm 1)\% \tag{2.28}$$

and the value of $\sigma(\text{sea})$ is obtained from eq.(2.26) with the quark structure functions obtained from deep inelastic scattering experiments. This way, Kleinknecht and Renk [5] obtain

$$|V_{cs}|^2 + |V_{cd}|^2 = 0.49 \sim 1.04 \,. \tag{2.29}$$

iv) On the other hand, opposite sign dimuon production in ν-reactions are expected to be produced by the reactions

$$\nu_\mu + d \to \mu^- c \to \mu^-\mu^+ X \qquad \sim |V_{cd}| \tag{2.30a}$$

and

$$\nu_\mu + s \to \mu^- c \to \mu^-\mu^+ X \qquad \sim |V_{cs}| \,. \tag{2.30b}$$

The first reaction can now be produced on a valence d-quark and on a sea quark. Therefore

$$\sigma(\nu_\mu N \to \mu^- c X) = |V_{cd}|^2 \sigma(\text{valence}) + (|V_{cd}|^2 + |V_{cs}|^2)\,\sigma(\text{sea}) \,, \tag{2.31}$$

with

$$\sigma(\text{valence}) = \frac{2}{\pi} G_F^2 M_N E \int dx \int dy \left(x' - \frac{m_c^2}{2M_N E}\right) V(x') \tag{2.32a}$$

and $V(x')$ the valence quark distribution

$$V(x) = u(x) + d(x) - 2\,\xi(x) \,. \tag{2.32b}$$

v) The difference of $\mu^+\mu^-$ production in ν and $\bar{\nu}$ reactions can then be parametrized as follows :

$$|V_{cd}|^2 = \frac{1}{\sigma(\text{valence})} \left[\sigma(\nu_\mu N \to \mu^-\mu^+ X) - \sigma(\bar{\nu}_\mu N \to \mu^+\mu^- X)\right] \frac{1}{BR(c \to \mu^+ X)} \,. \tag{2.33}$$

This is the procedure which leads to the results in eqs. (2.21,22) with errors which include the effect of using different parametrizations for the structure functions. The result

$$|S_1C_2| = 0.24 \pm 0.03 \tag{2.34}$$

combined with the value for C_1 given in eq. (2.11) leads to the constraints

$$C_2 \geqslant 0.89 \quad \text{and} \quad S_2 \leqslant 0.45 \ . \tag{2.35a,b}$$

It would be nice to have an independent determination of V_{cu} and V_{cs}. At long range, the most promising way should be via the analyses of D_{13} decays. From the theoretical point of view the difficulty here is the appearance of unknown form factors like the f^+ and f^- K_{13} formfactors. The behaviour of these form factors, however, can be rather constrained from general analyticity arguments and QCD along the lines discussed in ref. [7].

2.7 The Matrix Elements V_{cb} and V_{ub}

There is direct information on these matrix elements from the lifetime measurements of the B-particle reported by two experimental groups working at PEP:

MARK II [8],
$$\tau_B = \left(1.20 \begin{smallmatrix} +0.45 \\ -0.36 \end{smallmatrix} \pm 0.30\right) \times 10^{-12} \text{ sec.} \tag{2.36}$$

and

MAC [9],
$$\tau_B = (1.8 \pm 0.6 \pm 0.4) \times 10^{-12} \text{ sec.} \ ; \tag{2.37}$$

and from the CLEO [10] and CUSB [11] detectors which quote as 90 % c.l. upper limits

$$\frac{\Gamma(b \to u)}{\Gamma(b \to c)} < 5.0\ \% \qquad \text{CLEO} \tag{2.38}$$

and

$$\frac{\Gamma(b \to u)}{\Gamma(b \to c)} < 5.5\ \% \qquad \text{CUSB} \ . \tag{2.39}$$

The translation of these numbers into values for the C K M angles requires a phenomenological analysis of B-decays which we shall next review. For this purpose we shall take as experimental input the value

$$\tau_B = (1.4 + 0.7\,\tau) \times 10^{-12} \text{ sec}; \ -1 \leqslant \tau \leqslant 1. \tag{2.40}$$

At this stage, and for a comparison of orders of magnitude it is perhaps worthwhile to quote as well the experimental values of the lifetimes of D-particles (charm-up and charm-down bound states):

$$\tau_{D^\pm} = \left(9.2 \begin{smallmatrix} +1.7 \\ -1.2 \end{smallmatrix}\right) \times 10^{-13} \text{ sec.} \tag{2.41}$$

$$\tau_{D^0} = \left(4.4 \begin{smallmatrix} +0.8 \\ -0.6 \end{smallmatrix}\right) \times 10^{-13} \text{ sec.}; \tag{2.42}$$

of the F-particle (charm-strange bound state):

$$\tau_{F^\pm} = \left(1.9 \begin{smallmatrix} +1.3 \\ -0.7 \end{smallmatrix}\right) \times 10^{-13} \text{ sec.}; \tag{2.43}$$

and the well-known K-particles:

$$\tau_L = (5.183 \pm 0.040) \times 10^{-8} \text{ sec.} \qquad (2.44)$$

$$\tau_S = (0.8923 \pm 0.0022) \times 10^{-10} \text{ sec.} \qquad (2.45)$$

$$\tau_{K^\pm} = (1.2371 \pm 0.0026) \times 10^{-8} \text{ sec.} \qquad (2.46)$$

Remember also the mass values:

$$M_{B^\pm} = (5270.8 \pm 3.0) \text{ MeV}; \quad M_{B^o} = (5274.2 \pm 2.8) \text{ MeV} \qquad (2.47a,b)$$

$$M_{D^\pm} = (1869.4 \pm 0.6) \text{ MeV}; \quad M_{D^o} = (1864.7 \pm 0.6) \text{ MeV}$$

$$M_{F^\pm} = (1971 \pm 6) \text{ MeV} \qquad (2.48a,b,c)$$

$$M_{K^+} = (493.667 \pm 0.015) \text{ MeV}; \quad M_{K^o} = (497.67 \pm 0.13) \text{ MeV} . \qquad (2.49a,b)$$

The values quoted are the present PDG numbers [R.4].

The mere fact that B-particles have a longer lifetime than D-particles is already an indication that B-decays must be substantially supressed as compared to D-decays. Indeed, one expects the lifetime of a weak decaying particle to be characterized by a factor (effective four-fermion interaction)

$$\tau^{-1} \simeq \frac{G_F^2 M^5}{192\pi^3} \qquad (2.50)$$

times CKM like-factors, with M some effective mass which, for instance, in the quark spectator model would be the b-quark mass for B-decays or the C-quark mass for D-decays. In any case, CKM factors aside, one normally expects

$$\tau_B^{-1} \gg \tau_D^{-1} \qquad (2.51)$$

contrary to observation. Therefore we expect CKM factors to reduce quite considerably B-decay rates. Let us examine this quantitatively.

2.7a The B-Lifetime in the Spectator Model

For a heavy quark bound system, the spectator model, where the heavy quark decays freely in the presence of the light quark, should be a good approximation for a quantitative evaluation of the total decay rate. This is because the heavy quark provides a sufficiently large mass scale to treat the process as a short-distance phenomenon within the framework of perturbative QCD. The mechanism in terms of graphs goes as follows:

Fig. II.1

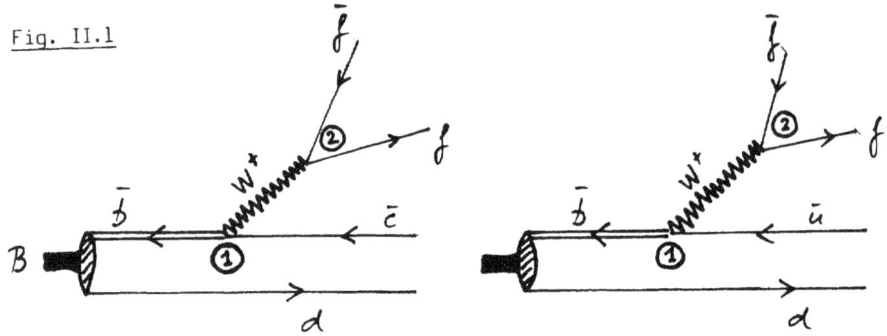

The outgoing $f\bar{f}$ at the ②-vertex can be

$$\nu_\ell,\ \bar{\ell} \quad \text{with } \ell = e, \mu, \tau\ ;$$

$$u, \bar{d};\quad u, \bar{s} \quad \text{and} \quad c, \bar{d};\ c, \bar{s}\ .$$

The CKM-factor at vertex-① can be

$$|V_{bc}|^2 \quad \text{and} \quad |V_{bu}|^2$$

and at vertex-②

$$|V_{ud}|^2,\ |V_{us}|^2;\quad |V_{cd}|^2,\ |V_{cs}|^2.$$

Then, with

$$m_{\nu_\ell} = m_e = m_\mu = m_d = m_u = 0\ , \tag{2.52}$$

the total decay rate

$$\Gamma_B = \Gamma(\bar{b} \to \bar{c}) + \Gamma(\bar{b} \to \bar{u}) \tag{2.53}$$

can be parametrized as follows:

$$\Gamma(\bar{b} \to \bar{c}) = \frac{G_F^2 m_b^5}{192\pi^3} |V_{bc}|^2 \Big\{ 2 f\Big(\frac{m_c}{m_b}\Big) + \varphi\Big(\frac{m_c}{m_b}, \frac{m_\tau}{m_b}\Big)$$

$$+ 3\eta_0 f\Big(\frac{m_c}{m_b}\Big)\,[c_1^2 + s_1^2 c_3^2] + 3\eta_0 h\Big(\frac{m_c}{m_b}\Big)\,[s_1^2 c_2^2 + (c_1 c_2 c_3 - s_2 s_3 \cos\delta)^2 + s_2^2 s_3^2 \sin^2\delta\,]\Big\}, \tag{2.54}$$

$$\Gamma(\bar{b} \to \bar{u}) = \frac{G_F^2 m_b^5}{192\pi^3} |V_{bu}|^2 \Big\{ 2 + f\Big(\frac{m_\tau}{m_b}\Big) +$$

$$+ 3\eta_0 (c_1^2 + s_1^2 c_3^2) + 3\eta_0 f\Big(\frac{m_c}{m_b}\Big)\,[s_1^2 c_2^2 + (c_1 c_2 c_3 - s_2 s_3 \cos\delta)^2 + s_2^2 s_3^2 \sin^2\delta\,]\Big\}, \tag{2.55}$$

where the functions $f(x)$, $\varphi(x,y)$ and $h(x)$ which are of kinematic origin correspond to phase-space factors for final states with masses:

$$
f(x) : \begin{array}{l} (m_{\nu_\ell}, m_\ell);(m_d, \underline{m_c}) \qquad \ell = e,\mu \\ (m_u, m_d(m_c)); (m_d, \underline{m_c}) \\ (m_{\nu_\tau}, \underline{m_\tau}); (m_u, m_d) \end{array} \quad ; \tag{2.56}
$$

$$
\varphi(x,y) : (m_{\nu_\tau}, \underline{m_\tau}); \; (m_d, \underline{m_c}) \quad ; \tag{2.57}
$$

$$
h(x) : (m_d(m_c), \underline{m_c}); \; (m_d, \underline{m_c}) \quad . \tag{2.58}
$$

Their exact expressions can be found, e.g. in ref. [12]. The factor η_0 accounts for gluonic corrections at the leading log approximation. Numerically, for the mass values

$$
m_c = 1.4 \text{ GeV}; \quad m_b = 4.6 \text{ GeV} \tag{2.59a,b}
$$

$$
M_W = 80 \text{ GeV} \quad \text{and} \quad \Lambda_{QCD} = 100 \text{ MeV} \tag{2.60a,b}
$$

one gets $\eta_0 \simeq 1.1$ and

$$
\begin{array}{l} f\left(\frac{m_c}{m_b}\right) = 0.51 \quad ; \quad f\left(\frac{m_\tau}{m_b}\right) = 0.33 \quad ; \\ \varphi\left(\frac{m_c}{m_b}, \frac{m_\tau}{m_b}\right) = 0.09 \quad ; \\ h\left(\frac{m_c}{m_b}\right) = 0.19 \quad . \end{array} \tag{2.61a}
$$

Then, with the further approximations

$$
c_1^2 + s_1^2 c_3^2 \simeq 1
$$

$$
s_1^2 c_2^2 + (c_1 c_2 c_3 - s_2 s_3 \cos\delta)^2 + s_2^2 s_3^2 \sin^2\delta \simeq 1 ,
$$

one obtains

$$
\Gamma(B \to all) = \frac{G_F^2 m_b^5}{192\pi^3} \left\{ 3.4 |V_{cb}|^2 + 7.3 |V_{ub}|^2 \right\} . \tag{2.61b}
$$

2.7b Phenomenological Determination of the CKM angles ϑ_2 and ϑ_3

The outcome of the analyses in the preceding subsection are the constraints

$$
\frac{|V_{cb}|^2}{0.29} + \frac{|V_{ub}|^2}{0.14} = \frac{192\pi^3}{G_F^2 m_b^5} \Gamma(B \to all) , \tag{2.62}
$$

and, from the experimental branching ratios quoted in (2.38,39)

$$|V_{ub}| \lesssim 0.16\ |V_{cb}| \quad . \tag{2.63}$$

The first constraint restricts $|V_{cb}|$ and $|V_{ub}|$ to be within a band which is limited by two ellipses in Fig. II.2. The second constraint restricts this domain to the dashed sector shown in the same figure. Here we have taken

$$m_b = 4.6\ \text{GeV}\ , \tag{2.64}$$

and we have combined the two lifetime measurements in eqs. (2.36,37) with a generous error, as indicated in eq. (2.40).

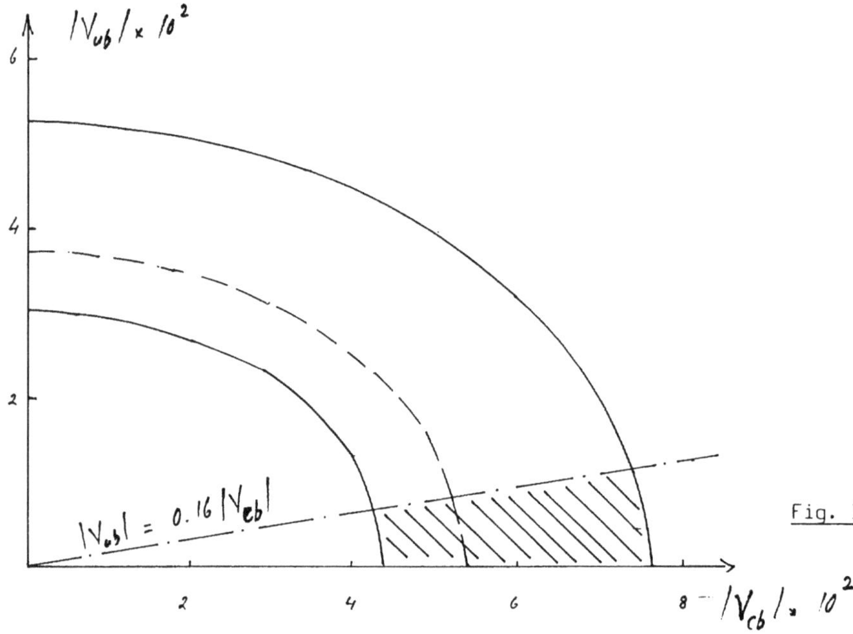

Fig. II.2

For numerical purposes, it is convenient to write the second constraint (2.63) as follows:

$$|V_{ub}| = 0.16\,\varkappa\,|V_{cb}|, 0 < \varkappa < 1. \tag{2.65}$$

Then, the constraints $|V_{cb}|$ and $|V_{ub}|$ read as follows:

$$|V_{cb}| = 0.054 \quad (1+0.5\,\tau)^{-1/2}(1+0.05\ \varkappa^2)^{-1/2} \tag{2.66a}$$

$$|V_{ub}| = 0.009\ \varkappa \left\{ (1+0.5\,\tau)\ (1+0.05\ \varkappa^2)\right\}^{-1/2} ; \tag{2.66b}$$

which can be approximated to

$$|V_{cb}| \simeq 0.054 \quad (1+0.5\,\tau)^{-1/2} \tag{2.67a}$$

$$|V_{ub}| \simeq 0.009\,\varkappa \ (1+0.5\,\tau)^{-1/2} . \tag{2.67b}$$

The last equation implies

$$/V_{ub}/ < 0.012 \qquad (2.68)$$

which in terms of the ϑ_3 angle ($V_{ub} = s_1 s_3$) says

$$s_3 < 0.05 \quad \text{and} \quad c_3 > 0.99 \quad . \qquad (2.69a,b)$$

Let us next discuss eq. (2.67a) in terms of CKM angles. It reads

$$(c_1^2 c_2^2 s_3^2 + s_2^2 c_3^2 + 2c_1 c_2 s_3 s_2 c_3 \cos\delta)^{1/2} = 0.054 \ (1+0.5\tau)^{-1/2}. \qquad (2.70)$$

Since both s_2 and s_3 are small quantities, to leading order in these variables we can approximate this equation to

$$s_2^2 + s_3^2 + 2s_2 s_3 \cos\delta = \frac{2.9}{1 + 0.5\tau} 10^{-3} \ . \qquad (2.71)$$

In the plane s_2, s_3 this equation restricts s_2 and s_3 to lie in an ellipse which for $\delta = \pi/2$ and $\delta = 3\pi/2$ becomes a circle, and for $\delta = 0, 2\pi$ and $\delta = \pi$ two lines. For $s_2 > 0$, and with the further constraint of eq. (2.68), the envelope of ellipses as δ varies from $\delta = 0$ to $\delta = 2\pi$ is the dashed domain shown in Fig. II.3. The "plus" area corresponds to the 1st and 4th quadrants for δ ; the minus area to the 2nd and 3rd.

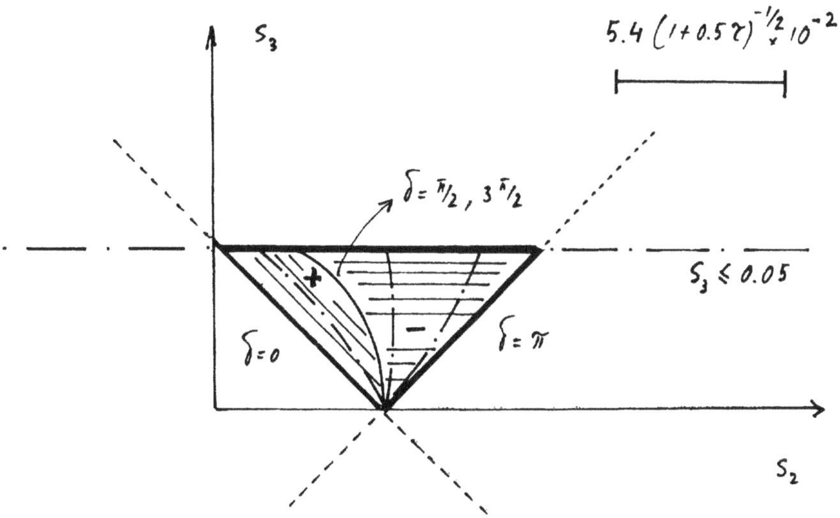

Fig. II.3

From eq. (2.70) we can find the limiting values for S_2 (the two-degenerate lines):

$$c_1 c_2 s_3 + s_2 c_3 = \frac{0.054}{(1+0.5\tau)^{1/2}} \quad \text{and} \qquad (2.72a)$$

$$-c_1 c_2 s_3 + s_2 c_3 = \frac{0.054}{(1+0.5\tau)^{1/2}} \ . \qquad (2.72b)$$

Then, from eq. (2.67b), and the fact we now ϑ_1,

$$s_1 s_3 = 0.009\, \varkappa\, (1 + 0.5\tau)^{-1/2} \quad . \tag{2.73}$$

Also, set $C_2 \simeq 1$ and we get

$$\frac{0.054 - 0.038\,\varkappa}{(1+0.5\tau)^{1/2}} \leqslant s_2 \leqslant \frac{0.054 + 0.038\,\varkappa}{(1+0.5\tau)^{1/2}} \quad , \tag{2.74}$$

i.e.,

$$0.013 \leqslant s_2 \leqslant 0.13, \tag{2.75}$$

a much more stringent limit than the limit from the analyses of ν-experiments [eq. (2.35b)].

The values for $|V_{cb}|$ and $|V_{ub}|$ inserted in Table I are those obtained in eqs. (2.67a,b) by letting $0 \leqslant \varkappa \leqslant 1$ and $-1 \leqslant \tau \leqslant 1$, i.e.

$$|V_{ub}| < 0.013 \tag{2.76}$$

$$|V_{cb}| = 0.054 \begin{array}{l} + 0.024 \\ - 0.011 \end{array} \quad . \tag{2.77}$$

The other CKM matrix elements in the 3-generation model are then determined by unitarity constraints. The values thus obtained are the ones which appear in Table I.

Table I.

	DOWN	STRANGE	BOTTOM
UP	c_1	$s_1 c_3$	$s_1 s_3$
	0.9735 ± 0.0015	0.221 ± 0.002	< 0.013
CHARM	$-s_1 c_2$	$c_1 c_2 c_3 - s_2 s_3 e^{i\delta}$	$c_1 c_2 s_3 + s_2 c_3 e^{i\delta}$
	-0.24 ± 0.03	$(0.96 \leq \quad \leq 0.98)$	$0.054 \begin{array}{l} + 0.024 \\ - 0.011 \end{array}$
TOP	$s_1 s_2$	$-c_1 s_2 c_3 - c_2 s_3 e^{i\delta}$	$-c_1 s_2 s_3 + c_2 c_3 e^{i\delta}$
	$0.003 \leq \quad \leq 0.03$	(≤ 0.17)	(> 0.99)

Note added in proof: There is a new measurement of τ_B (see ref. [14]):

$$\tau_B = \left(1.16 \pm {}^{0.37}_{0.94} \text{ (stat.)} \pm 0.23 \text{ (syst.)}\right) \times 10^{-12} \text{ sec.}$$

2.8 Effective Parametrization of the V-Matrix

The picture which emerges from the phenomenological determination of the CKM matrix elements is one where the angles ϑ_2 and ϑ_3 appear to be smaller than ϑ_1 ; in fact ϑ_2 and ϑ_3 are more like the square of the Cabibbo angle:

$$0.222 \leqslant s_1 \leqslant 0.235 \tag{2.78a}$$

$$0 \leqslant s_3 \leqslant 0.05 \tag{2.78c}$$

$$0.013 \leqslant s_2 \leqslant 0.13 \tag{2.78b}$$

which in terms of degrees for ϑ_1, ϑ_2 and ϑ_3 corresponds to

$$12.8^\circ \leqslant \vartheta_1 \leqslant 13.6^\circ \tag{2.79a}$$

$$0.7^\circ \leqslant \vartheta_2 \leqslant 7.5^\circ \tag{2.79b}$$

$$0 \leqslant \vartheta_3 \leqslant 3^\circ \tag{2.79c}$$

This suggests a simpler parametrization of the CKM matrix (see ref. [13]) which from the start incorporates the phenomenological feature that

$$\vartheta_1 \simeq \vartheta_{CABIBBO} \equiv \vartheta \quad \text{and} \quad \vartheta_{2,3} = O(\vartheta^2) . \tag{2.80}$$

For this purpose it is convenient to start with the definition

$$V = \exp(i\vartheta_2\lambda_7)\exp(i\vartheta_1\lambda_2)\begin{pmatrix} 1 & & \\ & 1 & \\ & & e^{i\delta} \end{pmatrix}\exp(i\vartheta_3\lambda_7) , \tag{2.81}$$

where

$$\lambda_7 = \begin{pmatrix} 0 & 0 & 0 \\ 0 & 0 & -i \\ 0 & i & 0 \end{pmatrix} \quad \text{and} \quad \lambda_2 = \begin{pmatrix} 0 & -i & 0 \\ i & 0 & 0 \\ 0 & 0 & 0 \end{pmatrix} . \tag{2.82a,b}$$

Then set

$$\left.\begin{aligned} \vartheta_1 &\equiv \vartheta \\ \vartheta_2 &\equiv \beta\vartheta^2 \\ \vartheta_3 &\equiv \gamma\vartheta^2 \end{aligned}\right\} \tag{2.83}$$

and expand the exponentials in powers of ϑ ; e.g.,

$$\exp(i\vartheta\lambda_2) = 1 + i\vartheta\lambda_2 + \frac{-1}{2!}\vartheta^2\lambda_2\lambda_2 + \frac{-i}{3!}\vartheta^3\lambda_2\lambda_2\lambda_2 + \dots \tag{2.84}$$

We arrive in this way at an effective parametrization in terms of powers of ϑ, the CP-violation parameter δ, and two parameters β and γ of O(1):

$$0.24 \leq \beta \leq 2.4 \qquad \text{and} \qquad 0 \leq \gamma \leq 1 \quad . \tag{2.85}$$

Developing to order ϑ^4, we obtain

$$V = \begin{pmatrix} 1-\frac{\vartheta^2}{2}+\frac{\vartheta^4}{24} & \vartheta-\frac{\vartheta^3}{6} & \gamma\vartheta^3 \\ -\vartheta+\frac{\vartheta^3}{6} & 1-\frac{\vartheta^2}{2}+\left(\frac{1}{24}-\frac{\gamma^2}{2}-\frac{\beta^2}{2}-\gamma\beta e^{i\delta}\right)\vartheta^4 & (\gamma+\beta e^{i\delta})\vartheta^2-\frac{\gamma}{2}\vartheta^4 \\ \beta\vartheta^3 & -(\beta+\gamma e^{i\delta})\vartheta^2+\frac{\beta}{2}\vartheta^4 & e^{i\delta}-\left[\beta\gamma+\left(\frac{\gamma^2}{2}+\frac{\beta^2}{2}\right)e^{i\delta}\right]\vartheta^4 \end{pmatrix} \tag{2.86}$$

This matrix is unitary up to and including terms of order $O(\vartheta^4)$.

We can see in this representation that transitions between quarks of the same generation are O(1). Transitions between the second and the first generations are $O(\vartheta)$, between the third and second generations of $O(\vartheta^2)$ and between the third and the first of $O(\vartheta^3)$. This is an extremely interesting pattern whose "raison d'être", which obviously has to be found beyond the standard model, is a challenge to theoreticians.

III. THE $\Delta S = 1$ EFFECTIVE HAMILTONIAN FOR NON-LEPTONIC DECAYS IN THE STANDARD MODEL

3.1 The $\Delta I = 1/2$ Rule for the Strangeness Changing Non-Leptonic Decays

Evidence for an approximate $|\Delta I| = 1/2$ selection rule for the non-leptonic strangeness changing amplitudes comes from several sources:

i) The branching ratio

$$\frac{\Gamma(K_S \to \pi^+\pi^-)}{\Gamma(K^+ \to \pi^+\pi^0)} \sim (21.2)^2 \quad . \tag{3.1}$$

Here the $\pi^+\pi^0$ system from the K^+ decay is in an I=2; i.e., the rate in the denominator is that of a pure $\Delta I = 3/2$ transition.

ii) The observation that $|\Delta I| = 1/2$ non-leptonic amplitudes are systematically larger for both kaons and baryons than the corresponding semileptonic amplitudes; e.g.,

$$\frac{\Gamma(K^+ \to \mu^+\nu)}{\Gamma(K_S \to 2\pi)} \sim \frac{1}{(12.2)^2} \quad ; \quad \frac{\Gamma(\Lambda \to pe\nu)}{\Gamma(\Lambda \to N\pi)} \sim \frac{1}{(27.7)^2} \quad . \tag{3.2a,b}$$

iii) The observation that the rates for $|\Delta I| = 3/2$ non-leptonic decays are of the same order of magnitude than the corresponding semileptonic decay rates; e.g.,

$$\frac{\Gamma(K^+\to\pi^+\pi^0)}{\Gamma(K^+\to\pi^0e^+\nu)}\sim 4.4 \quad ; \quad \frac{\Gamma(K^+\to\pi^+\pi^+\pi^-)}{\Gamma(K^+\to\pi^0e^+\nu)}\sim 1.2 \; . \tag{3.3}$$

From i), ii), and iii) one is led to the conclusion that, empirically, "the $|\Delta I|$ = 1/2 non-leptonic amplitudes are enhanced". This is the context of the so-called ΔI = 1/2 rule for ΔS = 1 non-leptonic decays, which in what follows we shall call the ΔI = 1/2 rule for short.

The explanation of the ΔI = 1/2 rule has been a continuous challenge to theorists throughout the development of the physics of weak interactions in particle physics. The question has been to know if this rule is due to intrinsic symmetry properties of some effective Hamiltonian or, more likely, to a dynamical interplay of the strong interactions in weak hadronic amplitudes. The first suggestion that the ΔI = 1/2 rule may be an effect of the short-distance behaviour of weak hadronic currents is due to K. Wilson [1]. Detailed calculations, however, had to await the development of QCD. The earlier calculations within the framework of the standard model were made by M.K. Gaillard and B.W. Lee [2];and, independently, by G. Altarelli and L. Maiani [3]. These authors showed that the leading logarithmic approximation (LLA) in QCD, goes in the right direction to explain the ΔI = 1/2 rule, but is not quantitatively large enough. Later on, the ITEP group M.A. Shifman, A.I. Vainshtein, and V.I. Zakharov suggested that a new type of diagrams - the so-called "penguin" diagrams - which appear when W-exchanges and gluon-exchanges are simultaneously combined with explicit quark-mass dependencies in the heavy charge 2/3 quarks (see Fig. III.5) - may be responsible for the required enhancement of the ΔI=1/2 rule [4]. The analysis of "penguin" diagrams from a general renormalization group point of view has been subsequently put on a firm basis (see refs. [5] and [6]); and several detailed calculations have been made,[7] and [8]. In what follows we shall briefly review our present theoretical wisdom on the ΔI = 1/2 rule within the framework of the standard model.

Throughout this section, we shall use the shorthand notation

$$(\bar{s}_L\gamma^\mu d_L) \equiv \sum_{i=\text{colour}} \bar{s}_i(x)\gamma^\mu \frac{1}{2}(1-\gamma_5)d_i(x) \quad ; \quad \text{and} \tag{3.4}$$

$$\lambda_q \equiv V_{qd}\,V_{qs}^* \qquad q = u,c,t \quad , \tag{3.5a}$$

where V_{qd} is the corresponding CKM matrix element. Unitarity of the V-matrix requires (for 3-generations)

$$\lambda_u + \lambda_c + \lambda_t = 0 \quad . \tag{3.5b}$$

3.2 The ΔS=1 Effective Hamiltonian for Non-Leptonic Decays in the limit $M_W^2 >> m_q^2$ for all Quarks

The relevant effective Hamiltonian induced by the exchange of a heavy W in the second-order Lagrangian of the standard model reads as follows:

$$\mathcal{H}_{eff}^{\Delta S=1}(x) = \frac{G_F}{\sqrt{2}} 4 \Big\{ \lambda_u (\bar{s}_L \gamma^\mu u_L)(\bar{u}_L \gamma_\mu d_L) + \\ + \lambda_c (\bar{s}_L \gamma^\mu c_L)(\bar{c}_L \gamma_\mu d_L) + \lambda_t (\bar{s}_L \gamma^\mu t_L)(\bar{t}_L \gamma_\mu d_L) \Big\} . \tag{3.6}$$

This corresponds to the picture already shown in Fig. I.1. The limit $M_W \to \infty$ is justified by the fact that $\mathcal{H}_{eff}^{\Delta S=1}(x)$ will only be applied to processes where all masses and momentum transfers involved are much smaller than M_W.

From the point of view of isospin properties it is convenient to rearrange $\mathcal{H}_{eff}^{\Delta S=1}$ in the following way:

$$\mathcal{H}_{eff}^{\Delta S=1}(x) = \frac{-G_F}{2\sqrt{2}} \Big\{ \lambda_c \big(O_c^{(+)} + O_c^{(-)} \big) + \lambda_t \big(O_t^{(+)} + O_t^{(-)} \big) \Big\} , \tag{3.7}$$

where (q = c,t)

$$O_q^{(\pm)}(x) \equiv 4 \Big\{ (\bar{s}_L \gamma^\mu u_L)(\bar{u}_L \gamma_\mu d_L) \pm (\bar{s}_L \gamma^\mu d_L)(\bar{u}_L \gamma_\mu u_L) \\ - (\bar{s}_L \gamma^\mu q_L)(\bar{q}_L \gamma_\mu d_L) \mp (\bar{s}_L \gamma^\mu d_L)(\bar{q}_L \gamma_\mu q_L) \Big\} . \tag{3.8}$$

Notice that in $O_q^{(-)}$ the term

$$(\bar{s}_L \gamma^\mu u_L)(\bar{u}_L \gamma_\mu d_L) - (\bar{s}_L \gamma^\mu d_L)(\bar{u}_L \gamma_\mu u_L)$$

is antisymmetric with respect to the exchange of the u_L and d_L fields, and therefore it transforms like a pure I = 1/2 operator. The operator $O_q^{(+)}$ contains pieces which transform both like I = 1/2 and I = 3/2. However, the operators $O^{(-)}$ and $O^{(+)}$ appear in the effective Hamiltonian, eq. (3.7), with the same weight. There is no hint of ΔI = 1/2 enhancement at this level.

The question we wish to discuss next is how this structure of the ΔS=1 non-leptonic Hamiltonian is modified in the presence of the color SU(3) corrections. We shall proceed in successive steps.

3.2a One-Loop Gluonic Corrections to $\mathcal{H}_{eff}^{\Delta S=1}(x)$

Let us consider the possible gluonic corrections to a typical 4-fermion term $(\bar{q}_{1L} \gamma^\mu q_{2L})(\bar{q}_{3L} \gamma_\mu q_{4L})$. As shown in Fig. III.1, there are two types of corrections: those which correspond to corrections within one vertex, like in Figs.

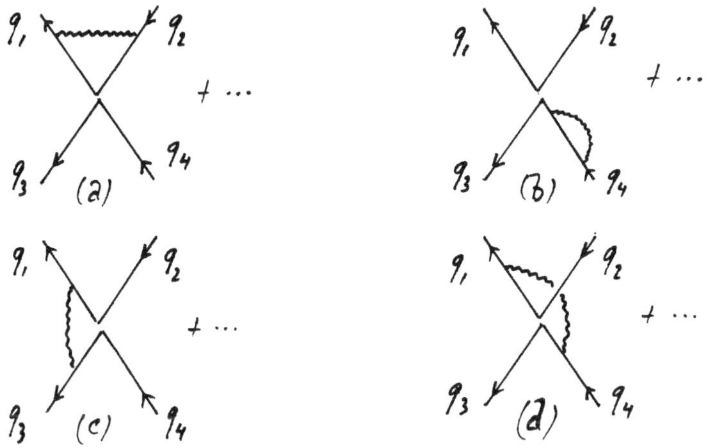

Fig. III.1

III.1a,b; and those where the gluon is exchanged between fermion lines belonging to different vertices, like in Figs. III.1c,d.

The first type of corrections only amounts to coupling constant renormalizations and wave function renormalization. They do not change the relative weight of the operators $O^{(+)}$ and $O^{(-)}$. By contrast, the exchange corrections like the ones in Figs. III.1c,d induce couplings of the type (*)

$$\sum_A (\bar{q}_{1L}\gamma^\mu \lambda^{(A)} q_{2L})(\bar{q}_{3L}\gamma_\mu \lambda^{(A)} q_{4L}) \quad , \tag{3.9}$$

i.e., color octet exchange. Here $(\lambda^{(A)})_{ij}$, A = 1,2,...8 are the SU(3)-color Gell-Mann matrices and

$$(\bar{q}_{1L}\gamma^\mu \lambda^{(A)} q_{2L}) \equiv \sum_{i,j} \bar{q}^{\,i}_{1L}\gamma^\mu (\lambda^{(A)})_{ij}\, q^j_{2L} \quad . \tag{3.10}$$

Now, combining the two-identities

Fierz-Michel:

$$[\gamma^\mu(1-\gamma_5)]_{\alpha\beta}\,[\gamma_\mu(1-\gamma_5)]_{\gamma\delta} = -[\gamma^\mu(1-\gamma_5)]_{\alpha\delta}\,[\gamma_\mu(1-\gamma_5)]_{\gamma\beta} \; ; \tag{3.12}$$

and the color algebra relation:

$$\sum_{A=1\ldots 8} \lambda^{(A)}_{ij}\,\lambda^{(A)}_{kl} = -\frac{2}{3}\delta_{ij}\delta_{kl} + 2\delta_{il}\delta_{kj} \; , \tag{3.13}$$

we can rewrite (remember Fermi statistics)

(*) For the benefit of the reader who tries to check this statement we recall the Dirac algebra identity

$$\bar{\psi}_1\gamma_\nu\gamma_\alpha\gamma^\mu\psi_2\;\bar{\psi}_3\gamma_\mu\gamma^\alpha\gamma^\nu\psi_4 = 4\bar{\psi}_1\gamma_\alpha\psi_2\;\bar{\psi}_3\gamma^\alpha\psi_4 \quad . \tag{3.11}$$

$$\sum_A (\bar{q}_{1L}\gamma^\mu \lambda^{(A)} q_{2L})(\bar{q}_{3L}\gamma_\mu \lambda^{(A)} q_{4L}) =$$

$$-\frac{2}{3}(\bar{q}_{1L}\gamma^\mu q_{2L})(\bar{q}_{3L}\gamma_\mu q_{4L}) + 2(\bar{q}_{1L}\gamma^\mu q_{4L})(\bar{q}_{3L}\gamma_\mu q_{2L}) , \tag{3.14}$$

which brings us back to the type of operators appearing in $O^{(+)}$ and $O^{(-)}$. However, because of the different color weights 2 and -2/3 the one loop gluonic corrections will clearly contribute differently to $O^{(+)}$ and $O^{(-)}$. In renormalization language this means that $O^{(+)}$ and $O^{(-)}$, although still multiplicatively renormalizable at the one-loop level with M_W the only large scale, have different anomalous dimensions.

At this stage we wish to comment further on the dynamics of this one loop calculation and present an argument, I believe originally due to E. Witten [9], which can be adapted via an inductive procedure to prove the existence of renormalization group equations for the coefficients of the operators $\mathcal{O}^{(\pm)}$ to leading order in the W-boson mass dependence and to all orders in the color coupling constant. For this purpose, let us consider a typical diagram - before the W-boson exchange is contracted to a point - like the one below

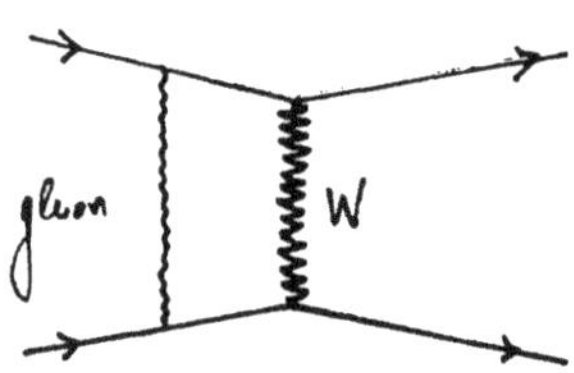

Fig. III.2

and let us fix the discussion in the Feynman - 't Hooft gauge; i.e., $\frac{-ig^{\mu\nu}}{k^2-M_W^2+i\epsilon}$ for the W-propagator. From power counting we see that in the diagram of Fig. III.2 one is not allowed to take $k^2 \to 0$ in the W-propagator and factorize a power $\frac{1}{M_W^2}$ before integration. This leads to a divergent integral. However, if we differentiate once with respect to the external momenta, or to one of the internal quark masses, then the improved convergence allows us to extract the leading power without impunity; i.e., to leading order in the heavy W-mass, the following diagrammatic equation is correct (the dot in the r.h.s. indicates differentiation with respect to the external momentum dependence of the corresponding propagator):

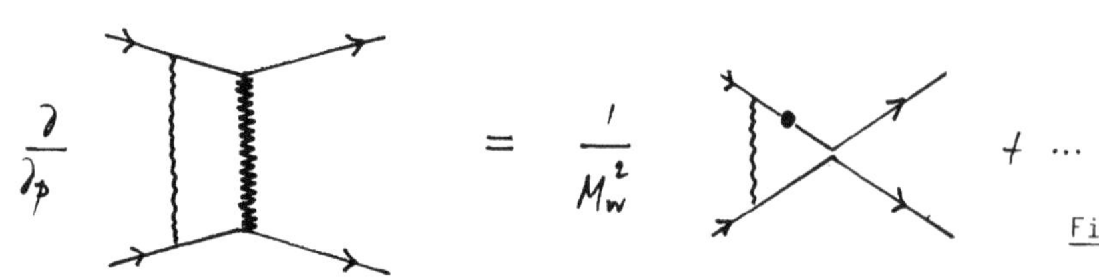
Fig. III.3

This implies that the diagram in Fig. III.2 must be equal to a leading term - independent of momenta and quark masses - with a coefficient F which possibly has log factors in M_W, plus a term of order $1/M_W^2$. The leading term, since it is independent of momenta, must be some local operator, which, by chirality conservation arguments, has to be of the $(\bar{q}_{1L}\gamma^\mu q_{2L})(\bar{q}_{3L}\gamma_\mu q_{4L})$ form, i.e., we must have

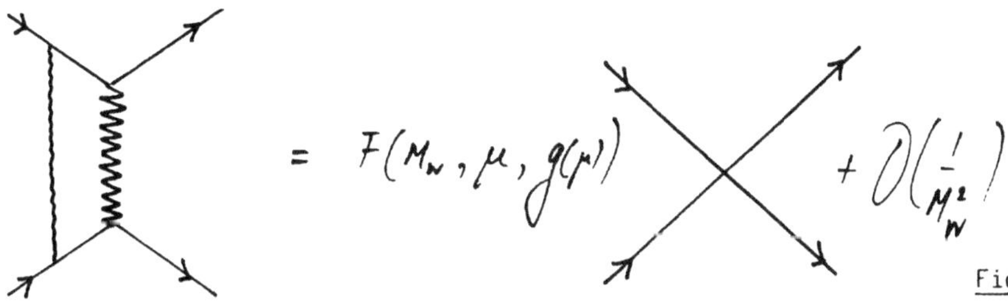

Fig. III.4

with μ some renormalization scale $\mu^2 < M_W^2$.

3.2b QCD Corrections to $\mathcal{H}_{eff}^{\Delta S=1}(x)$ at the Leading Logarithmic Approximation ($m_q^2 \ll M_W^2$ for all q)

The preceding argument can be generalized to all orders in perturbation theory in the color coupling constant $g(\mu)$. This means that to leading order in the heavy W vector boson mass - so far the only big mass scale, since all quark masses are neglected - the induced $\Delta S=1$ non-leptonic Hamiltonian, to all orders of perturbation theory in the color coupling, must be of the form

$$\mathcal{H}_{eff}^{\Delta S=1}(x) = \frac{-G_F}{2\sqrt{2}}\left\{\lambda_c\left(A_c^{(+)}O_c^{(+)} + A_c^{(-)}O_c^{(-)}\right) + c \leftrightarrow t\right\}, \tag{3.15}$$

and the coefficient functions $A_{q=c,t}^{(\pm)}$ must obey renormalization group equations of the type

$$\left\{\mu\frac{\partial}{\partial\mu} + \alpha_s\beta(\alpha_s)\frac{\partial}{\partial\alpha_s} - \gamma^{(\pm)}(\alpha_s)\right\}A_q^{(\pm)}\left(\frac{M_W}{\mu}, \alpha_s(\mu)\right) = 0\ , \tag{3.16}$$

where $\beta(\alpha_s)$ is the usual Callan-Symanzik beta function (see the references on textbooks; $N_c=3$, the number of colors; n_f is the number of flavors),

$$\beta(\alpha_s) = \left(-\frac{11}{6}N_c + \frac{1}{3}n_f\right)\frac{\alpha_s}{\pi} + O\left(\frac{\alpha_s}{\pi}\right)^2\ , \tag{3.17}$$

and $\gamma^{(\pm)}(\alpha_s)$ anomalous dimension functions which at the one loop level are simply

$$\gamma^{(+)}(\alpha_s) = \frac{\alpha_s}{\pi} + O\left(\frac{\alpha_s}{\pi}\right)^2 \tag{3.18a}$$

$$\gamma^{(-)}(\alpha_s) = -2\frac{\alpha_s}{\pi} + O\left(\frac{\alpha_s}{\pi}\right)^2\ . \tag{3.18b}$$

The general solution to eq. (3.16) can be cast in the following way:

$$A^{(\pm)}\left(\frac{M_W}{\mu},\alpha_s(\mu)\right) = A^{(\pm)}(1,\alpha_s(M_W))\exp\left\{-\int_{\alpha_s(\mu)}^{\alpha_s(M_W)} dx \frac{\gamma^{(\pm)}(x)}{x\beta(x)}\right\}, \tag{3.19}$$

with $\alpha_s(M_W)$ the running coupling constant, solution of the equation

$$\frac{1}{2}\log\frac{M_W^2}{\mu^2} = \int_{\alpha_s(\mu)}^{\alpha_s(M_W)} \frac{dx}{x\beta(x)} . \tag{3.20}$$

At the one loop level

$$\alpha_s(M_W) = \frac{\alpha_s(\mu)}{1-\frac{\alpha_s(\mu)}{\pi}\beta_1\frac{1}{2}\log\frac{M_W^2}{\mu^2}} = \frac{\pi}{-\beta_1/2\,\log\frac{M_W^2}{\Lambda^2}} \tag{3.21}$$

with $\beta = \frac{-11}{2}+\frac{1}{3}n_f$, and inserting eqs. (3.17) and (3.18) into eq. (3.19), we obtain (notice that in our normalization which factorizes λ_c and λ_t explicitly in the $\mathcal{H}_{eff}^{\Delta S=1}$ Hamiltonian, $A^{(\pm)}(1, \alpha_s(M_W))=1$)

$$A^{(\pm)}\left(\frac{M_W}{\mu},\alpha_s(\mu)\right) = \left(\frac{\alpha_s(M_W)}{\alpha_s(\mu)}\right)^{a^{(\pm)}}, \quad \text{with} \tag{3.22}$$

$$a^{(+)} = \frac{6}{33-2n_f}, \tag{3.23a}$$

$$a^{(-)} = \frac{-12}{33-2n_f}. \tag{3.23b}$$

This is the result of the LLA to the evaluation of the $A_c^{(\pm)}$ and $A_t^{(\pm)}$ coefficients in eq. (3.15). At this level, the number of quark flavors treated at the massless approximation (μ^2 must be taken $\gg m_q^2$ for all q) is clearly $n_f = 6$.

We see that from the point of view of the $\Delta I=1/2$ rule, the net effect of the LLA is to enhance the pure $\Delta I=1/2$ transitions governed by the $O^{(-)}$ operator by a factor

$$\left(\log\frac{M_W^2}{\Lambda^2}\Big/\log\mu^2/\Lambda^2\right)^{4/7}$$

while the transitions governed by the $O^{(+)}$ operator, which has both $\Delta I=1/2$ and $\Delta I=3/2$, is depressed by a factor $\left(\log\mu^2/\Lambda^2\Big/\log\frac{M_W^2}{\Lambda^2}\right)^{2/7}$.

These results go in the right direction to explain the phenomena i)-iii) we pointed out at the beginning but quantitatively they are not large enough.

3.2c Two-Loop Gluonic Corrections to $\mathcal{H}_W^{\Delta S=1}(x)$

The calculation of next-to-leading corrections (two-loops in QCD) to the effective ΔS =1 non-leptonic Hamiltonian has been done by Altarelli, Curci, Martinelli and Petrarca. This is a much more technically complicated calculation and I refer to the original papers [10] for details. The magnitude of the corrections which these authors find reinforces the trend of the results obtained at the LLA, but the same difficulty to get quantitative agreement persists if these short-distance enhancement factors are to be the only source of explanation of the Δ I=1/2 rule.

3.3 The ΔS=1 Effective Hamiltonian in the Presence of Heavy Quarks ("Penguin" diagrams)

The next step is to treat the t-quark as much heavier than the other quarks. We shall see how it is possible to eliminate the t-quark field from explicitly appearing in the effective Hamiltonian. However, in the process of removing the heavy t-quark field there appear new types of 4-fermion operators and also renormalization mixing among the various operators. This is due to the appearance of a new class of diagrams, the so-called "penguin" diagrams already mentioned in the introduction to section III. These diagrams, like the one shown in Fig. III.5

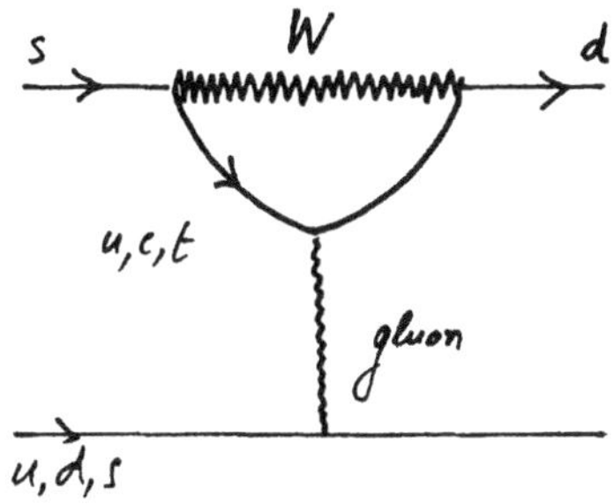

Fig. III.5

appear when one lets gluonic corrections operate in the second order weak Lagrangian, before removing the heavy W-field.

3.3a Explicit calculation of the Lowest Order "Penguin" Diagram

To fix the ideas, let us proceed to the evaluation of the effective vertex in Fig. III.6 responsible for the appearance of the penguin diagrams. To leading behaviour in $1/M_W^2$ powers, we have

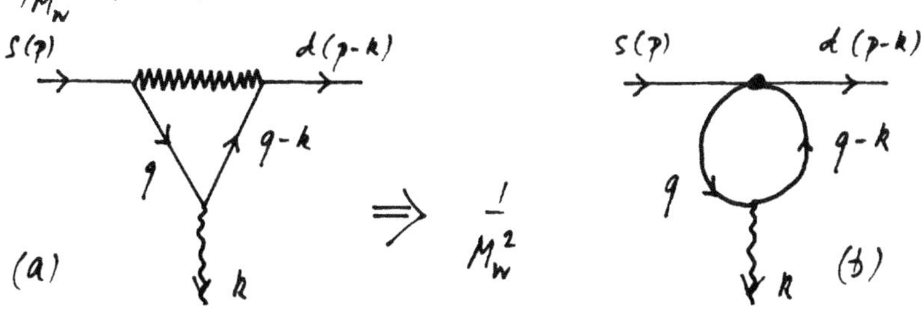

Fig. III.6

$$V_{eH}^{(A)}(p,k)_\mu = (-2i)\bar{d}_L(p-k)\,\bar{I}_\mu^A(k)\,s_L(p)\,\frac{g_W^2}{2M_W^2}\,, \tag{3.24}$$

with $I_\mu^A(k)$ the integral

$$\bar{I}_\mu^A(k) = \sum_{i=u,c,t}\lambda_i\int\frac{d^Dq}{(2\pi)^D}\,\gamma^\alpha\frac{i}{\not{q}-\not{k}-M_i+i\epsilon}\,\gamma_\mu\, ig\frac{\lambda^A}{2}\,\frac{i}{\not{q}-M_i+i\epsilon}\,\gamma_\alpha\,, \tag{3.25}$$

where M_i denotes the mass of the charge $2/3$ quark in the loop and $D=4-\epsilon$ the space-time dimension. In the limit $M_W^2 \gg M_i^2 \simeq 0$ for all i, this integral is inoperative because of the GIM-mechanism condition (unitarity of the V-matrix)

$$\sum_i \lambda_i = 0\,, \tag{3.26a}$$

which acts here like a regularization condition à la Pauli-Villars. This is the reason why "penguin" diagrams first appear in the process of distinguishing one large quark mass from the others. With $M_i \neq 0$, the part of the integral $I_\mu^A(k)$ which is proportional to an odd number of γ's will contribute to an effective vertex. Combining the two denominators with one Feynman parameter x and after a shift of the loop 4-momenta we find

$$\bar{I}_\mu^A(k) \doteq -ig\frac{\lambda^A}{2}\sum_i\lambda_i\int_0^1 dx\int\frac{d^Dq}{(2\pi)^D}\,\frac{1}{(q^2-R_i^2)^2}\;\times$$
$$\left\{-2\left[M_i^2(1-\epsilon/2)-\frac{1}{2}q^2(1-\frac{3}{2}\epsilon/2)\right]\gamma_\mu + 2x(1-x)\not{k}\gamma_\mu\not{k}\,(1-\epsilon/2)\right\}, \tag{3.26b}$$

where

$$R_i^2 \equiv M_i^2 - k^2x(1-x)\,. \tag{3.26c}$$

The evaluation of the integral over q leads to the result

$$\bar{I}_\mu^A(k) = -ig\frac{\lambda^A}{2}\,i\,\frac{\bar{\Lambda}^2}{(2\pi)^4}\,\nu^{-\epsilon}\,\frac{1}{\epsilon/2}\sum_i\lambda_i\int_0^1 dx\;\times$$
$$\left\{2x(1-x)\not{k}\gamma_\mu\not{k}\left[1+\epsilon/2\,(\log 4\pi-\gamma_E-1)-\epsilon/2\,\log\frac{R_i^2}{\nu^2}\right]\right.$$
$$-2M_i^2\left[1+\epsilon/2\,(\log 4\pi-\gamma_E-1)-\epsilon/2\,\log\frac{R_i^2}{\nu^2}\right]\gamma_\mu \tag{3.27}$$
$$\left.+2R_i^2\left[1+\epsilon/2\,(\log 4\pi-\gamma_E-3/2)+1/2\,\epsilon/2-\epsilon/2\,\log\frac{R_i^2}{\nu^2}\right]\gamma_\mu\right\}.$$

Most of the integrals over the Feynman parameter x are now trivial and we finally obtain

$$I_\mu^A(k) = -ig\frac{\lambda^A}{2}\, i\,\frac{\pi^2}{(2\pi)^4}\,\nu^{-\epsilon}\left\{\frac{1}{\epsilon/2}\sum_i \lambda_i\, 2/3\,(\not k k_\mu - k^2\gamma_\mu)\right.$$

$$+\sum_i \lambda_i\,(\log 4\pi - \gamma_E - 1)\, 2/3\,(\not k k_\mu - k^2\gamma_\mu)$$

$$\left. -\sum_i \lambda_i\, 2/3\,(\not k k_\mu - k^2\gamma_\mu)\int_0^1 \log\frac{M_i^2}{\nu^2}dx\right\}. \qquad (3.28)$$

As expected, the divergent term $(\sim 1/\epsilon)$ is flavor independent and its effect vanishes because of the GIM-symmetry condition $\sum \lambda_i = 0$. For the same reason, the constant term proportional to $\log 4\pi - \gamma_E - 1$ also disappears, and we are left with the term

$$I_\mu^A(k) = g\nu^{-\epsilon}\,\frac{\lambda^A}{2}\,\frac{1}{16\pi^2}\,(-2/3)\,(\not k k_\mu - k^2\gamma_\mu)\,\times$$

$$\times\int_0^1 dx\,\sum \lambda_i \log\frac{M_i^2 - k^2 x(1-x)}{\nu^2}dx\;. \qquad (3.29)$$

Remember that here one is interested only in terms which contribute to leading order in the heavy quark mass (M_t^2 for 3-generations; M_c^2 for 2-generations) and k - the gluon momentum transfer - is small with respect to heavy quark masses. Also, we always keep $M_W^2 \gg M_i^2$. Under these conditions, I_μ^A (k) reduces to the term

$$I_\mu^A(k) = g\nu^{-\epsilon}\,\frac{\lambda^A}{2}\,\frac{1}{16\pi^2}\,\frac{2}{3}\,(k^2\gamma_\mu - \not k k_\mu)\,\lambda_t \log\frac{M_t^2}{\nu^2}\;. \qquad (3.30)$$

The dependence on the factor $k^2\gamma_\mu - \not k k_\mu$ was to be expected on the grounds of gauge invariance. Also the log M_t^2/ν^2 factor, on the grounds of power counting. Only the factor 2/3 requires a detailed calculation...

The effective vertex I_μ^A (k) in eq. (3.30), when inserted in the full "penguin" diagram of Fig. III.5, has to be contracted - via the gluon propagator - with the point-like lower vertex. The factor k^2 cancels out and we are left with an effective 4-fermion interaction of the (V-A) x V type.

In coordinate space, the result for the effective vertex in eq. (3.30) corresponds to an effective Hamiltonian $(k_\mu \to -i\partial_\mu)$

$$\mathcal{H}_1(x) = -\frac{G_F}{\sqrt{2}}\,\frac{g_s}{12\pi^2}\log\frac{M_t^2}{\nu^2}\,2\,\bar d_L(x)\gamma^\mu\frac{\lambda^A}{2}s_L(x)\,\times \qquad (3.31)$$

$$\times(g_{\mu\nu}\partial^2 - \partial_\mu\partial_\nu)\,G_A^\nu(x) + h.c.,$$

where $G_A^\nu(x)$ denotes the gluon field A=1,2,...,8. In this result, one immediately recognizes the abelian part and the term of zeroth order in g_s of the covariant derivative

$$\left(g_{\mu\nu}\partial^2 - \partial_\mu\partial_\nu\right) G_A^\nu(x) = D^\nu G_{\nu\mu}^{(A)}(x) \Big/_{\text{Abelian};\, g_s \to 0} \tag{3.32}$$

where (see section I)

$$D_\nu \equiv \partial_\nu - i g_s \vec{T}\cdot\vec{G}_\nu(x) \quad ; \quad \text{and} \tag{3.33a}$$

$$\vec{G}_{\mu\nu}(x) = \partial_\mu \vec{G}_\nu - \partial_\nu \vec{G}_\mu + g_s \vec{G}_\mu \times \vec{G}_\nu \, . \tag{3.33b}$$

The question we have to investigate is whether the inclusion of other "Penguin"-like diagrams, always to order g_W^2, indeed leads to a corresponding effective Hamiltonian with $D^\nu G_{\nu\mu}^{(A)}(x)$ replacing $\left(g_{\mu\nu}\partial^2 - \partial_\mu\partial_\nu\right) G_A^\nu(x)$ in eq. (3.31). As we shall discuss in the next subsection, this is indeed the case. Then using the equations of motion of QCD, we shall be able to replace

$$D^\mu G_{\mu\nu}^{(A)}(x) = g_s\, \bar{u}(x)\gamma_\nu \frac{\lambda^A}{2} u(x) + \bar{d}(x)\gamma_\nu \frac{\lambda^A}{2} d(x) + \ldots \tag{3.34}$$

and the effective weak Hamiltonian reduces thus to a local gauge invariant four-fermion Hamiltonian.

3.3b "Penguin" diagrams to Higher Orders in QCD

In this subsection we shall show that the inclusion of further gluon corrections to the lowest order "penguin" diagram in Fig. III.5 indeed combine - at the approximation $M_W^2 \gg M_t^2 \gg$ - other m_q^2 to give the desired form of a local four-fermion effective Hamiltonian. The proof follows the diagrammatic analysis made by Wise and Witten in ref. [6]. The question is far from trivial, since diagrams like

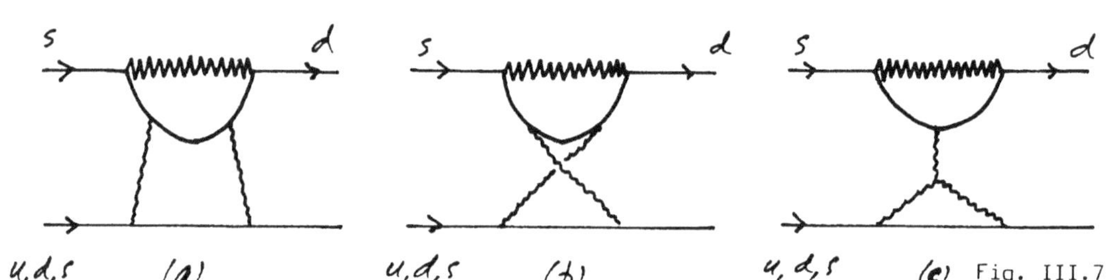

Fig. III.7

because of the non-local gluon structure, do not lead a priori to an interpretation in terms of a local four-fermion interaction. To see how, in spite of this apparent difficulty, the argument works let us proceed to an evaluation of the various types of one-loop vertex subdiagrams of the "Penguin" type.

First, let us evaluate the subdiagram

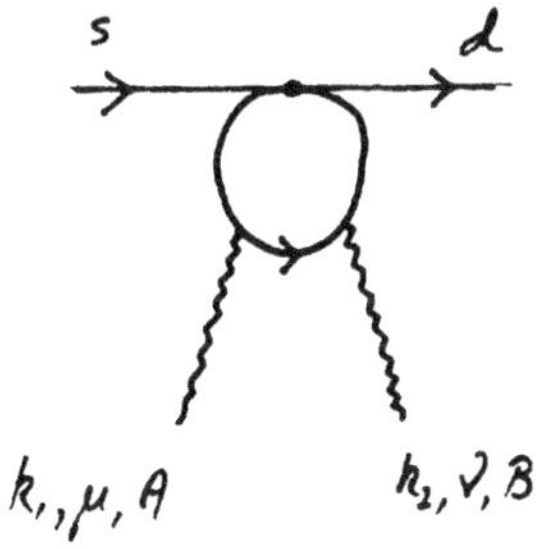

+ crossed graph

Fig. III.8

corresponding to the two-gluon emission in the one particle irreducible diagram contributing to the transition s $\rightarrow$ d + gluons at the one-loop level. Explicit calculation, much the same as we have carried out before for the one-gluon emission, gives the result

$$V_{eff}^{A,B}(p;k_1,k_2)_{\mu\nu} = (-2i)\,2\,\bar{d}_L(p-k_1-k_2)\,\bar{I}_{\mu\nu}^{A,B}(k_1k_2)\,s_L(p)\,\frac{g_w^2}{2M_W^2}, \tag{3.35}$$

with

$$\begin{aligned}\bar{I}_{\mu\nu}^{A,B}(k_1,k_2) = -i g_s^2 \nu^{-\epsilon}\,\frac{\lambda^C}{2}\,\frac{1}{16\pi^2}\cdot\frac{2}{3}\,\log\frac{M_t^2}{\nu^2}\,\cdot \\ f^{CAB}\left[(2k_2+k_1)_\mu\gamma_\nu - (2k_1+k_2)_\nu\gamma_\mu + (\not{k}_1-\not{k}_2)g_{\mu\nu}\right],\end{aligned} \tag{3.36}$$

where again the $1/\epsilon$ term is inoperative because of the GIM-condition $\sum\lambda_i = 0$.

This result is reproduced by an effective Hamiltonian

$$\begin{aligned}\mathcal{H}_2(x) = -\frac{G}{\sqrt{2}}\,\frac{g_s^2}{12\pi^2}\,\log\frac{M_t^2}{\nu^2}\,2\,\bar{d}_L(x)\gamma^\mu\frac{\lambda^C}{2}s_L(x)\,\cdot \\ f^{CAB}\left[2G_\nu^A(x)\partial^\nu G_\mu^B(x) + G_\mu^B(x)\partial^\nu G_\nu^A(x) + G^{\nu B}(x)\partial_\mu G_\nu^A(x)\right].\end{aligned} \tag{3.37}$$

We recognize in the r.h.s. the piece coming from the quadratic gluon field terms in the covariant derivativd $D^\nu G_{\mu\nu}^{(A)}(x)$.

Now, up to a sign, the effective vertex in eqs. (3.35) and (3.36) is precisely the same as the one from the subdiagram

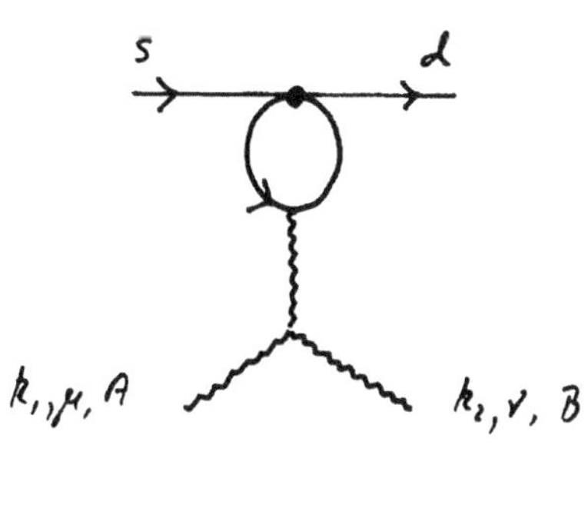

Fig. III.9

which is not one particle irreducible. This can be readily seen from the explicit result in eq. (3.30). The k^2 factor from the effective upper vertex cancels the $\frac{1}{k^2}$ from the gluon propagator and we are left with the 2-gluon vertex. Therefore, if we could replace Fig. III.9 as a subdiagram of Fig. III.7c, and Figs. III.8 as subdiagrams of Figs. III.7a and III.7b there would then be a complete cancellation between the diagrams in Fig. III.7. In practice, the cancellation is only exact at the soft gluon limit when k_1 and k_2 are small with respect to the heavy quark mass. There will be factors left out from the short-distance region in the one-loop evaluation of $\mathcal{H}_1$, - from the lower loop in the diagram of Fig. III.7c - and in the one-loop evaluation of $\mathcal{H}_2$ - from the lower loops in the diagrams of Figs. III.7a and III.7b; but using the external differentiation technique we described in the previous subsection, one can easily observe that the short-distance contributions can be factorized as Wilson coefficients of overall effective local four-fermion-like operators.

Other two-loop diagrams like

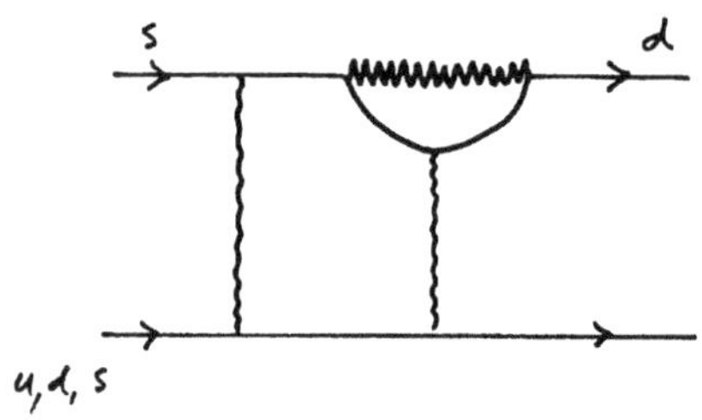

Fig. III.10

can be easily shown to factorize in the same way, using the external momenta differentiation technique. They contribute both as a correction to the matrix elements of $\mathcal{H}_1$, and to the Wilson coefficient of other local four-fermion operators.

The successful analysis of two-loop diagrams was based on the crucial role played by the one particle irreducible two-gluon emission diagrams. Let us next examine the case of three-gluon emission attached to the heavy quark line, i.e. the 6 diagrams

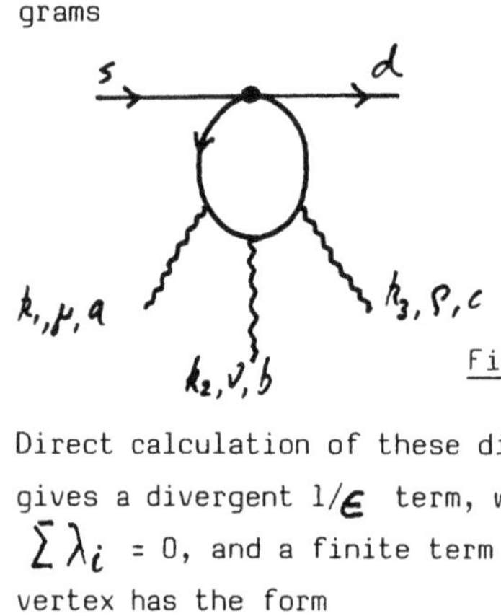

+ 5 crossed graphs .

Fig. III.11

Direct calculation of these diagrams in the large quark-mass and soft-gluon limit gives a divergent $1/\epsilon$ term, which again is inoperative because of the GIM-condition $\sum \lambda_i = 0$, and a finite term with a $\log M_t^2$ factor. The corresponding effective vertex has the form

$$V_{eff}^{A,B,C}(p;k_i)_{\mu\nu\rho} = (-2i)\frac{g_w^2}{2M_W^2}\,\bar d_L(p-k_1-k_2-k_3)\,\bar T^{A,B,C}_{\mu\nu\rho}(k_i)\,s_L(p) \;, \tag{3.38}$$

$$\bar T^{A,B,C}_{\mu\nu\rho}(k_1,k_2,k_3) = g_s^3\,\nu^{-\epsilon}\,\frac{1}{16\pi^2}\,\frac{2}{3}\,\frac{\lambda^D}{2}\,\log\frac{M_t^2}{\nu^2}\,\times$$

$$\Big[\left(f^{EAB}f^{ECD} - f^{EBC}f^{EAD}\right)g_{\rho\mu}\gamma_\nu - \left(f^{EAB}f^{ECD} + f^{EAC}f^{EBD}\right) \tag{3.39}$$

$$\times\, g_{\rho\nu}\gamma_\mu + \left(f^{EAC}f^{EBD} + f^{EBC}f^{EAD}\right)g_{\mu\nu}\gamma_\rho\Big] \,.$$

This effective vertex is reproduced by the following effective Hamiltonian

$$\mathcal{H}_3(x) = -\frac{G}{\sqrt 2}\,\frac{g_s^3}{12\pi^2}\,\log\frac{M_t^2}{\nu^2}\,2\,\bar d_L(x)\gamma^\mu\,\frac{\lambda^D}{2}\,s_L(x)\,\times \tag{3.40}$$

$$\times\, f^{EBC}f^{EAD}\left[G_\nu^A(x), G_C^\nu(x)\right]G_\mu^B(x) \,.$$

We recognize in the r.h.s. the cubic piece of the covariant product $D_\nu G_D^{\nu\mu}(x)$ coming from the product of the $-ig_s\vec\lambda\cdot\vec G_\mu$ term and the non-abelian term in $G_{\nu\mu}(x)$. By an argument entirely analogous to the one discussed for the two-loop diagrams in Fig. III.7 it can be seen that the total contribution from the three-loop diagrams in the Fig. III.12 below

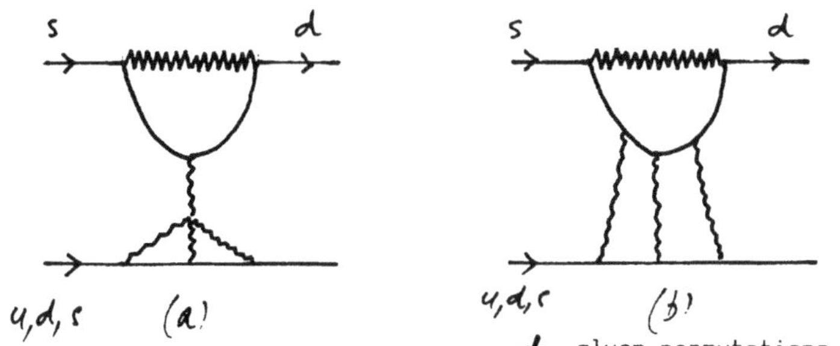

+ gluon permutations . Fig. III.12

cancels, up to terms independent of the external momenta, which are proportional to the tree approximation of local four-quark operators. The net effect of the three-loop diagrams is to give therefore $O(g_s^6)$ contributions to the Wilson coefficients of local four-fermion operators.

Notice that in the large quark-mass and soft-gluon limit the other one particle irreducible diagrams where gluon fields are attached to the heavy quark loop are already convergent - even in the absence of the GIM-mechanism - and therefore cannot produce a log M_t^2/ν^2 factor. They correspond to non-leading operators.

The effective leading weak Hamiltonian, at the one-loop level, corresponding to the sum of diagrams contributing to the process $s \to$ d + (gluons) with a heavy quark in the loop has the form

$$\mathcal{H}_{PENGUIN}(x) = \mathcal{H}_1(x) + \mathcal{H}_2(x) + \mathcal{H}_3(x) =$$

$$= -\frac{G}{\sqrt{2}} \frac{g_s}{12\pi^2} \log\frac{M_t^2}{\nu^2}\, 2\bar{d}_L(x)\gamma^\mu \frac{\lambda^A}{2} s_L(x) \left(D^\nu G_{\nu\mu}(x)\right)^A =$$

$$= -\frac{G}{\sqrt{2}} \frac{g_s^2}{12\pi^2} \log\frac{M_t^2}{\nu^2}\, 2\bar{d}_L(x)\gamma^\mu \frac{\lambda^A}{2} s_L(x) \times$$

$$\times \sum_{\text{light flavours}} \bar{q}_i(x)\gamma_\mu \frac{\lambda^A}{2} q_i(x) . \tag{3.41}$$

We conclude from this analys that the sum of all penguin-type diagrams - with arbitrary gluon insertions - can be represented by a sum of Wilson coefficients times matrix elements of local four-fermion operators.

3.3c The Δ S=1 Effective Hamiltonian Without Explicit Dependence on the Heavy t-Quark Field

We are now in the position to carry through the procedure of eliminating the t-quark field from explicitly appearing in the effective Hamiltonian. To order $1/M_t^2$ we have seen in the previous subsections that all the dependence of amplitudes on the heavy t-quark mass can be reabsorbed into renormalization effects of four-fermion operators. What we now need is a basis of independent four-fermion operators which close under one-loop renormalization and expand the $O_t^{(\pm)}$ and $O_c^{(\pm)}$ operators into this new basis. Clearly, for $O_c^{(\pm)}$ there is no problem since the t-quark field does not explicitly appear in $O_c^{(\pm)}$. At this level, the $O_c^{(\pm)}$ operators are still multiplicatively renormalizable,

$$\langle O_c^{(\pm)}\rangle = B^{(\pm)}\left(\frac{M_t}{\mu}, \alpha_s(\mu)\right) \langle O_c^{(\pm)}\rangle' , \tag{3.42}$$

where $\langle O_c^{(\pm)}\rangle'$ means that matrix elements of the new operator must be evaluated in an effective five-quark theory with a new coupling $g_s'(M_t/\mu, g_s)$ and new mass parameters m_i', i = u, d, s, c, b; i.e., matrix elements of $O'^{(\pm)}_c$ obey the renormalization group equation

$$\mu\frac{d}{d\mu}\langle O_c^{(\pm)}\rangle' \equiv$$

$$\left(\mu \frac{\partial}{\partial \mu} + \alpha_s' \beta'(\alpha_s') \frac{\partial}{\partial \alpha_s'} + \sum_q \gamma_q'(\alpha_s') m_q' \frac{\partial}{\partial m_q'} \right) \langle O_c^{(\pm)} \rangle' =$$

$$- \gamma'^{(\pm)}(\alpha_s') \langle O_c^{(\pm)} \rangle' \tag{3.43}$$

where $\beta'(\alpha_s')$, $\gamma_q'(\alpha_s')$ are the β-function and the anomalous mass dimension functions of a theory with five-flavours, and $\gamma'^{(\pm)}(\alpha_s')$ is the anomalous dimension of the effective $O'^{(\pm)}$ operator in a theory with five-flavours $\mu^2 \gg m_q'^2$. On the other hand,

$$\mu \frac{d}{d\mu} \langle O_c^{(\pm)} \rangle \equiv$$

$$\left(\mu \frac{\partial}{\partial \mu} + \alpha_s \beta(\alpha_s) \frac{\partial}{\partial \alpha_s} + \gamma_t(\alpha_s) M_t \frac{\partial}{\partial M_t} \right) \langle O_c^{(\pm)} \rangle =$$

$$= - \gamma^{(\pm)}(\alpha_s) \langle O_c^{(\pm)} \rangle \quad , \tag{3.44}$$

where $\beta(\alpha_s)$, $\gamma_t(\alpha_s)$ and $\gamma^{(\pm)}(\alpha_s)$ are quantities defined in the full six-quark theory. Furthermore, from eq. (3.42), it follows that

$$\mu \frac{d}{d\mu} \langle O_c^{(\pm)} \rangle = \left(\mu \frac{d}{d\mu} B^{(\pm)} \right) \langle O_c^{(\pm)} \rangle' + B^{(\pm)} \mu \frac{d}{d\mu} \langle O_c^{\pm} \rangle' \quad , \tag{3.45}$$

i.e.,

$$- \gamma^{(\pm)} B^{(\pm)} \langle O_c^{(\pm)} \rangle' = \left(\mu \frac{d}{d\mu} B^{(\pm)} \right) \langle O_c^{(\pm)} \rangle' - \gamma'^{(\pm)} B^{(\pm)} \langle O_c^{(\pm)} \rangle' \quad , \tag{3.46}$$

and therefore, the coefficient $B^{(\pm)}$ obeys the renormalization group equation

$$\left\{ \mu \frac{\partial}{\partial \mu} + \alpha_s \beta(\alpha_s) \frac{\partial}{\partial \alpha_s} + \gamma_t(\alpha_s) M_t \frac{\partial}{\partial M_t} + \gamma^{(\pm)}(\alpha_s) - \gamma'^{(\pm)}(\alpha_s') \right\} B^{(\pm)}\left(\frac{M_t}{\mu}, \alpha_s \right) = 0 . \tag{3.47}$$

Since $B^{(\pm)}\left(\frac{M_t}{\mu}, \alpha_s \right)$ is dimensionless, it is convenient to introduce the variable

$$y = \frac{1}{2} \log M_t^2 / \mu^2 \quad , \tag{3.48}$$

and rewrite eq. (3.47) in the following way (using Euler's homogeneity equation):

$$\left\{ \frac{-\partial}{\partial y} + \frac{\alpha_s \beta(\alpha_s)}{1 - \gamma_t(\alpha_s)} \frac{\partial}{\partial \alpha_s} + \frac{\gamma^{(\pm)}(\alpha_s) - \gamma'^{(\pm)}(\alpha_s')}{1 - \gamma_t(\alpha_s)} \right\} B^{(\pm)}\left(\frac{M_t}{\mu}, \alpha_s \right) = 0 . \tag{3.49}$$

This equation can be solved, via the introduction of a new running coupling constant $\bar{\alpha}_s(y,\alpha_s)$, solution of the ordinary differential equation (characteristic equation)

$$\frac{d\bar{\alpha}(y,\alpha)}{dy} = \frac{\alpha_s \beta(\alpha_s)}{1-\gamma_t(\alpha_s)} , \qquad (3.50)$$

with the boundary condition $\bar{\alpha}(0,\alpha_s) = 1$; i.e.,

$$y = \tfrac{1}{2}\log\frac{M_t^2}{\mu^2} = \int_{\alpha_s}^{\bar{\alpha}(y,\alpha_s)} \frac{1-\gamma_t(x)}{x\beta(x)}\, dx . \qquad (3.51)$$

The general solution of eq. (3.49) can then be cast in the following way

$$\mathcal{B}^{(\pm)}\left(\frac{M_t}{\mu},\alpha_s(\mu)\right) = \mathcal{B}^{(\pm)}\left[1,\bar{\alpha}(y,\alpha_s)\right] \times$$

$$\exp\left\{\int_{\alpha_s(\mu)}^{\bar{\alpha}(y,\alpha_s)} dx\, \frac{\gamma^{(\pm)}(x)}{x\beta(x)}\right\} \exp\left\{-\int_{\alpha'(y,\alpha_s(\mu))}^{\alpha'(0,\bar{\alpha}(y,\alpha_s))} dx\, \frac{\gamma'^{(\pm)}(x)}{x\beta'(x)}\right\} , \qquad (3.52)$$

where $\alpha'(\log M_t/\mu, \alpha_s(\mu))$ is the running coupling constant of an effective theory with five-flavours, i.e. it obeys the ordinary differential equation

$$\mu\frac{d\alpha'(\log M_t/\mu, \alpha_s(\mu))}{d\mu} = \alpha'\beta'(\alpha') , \qquad (3.53)$$

with

$$\beta'(\alpha') = \left(-\frac{11}{6}N_c + \frac{1}{3}n_f\right)\frac{\alpha'}{\pi} + \ldots \qquad (3.54)$$

and $n_f = 5$. Notice that the auxiliary running coupling constant defined in eq. (3.51) differs from the usual definition by the term $\gamma_t(x)/x\beta(x)$ in the integrand. It has, however, the same leading behaviour for small x as the usual definition, which for practical purposes means it will coincide with the usual definition in the LLA approximation. In this approximation we simply find

$$\mathcal{B}^{(\pm)}\left(\frac{M_t}{\mu},\alpha_s(\mu)\right) = \left(\frac{\alpha_s(M_t)}{\alpha_s(\mu)}\right)^{-a^{(\pm)}} \left(\frac{\alpha_s(M_t)}{\alpha_s(\mu)}\right)^{a'^{(\pm)}} \qquad (3.55)$$

where $a'^{(\pm)}$ are the coefficients in eq. (3.23) for $n_f = 5$.

The process of reducing the t-quark field from explicitly appearing in the operators $O_t^{(\pm)}$ is slightly more complicated due to the fact that these operators are no longer multiplicatively renormalizable. A choice of a basis of four-fermion operators which closes under renormalization at the one-loop approximation is provided by the following set of operators (summation over repeated color indices i and j is understood):

$$O_1 = 4(\bar{s}_L \gamma^\mu d_L)(\bar{u}_L \gamma_\mu u_L) \tag{3.56a}$$

$$O_2 = 4\, \bar{s}_L^i \gamma^\mu d_L^j\, \bar{u}_L^j \gamma_\mu u_L^i \tag{3.56b}$$

$$O_3 = 4(\bar{s}_L \gamma^\mu d_L) \sum_{u,d,s,c,b} (\bar{q}_L \gamma_\mu q_L) \tag{3.56c}$$

$$O_4 = 4\, \bar{s}_L^i \gamma^\mu d_L^j \sum_{u,d,s,c,b} \bar{q}_L^j \gamma_\mu q_L^i \tag{3.56d}$$

$$O_5 = 4(\bar{s}_L \gamma^\mu d_L) \sum_{u,d,s,c,b} (\bar{q}_R \gamma_\mu q_R) \tag{3.56e}$$

$$O_6 = 4\, \bar{s}_L^i \gamma^\mu d_L^j \sum_{u,d,s,c,b} \bar{q}_R^j \gamma_\mu q_R^i \, , \quad \text{where} \tag{3.56f}$$

$$q_R \equiv \frac{1}{2}(1+\gamma_5)\, q(x) \, . \tag{3.57}$$

Up to terms of order $\frac{1}{M_t^2}$, the dependence of amplitudes on the t-quark mass can be explicitly factorized as follows

$$\langle O_t^{(\pm)} \rangle = \sum_{I=1}^{6} B_I^{(\pm)}\left(\frac{M_t}{\mu}, \alpha_s(\mu)\right) \langle O_I \rangle' \, , \tag{3.58}$$

where $\langle O_I \rangle'$ means that the matrix elements of the O_I operators have to be evaluated in an effective five-quark theory with a new coupling $g_s'(M_t/\mu, g_s)$ and new mass parameters m_j', i = u, d, s, c, b, much the same as already discussed for the $O_c^{(\pm)}$ matrix elements. From the fact that the operators O_1 and O_2 already appear to zeroth order in α_s, and the others are absent, we have the boundary conditions

$$B_1^{(\pm)}(1, 0) = \pm 1 \tag{3.59a,b}$$

$$B_2^{(\pm)}(1, 0) = +1, \quad \text{and} \tag{3.60a,b}$$

$$B_I^{(\pm)}(1, 0) = 0 \quad , \quad I = 3, 4, 5, 6. \tag{3.61 a,b,c,d}$$

The operators $O_t^{(\pm)}$ have anomalous dimensions $\gamma_t^{(\pm)}$, which can be obtained from the logarithmic dependence on the renormalization scale μ of matrix elements $\langle O_t^{(\pm)} \rangle$

evaluated in the six-quark theory

$$\mu \frac{d}{d\mu} \langle O_t^{(\pm)} \rangle = - \gamma^{(\pm)}(\alpha_s) \langle O_t^{(\pm)} \rangle \quad , \tag{3.62}$$

while the operators O_I have an anomalous dimension matrix $\gamma'_{IJ}(\alpha_s')$ which can be evaluated from the logarithmic μ dependence of their matrix elements in the 5-quark theory

$$\mu \frac{d}{d\mu} \langle O_I \rangle' = - \gamma'_{IJ}(\alpha_s') \langle O_J \rangle' \quad . \tag{3.63}$$

In the language of renormalization of composite operators [11], the operators $O_I^{(\text{bare})}$ expressed in terms of bare fields and the renormalized operators O_I are related by a matrix of renormalization constants

$$O_I^{\text{bare}} = Z_{IJ}\, O_J \quad , \tag{3.64}$$

which are functions of the renormalized coupling g_s', and

$$\gamma'_{IJ}(\alpha_s') = Z^{-1}_{IK}\, \mu \frac{d}{d\mu} Z_{KJ} \quad . \tag{3.65}$$

At the one-loop level, and setting the masses of the light up, down and strange quark to zero, one finds in the Landau gauge

$$\gamma'_{IJ}(\alpha_s) = \frac{\alpha_s'}{\pi} \frac{1}{2} \begin{pmatrix} -1 & 3 & 0 & 0 & 0 & 0 \\ 3 & -1 & -1/9 & 1/3 & -1/9 & 1/3 \\ 0 & 0 & -11/9 & 11/3 & -2/9 & 2/3 \\ 0 & 0 & 22/9 & 2/3 & -5/9 & 5/3 \\ 0 & 0 & 0 & 0 & 0 & -3 \\ 0 & 0 & -5/9 & 5/3 & -5/9 & -19/3 \end{pmatrix} \quad . \tag{3.66}$$

One can derive renormalization group equations for the coefficients $B_I^{(\pm)}\left(\frac{M_t}{\mu}, \alpha_s(\mu)\right)$ in much the same way as we did for $B^{\pm}$ in eq. (3.47). Here, however, because of the matrix equation (3.63), we end up with a system of renormalization group equations:

$$\sum_{J=1}^{6} \left\{ \left(\mu \frac{\partial}{\partial \mu} + \alpha_s \beta(\alpha_s) \frac{\partial}{\partial \alpha_s} + \gamma_t(\alpha_s) M_t \frac{\partial}{\partial M_t} + \gamma^{(\pm)}(\alpha_s) \right) \delta_{IJ} - \gamma'^{T}(\alpha_s)_{IJ} \right\} B_J^{(\pm)}\left(\frac{M_t}{\mu}, \alpha_s(\mu)\right) = 0 \quad , \quad I = 1, 2, \ldots, 6 , \tag{3.67}$$

where γ'^{T} denotes the transpose of the anomalous dimension matrix (3.66). This matrix can be diagonalized by the transformation

$$W^{-1}_{IL}\,\gamma'^{T}_{LK}(\alpha_s')\,W_{KJ} = \gamma'_I(\alpha_s')\,\delta_{IJ} \quad , \tag{3.68}$$

which defines a set of anomalous dimensions eigenvalues γ'_I . They correspond to the anomalous dimension of the set of eigenoperators which are multiplicatively renormalizable. Their corresponding coefficients $\tilde{B}^{(\pm)}_I\left(\frac{M_t}{\mu}, \alpha_s(\mu)\right)$ are related to the $B^{(\pm)}_I\left(\frac{M_t}{\mu}, \alpha_s(\mu)\right)$ by the W^{-1} transformation

$$\tilde{B}^{(\pm)}_I = \sum_J W^{-1}_{IJ}\, B^{(\pm)}_J \quad . \tag{3.69}$$

Numerical values of the matrix W_{IJ} can be found in appendix III.A. The diagonal coefficients $\tilde{B}^{(\pm)}_I(M_t/\mu, \alpha_s(\mu))$ obey renormalization group equations decoupled for each value of I = 1, 2, ... 6:

$$\left\{\mu\frac{\partial}{\partial\mu} + \alpha_s\beta(\alpha_s)\frac{\partial}{\partial\alpha_s} + \gamma_t(\alpha_s)M_t\frac{\partial}{\partial M_t} + \gamma^{(\pm)}(\alpha_s) - \gamma'_I(\alpha_s')\right\}\tilde{B}^{(\pm)}_I\left(\frac{M_t}{\mu}, \alpha_s(\mu)\right) = 0 . \tag{3.70}$$

These equations can be solved in exactly the same way as eq. (3.47), i.e.

$$\tilde{B}^{(\pm)}_I\left(\frac{M_t}{\mu}, \alpha_s(\mu)\right) = \tilde{B}^{(\pm)}_I\left(1, \bar{\alpha}(y,\alpha_s)\right)\cdot$$

$$\exp\left\{\int_{\alpha_s(\mu)}^{\bar{\alpha}(y,\alpha_s)} dx\,\frac{\gamma^{(\pm)}(x)}{x\beta(x)}\right\}\exp\left\{-\int_{\alpha'(y,\alpha_s(\mu))}^{\alpha'(0,\bar{\alpha}(y,\alpha_s))} dx\,\frac{\gamma'_I(x)}{x\beta'(x)}\right\} . \tag{3.71}$$

At the one-loop approximation

$$\frac{\gamma'_I(x)}{x\beta'(x)} = \frac{2a'_I}{x} \quad . \tag{3.72}$$

Numerical values for the a'_I can be found in appendix III.A. The LLA for the $\tilde{B}^{(\pm)}$ coefficients can then be easily obtained :

$$\hat{B}^{(\pm)}_I\left(\frac{M_t}{\mu}, \alpha_s(\mu)\right) = \tilde{B}^{(\pm)}_I\left(1, \bar{\alpha}(y,\alpha_s)\right)\cdot$$

$$\left(\frac{\alpha_s(M_t)}{\alpha_s(\mu)}\right)^{-a^{(\pm)}}\left(\frac{\alpha_s(M_t)}{\alpha_s'(\mu)}\right)^{a'_I} , \tag{3.73}$$

and using the linear transformation between B's and $\tilde{B}$'s we finally have

$$B_I^{(\pm)}\left(\frac{M_t}{\mu}, \alpha_s(\mu)\right) = \left(\frac{\alpha_s(M_t)}{\alpha_s(\mu)}\right)^{-a^{(\pm)}} W_{IK} \left(\frac{\alpha_s(M_t)}{\alpha_s'(\mu)}\right)^{a_K'} \cdot$$

$$W_{KJ}^{-1}\, \tilde{B}_J^{(\pm)}\left(1, \bar{\alpha}(y, \alpha_s(\mu))\right) . \qquad (3.74)$$

We are now in the position to give the effective $\Delta S=1$ non-leptonic Hamiltonian at the LLA once the heavy W-field and the heavy t-quark field have been removed from explicitly appearing :

$$\mathcal{H}_{eff}^{\Delta S=1} = -\frac{G_F}{\sqrt{2}}\left\{ \lambda_c \left[\left(\frac{\alpha_s(M_W)}{\alpha_s(M_t)}\right)^{a^{(+)}} \left(\frac{\alpha_s(M_t)}{\alpha_s'(\mu)}\right)^{a'^{(+)}} O_c^{(+)} + (+) \leftrightarrow (-)\right]\right.$$

$$+ \lambda_t \left(\sum_{I=1}^{6} \left[\sum_{J,K} W_{IK} \left(\frac{\alpha_s(M_t)}{\alpha_s'(\mu)}\right)^{a_K'} W_{KJ}^{-1}\, B_J^{(+)}(1, \bar{\alpha}) \right] \left(\frac{\alpha_s(M_W)}{\alpha_s(M_t)}\right)^{a^{(+)}}\right.$$

$$\left.\left. + \sum_{I=1}^{6} \left[(+) \leftrightarrow (-) \text{ term} \right] \right) O_I \right\} . \qquad (3.75)$$

All matrix elements of the operators in the r.h.s. are understood to be evaluated in the effective five-quark theory.

3.3d The ΔS=1 Effective Hamiltonian with Explicit Dependence on Light-Quark Fields Only

The procedure of eliminating the b-quark field from explicitly appearing in the effective Hamiltonian of eq. (3.76) is entirely analogous to the one already described for the elimination of the t-quark field in the previous section. One has to introduce a new set of operators (with summation over color indices i and j understood):

$$P_1 = 4\, (\bar{s}_L \gamma^\mu d_L)(\bar{u}_L \gamma_\mu u_L) \qquad (3.76a)$$

$$P_2 = 4\, \bar{s}_L^i \gamma^\mu d_L^j\, \bar{u}_L^j \gamma_\mu u_L^i \qquad (3.76b)$$

$$P_3 = 4\, (\bar{s}_L \gamma^\mu d_L) \sum_{u,d,s,c} (\bar{q}_L \gamma_\mu q_L) \qquad (3.76c)$$

$$P_4 = 4\, \bar{s}_L^i \gamma^\mu d_L^j \sum_{u,d,s,c} \bar{q}_L^j \gamma_\mu q_L^i \tag{3.76d}$$

$$P_5 = 4\, (\bar{s}_L \gamma^\mu d_L) \sum_{u,d,s,c} (\bar{q}_R \gamma_\mu q_R) \tag{3.76e}$$

$$P_6 = 4\, \bar{s}_L^i \gamma^\mu d_L^j \sum_{u,d,s,c} \bar{q}_R^j \gamma_\mu q_R^i \quad , \tag{3.76f}$$

and express the matrix elements $\langle O_I \rangle'$ with explicit factorization of their M_b' dependence, i.e.,

$$\langle O_I \rangle' = \sum_{J=1}^{6} C_{IJ} \left(\frac{M_b'}{\mu}, \alpha_s'(\mu) \right) \langle P_J \rangle'' \quad , \tag{3.77}$$

where $\langle P_J \rangle''$ means that matrix elements of the operators P_J have to be evaluated in an effective four-quark theory with a new coupling contact g_s'' $(\frac{M_b'}{\mu}, g_s' \qquad)$ and new mass parameters m_i'', i = u, d, s, e. The operators P_I, which close under one-loop renormalization, have an anomalous dimension matrix $\gamma_{IJ}''(\alpha_s'')$ which has to be evaluated much the same as $\gamma_{IJ}'(\alpha_s')$ in the previous subsection. At the one-loop level

$$\gamma_{IJ}''(\alpha_s'') = \frac{\alpha_s''}{\pi}\frac{1}{2} \begin{vmatrix} -1 & 3 & 0 & 0 & 0 & 0 \\ 3 & -1 & -1/9 & 1/3 & -1/9 & 1/3 \\ 0 & 0 & -11/9 & 11/3 & -2/9 & 2/3 \\ 0 & 0 & 22/9 & 1/3 & -4/9 & 4/3 \\ 0 & 0 & 0 & 0 & 1 & -3 \\ 0 & 0 & -4/9 & 4/3 & -4/9 & -20/3 \end{vmatrix} . \tag{3.78}$$

The new coefficient functions C_{IJ} can be shown to obey a matrix renormalization group equation

$$\sum_{I,J} \Big[\Big(\mu\frac{\partial}{\partial\mu} + \alpha_s'\beta(\alpha_s')\frac{\partial}{\partial\alpha_s'} + \gamma_f'(\alpha_s') M_b' \frac{\partial}{\partial M_b'} \Big) \delta_{KI}\, \delta_{LJ}$$

$$+ \gamma_{KI}'(\alpha_s')\delta_{LJ} - \delta_{KI}\gamma_{LJ}''^{T}(\alpha_s'') \Big] C_{IJ} = 0 \, . \tag{3.79}$$

The diagonalization of γ''^{T} requires the introduction of a new transformation W'

$$W_{IL}'^{-1} \gamma''^{T}(\alpha_s'')_{LK} W_{KJ}' = \gamma_I''(\alpha_s'')\, \delta_{IJ} \quad . \tag{3.80}$$

One can then solve the renormalization group equation matrix for the C-coefficients much the same as in the case already discussed for the t-quark, with boundary conditions

$$C_{IJ}(1,0) = \delta_{IJ} . \tag{3.81}$$

The resulting expression for the $\Delta S=1$ effective Hamiltonian at the LLA with the heavy W-field, the heavy t-quark field and the heavy b-quark field removed from explicitly appearing, is then

$$\mathcal{H}_{eff}^{\Delta S=1}(x) = -\frac{G_F}{2\sqrt{2}}\Big\{\lambda_c\Big[\Big(\frac{\alpha_s(M_W)}{\alpha_s(M_t)}\Big)^{a^{(+)}}\Big(\frac{\alpha_s(M_t)}{\alpha_s'(M_b')}\Big)^{a'^{(+)}}\Big(\frac{\alpha_s'(M_b')}{\alpha_s''(\mu)}\Big)^{a''^{(+)}}O_c^{(+)}$$
$$+ (+)\Leftrightarrow(-)\ \text{term}\Big]_- + \lambda_t\sum_{J,I=1}^{6}\Big[\sum_{K,L=1}^{6} W'_{JK}\Big(\frac{\alpha_s'(M_b')}{\alpha_s''(\mu)}\Big)^{a''_K} *$$
$$* W'_{KL}C_{LI}(1,\bar\alpha')\Big]\Big(\Big[\sum_{K,L=1}^{6} W_{IK}\Big(\frac{\alpha_s(M_t)}{\alpha_s'(M_b')}\Big)^{a'_K} W^{-1}_{KL}B_L^{(+)}(1,\bar\alpha)\Big]_-\Big(\frac{\alpha_s(M_W)}{\alpha_s(M_t)}\Big)^{a^{(+)}}$$
$$+ (+)\Leftrightarrow(-)\ \text{term}\Big)P_J , \tag{3.82}$$

where all the matrix elements of the operators in the r.h.s. are understood to be evaluated in an effective four-quark theory. Numerical values of the coefficients a'_I and the matrix elements W' can be found in appendix III.A.

The final step is that of eliminating the c-quark field from explicitly appearing in the effective Hamiltonian of eq. (3.82). For that purpose, one has to introduce the set of operators

$$Q_1 = 4\,(\bar s_L\gamma^\mu d_L)(\bar u_L\gamma_\mu u_L) \tag{3.83a}$$

$$Q_2 = 4\,\bar s_L^i\gamma^\mu d_L^j\ \bar u_L^j\gamma_\mu u_L^i \tag{3.83b}$$

$$Q_3 = 4\,(\bar s_L\gamma^\mu d_L)\,[(\bar u_L\gamma_\mu u_L)+(\bar d_L\gamma_\mu d_L)+(\bar s_L\gamma_\mu s_L)] \tag{3.83c}$$

$$Q_4 = 4\,\bar s_L^i\gamma^\mu d_L^j\,[\bar u_L^j\gamma_\mu u_L^i+\bar d_L^j\gamma_\mu d_L^i+\bar s_L^j\gamma_\mu s_L^i] \tag{3.83d}$$

$$Q_5 = 4\,(\bar s_L\gamma^\mu d_L)\,[(\bar u_R\gamma_\mu u_R)+(\bar d_R\gamma_\mu d_R)+(\bar s_R\gamma_\mu s_R)] \tag{3.83e}$$

$$Q_6 = 4\,\bar s_L^i\gamma^\mu d_L^j\,[\bar u_R^j\gamma_\mu u_R^i+\bar d_R^j\gamma_\mu d_R^i+\bar s_R^j\gamma_\mu s_R^i] . \tag{3.83f}$$

These operators close under one-loop renormalization, but they are linearly dependent,

$$Q_1 + Q_4 = Q_2 + Q_3 . \tag{3.84}$$

It has become conventional to use the five operators

$$Q_1,\ Q_2,\ Q_3,\ Q_5 \text{ and } Q_6$$

as a complete basis to express the matrix elements $\langle P_I \rangle''$ with explicit factorization of their M_s'' dependence, i.e.,

$$\langle P_I \rangle'' = \sum_{R=1,2,3,5,6} D_{IR}\left(\frac{M_c''}{\mu}, \alpha_s''(\mu)\right) \langle Q_R \rangle''' , \tag{3.85}$$

where $\langle Q_R \rangle'''$ means that matrix elements of the Q_R operators have to be evaluated in an effective three-quark theory with a new coupling constant $g_s'''\left(\frac{M_c'}{\mu}, g_s''\right)$ and new mass parameters m_u''', m_d''' and m_s'''. The operators Q_R have an anomalous dimension matrix $\gamma'''_{RS}(\alpha_s''')$ which at the one-loop level has the form

$$\gamma'''_{RS}(\alpha_s''') = \frac{\alpha_s'''}{\pi}\,\frac{1}{2}\begin{vmatrix} -1 & 3 & 0 & 0 & 0 \\ 9/3 & -2/3 & 2/9 & -1/9 & 1/3 \\ -11/3 & 11/3 & 22/9 & -2/9 & 2/3 \\ 0 & 0 & 0 & 1 & -3 \\ -1 & 1 & 2/3 & -1/3 & -7 \end{vmatrix} . \tag{3.86}$$

The new coefficient functions D_{IR} obey matrix renormalization group equations, much the same as the C_{IJ} coefficient functions in eq. (3.79). Their diagonalization requires the introduction of a third transformation :

$$\sum_{R,S} W''^{-1}_{UR}\, \gamma'''^{T}_{RS}(\alpha_s''')\, W''_{ST} = \gamma_T''(\alpha_s''')\,\delta_{UT} . \tag{3.87}$$

The boundary condition for the $D_{IR}(1, \bar{\alpha}_s'')$ coefficient functions at the LLA are

$$D_{IR}(1,0) = \delta_{IR} \qquad I,R = 1,\ 2,\ 3,\ 5,\ 6 \quad \text{and} \tag{3.88a}$$

$$D_{4\,1}(1,0) = -1;\ D_{4\,2}(1,0) = D_{4\,3}(1,0) = 1 \quad ;$$
$$D_{4\,5}(1,0) = D_{4\,6}(1,0) = 0 \quad . \tag{3.88b}$$

Another new feature in the reduction of the c-quark field is the fact that the $\mathcal{O}_c^{(\pm)}$ are no longer multiplicative renormalizable. One also has to expand

$$\langle \mathcal{O}_c^{(\pm)} \rangle = D_R^{(\pm)}\left(\frac{M_c''}{\mu}, \alpha_s''\right) \langle Q_R \rangle''' . \tag{3.89}$$

The boundary condition for the $D_R^{(\pm)}(1, \bar{\alpha}_s'')$ coefficient functions at the LLA are

$$D_1^{(\pm)}(1,0) = \pm 1 \quad , \quad D_2^{(\pm)}(1,0) = +1 \text{ and } D_R^{(\pm)}(1,0) = 0 \text{ for } R = 3,5,6. \tag{3.90}$$

We are now in the position to give the final form of the $\Delta S=1$ effective Hamiltonian with only the light-quark fields u, d, s explicitly appearing in the operators Q_R:

$$\mathcal{H}_{eff}^{\Delta S=1}(x) = -\frac{G_F}{2\sqrt{2}}\Big\{\lambda_c \sum_R \Big(\Big[\sum_{S,T} W''_{RS}\Big(\frac{\alpha_s''(M_c'')}{\alpha_s'''(\mu)}\Big)^{a_S'''} W''^{-1}_{ST} D_T^{(+)}\Big]\cdot$$

$$\cdot\Big(\frac{\alpha'(M_b')}{\alpha''(M_c'')}\Big)^{a''^{(+)}}\Big(\frac{\alpha(M_t)}{\alpha'(M_b')}\Big)^{a'^{(+)}}\Big(\frac{\alpha(M_W)}{\alpha(M_t)}\Big)^{a^{(+)}} + \sum (+)\leftrightarrow(-)\ \text{terms}\Big)Q_R$$

$$+\lambda_t \sum_{I,J,R}\Big[\sum_{S,T} W''_{RS}\Big(\frac{\alpha''(M_c'')}{\alpha'''(\mu)}\Big)^{a_S'''} W''^{-1}_{ST} D_{TJ}\Big]\cdot$$

$$\Big[\sum_{K,L} W'_{JK}\Big(\frac{\alpha'(M_b')}{\alpha''(M_c'')}\Big)^{a_K''} W'^{-1}_{KL} C_{LI}\Big]\cdot$$

$$\Big[\sum_{K,L} W_{IK}\Big(\frac{\alpha(M_t)}{\alpha'(M_b')}\Big)^{a_K'} W^{-1}_{KL}\Big(B_L^{(+)}\Big(\frac{\alpha_s(M_W)}{\alpha_s(M_t)}\Big)^{a^{(+)}} + (+)\leftrightarrow(-)\ \text{terms}\Big)Q_R\Big\}, \tag{3.91}$$

where the summations over the indices R, S and T run only through 1, 2, 3, 5 and 6, and the matrix elements of the Q_R operators in the r.h.s. are understood to be evaluated in an effective three-quark theory. The numerical values of the coefficients a''' and the matrix elements W''_{RS} can be found in appendix III.A.

3.4 Decoupling of Quarks Much Heavier than the W

The derivation of $\mathcal{H}_{eff}^{\Delta S=1}(x)$ in eq. (3.91) relies on the successive approximations

$$M_W^2 \gg M_t^2 \gg M_b^2 \gg M_c^2 \gg m_q^2 \qquad q = u,d,s.$$

One may wonder how the structure of the $\Delta S=1$ effective Hamiltonian would change in the presence of quark generations with masses much higher than the charged intermediate vector boson W. We shall show below that heavy quarks with charge 2/3 and mass $M_{2/3} \gg M_W$ contribute terms of order $(1/M_W^2)(M_W^2/M_{2/3}^2)$ as compared to the contribution $(1/M_W^2)\log M_W^2/M_i^2$ for i = u,c,t and can be safely ignored for phenomenological purposes. In the language of effective field theory, they decouple with factors $1/M_{2/3}^2$. To see this, one only has to compute the leading "penguin" diagram in Fig. III.6(a) without reduction of the W-field. The exact calculation of this diagram also helps one to understand the passage from Fig. III.6(a) to Fig. III.6(b). Therefore we consider it worthwhile to do it explicitly below.

The Feynman rules for the vertex of Fig. III.6(a) give the expression

$$I_\mu^A(p,k) = i g_s \sum_i \lambda_i \int \frac{d^D q}{(2\pi)^D} \gamma^\alpha i g_w \frac{i(\not q - \not k + M_i)}{(q-p)^2 - M_i^2 + i\epsilon} \gamma_\mu T^{(A)} \times$$
$$\frac{i(\not q + M_i)}{q^2 - M_i^2 + i\epsilon} \gamma_\alpha i g_w \frac{i}{(p-q)^2 - M_W^2 + i\epsilon} , \tag{3.92}$$

where $D = 4-\epsilon$. Combining the denominators with two Feynman parameters, and after a shift of the integration momentum

$$q \rightarrow q + kxy + p\,(1-x),$$

and neglecting terms $\sim$ p, the external quark momentum, we are left with the following expression:

$$I_\mu^A(p,k) = g_w^2 g_s T^{(A)} \sum \lambda_i \, 2 \int_0^1 dx\, x \int_0^1 dy \int \frac{d^D q}{(2\pi)^D} \frac{1}{[q^2 - \mathbb{R}_i^2 + i\epsilon]^3} \times$$
$$\Big[-\frac{2}{D}(2-D) q^2 \gamma_\mu - (D-4)\frac{1}{D}(2-D) q^2 \gamma_\mu + (2-D)\gamma_\mu M_i^2 +$$
$$+ 2xy(1-xy)\left(1 + \frac{D-4}{2}\right) \not k \gamma_\mu \not k \Big] , \tag{3.93a}$$

where

$$\mathbb{R}_i^2 = M_i^2 x + M_W^2 (1-x) - k^2 xy(1-xy) . \tag{3.93b}$$

The integral over the loop momentum q can be easily done and we find

$$I_\mu^A(p,k) = - g_w^2 g_s T^{(A)} \frac{i\pi^2}{(2\pi)^4} \, 2 \int_0^1 dx\, x \int_0^1 dy \sum \lambda_i \Big\{ \gamma_\mu \log \mathbb{R}_i^2$$
$$+ [2xy(1-xy) \not k \gamma_\mu \not k - 2\gamma_\mu M_i^2] \frac{1}{2} \frac{1}{\mathbb{R}_i^2} \Big\} . \tag{3.94}$$

Since we are only interested in terms which contribute to leading order in heavy quark masses and heavy W-mass, we shall treat k^2 as small compared to M_i^2 and M_W^2 and keep only terms quadratic in the k-momentum. Then the integrals over y and x are easily done. There are also terms with no k^2 factor, proportional to γ^μ. These are cancelled out by quark wave-function renormalization from the diagrams in Fig. III.13 :

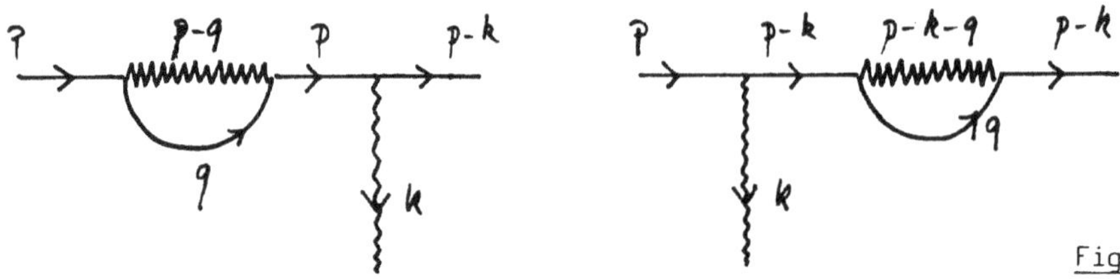

Fig. III.13

The result we find is

$$I_\mu^A(p,k) = -g_W^2 g_s T^{(A)} i \frac{\pi^2}{(2\pi)^4} 2(k_\mu k_\nu - k^2\gamma_\mu) \sum \lambda_i \left(I_2 - \frac{2}{3} I_3\right), \qquad (3.95)$$

where

$$I_n = \int_0^1 dx\, x^N \frac{1}{M_i^2 x + M_W^2(1-x)} \qquad (3.96)$$

$$I_1 = \frac{1}{M_i^2 - M_W^2}\left(1 - \frac{M_W^2}{M_i^2 - M_W^2} \log \frac{M_i^2}{M_W^2}\right) \qquad (3.97a)$$

$$I_2 = \frac{1}{2}\frac{1}{M_i^2 - M_W^2} - \frac{M_W^2}{M_i^2 - M_W^2} I_1 \qquad (3.97b)$$

$$I_3 = \frac{1}{3}\frac{1}{M_i^2 - M_W^2} - \frac{M_W^2}{M_i^2 - M_W^2} I_2 \; . \qquad (3.97c)$$

From these expressions, the following asymptotic behaviours emerge :

$$I_2 - \frac{2}{3} I_3 \simeq \begin{cases} -\frac{1}{M_W^2}\frac{1}{3}\log\frac{M_i^2}{M_W^2} \;\ldots\; \text{for } M_i^2 \ll M_W^2 & (3.98a) \\ \frac{1}{M_W^2}\frac{1}{6} \;\ldots\; \text{for } M_i^2 = M_W^2 & (3.98b) \\ \frac{1}{M_W^2}\frac{M_W^2}{M_i^2}\frac{5}{18} \;\ldots\; \text{for } M_i^2 \gg M_W^2 \, . & (3.98c) \end{cases}$$

The behaviour for $M_i^2 << M_W^2$ corresponds to the one we already found in eq. (3.30). It differs only by a $\log \mu^2$ factor which cancels because of the GIM condition $\sum \lambda_i = 0$. For $M_i^2 >> M_W^2$ we see the decoupling factor $1/M_i^2$ explicitly appearing.

3.5 Numerical Results and Phenomenological Implications

To recapitulate the context of the previous sections it is instructive to compare the final expression for $\mathcal{H}_{eff}^{\Delta S=1}$ (x) in eq. (3.91) with the initial expression in eq. (3.6) before QCD corrections and heavy quark mass effects have been taken into account. When only the fields of the light quarks u, d, and s are kept explicitly we see that eq. (3.6) corresponds to the case where only the operator Q_2 is kept, with coefficient unity, and the coefficients of all the other operators are put to zero. The QCD corrections have brought in the operator Q_1, changed the coefficient of Q_2, and brought in the "Penguin" diagrams, which appear when heavy quarks are successively removed from explicitly appearing in the Lagrangian. This results in new operators Q_3, Q_5 and Q_6 and in further changes in the structure of the Wilson coefficients.

Although $\mathcal{H}_{eff}^{\Delta S=1}$ (x) in eq. (3.91) has only u, d, and s quark fields explicitly appearing, there are CP-violating terms induced by the procedure of removing the heavy t-and b-fields via the "penguin" diagrams. Since "penguins" only generate I = 1/2 operators and renormalization does not mix I = 1/2 operators with I = 3/2 operators, the CP-violation coefficient of the Q_1-operator must be equal in magnitude and opposite in sign to the CP-violating coefficient of the Q_2 operator in order that the $Q_1 + Q_2$ combination which contains a I = 3/2 piece be purely CP-conserving. In this respect it is convenient to trade the two CKM-coefficients (see eq. (3.5a))

$$\lambda_c = -s_1 c_1 c_3 \left(c_2^2 - \frac{s_2 c_2 s_3}{c_1 c_3} e^{-i\delta} \right) \tag{3.99a}$$

and

$$\lambda_t = -s_1 c_1 c_3 \left(s_2^2 + \frac{s_2 c_2 s_3}{c_1 c_3} e^{-i\delta} \right) \tag{3.99b}$$

by their sum $\lambda_c + \lambda_t = -S_1 C_1 C_3$, which is an overall factor in $\mathcal{H}_{eff}^{\Delta S=1}$ (x), and the difference

$$\lambda_c - \lambda_t = (\lambda_c + \lambda_t)(1 - 2\tau), \quad \text{with} \tag{3.100a}$$

$$\tau = s_2^2 + \frac{s_2 c_2 s_3}{c_1 c_3} e^{-i\delta} \tag{3.100b}$$

where the CP-violation phase appears.

The only arbitrary scale appearing in $\mathcal{H}_{eff}^{\Delta S=1}$ (x) is now the renormalization scale μ^2. Because of the last step of removing the charm quark we must take $\mu^2 < m_c^2$ which brings us to an application of QCD renormalization to the GeV region. This is a point of concern for phenomenology, since long distance effects which have not been taken into account in the approach which has been used to derive eq. (3.91) may be rather inportant at such low μ^2 values. In any case, in a systematic short distance approach like the one which has been discussed, the μ^2 dependence of the coefficients in the effective hamiltonian should be cancelled with the μ^2 dependence in the evaluation of the operator matrix elements for a given process. In practise, however, the evaluation of matrix elements is done in hadronic models (bag models; harmonic oscillator models etc. ...) where the appearance of a QCD-scale like μ^2 is far from obvious and one has to resort to an arbitrary choice of $\alpha_s(\mu^2)$.

Numerical evaluations of the coefficients of the operator Q_R in eq. (3.91) for various ranges of mass parameters has been made by various groups (see, e.g. refs. [7] and [8]). Here we quote a recent evaluation by Buras and Słominski [12] which uses as input parameters the following values: the W-mass is fixed, according to the later values published by the UA2-group at CERN, to

$$M_W = 83 \text{ GeV} . \tag{3.101}$$

The top-quark mass is chosen within the range of variation

$$30 \text{ GeV} \lesssim m_t \lesssim 60 \text{ GeV} , \tag{3.102}$$

as suggested by the recent analyses of candidate events at the p-$\bar{p}$ collider, and the b-quark and c-quark masses as follows:

$$m_b = 4.6 \text{ GeV} \quad \text{and} \quad m_c = 1.4 \text{ GeV} . \tag{3.103}$$

Buras and Słominski [12] have chosen to present their results in the basis of multiplicatively renormalizable operators. As discussed in their paper, this has some phenomenological advantages. Their result reads as follows:

$$\mathcal{L}_{eff}^{\Delta I=1} = \frac{G_F}{\sqrt{2}}(\lambda_c+\lambda_t)\sum_{k=1}^{5}\left(\alpha_s'''(\mu^2)\right)^{p_k}(\alpha_k+\tau\beta_k)\,\tilde{Q}_k , \tag{3.104}$$

where $\tilde{Q}_k$ denotes the diagonal operators, and α_k, β_k and p_k are numerical coefficients. The $\tilde{Q}_R$-operators are linear combinations of the Q_R-operators defined in eq. (3.83):

$$\tilde{Q}_1 = -0.215\,(Q_2-Q_1) - 0.064\,Q_3 + 0.036\,Q_5 + Q_6 \tag{3.105a}$$

$$\tilde{Q}_2 = Q_2-Q_1 - 0.043\,Q_3 + 0.026\,Q_5 + 0.070\,Q_6 \tag{3.105b}$$

$$\tilde{Q}_3 = 0.052\,(Q_2-Q_1) + 0.165\,Q_3 + Q_5 - 0.356\,Q_6 \tag{3.105c}$$

$$\tilde{Q}_4 = Q_1 + \frac{2}{3}Q_2 - \frac{1}{3}Q_3 \tag{3.105d}$$

$$\tilde{Q}_5 = 0.602\,(Q_2-Q_1) + Q_3 - 0.202\,Q_5 + 0.152\,Q_6 . \tag{3.105e}$$

The numerical values for the powers p_k are

$$p_1 = 0.802,\quad p_2 = 0.417,\quad p_3 = -0.120,\quad p_4 = -2/9 \quad \text{and} \quad p_5 = -0.298.$$

The values of the coefficients for $\Lambda = 0.1$ GeV and various choices of m_t can be found in Table III.1 in the appendix.

The picture which emerges from this numerical evaluation is that the coefficient of Q_2-Q_1 is enhanced, while the coefficient for Q_2+Q_1 is suppressed. As mentioned before, this goes in the right direction to explain the $\Delta I = 1/2$ rule but it is not big enough. The "penguin" operators, which are $I = 1/2$, have coefficients which are smaller by a factor of roughly one order of magnitude. An explanation of the $\Delta I = 1/2$ rule via "penguins" necessarily has to resort therefore to enhancement factors in the matrix elements.

The possibility of an enhancement of the matrix elements of the (V-A) x (V+A) part of the "penguin" operators was emphasized by Shifman, Vainshtein and Zakharov in their original paper. The question has been recently reexamined in the language of effective chiral Lagrangians [13]. The quantitative details of these matrix elements as a complementary explanation of the $\Delta I = 1/2$ rule are, however, still controversial and we refer to the literature for details (see refs. [4], [7], [8], and [12], to [20]).

The appearance of an intrinsic CP-violation in $\mathcal{H}_{eff}^{\Delta S=1}$ opens the interesting possibility of a new test of the origin of CP-violation in the KM-phase δ. This we shall discuss in detail in the next section.

APPENDIX III.A

Numerical Values for the Transformations W, W', W'' and the Coefficients a', a'' and a'''

The numerical values given below have been extracted from Gilman and Wise, ref. [7].

The matrix elements of the transformation W introduced in eq. (3.68) are

$$W_{IJ} = \begin{pmatrix} 0 & -0.6948 & 0 & 0 & 0.7058 & 0 \\ 0 & 0.6948 & 0 & 0 & 0.7058 & 0 \\ 0.1504 & 0.2316 & -1.253 & 0.1668 & -0.1008 & 0.4268 \\ -0.2089 & -0.2316 & 1.084 & 0.08120 & -0.1008 & 0.8241 \\ 0.0329 & 0 & 0.1043 & 0.9392 & 0 & -0.3322 \\ 0.6169 & 0 & 0.2132 & -0.3451 & 0 & 0.2804 \end{pmatrix} . \tag{A.1}$$

The coefficients a'_I introduced in eq. (3.72) are

$$\begin{aligned} a'_1 &= -0.8994;\ a'_2 = -12/23;\ a'_3 = -0.4230 \\ a'_4 &= 0.1456;\ a'_5 = 6/23;\ a'_6 = 0.4086 \ . \end{aligned} \tag{A.2}$$

The matrix elements of the transformation W' introduced in eq. (3.80) are

$$W'_{IJ} = \begin{pmatrix} 0 & 0.6755 & 0 & 0 & 0.7060 & 0 \\ 0 & -0.6755 & 0 & 0 & 0.7060 & 0 \\ -0.1301 & -0.3378 & -1.209 & 0.1407 & -0.1177 & 0.4725 \\ 0.1827 & 0.3378 & 1.104 & 0.06713 & -0.1177 & 0.8020 \\ -0.02959 & 0 & 0.06412 & 0.9633 & 0 & -0.3002 \\ -0.6532 & 0 & 0.1497 & -0.3486 & 0 & 0.2391 \end{pmatrix} . \quad (A.3)$$

The coefficients a''_K in eq. (3.82) are

$$a''_1 = -0.8451; \; a''_2 = -12/25; \; a''_3 = -0.4201$$
$$a''_4 = 0.1317; \; a''_5 = 6/25; \; a''_6 = 0.3469 \quad . \quad (A.4)$$

The matrix elements of the transformation W" introduced in eq. (3.87) are

$$W''_{RS} = \begin{pmatrix} 0.1687 & -0.7144 & 0.05263 & 0.8485 & 0.6909 \\ -0.1687 & 0.7144 & -0.05263 & 0.5657 & -0.6909 \\ -0.05016 & -0.03095 & -0.1655 & -0.2828 & -1.148 \\ 0.02813 & 0.01873 & -1.004 & 0 & 0.2323 \\ 0.7836 & 0.04972 & 0.3573 & 0 & -0.1749 \end{pmatrix} . \quad (A.5)$$

The coefficients a'''_s in eq. (3.91) are

$$a_1 = -0.8025; \; a_2 = -0.4173; \; a_3 = 0.1196$$

$$a'''_5 = 6/27; \quad a'''_6 = 0.2977 \quad (A.6)$$

TABLE III.1 Numerical Values of the Coefficients α_k and β_k in Eq. (3.104) Obtained by Buras and Słominski [12] (Λ = 0.1 GeV)

α_k, β_k	m_t = 30 GeV	m_t = 40 GeV	m_t = 60 GeV
α_1	-0.197	-0.197	-0.197
$\beta_1 10^2$	-9.53	-10.31	-11.39
α_2	1.25	1.25	1.25
$\beta_2 10^2$	-4.62	-4.94	-5.37
α_3	0.0175	0.0176	0.0176
$\beta_3 10^2$	0.590	0.621	0.662
α_4	0.362	0.362	0.362
$\beta_4 10^2$	0	0	0
α_5	0.126	0.126	0.126
$\beta_5 10^2$	-1.56	-1.63	-1.73

IV. VIOLATION OF CP-INVARIANCE IN THE STANDARD MODEL

4.1 Phenomenological Overview of the K^o-$\bar{K}^o$ System

We shall be concerned with the complex mass matrix of the K^o-$\bar{K}^o$ system; more precisely with the complex off-diagonal term M_{12} responsible for the mixing of K^o and $\bar{K}^o$:

$$\begin{array}{cc} K^o & \bar{K}^o \end{array}$$
$$\begin{pmatrix} \mathcal{M}_{11} & \mathcal{M}_{12} \\ \mathcal{M}_{21} & \mathcal{M}_{22} \end{pmatrix} \begin{array}{c} K^o \\ \bar{K}^o \end{array} \quad . \tag{4.1}$$

In the standard model, the transition M_{12} is allowed only as a higher order process of the type (J represents the charged current; $\overline{WW}$ the W-propagator)

$$J\,\overline{W\,W}\,J\;J\,\overline{W\,W}\,J \quad , \tag{4.2}$$

thanks to the explicit breaking of flavor symmetry by the different charge 2/3 quark masses. As a result, the states K^o and $\bar{K}^o$ which are eigenstates of strangeness with S = +1 and -1 respectively are unstable, and the states obeying an exponential decay law K_S and K_L are linear superpositions of K^o and $\bar{K}^o$. Historically the early phenomenological description of K^o-$\bar{K}^o$ mixture was introduced by Gell-Mann and Pais [1], and was further developed to introduce possible violation of CP-invariance and CPT-invariance by Lee, Oehme and Yang [2], and by Sachs [3]. Soon after the discovery by Christenson, Cronin, Fitch and Turlay [4] that $K_L \to \pi^+\pi^-$, there were new phenomenological analyses by Wu and Yang [5], Bell and Steinberger [6], and Lee and Wu [7]. Here we shall limit ourselves to reproducing the essentials of the formalism in order to establish definitions and notations, and we refer to the excellent lectures of J. Steinberger in ref. [8] and the review article by Kleinknecht [R-15] for further details. (See also the lectures of Anne Marie Lutz in these proceedings.)

Assuming CPT-invariance alone, an assumption which in any case is built in in the standard model, the states $|K_L\rangle$ and $|K_S\rangle$ which diagonalize the K^o-$\bar{K}^o$ mass matrix are the linear combinations

$$|K_L\rangle = \frac{1}{\sqrt{|p|^2+|q|^2}} \left(p|K^o\rangle + q|\bar{K}^o\rangle\right) , \tag{4.3a}$$

$$|K_S\rangle = \frac{1}{\sqrt{|p|^2+|q|^2}} \left(p|K^o\rangle - q|\bar{K}^o\rangle\right) . \tag{4.3b}$$

These are the states with simple exponential decay laws (τ is the proper time):

$$|K_L\rangle \to e^{-iM_L\tau}|K_L\rangle \; , \; |K_S\rangle \to e^{-iM_S\tau}|K_S\rangle \; , \tag{4.4}$$

where the real part of $M_{L,S}$ is the mass $m_{L,S}$ and the imaginary part of $M_{L,S}$ is half of the total decay rate $\Gamma_{L,S}$ of the long-lived (L) and short-lived (S) particles:

$$M_{L,S} = m_{L,S} - \frac{i}{2}\Gamma_{L,S} \,. \tag{4.5}$$

Experimentally,

$$\Gamma_L = (1.929 \pm 0.015)\, 10^7 \ \text{sec}^{-1} \quad \text{and} \quad \Gamma_S = (1.121 \pm 0.003) \quad 10^{10}\text{sec}^{-1} \tag{4.6a,b}$$

$$M_{K^o} = (497.67 \pm 0.13) \ \text{MeV} \tag{4.7}$$

and the K_L-K_S mass difference is

$$m_L - m_S \equiv \Delta m = (3.521 \pm 0.014) \times 10^{-12} \ \text{MeV} \,. \tag{4.8}$$

These are the values quoted in the latter edition of the PDG tables [R-4]. Notice that empirically Δm is very close to $\Gamma_s/2$:

$$\frac{\Delta m}{\Gamma_s} = 0.477 \pm 0.001 \quad , \tag{4.9}$$

and that Γ_S is about 600 times larger than Γ_L.

CP-invariance requires p=q and then

$$|K_L\rangle \longrightarrow |K_2^o\rangle \quad , \quad |K_s\rangle \longrightarrow |K_1^o\rangle$$

where K_1^o and K_2^o are eigenstates of CP, with

$$CP\,|K_1^o\rangle = +\,|K_1^o\rangle \quad , \quad CP\,|K_2^o\rangle = -\,|K_2^o\rangle \tag{4.10}$$

and where we have used phase conventions such that

$$CP\,|K^o\rangle = -\,|\bar{K}^o\rangle \,. \tag{4.11}$$

Two pions, either $\pi^+\pi^-$ or $\pi^o\pi^o$, in an angular momentum state J = 0, have CP = +1; therefore if CP-invariance had been a good symmetry, only one component of the K^o-$\bar{K}^o$ system, the short-lived component, would decay into 2π. Experimentally it is found that [R-4]

$$\eta_{+-} = \frac{A(K_L \to \pi^+\pi^-)}{A(K_S \to \pi^+\pi^-)} = (2.274 \pm 0.022) \cdot 10^{-3}\, e^{i(44.6 \pm 1.2)^o} \,, \tag{4.12}$$

and

$$\eta_{oo} = \frac{A(K_L \to \pi^o\pi^o)}{A(K_S \to \pi^o\pi^o)} = (2.33 \pm 0.08) \cdot 10^{-3}\, e^{i(54 \pm 5)^o} \,, \tag{4.13}$$

indicating that CP-invariance is violated. Furthermore, it has also been found that the semi-leptonic decay rates have a charge asymmetry:

$$\delta = \frac{\Gamma(K_L \to \nu \ell^+ \pi^-) - \Gamma(K_L \to \bar{\nu} \ell^- \pi^+)}{\Gamma(K_L \to \nu \ell^+ \pi^-) + \Gamma(K_L \to \bar{\nu} \ell^- \pi^+)} = (0.330 \pm 0.012)\% \,. \tag{4.14}$$

In general, $|K_L\rangle$ and $|K_S\rangle$ are defined up to arbitrary phases, and the relative phase of the $|K^0\rangle$ and $|\bar{K}^0\rangle$ states - which are in different sectors of the strong interaction Hamiltonian - is also arbitrary. Then the complex quantity

$$\epsilon = \frac{p-q}{p+q} , \tag{4.15}$$

which is often used in the literature, is phase convention dependent; however, the $|K_L\rangle$, $|K_S\rangle$ non orthogonality(*)

$$\langle K_L | K_S \rangle = \frac{|p|^2 - |q|^2}{|p|^2 + |q|^2} = \frac{2\,\mathrm{Re}\,\epsilon}{1+|\epsilon|^2} , \tag{4.16}$$

is a physical parameter which has been determined,e.g., from the measurements of the charge asymmetry in K_{l3}-decays :

$$\mathrm{Re}\,\epsilon = (1.621 \pm 0.088) \times 10^{-3} . \tag{4.17}$$

Inverting the $|K_L\rangle$ and $|K_S\rangle$ system of equations, we get

$$|K^0\rangle = \frac{\sqrt{|p|^2+|q|^2}}{2p}\Big(|K_L\rangle + |K_S\rangle \Big) , \tag{4.18a}$$

$$|\bar{K}^0\rangle = \frac{\sqrt{|p|^2+|q|^2}}{2q}\Big(|K_L\rangle - |K_S\rangle \Big) . \tag{4.18b}$$

Then, in a period τ, a state initially pure $|K^0\rangle$ evolves to

$$\begin{aligned} |K^0\rangle \rightarrow & \frac{\sqrt{|p|^2+|q|^2}}{2p}\Big(e^{-iM_L\tau}|K_L\rangle + e^{-iM_S\tau}|K_S\rangle \Big) = \\ = & \frac{1}{2}\Big(e^{-iM_L\tau} + e^{-iM_S\tau}\Big)|K^0\rangle + \\ & + \frac{q}{2p}\Big(e^{-iM_L\tau} - e^{-iM_S\tau}\Big)|\bar{K}^0\rangle , \end{aligned} \tag{4.19a}$$

and, likewise,

$$\begin{aligned} |\bar{K}^0\rangle \longrightarrow & \frac{p}{2q}\Big(e^{-iM_L\tau} - e^{-iM_S\tau}\Big)|K^0\rangle \\ & + \frac{1}{2}\Big(e^{-iM_L\tau} + e^{-iM_S\tau}\Big)|\bar{K}^0\rangle . \end{aligned} \tag{4.19b}$$

(*) Notice that, under the assumption of CPT-invariance alone, $\langle K_L | K_S \rangle$ is a real quantity. However, under the assumption of T-invariance alone, $\langle K_L | K_S \rangle$ is purely imaginary. Experimental evidence against the latter possibility can be obtained from the unitarity equation below (see eq.(4.42)) and the experimental measurements of η_{+-} and η_{00}. See ref. [8] for a detailed discussion of this important point.

For a small period of time $\delta\tau$ we shall have

$$|K^0\rangle \rightarrow |K^0\rangle - i\delta\tau\left(\mathcal{M}_{11}|K^0\rangle + \mathcal{M}_{21}|\bar{K}^0\rangle\right) , \qquad (4.20a)$$

$$|\bar{K}^0\rangle \rightarrow |\bar{K}^0\rangle - i\delta\tau\left(\mathcal{M}_{12}|K^0\rangle + \mathcal{M}_{22}|\bar{K}^0\rangle\right) , \qquad (4.20b)$$

where

$$\mathcal{M}_{ij} = \frac{1}{2}\begin{pmatrix} M_L + M_S & \frac{p}{q}(M_L - M_S) \\ \frac{q}{p}(M_L - M_S) & M_L + M_S \end{pmatrix} . \qquad (4.21)$$

This is the complex mass matrix of the K^0-$\bar{K}^0$ system introduced in eq. (4.1):

$$\mathcal{M}_{ij} = M_{ij} - i\frac{\Gamma_{ij}}{2} , \qquad (4.22)$$

where Γ_{ij} denotes the absorptive part and M_{ij} the dispersive part, i.e.,

$$\Gamma_{12}(M_K^2) = \sum_{\Gamma}\int d\Gamma\left(\langle\Gamma|T|\bar{K}^0\rangle\right)^*\langle\Gamma|T|K^0\rangle \qquad (4.23)$$

with summation extended to all possible states Γ allowed by phase space, and

$$M_{12} = \langle\bar{K}^0|T|K^0\rangle + \frac{1}{\pi}P\int_0^\infty ds\,\frac{1}{M_K^2 - s}\,\Gamma_{12}(s) . \qquad (4.24)$$

Here, T denotes the transition matrix which in the standard model has to be evaluated, via perturbation theory, using the $SU(3)_c \times SU(2)_L \times U(1)$ interaction Lagrangian at the required order. In the explicit evaluation of M_{12}, where usually a short-distance evaluation of the effective T-operator is made, care must be taken so as not to double count dispersive contributions in the second term. We shall come back to this point later on.

From eq. (4.21) and (4.15) we then have

$$\frac{1-\epsilon}{1+\epsilon} = \frac{q}{p} = \frac{1}{2}\,\frac{\Delta m + i\,\Delta\Gamma/2}{M_{12} - i\,\Gamma_{12}/2} = \frac{M_{21} - i\,\Gamma_{21}/2}{\frac{1}{2}(\Delta m + i\,\Delta\Gamma/2)} \qquad (4.25)$$

where (beware that with our definitions Δm and $\Delta\Gamma$ are positive)

$$\Delta m \equiv m_L - m_S \quad \text{and} \quad \Delta\Gamma \equiv \Gamma_S - \Gamma_L , \qquad (4.26a,b)$$

and, because of CPT-invariance,

$$M_{21} = M_{12}^* , \qquad \Gamma_{21} = \Gamma_{12}^* . \qquad (4.27a,b)$$

To proceed any further we need to find out how ϵ appears in the amplitudes observed in $K \rightarrow \pi\pi$ decays, i.e. in the quantities conventionally called

$$\eta_{+-} = \frac{A(K_L \to \pi^+\pi^-)}{A(K_S \to \pi^+\pi^-)} \quad \text{and} \quad \eta_{00} = \frac{A(K_L \to \pi^0\pi^0)}{A(K_S \to \pi^0\pi^0)}, \tag{4.28a,b}$$

which experimentally have the values indicated in eqs. (4.12) and (4.13).

Because of Bose statistics the $\pi^+\pi^-$ state and the $\pi^0\pi^0$ state from K-decays can be in an isospin I = 0 and I = 2 only. It is then convenient to introduce the amplitudes (the i factor comes because of our convention CP $|K^0\rangle = -|\bar{K}^0\rangle$)

$$\langle(\pi\pi)_I | T | K^0\rangle = i A_I e^{i\delta_I}, \quad I = 0,2, \tag{4.29a,b}$$

where δ_I denotes the S-wave $\pi\pi$ phase shifts with isospin I. Then CPT-invariance plus Watson's theorem on final state interactions imply that

$$\langle(\pi\pi)_I | T | \bar{K}^0\rangle = -i A_I^* e^{i\delta_I}, \quad I = 0,2. \tag{4.30a,b}$$

Notice that CP-invariance requires, furthermore, the relative reality of A_0 and A_2, i.e., $A_2 = \lambda A_0$ with λ <u>real</u>. In general A_0 and A_2 are two independent complex numbers. It is now natural to introduce a complete system of parameters to describe the K_L and K_S transitions into two pions in terms of these isospin amplitudes. We shall call them (this is not always a conventional notation)

$$\epsilon_0 = \frac{A[K_L \to (\pi\pi)_{I=0}]}{A[K_S \to (\pi\pi)_{I=0}]} = \frac{(1+\epsilon)A_0 - (1-\epsilon)A_0^*}{(1+\epsilon)A_0 + (1-\epsilon)A_0^*}, \tag{4.31}$$

$$\epsilon_2 = \frac{A[K_L \to (\pi\pi)_{I=2}]}{A[K_S \to (\pi\pi)_{I=0}]} = \frac{(1+\epsilon)A_2 - (1-\epsilon)A_2^*}{(1+\epsilon)A_0 + (1-\epsilon)A_0^*} e^{i(\delta_2-\delta_0)}, \tag{4.32}$$

and

$$\omega = \frac{A[K_S \to (\pi\pi)_{I=2}]}{A[K_S \to (\pi\pi)_{I=0}]} = \frac{(1+\epsilon)A_2 + (1-\epsilon)A_2^*}{(1+\epsilon)A_0 + (1-\epsilon)A_0^*} e^{i(\delta_2-\delta_0)}. \tag{4.33}$$

They can also be written in the following way, perhaps more transparently,

$$\epsilon_0 = \frac{\epsilon + i\dfrac{\mathrm{Im}A_0}{\mathrm{Re}A_0}}{1 + i\epsilon\dfrac{\mathrm{Im}A_0}{\mathrm{Re}A_0}}, \tag{4.34}$$

$$\epsilon_2 = \frac{i\dfrac{\mathrm{Im}A_2}{\mathrm{Re}A_0} + \epsilon\dfrac{\mathrm{Re}A_2}{\mathrm{Re}A_0}}{1 + i\epsilon\dfrac{\mathrm{Im}A_0}{\mathrm{Re}A_0}} e^{i(\delta_2-\delta_0)}, \tag{4.35}$$

and

$$\omega = \frac{\frac{\mathrm{Re}A_2}{\mathrm{Re}A_0} + i\epsilon \frac{\mathrm{Im}A_2}{\mathrm{Re}A_0}}{1 + i\epsilon \frac{\mathrm{Im}A_0}{\mathrm{Re}A_0}} e^{i(\delta_2 - \delta_0)} . \tag{4.36}$$

In terms of these parameters, the observable η_{+-} and η_{00} in eqs. (4.28) read as follows:

$$\eta_{+-} = \frac{\epsilon_0 + \frac{1}{\sqrt{2}}\epsilon_2}{1 + \frac{\omega}{\sqrt{2}}} , \quad \eta_{00} = \frac{\epsilon_0 - \sqrt{2}\,\epsilon_2}{1 - \omega\sqrt{2}} . \tag{4.37a,b}$$

In order to further separate different possible sources of CP-violation it is convenient to introduce yet another parameter conventionally called ϵ':

$$\epsilon' = \frac{1}{\sqrt{2}}\,\omega\left(\frac{\epsilon_2}{\omega} - \epsilon_0\right) . \tag{4.38}$$

From the previous expressions, it can be seen after some straightforward algebra that

$$\epsilon' = \frac{i}{\sqrt{2}} \frac{(1-\epsilon^2)e^{i(\delta_2-\delta_0)}}{(\mathrm{Re}A_0 + i\epsilon\,\mathrm{Im}A_0)^2}\left(\mathrm{Im}A_2\,\mathrm{Re}A_0 - \mathrm{Im}A_0\,\mathrm{Re}A_2\right) , \tag{4.39}$$

i.e., ϵ' is proportional to the lack of relative reality of the isospin amplitudes A_0 and A_2. Observation of $\epsilon' \neq 0$ implies

$$\mathrm{Im}A_2\,\mathrm{Re}A_0 \neq \mathrm{Im}A_0\,\mathrm{Re}A_2$$

and hence, CP-violation in the weak $\Delta S=1$ non-leptonic Hamiltonian. The amplitudes η_{+-} and η_{00} when expressed in terms of ϵ' read as follows:

$$\eta_{+-} = \epsilon_0 + \epsilon' \frac{1}{1 + \frac{\omega}{\sqrt{2}}} , \quad \text{and} \tag{4.40a}$$

$$\eta_{00} = \epsilon_0 - 2\epsilon' \frac{1}{1 - \omega\sqrt{2}} . \tag{4.40b}$$

All these phenomenological parameters are constrained by the Bell-Steinberger [6] unitarity requirements(*)

(*) These equations follow from the fact that the total probability for the decay of an arbitrary K-meson state

$$|\psi\rangle = a|K_S\rangle + b|K_L\rangle$$

with a and b arbitrary complex parameters, must be compensated by a decrease of the norm of the ψ state:

$$\sum_f |\langle f|T|\psi\rangle|^2 = -\frac{d}{d\tau}|\psi|^2 .$$

$$\Gamma_L = \sum_{\Gamma} \int d\Gamma \, |\langle \Gamma | T | K_L \rangle|^2 , \tag{4.41a}$$

$$\Gamma_S = \sum_{\Gamma} \int d\Gamma \, |\langle \Gamma | T | K_S \rangle|^2 \tag{4.41b}$$

and

$$-i\left(M_L^* - M_S\right) \langle K_L | K_S \rangle = \sum_{\Gamma} \int d\Gamma \left(\langle \Gamma | T | K_L \rangle\right)^* \langle \Gamma | T | K_S \rangle \, . \tag{4.42}$$

The latter is particularly interesting. Using the Schwartz inequality it tells us that

$$\left| \frac{\Gamma_S + \Gamma_L}{2} - i\Delta m \right| \frac{2 \operatorname{Re} \epsilon}{1+|\epsilon|^2} \leq \Gamma_L^{1/2} \Gamma_S^{1/2} , \tag{4.43}$$

an interesting inequality which can also be applied to D^0-$\bar{D}^0$ mixing and B^0-$\bar{B}^0$ mixing. In the case of the K^0-$\overline{K^0}$ system one can get more useful information from eq. (4.42) by explicitly writing the contribution of 2π-states, which is the dominant one, 3π-decays, etc. (*):

$$\left(\frac{\Gamma_S + \Gamma_L}{2} - i\Delta m \right) \frac{2 \operatorname{Re} \epsilon}{1+|\epsilon|^2} = \left(\epsilon_0^* + \epsilon_2^* \omega\right) \Gamma_S(I=0) + \ldots \quad . \tag{4.44}$$

A further constraint on ω is provided by the branching ratio

$$\frac{\Gamma(K_S \to \pi^+\pi^-)}{\Gamma(K_S \to \pi^0\pi^0)} = 2 \left| \frac{1 + \frac{1}{\sqrt{2}}\omega}{1 - \omega\sqrt{2}} \right|^2 P.S. \left(1 + R.C.\right) , \tag{4.45}$$

where P.S. denotes phase-space ratio, and R.C. radiative corrections of $O\left(\frac{\alpha}{\pi}\right)$ (**). Experimentally,

$$\frac{\Gamma(K_S \to \pi^+\pi^-)}{\Gamma(K_S \to \pi^0\pi^0)} = 2.19 \pm 0.02 , \tag{4.46}$$

consistent with the $\Delta I = 1/2$ prediction $\omega = 0$.

All the equations derived so far are exact, and, except for the choice $CF|K^0\rangle = -|\bar{K}^0\rangle$ in eq. (4.11), phase convention independent. In principle, the $K \to 2\pi$ system can be characterized by three phenomenological complex parameters: ϵ_0, ϵ' and ω. Two complex constraints are known experimentally: η_{+-} and η_{00}, and a further constraint on ω: eq. (4.45). The unitarity condition in eq. (4.44) provides an overall constraint. It is possible to argue from these constraints that ϵ,

---

(*) For an early application to the phenomenology of K-decays see, e.g. ref. [9].

(**) For an evaluation of the radiative corrections, prior to the development of the standard model, see ref. [10].

ϵ_0, ϵ' and ω must be small parameters, and, to a good approximation, one can write

$$\eta_{+-} \simeq \epsilon_0 + \epsilon' \quad , \quad \eta_{00} \simeq \epsilon_0 - 2\epsilon' \quad . \tag{4.47a,b}$$

Traditionally, since the early paper by Wu and Yang [5], it has become customary to freeze the freedom on the choice of the relative phase of $|K^0\rangle$ and $|\bar{K}^0\rangle$ by fixing the isospin amplitude A_0 to be real. In the formulation of the standard model, however, the convention which is implicitly made by the way that the arbitrary phases have been fixed in the CKM matrix is the one where, on the contrary, A_2 is real. This is the convention we shall adopt from here onwards. Then, we can write

$$\epsilon_0 \simeq \epsilon + i \frac{\mathrm{Im} A_0}{\mathrm{Re} A_0} \quad , \quad \text{and} \tag{4.48}$$

$$\epsilon' \simeq \frac{-i}{\sqrt{2}} e^{i(\delta_2 - \delta_0)} \frac{A_2}{\mathrm{Re} A_0} \frac{\mathrm{Im} A_0}{\mathrm{Re} A_0} \quad , \tag{4.49}$$

$$1 - 2\epsilon \simeq \frac{\mathrm{Re} M_{12} - i \mathrm{Re} \frac{\Gamma_{12}}{2} - i\left(\mathrm{Im} M_{12} - i \mathrm{Im} \frac{\Gamma_{12}}{2}\right)}{\frac{1}{2}\left(\Delta m + i \frac{\Delta\Gamma}{2}\right)} \quad . \tag{4.50}$$

Again, empirically, Γ_{12} is dominated by the $(2\pi)_{I=0}$ intermediate state, and to a good approximation

$$\Gamma_{12} \simeq \left(i A_0^* e^{i\delta_0}\right)^* i A_0 e^{i\delta_0} = -(\mathrm{Re} A_0 + i \mathrm{Im} A_0)^2 \quad ; \tag{4.51}$$

therefore

$$\frac{\mathrm{Im} \Gamma_{12}}{\mathrm{Re} \Gamma_{12}} \simeq 2 \frac{\mathrm{Im} A_0}{\mathrm{Re} A_0} \quad . \tag{4.52}$$

To a first approximation, we have from eq. (4.50)

$$\mathrm{Re} M_{12} \simeq \frac{1}{2} \Delta m \quad \text{and} \quad \mathrm{Re} \Gamma_{12} \simeq -\frac{\Delta\Gamma}{2} \quad , \tag{4.53a}$$

and empirically (see eq. (4.9)) we know that

$$\Delta m \simeq \frac{\Gamma_S}{2} \quad \text{and} \quad \Gamma_L \ll \Gamma_S \quad . \tag{4.54a,b}$$

These approximations incorporated in eq. (4.50) result in a simple compact expression for ϵ:

$$\epsilon = \frac{1}{1+i}\left[i \frac{\mathrm{Im} M_{12}}{2 \mathrm{Re} M_{12}} + \frac{\mathrm{Im} A_0}{\mathrm{Re} A_0}\right] \quad . \tag{4.55}$$

With this expression inserted in eq. (4.48) we have

$$\epsilon_0 = \frac{1}{\sqrt{2}} e^{i\pi/4} \left(\frac{\mathrm{Im}\, M_{12}}{2\,\mathrm{Re}\, M_{12}} + \frac{\mathrm{Im}\, A_0}{\mathrm{Re}\, A_0} \right) . \qquad (4.56)$$

Solving eqs. (4.47) for ϵ_0 and ϵ' gives

$$\epsilon_0 = \frac{1}{3} \eta_{00} + 2\eta_{+-} \quad \text{and} \quad \epsilon' = \frac{1}{3} (\eta_{+-} - \eta_{00}) , \qquad (4.57a,b)$$

to be confronted with the phenomenological expressions

$$\epsilon_0 = \frac{1}{\sqrt{2}} e^{i\pi/4} \frac{1}{\Delta m} \left(\mathrm{Im}\, M_{12} + 2 \frac{\mathrm{Im}\, A_0}{\mathrm{Re}\, A_0} \mathrm{Re}\, M_{12} \right) , \qquad (4.58)$$

and

$$\epsilon' = \frac{1}{\sqrt{2}} e^{i(\delta_2 - \delta_0 - \frac{\pi}{2})} \frac{A_2}{\mathrm{Re}\, A_0} \frac{\mathrm{Im}\, A_0}{\mathrm{Re}\, A_0} . \qquad (4.59)$$

The experimental value of the π-π phase shift difference which appears in ϵ' is known experimentally. The most recent result [11] is

$$\delta_0 - \delta_2 = 29.2^\circ \pm 3^\circ . \qquad (4.60)$$

There are several experiments in progress which can measure the ratio ϵ'/ϵ_0 to an accuracy of a few parts in a thousand. This brings in the interesting possibility to resolve the question of a superweak mechanism [12] versus a milliweak mechanism of CP-violation. The standard model falls into the latter class and, as we shall discuss in the following, it is possible to make predictions about the ratio ϵ'/ϵ_0.

4.2 Calculation of the Off-Diagonal K^0-$\bar{K}^0$ Mass Matrix Element in the Standard Model

Here we shall be concerned with the calculation of the matrix element M_{12} (see eqs. (4.1), (4.22), and (4.24)) within the framework of the standard model. To lowest order in the weak coupling, this matrix element is generated by the Feynman diagrams shown below (the so-called box diagrams):

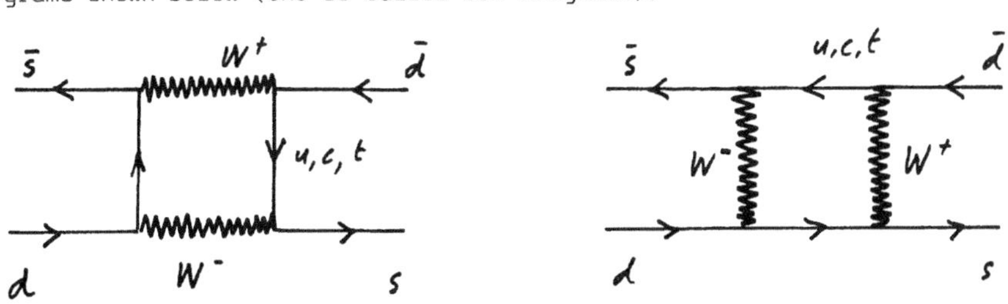

Fig. IV-I

In principle, the calculation of M_{12} involves the evaluation of these diagrams in the presence of gluonic exchanges between the quarks with the $\bar{s}d$ and $s\bar{d}$ quarks confined in the K^0 and $\bar{K}^0$ wave functions. Clearly, the calculation is a formidable task and one is obliged to split it into various pieces:

i) short-distance contribution (free quark model)

ii) QCD-corrections to i)

iii) evaluation of the matrix element of the effective $\Delta S=2$ Hamiltonian resulting from steps i) and ii) - this is the (in)famous B-factor -

iv) corrections due to long-distance hadronic effects, NOT already included in the previous steps.

In what follows, we shall first review the present state of the art in the evaluation of these successive steps, then discuss the phenomenological implications of the results in the next subsection.

4.2a The Short-Distance Contribution to M_{12}

The calculation of an effective $\Delta S=2$ Hamiltonian from the GSW-Lagrangian in the free quark approximation and for two-generations was made by Gaillard and Lee [13]. It corresponds to the calculation of the box diagrams shown in Fig. IV.1. Here, the short-distance scale is provided by the large W-mass and the breaking of the GIM-cancellation by the difference between the charm-quark mass and the up-quark mass. This calculation, extended to 3-generations, and keeping the exact mass dependence of the virtual propagators, has been made by Buras [14], and by Inami and Lim[15] with the following result:

$$M_{12} = \frac{G_F^2 M_W^2}{16\pi^2} \sum_{i,j=u,c,t} \lambda_i \lambda_j A_{ij} \left(x_i \equiv \frac{m_i^2}{M_W^2},\ x_j \equiv \frac{m_j^2}{M_W^2} \right) \times$$

$$\frac{1}{2M_K} \langle \bar{K}^0 | \bar{s}_\alpha \gamma^\mu (1-\gamma_5) d_\alpha\ \bar{s}_\beta \gamma_\mu (1-\gamma_5) d_\beta | K^0 \rangle , \qquad (4.61)$$

where summation over the repeated color indices α and β is understood. It can be seen directly from the diagrams in Figs. IV.I that the dependence on the CKM matrix elements V_{ij} must be a quadratic form in the quantities

$$\lambda_i \equiv V_{id} V_{is}^* \quad , \quad i = u,c,t \ . \qquad (4.62)$$

The dynamics is contained in the coefficients A_{ij} which are functions of the mass ratios

$$x_i \equiv \frac{m_i^2}{M_W^2} \quad , \quad i = up,\ charm,\ top \ . \qquad (4.63)$$

When i=j, there is only one type of charge 2/3 quark in the loop and

$$A_{ii} = \frac{x_i}{(1-x_i)^2}\left[3 - \frac{19}{4}x_i + \frac{1}{4}x_i^2\right] + 2\left[1 - \frac{3x_i^2}{4(1-x_i)}\right]\frac{x_i \log x_i}{(1-x_i)^2} \quad . \tag{4.64}$$

For i = j, the coefficient function A_{ij} depends on two-mass ratios

$$A_{ij} = \frac{x_i + x_j - \frac{11}{4}x_i x_j}{(1-x_i)(1-x_j)} + \frac{1}{x_i - x_j}\left[\frac{x_i^2(1-2x_j+\frac{1}{4}x_i x_j)}{(1-x_i)^2}\log x_i - i \leftrightarrow j\right] . \tag{4.65}$$

For $X_i \ll 1$, these expressions reduce to the simple form

$$A_{ii} = 3x_i + 2x_2 \log x_i \quad , \tag{4.66}$$

and

$$A_{ij} = x_i + x_j + \frac{1}{x_i - x_j}\left(x_i^2 \log x_i - x_j^2 \log x_j\right) \quad . \tag{4.67}$$

For two-generations, and with $x_u = 0$, we simply have

$$A_{uu} = 0 \quad ; \quad A_{cc} = 3x_c + 2x_c \log x_c \; ; \tag{4.68a}$$

$$A_{uc} = A_{cu} = X_c + X_c \log x_c \quad . \tag{4.68b}$$

Furthermore, in this case there is only one mixing angle, the Cabbibo angle, and

$$\lambda_u \lambda_u = \lambda_c \lambda_c = (\sin\vartheta_c \cos\vartheta_c)^2 \tag{4.69a}$$

$$\lambda_u \lambda_c = \lambda_c \lambda_u = -(\sin\vartheta_c \cos\vartheta_c)^2 . \tag{4.69b}$$

Then

$$M_{12} = \frac{G_F^2 M_W^2}{16\pi^2}(\sin\vartheta_c \cos\vartheta_c)^2 \frac{m_c^2}{M_W^2} \times$$

$$\frac{1}{2M_K}\langle \bar{K}^0 | \bar{s}_\alpha \gamma^\mu (1-\gamma_5) d_\alpha \, \bar{s}_\beta \gamma_\mu (1-\gamma_5) d_\beta | K^0 \rangle \quad . \tag{4.70}$$

Since M_{12} is related to the K_L-K_S mass difference by eq. (4.53a),

$$m_L - m_S = 2\mathrm{Re}\, M_{12} \quad ,$$

eq. (4.70) can be used as an estimate of the charm quark mass, provided one knows the K^0-$\bar{K}^0$ matrix element of the effective $\Delta S=2$ operator. In their pioneering paper Gaillard and Lee [13] used the vacuum saturation assumption (see section 4.2c) to estimate the latter and concluded from their analyses that m_c should not be too large:

$$m_c \text{ (Gaillard-Lee)} \simeq 1.2 \text{ GeV} \quad .$$

The merit of this "educated guess" lies in the fact that it was made prior to the discovery of charm!

We can also see that in the two-generation case M_{12} is real, i.e. the CKM-matrix, when reduced to the Cabibbo matrix, cannot provide an imaginary part to M_{12}.

Let us next discuss M_{12} in the case of three-generations. We can use the unitarity constraint

$$\sum_{i=u,c,t} \lambda_i \equiv V_{id} V_{is}^* = 0$$

to eliminate one of the λ_i-parameters. Keeping in mind that eventually we wish to discuss gluonic corrections, the most convenient choice is to eliminate λ_u, as we did in the discussion of the effective $\Delta S=1$ Hamiltonian in section III, i.e., we replace

$$\lambda_u = -\lambda_c - \lambda_t$$

in eq. (4.61) which results in

$$M_{12} = \frac{G_F^2 M_W^2}{16\pi^2} \Big\{ (\lambda_c)^2 (A_{uu} - 2A_{uc} + A_{cc}) + (\lambda_t)^2 (A_{tt} - 2A_{ut} + A_{uu}) + 2\lambda_c\lambda_t (A_{ct} + A_{uu} - A_{ut} - A_{uc}) \Big\} \cdot \frac{1}{2M_K} \langle \bar{K}^0 | \bar{s}_\alpha \gamma^\mu (1-\gamma_5) d_\alpha \, \bar{s}_\beta \gamma_\mu (1-\gamma_5) d_\beta | K^0 \rangle \quad . \tag{4.71}$$

Notice that the leading behaviour, for $X_i \ll 1$, of these three combinations of A_{ij} amplitudes becomes (we always let $X_u = 0$)

$$A_{uu} - 2A_{uc} + A_{cc} \rightarrow X_c \tag{4.72a}$$

$$A_{tt} - 2A_{ut} + A_{uu} \rightarrow X_t \quad \text{and} \tag{4.72b}$$

$$A_{ct} + A_{uu} - A_{uc} - A_{ut} \rightarrow X_c \log \frac{X_t}{X_c} \quad . \tag{4.72c}$$

The 3-generation expression for M_{12} has now both a real part, which contributes to the K_L-K_S mass difference, and an imaginary part which induces CP-violation in the K^o-$\bar{K}^o$ system. There are three sources of contributions to Im M_{12}:

$$\mathrm{Im}(\lambda_c)^2 \simeq 2\gamma\beta\,\theta^6 \sin\delta \,, \tag{4.73a}$$

$$\mathrm{Im}(\lambda_t)^2 \simeq -2\gamma^2\beta\,\theta^{10} (\gamma + \beta\cos\delta) \sin\delta \,, \tag{4.73b}$$

$$\mathrm{Im}\, 2\lambda_c\lambda_t \simeq 2\gamma\beta\,\theta^6 \sin\delta \,, \tag{4.73c}$$

where we have used the Cabibbo expanded version of the CKM matrix. The induced CP-violation is a small effect ($\sim \theta^6$) and, regardless of the CP-violating phase value, vanishes if either γ or β is null. We keep the term $\sim \theta^{10}$ from $\mathrm{Im}(\lambda_t)^2$, because in the expression of Im M_{12} it appears multiplied by m_t^2 which is a big factor.

The short-distance contribution to Re M_{12} is modulated by the coefficients

$$\mathrm{Re}(\lambda_c)^2 \simeq \theta^2\left(1-\frac{5}{6}\theta^2\right), \tag{4.74a}$$

$$\mathrm{Re}(\lambda_t)^2 \simeq \gamma^2\theta^{10}\left(\gamma^2+2\gamma\beta\cos\delta+\beta^2\cos 2\delta\right), \tag{4.74b}$$

$$\mathrm{Re}\,\lambda_c\lambda_t \simeq 2\gamma\theta^6\left(\gamma+\beta\cos\delta\right). \tag{4.74c}$$

Finally, we wish to make a remark concerning the short-distance contribution to M_{12} from charge 2/3 quarks much heavier than the W-mass. It can be readily seen from the general expressions for A_{ii} and A_{ij} in eq. (4.64) and (4.65) that contrary to what happens in the case of "penguin" contributions to the $\Delta I=1/2$ (see section (III.4)), here heavy quarks give contributions proportional to their mass squared, with, say, $M_T^2 \gg M_W^2$,

$$A_{TT} \longrightarrow \frac{1}{4}\frac{M_T^2}{M_W^2}+\frac{3}{2}\lg\frac{M_T^2}{M_W^2}. \tag{4.75}$$

This is another interesting case of non-decoupling of ultraheavy fermions in the standard model of the type first discussed by Veltman [16]. In this case, however, the contribution is modulated by the CKM-factor λ_T^2, which by unitarity is constrained to be rather small.

4.2b QCD-Corrections to the Short-Distance Evaluation of M_{12}.

The development of QCD gave a new dimension to the Gaillard and Lee [13] calculation of the K^o-$\bar{K}^o$ mixing. It was clear that their result should correspond to the asymptotic freedom behaviour only, and there should be calculable gluonic corrections to this behaviour. Early calculations by Vainshtein et al. [17], and Novikov et al. [18] have been put on a firm basis after Witten's analyses of the short-distance behaviour of weak interactions [19]. In the case of the K^o-$\bar{K}^o$ mixing with three-generations this has been carried through explicitly in papers by Gilman and Wise [20], [21]. The technique is rather similar to the calculation of the effective $\Delta S=1$ non-leptonic Hamiltonian which we reviewed in detail in the previous section. Therefore, here we shall limit ourselves to a descriptive overview of the calculations.

The procedure consists in successively removing the heavy W-field from explicity appearing in the $\Delta S=2$ Hamiltonian; then the heavy t-quark field; the b-quark field and finally the c-quark field. This is done at the one-loop level. Successive correction factors

$$\left(\frac{\alpha_s(M_W)}{\alpha_s(M_t)}\right)^a ; \left(\frac{\alpha_s(M_t)}{\alpha_s'(M_b')}\right)^b ; \left(\frac{\alpha_s'(M_b')}{\alpha_s''(M_c'')}\right)^c ; \left(\frac{\alpha_s''(M_c'')}{\alpha_s'''(\mu)}\right)^d$$

appear, as solutions of the various renormalization group equations of the $\Delta S=2$ operator coefficients. Here $\alpha_s(M)$ means that the running α_s of QCD has to be evaluated in an effective 6-quark theory; α' and M' mean that they have to be evaluated in an effective 5-quark theory; double primes in an effective 4-quark theory; and so on. The exponents a, b, c and d are governed by the one-loop coefficient of the anomalous dimension of the successive $\Delta S=2$ operators.

In fact, Gilman and Wise [21], in the process of extracting the heavy t-quark field from explicitly appearing in the $\Delta S=2$ Hamiltonian, find eight operators which mix under one-loop renormalization. Four of these operators are induced by the strong interactions and have coefficients which vanish in the limit when the strong interactions are switched off. They are dismissed, precisely because of this reason. Of the four operators left, two of them produce large m_c^2 factor; the others do not, and they do not mix directly with the operators which produce large factors. They are also dismissed on these grounds. Finally, of the two operators which remain, only one is left when the charm field is removed from explicitly appearing in the Lagrangian.

The approximations of the Gilman and Wise calculation sound reasonable, provided one is concentrating on the short-distance contribution to the $\Delta S=2$ Hamiltonian, i.e., the approximation of keeping leading m_q^2 factors only. The operators left out although non-leading in heavy quark masses may well give sizable contributions when evaluated between the K^0 and $\bar{K}^0$ states, i.e., long distance contributions to non-leading short-distance operators. This is a possible loophole to keep in mind when confronting the Gilman and Wise results to phenomenology.

To get an idea of the QCD-corrections to the free quark expression in eq. (4.71) - with the leading short-distance expressions eq. (4.72) incorporated - we can rewrite $M_{12}^{(QCD)}$ in the following way:

$$M_{12}^{(QCD)} = \frac{G_F^2 M_W^2}{16\pi^2}\left\{\lambda_c^2 \eta_1 x_c + \lambda_t^2 \eta_2 x_t + 2\lambda_c\lambda_t \eta_3 x_c \log\frac{x_t}{x_c}\right\} \times \frac{1}{2M_K} \langle \bar{K}^0 | \bar{s}_\alpha \gamma^\mu (1-\gamma_5) d_\alpha \, \bar{s}_\beta \gamma_\mu (1-\gamma_5) d_\beta | K^0 \rangle , \qquad (4.76)$$

where η_1, η_2 and η_3 are QCD-corrective factors. In view of the approximations which lead to these factors we do not think it is consistent to apply them to the full exact expression, in terms of masses, of the box calculation. Numerically, with M_W = 80 GeV, m_c = 1.5 GeV and m'_b = 4.5 GeV, Gilman and Wise find - with m_t = 30 GeV -

$$\eta_1 = 0.69 \; , \quad \eta_2 = 0.59 \quad \text{and} \quad \eta_3 = 0.41 \; , \tag{4.77a}$$

for Λ''^2 (the Λ of two-generations) = 0.01 GeV², and

$$\eta_1 = 0.99 \; , \quad \eta_2 = 0.60 \quad \text{and} \quad \eta_3 = 0.40 , \tag{4.77b}$$

for Λ''^2 = 0.1 GeV².

4.3 The "Infamous" B-Factor

In the short-distance evaluation of M_{12} one is finally confronted with the problem of calculating the matrix element of the 4-fermion $\Delta S=2$ operator - $(\bar{s}_L \gamma^\mu d_L) \times (\bar{s}_L \gamma_\mu d_L)$ - between K^0 and $\bar{K}^0$ states. In their early paper, Gaillard and Lee [13] estimated this matrix element via the assumption of vacuum saturation. The assumption consists in inserting the vacuum state in all possible ways in the matrix element:

$$\langle \bar{K}^0 | (\bar{s}_L \gamma^\mu d_L)(\bar{s}_L \gamma_\mu d_L) | K^0 \rangle \Big|_{V.S.} = \frac{1}{4} 2 \left(1 + \frac{1}{3}\right) \sqrt{2} f_K P_K^\mu \sqrt{2} f_K P_{K\mu} \; . \tag{4.78}$$

This result comes as follows: the factor $\frac{1}{4}$ comes from $d_L = \frac{1}{2}(1-\gamma_5)\psi_d(x)$; by Fierz rearrangement there appears a term with a $\frac{1}{N}$ factor - the $\frac{1}{3}$ term -; each term admits two distinct ways of inserting the vacuum - the factor of 2 -. Then, one is left with the evaluation of

$$\langle 0 | \bar{s} \gamma^\mu \gamma_5 d | K^0 \rangle = \sqrt{2} f_K P_K^\mu \; , \tag{4.79}$$

which is the definition of f_K - known from $K \to \mu\nu$ decay - f_K = 0.119 GeV; and $P_K^\mu \cdot P_{K\mu} = M_K^2$, since K's are on shell. This is the way that Gaillard and Lee obtained a good estimate of the charm-quark mass, as mentioned in section 2a.

Serious doubts on the reliability of the vacuum saturation estimate were first spelled out by Schrock and Treiman [22]. These authors examined the possibility of evaluating the matrix element in question by inserting a complete sum of intermediate physical hadronic states

$$\langle \bar{K}^0 | O_{\Delta S=2} | K^0 \rangle = \sum_{N = \text{HADRONS}} \langle \bar{K}^0 | (\bar{s}_L \gamma^\mu d_L) | N \rangle \langle N | (\bar{s}_L \gamma_\mu d_L) | K^0 \rangle \; . \tag{4.80}$$

In particular, there will be a contribution from the π^0 intermediate state, which formally is given by the integral

$$\langle \bar{K}^0 | O_{\Delta S=2} | K^0 \rangle \Big|_{\pi^0} = \frac{-1}{16\pi^2} \frac{1}{4M_K^2} \int_{-\infty}^{(M_K - m_\pi)^2} \frac{dt}{t} \lambda^{1/2}(t, M_K^2, m_\pi^2) \left[|d|^2 - \lambda(t, M_K^2, m_\pi^2) |f_+|^2 \right] , \tag{4.81}$$

where $d(t)$ and $f_+(t)$ are the scalar and vector K_{l3} - form factors

$$\langle \pi^0 | (\bar{s}\gamma^\mu d) | K^0 \rangle = (p_K + p_\pi)^\mu f_+ + (p_K - p_\pi)^\mu f_- \ , \tag{4.82}$$

with

$$d(t) = (M_K^2 - m_\pi^2) f_+(t) + t f_-(t) \ ; \tag{4.83}$$

and $\lambda(a, b, c)$ the usual kinetic function $\lambda(a, b, c) = a^2+b^2+c^2 - 2ab - 2bc - 2ca$. Unfortunately, the integration domain demands the knowledge of these form factors in a region where they are not experimentally measured. With a reasonable extrapolation of the form factors$^{(*)}$, Schrock and Treiman find

$$\langle \bar{K}^0 | \mathcal{O}_{\Delta S=2} | K^0 \rangle \Big|_{\pi^0} \simeq - \langle \bar{K}^0 | \mathcal{O}_{\Delta S=2} | K^0 \rangle \Big|_{\text{VACUUM}} \ , \tag{4.84}$$

i.e., a contribution of the order of that coming from the hadronic vacuum intermediate state - and opposite in sign -! Successive evaluation of other hadronic intermediate states 2π , 3π , etc. is a formidable task and one is obliged to devise other methods of evaluation.

The latest MIT-bag model calculation [24], which corrects previous evaluations, gives

$$\langle \bar{K}^0 | \mathcal{O}_{\Delta S=2} | K^0 \rangle \simeq -0.4 \ \frac{4}{3} f_K^2 M_K^2 \ . \tag{4.85}$$

If taken seriously, this would imply either that long distance contributions not included in the evaluation of M_{12} are very important, or that the three-generation standard model fails to reproduce the observed K_L-K_S mass difference! Most likely, what the calculation shows is that the bag model is inadequate for this application.

In a critical phenomenological analysis of the situation, Wolfenstein [25] pointed out two sources of unknowns in the theoretical evaluations of M_{12}: long distance effects and what he called the B-factor

$$\langle \bar{K}^0 | \mathcal{O}_{\Delta S=2} | K^0 \rangle = B \ \frac{4}{3} f_K^2 M_K^2 \ , \tag{4.86}$$

and urged caution with phenomenological applications which use the short-distance evaluation of $\mathrm{Re}\, M_{12}$ only.

Since then, there has been an interesting evaluation of B, by Donoghue, Golowich and Holstein [26] which uses SU(3) symmetry to relate the matrix elements

$$\langle \bar{K}^0 | \mathcal{O}_{\Delta S=2} | K^0 \rangle = \sqrt{2} \ \langle \bar{K}^0 | \mathcal{O}^{(27)}_{\Delta I=3/2} | \pi^0 \rangle \ , \tag{4.87}$$

(*) In fact it is possible to obtain rigorous upper and lower bounds of the integral in eq. (4.81), using only the experimental information on K_{l3} form factors. The procedure has been described in connection with another application, in ref. [23].

and then PCAC to relate the matrix element in the r.h.s. to the pure $\Delta I = 3/2$ $K^- \to \pi^- \pi^0$ amplitude which is extracted - in absolute value - from the experimentally observed decay rate. In this way, these authors obtain the result

$$|B| \simeq 0.33 \; . \tag{4.88}$$

A similar result has been obtained by Gilman and Wise [27] who have done the same calculation in the language of chiral perturbation theory.

These are nice calculations. It would be nice, however, to see how they are modified in a systematic treatment of mass expansions in chiral perturbation theory, say à la Gasser and Leutwyler [28].

Ultimately, the B-factor may well be obtained to a sufficient accuracy from lattice gauge calculations. A first trial by Cabibbo, Martinelli and Petronzio [29] has resulted in the value

$$B = 1.2 \pm \text{error (?)} \; . \tag{4.89}$$

Other recent estimates, using various types of hadronic models, give results

$$1.4 \lesssim B \lesssim 2.8 \quad \text{, quark harmonic oscillator models, ref. [30]} \tag{4.90}$$

$$-0.5 \lesssim B \lesssim 1.6, \text{ static quark model; } \; B \sim 1.5 - 2, \text{ other models, Wagner[31]} \tag{4.91}$$

$$1.8 < B < 2.5 \quad \text{, Lea, Nardulli and Preparata [32],} \tag{4.92}$$

significantly larger than the SU(3)-PCAC calculation, eq. (4.88).

4.3a An Upper Bound on |B| from Hadronic Sum Rules

In view of the disparity of results obtained in model calculations, it would be nice to get, at least, bounds on B. It is indeed possible to obtain an upper bound on B following the procedure discussed in ref. [33] which we next review. We shall also add a few comments on this bound. The derivation proceeds as follows:

First one observes that the matrix element $\langle \bar{K^0} | \mathcal{O}_{\Delta S=2} | K^0 \rangle$ can be viewed as the value at t=0 of a scalar form factor which is a real analytic function in the complex t-plane with a branch cut along the real axis $4M_{K^0}^2 \lesssim t < \infty$.

Second observation is that the modulus squared of this form factor $|F(t)|^2$ for $4M_{K^0}^2 \lesssim t < \infty$ governs the lowest hadronic intermediate state contributing to the absorption part of the two-point function

$$\psi(q^2) = i \int dx \, e^{iq.x} \langle 0 | T(\mathcal{O}_{\Delta S=2}(x) \, \mathcal{O}^{+}_{\Delta S=2}(0)) | 0 \rangle \; , \tag{4.93}$$

with $\mathcal{O}_{\Delta S=2}$ the same local operator $:(\bar{s}_L \gamma^\mu d_L)(\bar{s}_L \gamma_\mu d_L):$ which defines the effective $\Delta S=2$ Hamiltonian at short distances. More precisely, the absorptive part of $\psi(q^2)$ is the spectral function

$$\frac{1}{\pi} \mathrm{Im}\, \psi(t) = \frac{1}{2\pi} \sum_{\Gamma} \int \langle 0 | \mathcal{O}_{\Delta S=2}(0) | \Gamma \rangle \langle \Gamma | \mathcal{O}^{+}_{\Delta S=2}(0) | 0 \rangle (2\pi)^4 \delta^{(4)}(p - \Sigma p_\gamma) \tag{4.94}$$

where the sum is extended to all possible hadronic intermediate states with strangeness 2, with integration over phase space. The K^0K^0 is the lowest intermediate state and its contribution, which depends on $|F(t)|^2$, sets a lower bound to $\frac{1}{\pi} \operatorname{Im} \psi(t)$ which by definition is a sum of positive contributions:

$$\frac{1}{\pi} \operatorname{Im} \psi(t) \geq \frac{1}{16\pi^2} \left(1 - \frac{4M_{K^0}^2}{t}\right)^{1/2} \frac{1}{2} |F(t)|^2 \, \theta(t - 4M_{K^0}^2) \,, \tag{4.95}$$

and F(t) is precisely the form factor whose value at t=0 we are looking for.

Equation (4.95) is the starting point for the derivation of an upper bound on $|F(0)|$. The interest of relating $F(t)$ to $\psi(q^2)$ lies in the fact that, quite generally, two-point functions are objects much more easy to handle in QCD than direct calculation of low energy parameters.

Before we enter the formal derivation of the bound, let us investigate the properties of $\psi(q^2)$ in QCD. The behaviour of $\psi(q^2)$ in the deep euclidean region, $Q^2 = -q^2$ very large, is governed by the asymptotic freedom behaviour of QCD, i.e., by the diagrams shown in Fig. IV-2. QCD also tells us that $\psi(q^2)$ obeys a dispersion relation defined up to an arbitrary polynomial in Q^2 of degree four at most. To eliminate this arbitrariness (external subtractions) we are obliged to work with a minimum of five derivatives of $\psi(q^2)$ with respect to q^2. We can then transform the inequality in eq. (4.95) into an inequality for the 5-derivative of $\psi(q^2)$ as follows:

$$\mathcal{F}(Q^2) \equiv -\frac{\partial^5 \psi(q^2)}{(\partial Q^2)^5} = 5! \int_0^\infty dt \, \frac{1}{(t+Q^2)^6} \, \frac{1}{\pi} \operatorname{Im} \psi(t) \geq$$

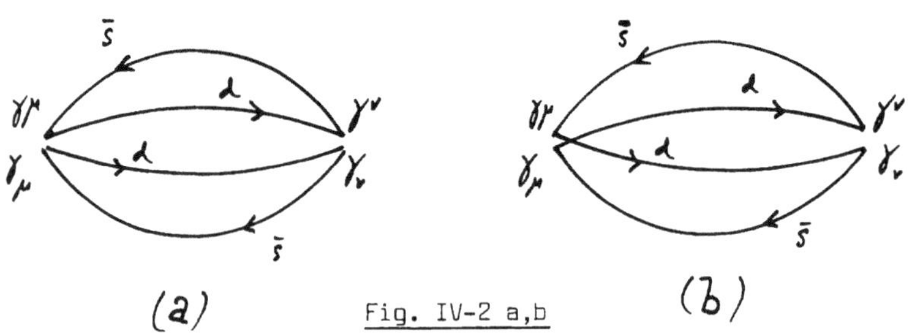

Fig. IV-2 a,b

Feynman diagram contributing to the lowest order evaluation of $\psi(q^2)$ in eq. (4.93).

$$\geq \frac{15}{4\pi^2} \int_{4M_K^2}^\infty dt \, \frac{1}{(t+Q^2)^6} \left(1 - \frac{4M_K^2}{t}\right)^{1/2} |F(t)|^2 \,. \tag{4.96}$$

The calculation of the contribution from the diagrams in Fig. IV-2 to $\mathcal{F}(Q^2)$, i.e., the asymptotic freedom contribution is rather straightforward and gives [33]

$$\mathcal{F}(Q^2)_{\text{All QCD}} = \frac{1}{16}\frac{1}{(16\pi^2)^3}\, 3\times 3\times 2\left(1+\frac{1}{3}\right)\frac{32}{15}\frac{1}{Q^2}\,, \tag{4.97}$$

where we have exhibited explicitly the color factors of 3 and $\frac{1}{3}$, the 2π factors and the factor 2 of time ordering. Only the factor $\frac{32}{15}$ requires an explicit calculation.

It is interesting to point out that the diagram in Fig. IV-2(b) which is suppressed by the color factor $\frac{1}{3}$ gives the same result, but for this color factor, than the one is Fig. IV-2(a).

An interesting question to ask is what about the gluonic corrections to eq. (4.97)? Here, the question of renormalization invariance of the $\mathcal{O}_{\Delta S=2}$ operator enters! Calculations of the gluonic corrections to the Wilson coefficient of the operator $\mathcal{O}_{\Delta S=2}$, which we have reviewed in section 4.26, have been done only to the leading log approximation (LLA). From this, one is left with an arbitrary μ-dependence in the final expression of the Wilson coefficient - an overall $(\alpha_s(\mu))^{-2/9}$ factor - which is determined from the lowest order anomalous dimension of the $\mathcal{O}_{\Delta S=2}$ operator.

The calculation of the two-point function $\mathcal{F}(Q^2)$ - at the same LLA - also exhibits a μ-dependence, with the result

$$\mathcal{F}(Q^2)_{QCD} = \frac{1}{16}\frac{1}{(16\pi^2)^3}\, 3\times 3\times 2\left(1+\frac{1}{3}\right)\frac{32}{15}\frac{1}{Q^2}\left(\frac{\alpha_s(\mu)}{\alpha_s(Q)}\right)^{4/9}\left\{1+A\frac{\alpha_s(Q)}{\pi}+\cdots\right\}. \tag{4.98}$$

Here A is a <u>numerical</u> coefficient which we have not been brave enough to compute explicitly (but it is not too difficult to do ...). The important point to emphasize is that the ambiguity in the μ^2-dependence of the Wilson coefficient in the short-distance evaluation of M_{12} in section 4.2b is completely reabsorbed in $\mathcal{F}(Q^2)_{QCD}$ - if calculated in the LLA - and hence, as well shall see, in a QCD bound for $|B|$.

An upper bound on $|B|$ from the inequality in eq. (4.96) can be derived using two standard inequalities. The first is a generalization for continuous positive functions of the well known fact that with $x_i \geqslant 0$,

$$\frac{1}{N}\sum_{i=1}^{N} x_i \geqslant \left(\prod_{i=1}^{N} x_i\right)^{1/N} = \exp\left(\log\left(\prod_{i=1}^{N} x_i\right)^{1/N}\right) = \exp\frac{1}{N}\sum_{i=1}^{N}\log x_i\,. \tag{4.99}$$

To apply it in our case, let us introduce a measure

$$d\mu(t) = \frac{1}{\pi}\sqrt{\frac{t_0}{t-t_0}}\,\frac{dt}{t}\,, \qquad t_0 \equiv 4M_K^2\,, \tag{4.100}$$

with the property that

$$\int_0^\infty d\mu(t) = 1\,. \tag{4.101}$$

Then, with

$$\rho(t,Q^2) \equiv \frac{15}{4\pi^2}\,\pi t\left(\frac{t-t_0}{t_0}\;\frac{t-t_0}{t}\right)^{1/2}\frac{1}{(t+Q^2)^6}\,, \tag{4.102}$$

which is positive for $t \geq t_0 \equiv 4M_K^2$, we can rewrite the r.h.s. of eq. (4.96) in the following way:

$$\int_{t_0}^{\infty} d\mu(t)\,\rho(t,Q^2)\,|F(t)|^2 \geq \exp\int_{t_0}^{\infty} d\mu(t)\,\log\left[\rho(t,Q^2)\,|F(t)|^2\right] =$$

$$= \left(\exp\int_{t_0}^{\infty} d\mu(t)\,\log\rho(t,Q^2)\right)\left(\exp\int_{t_0}^{\infty} d\mu(t)\,\log|F(t)|^2\right). \tag{4.103}$$

The second inequality we need is Jensen's inequality for analytic functions (see, e.g. ref. [34]), which states that

$$\exp\int_{t_0}^{\infty} d\mu(t)\,\log|F(t)|^2 \geq |F(0)|^2 \,. \tag{4.104}$$

Here, the inequality sign becomes an inequality in the case of an analytic function with no zeros in the complex plane.

The integral $\exp\int_{t_0}^{\infty} d\mu(t)\,\log\rho(t,Q^2)$ can be computed explicitly, and combining eqs. (4.96), (4.103) and (4.104) we arrive at the result

$$|F(0)| \leq \frac{4}{3}f_K^2M_K^2\left(t_0\mathcal{F}(Q^2)\right)^{1/2}\left(\frac{6\pi}{5}\right)^{1/2}\frac{t_0}{f_K^2}\left[1+\left(1+\frac{Q^2}{t_0}\right)^{1/2}\right]^6, \tag{4.105}$$

where, for convenience, the vacuum saturation value $\frac{4}{3}f_K^2M_K^2$ has been factored out explicitly, i.e.,

$$|B| \leq \sqrt{\frac{6\pi}{5}}\;\frac{t_0}{f_K^2}\left(1+\sqrt{1+\frac{Q^2}{t_0}}\right)^6\left(t_0\mathcal{F}(Q^2)\right)^{1/2}. \tag{4.106}$$

This is an upper bound for B which explicitly shows the $\alpha_s(\mu)$-dependence which will be cancelled out with the corresponding μ-dependence of the Wilson coefficient.

For large Q^2, $\mathcal{F}(Q^2) \longrightarrow \mathcal{F}(Q^2)|_{QCD}$ - if we believe QCD to correctly describe strong interactions at short distances- and the bound grows as $(Q^2)^{5/2}$. Therefore, it is not very interesting as a phenomenological constraint; e.g. at $Q^2 \simeq 1\ \mathrm{GeV}^2$, which can hardly be considered a safe asymptotic value for eq. (4.99) to be an accurate approximation to the physical $\mathcal{F}(Q^2)$, one gets $|B| < 25$. The bound in eq. (4.106) has the merit, however, of being rigorous and shows explicitly how the arbitrary $\alpha_s(\mu^2)^{-2/9}$ factor in the Wilson coefficient can be reabsorbed in the LLA evaluation of $\mathcal{F}(Q^2)|_{QCD}$.

4.3b Phenomenological Estimates of the Upper Bound for $|B|$

There are two sources of possible improvement of the QCD-bound in eq. (4.106). One is to have an estimate of $\hat{F}_1(Q^2)$ at non-asymptotic Q^2-values; another to know more properties of the form factor F(t) itself. This we examine next.

One possible improvement on $\hat{F}_1(Q^2)$ would be to incorporate non-perturbative short-distance corrections à la Shifman, Vainshtein and Zakharov [35]. The leading corrections go as $m\langle\bar{\psi}\psi\rangle/Q^4$ and $\alpha_s\langle G^2\rangle/Q^4$. From a phenomenological point of view, this would allow one to apply the bound to Q^2 values around 1 GeV^2 with more reliability, but not much more. Unfortunately, the factor $\left(1+\sqrt{1+Q^2/t_0}\right)^6$ is a big factor.

Another possibility is to make a phenomenological ansatz for the spectral function $\frac{1}{\pi}\mathrm{Im}\,\psi(t)$ itself in eq. (4.94), and then calculate $\hat{F}_1(Q^2)$, i.e.,

$$\frac{1}{\pi}\mathrm{Im}\,\psi(t) = \sum\left(\text{low energy hadronic contributions}\right) + \text{QCD-continuum.}$$

From the QCD-calculation of $\psi(Q^2)$, we know the asymptotic behaviour of $\frac{1}{\pi}\mathrm{Im}\,\psi(t)$:

$$\frac{1}{\pi}\mathrm{Im}\,\psi(t)\Big|_{QCD} = \frac{1}{(16\pi^2)^3}\,\frac{1}{10}\left(1+\frac{1}{3}\right)t^4\left\{1+O\left(\frac{\alpha_s}{\pi}\right)\right\}. \tag{4.107}$$

On the other hand there is a large class of hadronic contributions - the class which is graphically represented in Fig. IV-3 -

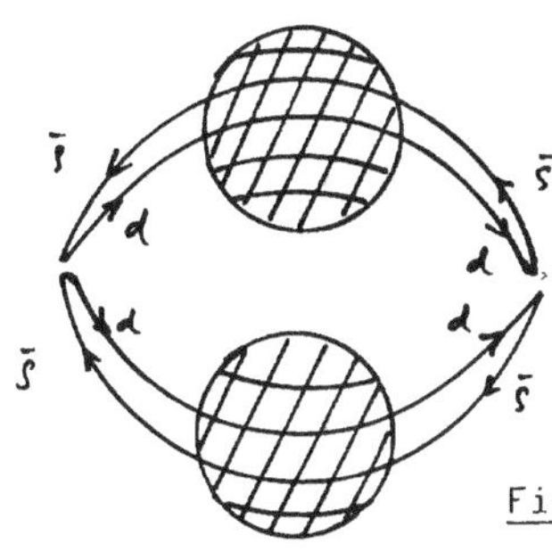

Fig. IV-3

which can be phenomenologically parametrized to a good accuracy. It corresponds to the approximation

$$\frac{1}{\pi}\mathrm{Im}\,\psi(t) \simeq \frac{1}{2\pi}\sum_{\Gamma,\Gamma'}\int\langle 0|(\bar{s}_L\gamma^\mu d_L)|\Gamma\rangle\langle 0|(\bar{s}_L\gamma_\mu d_L)|\Gamma'\rangle \times$$
$$\langle\Gamma|(\bar{s}_L\gamma^\nu d_L)^\dagger|0\rangle\langle\Gamma'|(\bar{s}_L\gamma_\nu d_L)^\dagger|0\rangle\,(2\pi)^4\delta^{(4)}(p-p_\Gamma-p_{\Gamma'}). \tag{4.108a}$$

The lowest hadronic intermediate states are K^0,K^0; K^0,K^{0*}; and $K^{0*}K^{0*}$. Their contribution to $\hat{F}_1$ (Q^2) can be written as a convolution over the spectral functions

$$\frac{1}{\pi}\mathrm{Im}\,\Pi^{(0)}_{AA} = 2f_K^2\,\delta(t-M_K^2), \tag{4.108b}$$

and

$$\frac{1}{\pi}\,\mathrm{Im}\,\Pi^{(1)}_{VV}(t) = \frac{1}{48\pi^2}\,\frac{1}{4}\,\frac{M_{K^*}^4\left(1+\Gamma_{K^*}^2/M_{K^*}^2\right)}{(t-M_{K^*}^2)+M_{K^*}^2\Gamma_{K^*}^2}\left[1-2\,\frac{M_K^2+m_\pi^2}{t}\right]\times$$

$$\times\,\lambda^{1/2}(M_{K^*}^2, M_K^2, t)\ . \tag{4.109}$$

More precisely, one finds (see ref. [33] for details)

$$\tilde{F}(Q^2)_{Hadrons} = \frac{1}{128\pi^2}\sum_{i,j=0,1}\int_0^1 dx\,\frac{x^3(1-x)^3}{\Delta^3}\times$$

$$\int_0^\infty dt\left(\frac{1}{\pi}\mathrm{Im}\,\Pi^{(i)}_{VV}(t)+\frac{1}{\pi}\mathrm{Im}\,\Pi^{(i)}_{AA}(t)\right)\int_0^\infty dt'\left(\frac{1}{\pi}\mathrm{Im}\,\Pi^{(j)}_{VV}(t')+\frac{1}{\pi}\mathrm{Im}\,\Pi^{(j)}_{AA}(t')\right)\times$$

$$\left\{A_{ij}(t,t')\,\frac{6x^2(1-x)^2}{\Delta^2}-3(t+t')\,\frac{x(1-x)}{\Delta}\left[1+4\,\frac{tx+t'(1-x)}{\Delta}\right]+\right.$$

$$\left.1+3\,\frac{[tx+t'(1-x)]}{\Delta}+6\,\frac{[tx+t'(1-x)]^2}{\Delta}\right\}, \quad \text{where} \tag{4.110a}$$

$$\Delta \equiv Q^2x(1-x)+tx+t'(1-x)\ ; \quad \text{and} \tag{4.110b}$$

$$\left.\begin{aligned} A_{11}(t,t') &= 8tt'+(t+t')^2\\ A_{10}(t,t') &= A_{01}(t',t) = (t-t')^2\\ A_{00}(t,t') &= (t+t')^2 \end{aligned}\right\}. \tag{4.110c}$$

To this representation, we add the contribution induced by the asymptotic QCD spectral function in eq. (4.107) from a certain threshold $\hat{t}$ onwards:

$$\tilde{F}(Q^2)_{Asy.\,QCD} = 5!\int_{\hat{t}}^\infty\frac{dt}{(t+Q^2)^6}\,\frac{1}{(16\pi^2)^3}\,\frac{1}{10}\left(1+\frac{1}{3}\right)t^4\,\theta(t-\hat{t})\times$$

$$= \frac{1}{(16\pi^2)^3}\,\frac{12}{5}\,\frac{1}{\hat{t}+Q^2}\left\{1+\sum_{i=1}^4\left(\frac{\hat{t}}{\hat{t}+Q^2}\right)^i\right\}. \tag{4.111}$$

The final phenomenological expression we propose for $\tilde{F}(Q^2)$ is then

$$\tilde{F}(Q^2) = \tilde{F}(Q^2)_{HADRONS}+\tilde{F}(Q^2)_{Asy.\,QCD}\ . \tag{4.112}$$

Several remarks are in order:

i) There is a correspondence between the approximation which consists in considering only the class of hadronic contributions in Fig. IV-3, i.e., eq. (4.107), and the approximation in the evaluation of $\langle \bar{K}^0 | \mathcal{O}_{\Delta S=2} | K^0 \rangle$ via the insertion of a complete set of intermediate physical hadronic states, i.e., eq. (4.80) . They both use the factorization of the $\mathcal{O}_{\Delta S=2}$ operator into a product of (V-A) $\Delta S=1$ currents.

ii) From the point of view of QCD, the class of contributions in Fig. IV-3 contains all the diagrams which are leading in color and <u>some</u> subleading ones, but not all the subleading.

iii) The choice of $\hat{t}$ in eq. (4.111), i.e., the onset of the continuum, must be sufficiently large so as to avoid double counting with the K^0 and K^{0*} contributions which have been already taken into account in $\mathcal{F}(Q^2)|_{\text{hadronic}}$, but not so large that we underestimate the asymptotic hadronic continuum. In our numerical evaluations we have taken $\hat{t}$ within a range

$$(2.8\,\mathrm{GeV})^2 = (2M_{K^*}+2M_K)^2 \lesssim \hat{t} \lesssim (4M_{K^*})^2 = (3.6\,\mathrm{GeV})^2 . \tag{4.113}$$

iv) An unfortunate feature about this estimate of $\mathcal{F}(Q^2)$ is that, because of the factorization involved in the evaluation of the hadronic low energy part, we have lost track of the way to incorporate the arbitrary $\alpha_s(\mu)^{-2/9}$ term in the Wilson coefficient.

With $\mathcal{F}(Q^2)$ given by eq. (4.112), the optimal bound for B is attained at $Q^2 \sim 0$. and the value is

$$|B| < 11 , \tag{4.114}$$

i.e., an order of magnitude bigger than the vacuum saturation value.

The other possible improvement of the bound on $|B|$ requires some input knowledge on the form factor $F(t)$ itself. To examine the kind of improvement one can have, let us assume that we know something about the relative slope of the form factor at the origin, i.e., the dimensionless quantity

$$\rho_1 \equiv -4t_0 \frac{1}{F(0)} \frac{d}{dt} F(t)\Big|_{t=0} . \tag{4.115}$$

Following the techniques described, e.g. in ref. [23] , it can be shown that information on ρ_1 improves the general expression for the bound in eq. (4.106) as follows:

$$|B| \leq \sqrt{t_0 \mathcal{F}(Q^2)} \sqrt{\frac{6\pi}{5}} \frac{t_0}{f_K^2} \left(1+\sqrt{1+\frac{Q^2}{t_0}}\right)^6 \frac{1}{\left[1+(\rho_1+\lambda_1(Q^2))^2\right]^{1/2}} \tag{4.116}$$

where $\lambda_1(Q^2)$ is the function

$$\lambda_1(Q^2) = \frac{5}{2} - \frac{12}{1+\sqrt{1+\frac{Q^2}{t_0}}} \quad . \tag{4.117}$$

Further knowledge, like the relative second derivative at the origin, say knowledge on

$$\rho_2 = 16 t_0^2 \frac{1}{F(0)} \frac{d}{dt^2} F(t) \Big|_{t=0} + 4\rho_1 \tag{4.118}$$

would improved the bound to

$$|B| \leq \sqrt{t_0 \bar{F}(Q^2)} \sqrt{\frac{6\pi}{5}} \frac{t_0}{f_K^2} \left(1+\sqrt{1+\frac{Q^2}{t_0}}\right)^6 \times$$

$$\left\{1+(\rho_1+\lambda_1(Q^2))^2 + \frac{1}{4}\left[\rho_2+2\rho_1\lambda_1(Q^2)+\lambda_2(Q^2)\right]^2\right\}^{-1/2} \tag{4.119}$$

where $\lambda_2(Q^2)$ is the function

$$\lambda_2(Q^2) = \left(\lambda_1(Q^2)\right)^2 + \frac{1}{2} - 24 \frac{\sqrt{1+Q^2/t_0}}{\left(1+\sqrt{1+Q^2/t_0}\right)^2} \quad . \tag{4.120}$$

For a form factor in an exotic channel like K^0K^0 in an S-wave it is very likely that there is very little structure and it is reasonable to expect that ρ_1 and ρ_2 are small parameters. In order to obtain an educated guess about the size of ρ_1 and ρ_2 we shall adopt for F(t) the form factor representation [36],

$$\frac{F(q^2)}{F(0)} = \exp\left(\frac{1}{\pi} q^2 P\int_{t_0}^{\infty} \frac{dt}{t} \frac{1}{t-q^2} \delta(t)\right) , \tag{4.121}$$

where $\delta(t)$ is the phase of the form factor, which in the elastic region coincides with the K^0K^0 J=0, I=1 phase shift. This representation follows under the requirements (see, e.g. ref. [37] where earlier references can be found) that i) $F(q^2)$ is polynomially bounded and ii) $F(q^2)$ has no zeros in the complex plane. The first requirement follows from QCD. Unfortunately we have no control on the second requirement. We assume that this is the case and find

$$\frac{F'(0)}{F(0)} = \frac{1}{\pi}\int_{t_0}^{\infty} dt \frac{1}{t^2} \delta(t) \quad , \tag{4.122}$$

and

$$\frac{F''(0)}{F(0)} = \left(\frac{F'(0)}{F(0)}\right)^2 + \frac{2}{\pi}\int_{t_0}^{\infty} dt \, \frac{1}{t^3} \, \delta(t) \quad . \tag{4.123}$$

These integrals are dominated by the low energy behaviour of the phase $\delta(t)$, and, to a good approximation, we can replace $\delta(t)$ by the phase shift $\delta^{J=1}_{K^0K^0}(t)_{J=0}$. This phase shift has not been measured, but it can be related to the exotic $\pi^0\pi^0$ I=2, J=0 phase shifts, which are experimentally known [38], using SU(3)-symmetry and a phase space correction

$$|\vec{q}_{\pi\pi}|_{CM} \Leftrightarrow |\vec{q}_{KK}|_{CM} \quad .$$

This way we obtain the estimates

$$\rho_1 = 0.4 \pm 0.4 \quad , \quad \rho_2 = 0.5 \pm 0.5 \quad , \tag{4.124}$$

which have been rounded with a 100 % uncertainty to compensate a possible large margin of error.

With the central values for ρ_1 and ρ_2 in eqs. (4.124) inserted into eq. (4.119) the minimum of the bound is attained at $Q^2 = -\ 0.5\ \text{GeV}^2$. Allowing for a 100 % error in ρ_1 and ρ_2, and for a variation in the choice of the continuous threshold $\dot{t}$, we could get an increase by 25% at most at this minimum, hence the final result

$$|B| \lesssim (2.0 \pm 0.5) \quad . \tag{4.125}$$

Note added in proof: It may be possible to do a calculation of B by combining the effective Lagrangian technique developed in ref. [28] with the QCD-duality approach of ref. [54] . This is at present under investigation.

4.4 Long Distance Contributions to M_{12}

The problem here is to evaluate the contribution from the dispersive integral in eq. (4.24), i.e.,

$$\frac{1}{\pi} P \int_0^{\infty} ds \, \frac{1}{s - M_K^2} \sum_{\Gamma} \int d\Gamma \left(\langle\Gamma|T_{\Delta S=1}|\bar{K}^0\rangle\right)^* \left(\langle\Gamma|T_{\Delta S=1}|K^0\rangle\right) , \tag{4.126}$$

over intermediate hadronic states Γ, without double counting with the previous short distance evaluation of the term $\langle\bar{K}^0|T_{\Delta S=2}|K^0\rangle$. The importance of such long distance contributions (LDC) in the phenomenological implications of the standard model was emphasized in early papers by Wolfenstein [39] and Hill [40], and it has triggered a lot of attention more recently (see refs. [41] to [45]).

The importance of contributions to the integral in eq. (4.126) from individual hadronic states is known from early work prior to the development of QCD and the standard model (see refs. [46,47,48]). The problem now is to extract the part of these contributions not already contained in the short-distance evaluation of M_{12}. This question is particularly pertinent with regard to a proper evaluation of Re M_{12}. Although this is still a controversial subject at present, we shall critically summarize some of the attempts exposed in the literature and indicate possible lines of improvement.

Let us first discuss LDC to Re M_{12}. At the limit of CP-invariance - certainly a good approximation to estimate Re M_{12} - the integral in eq. (4.126) can be written as a difference of CP = -1 and CP = +1 contributions:

$$\mathrm{Re}\, M_{12}^{\mathrm{disp.}} = \frac{1}{2\pi} P\int_0^\infty ds \frac{1}{M_K^2 - s} \left(\sum_r \int d\rho \left(|\langle r | T_{\Delta S=1} | K_2^0 \rangle|^2 - K_2^0 \leftrightarrow K_1^0 \right) \right). \qquad (4.127)$$

In particular, the hadronic vacuum (CP = +1) will contribute negatively, while the one pion state (CP = -1) will give a positive contribution. Other one particle states more massive than the K-mass, like the η , will contribute negatively. The 2π - intermediate state will give negative contributions for $s < M_K^2$ and positive for $s > M_K^2$.

The most powerful technique we have at present to evaluate these contributions is by the method of chiral symmetric effective Lagrangians (see refs.[28,49] and [50]). An attempt on this line has been recently suggested by Donoghue, Golowich and Holstein [42] which we next review. These authors take as a starting point the ΔS=1 effective Lagrangean suggested by Cronin [49] many years ago:

$$\mathcal{L}_{\Delta S=1} = g\, \mathrm{Tr}\left(\partial_\mu U^\dagger \partial^\mu U \lambda_6\right) , \qquad (4.128a)$$

where U is

$$U = \exp\left(i\frac{\sqrt{2}}{f_\pi} \Phi(x) \right) , \qquad (4.128b)$$

and $\Phi(x)$ the SU(3) octet of pseudoscalar Goldstone meson fields ,

$$\Phi(x) \equiv \frac{1}{\sqrt{2}} \vec{\lambda}\cdot\vec{\Phi}(x) = \begin{pmatrix} \frac{\pi^0(x)}{\sqrt{2}} + \frac{\eta(x)}{\sqrt{6}} & \pi^+(x) & K^+(x) \\ \pi^-(x) & -\frac{\pi^0(x)}{\sqrt{2}} + \frac{\eta(x)}{\sqrt{6}} & K^0(x) \\ K^-(x) & \bar{K}^0(x) & -2\frac{\eta(x)}{\sqrt{6}} \end{pmatrix} . \qquad (4.128c)$$

The coupling constant g is a free parameter, which has to be determined phenomenologically, e.g. from the $K_S \to \pi\pi$ decay rate, and f_π is the pion to vacuum coupling constant

$$f_\pi = 93.3 \text{ MeV}. \qquad (4.128d)$$

When expanded in powers of Φ this effective Lagrangian allows the calculation of ΔS=1 transitions K -> vacuum; $K \to \pi$; $K \to \eta$; $K \to 2\pi$; and $K \to 3\pi$

amplitudesin a way consistent with the current algebra and PCAC predictions, in terms of one parameter only. In fact the K $\rightarrow$ vacuum amplitude vanishes in this description. The contributions from $\Gamma = \pi^0$ and η in the dispersive integral are proportional to a factor

$$4M_K^2 - 3M_\eta^2 - m_\pi^2$$

which, for $\eta \equiv \eta_8$, as one should have to be consistent with the effective Lagrangean approach, vanishes by the Gell-Mann Okubo formula.

One is left with $\Gamma = 2\pi$ as the bulk of LD contributions to Re M_{12}^{Disp}. The 3π contributions will give a mass shift proportional to $\Gamma_L \ll \Gamma_S$, and can be dismissed at the level of accuracy one is working. The problem now is that the 2π -contribution in eq. (4.127) involves a divergent loop integral. The authors of ref. [42] have suggested to cut-off the loop integral (or the dispersive integral) to some cut-off Λ^2 beyond which it is argued that the contribution goes effectively to the short-distance piece already evaluated. Fixing the cut-off to Λ = 0.7 GeV, they find

$$\Delta m / \Gamma_r \simeq 0.32 \qquad (4.129)$$

and conclude that "the long-range dispersive effect is both large and of the correct sign".

The objection we wish to raise to this calculation is that not only it is very sensitive to the choice of the cut-off (their result is quadratically divergent!), but also to the regularization prescription. If,e.g., one uses instead a dimensional regularization of the divergent loop - which incidentally has the virtue of preserving chiral symmetry - we find that the quadratic divergence appears as a $1/\epsilon$ pole with a residue proportional to $M_K^2 m_\pi^2$ - not $M_K^2\Lambda^2$- showing that very little reliability can be expected from fixing cut-offs arbitrarily.

In fact, the method of how to deal with loops in the framework of effective Lagrangians was outlined in a paper by Weinberg [50] , and has been recently developed in a systematic way in papers by Leutwyler and Gasser [28] . It may be possible to apply these techniques both to the evaluation of the B-factor and the long distance pieces of the Δm contribution in a systematic way. This is under study.

We next turn our attention to the question of LDC to Im M_{12}. Here, it has been noticed by Ginsparg, Glashow and Wise [51] that although Im M_{12} acquires contributions from the 2π-intermediate states in the dispersive integral, the contribution is such that it cancels out in the combination

$$\text{Im}\, M_{12} + 2\,\frac{\text{Im} A_0}{\text{Re} A_0}\,\text{Re}\, M_{12}$$

which determines ϵ_0(see eq. (4.58)). There is still the question of the contribution from other intermediate states, the η' in particular, which is under debate.

For recent literature on the subject see refs. [52] and [53]. Again, we believe that a self-consistent calculation could be done within the framework of chiral symmetric effective Lagrangians and QCD.

4.5 Analysis of predictions on ϵ'/ϵ_0

The basic equations derived in the previous sections for phenomenological applications are eqs. (4.58) and (4.59):

$$\epsilon_0 = \frac{1}{\sqrt{2}} e^{i\pi/4} \frac{1}{\Delta m} \left(\mathrm{Im}\, M_{12} + 2 \frac{\mathrm{Im} A_0}{\mathrm{Re} A_0} \mathrm{Re}\, M_{12} \right) , \tag{4.58}$$

and

$$\epsilon' = \frac{1}{\sqrt{2}} e^{i(\delta_2 - \delta_0 - \pi/2)} \frac{A_2}{\mathrm{Re} A_0} \frac{\mathrm{Im} A_0}{\mathrm{Re} A_0} . \tag{4.59}$$

We have also discussed the short-distance evaluation of M_{12}, with the result

$$M_{12\,\mathrm{S.D.}} = \frac{G_F^2 M_W^2}{4\pi^2} \left\{ \lambda_c^2 \eta_1 x_c + \lambda_t^2 \eta_2 x_t + 2\lambda_c \lambda_t x_t \log \frac{x_c}{x_t} \right\} \frac{1}{2M_K} B \frac{4}{3} f_K^2 M_K^2 . \tag{4.130}$$

Here, the real parts of λ_c^2, λ_t^2 and $2\lambda_c\lambda_t$ can be read in eqs. (4.74); the corresponding imaginary parts in eqs. (4.73), η_1, η_2 and η_3 are the perturbative QCD enhancement factors, calculated in the LLA, $x_i \equiv m_i^2/M_W^2$, i=c and t, and B is the factor discussed at length in section 3. Because of the present uncertainty in the evaluation of B, and also because of the uncertainty about the size of long-distance contributions to Re M_{12} - the so-called Re M_{12}(Disp) in eq. (4.127) - it seems reasonable to avoid the relation $(\Delta m)_{\mathrm{exp.}}$ to $(\mathrm{Re}\, M_{12})_{\mathrm{theory}}$ as a further piece of information. Therefore Δm in eq. (4.58) is usually taken as an experimental input.

The quantity $\mathrm{Im} A_0/\mathrm{Re} A_0$, which appears both in eqs. (4.58) and (4.59), is governed by the CP-violating part of the effective $\mathcal{H}_{\mathrm{eff}}^{\Delta S=1}$ Hamiltonian discussed in chapter III. Here, the standard approximation is to attribute the bulk of the contribution to the effect of the "penguin" operator Q_6 in eq. (3.83f), i.e.,

$$\frac{\mathrm{Im} A_0}{\mathrm{Re} A_0} = \frac{\mathrm{Im} \langle (\pi\pi)_{I=0} | \mathcal{H}_{\mathrm{eff}}^{\Delta S=1} | K^0 \rangle}{\mathrm{Re} A_0} \simeq$$

$$\simeq -\frac{G_F}{\sqrt{2}} (\lambda_c + \lambda_t) \frac{s_2 c_2 s_3}{c_1 c_3} \sin\delta \, C_6(\mu) \frac{1}{\mathrm{Re} A_0} \langle (\pi\pi)_{I=0} | Q_6 | K^0 \rangle \equiv H s_2 c_2 s_3 \sin\delta \tag{4.131}$$

where H is defined as

$$H = \frac{G_F}{\sqrt{2}} s_1 \frac{1}{\mathrm{Re} A_0} C_6(\mu) \langle (\pi\pi)_{I=0} | Q_6 | K^0 \rangle . \tag{4.132}$$

Calculations of H are again uncertain - though perhaps not as much as those of B- ranging in values from $H \simeq -0.4$ to $H \simeq -1.0$ (see refs. of section III). Again this is another case where a combination of the effective Lagrangian technique developed by Gasser and Leutwyler [28] with the QCD-duality equations of ref. [54] may considerably improve the reliability of these calculations.

Finally, let me also remember that $\delta_2 - \delta_0$ and $A_2/\mathrm{Re}A_0$ in eq. (4.59) are both taken from experiment, eqs. (4.60) and (4.45); and, of course the Kobayashi- Mascawa angles are constrained as discussed in chapter II.

There is a wealth of phenomenological analyses in the literature based on these equations which relate mixing angles to ϵ'/ϵ_0 and to the hitherto unknown mass of the t-quark. For a comprehensive updated review, where many other references can be found, we recommend the talk of Buras at the Freiburg workshop [55]. Here we shall limit ourselves to some semiquantitative features which seem to clearly emerge from these analyses:

i) The quantity ϵ'/ϵ_0 appears to be nearly real; and because of the systematic negative sign in the various calculations of H, ϵ'/ϵ_0 is predicted to be positive.

ii) The uncertainties on the calculations of the factors B and H are reflected in the predictions for ϵ'/ϵ_0 as follows: a larger B makes ϵ'/ϵ_0 smaller; a larger H makes ϵ'/ϵ_0 bigger. Also ϵ'/ϵ_0 decreases as the mass of the t-quark grows.

iii) A nice lower bound on ϵ'/ϵ_0 as a function of m_t for a fixed B-value and a fixed B-lifetime value has been derived by Gilman and Hagelin. We reproduce their numerical results in Fig. IV (extracted from ref. [56]).

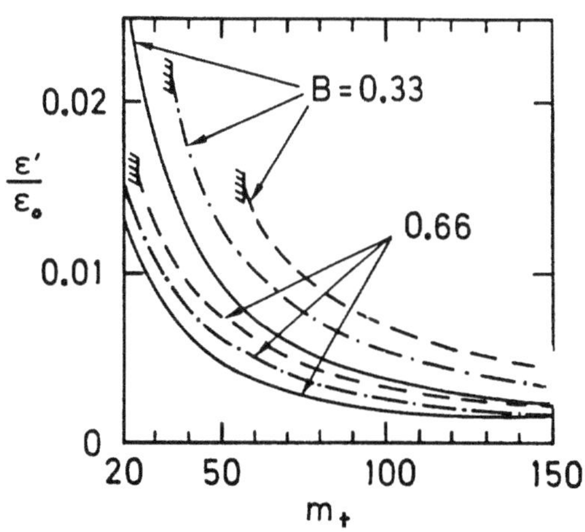

Fig. IV-IV
Lower bound on $\varepsilon/\varepsilon_0$ as a function of m_t for B = 1/3 and B = 2/3; and for $\tau_B = 0.6 \times 10^{-12}$sec (-); $\tau_B = 0.9 \times 10^{-12}$sec (-.-.-); and $\tau_B = 1.2 \times 10^{-12}$sec (----)

iv) Barring the possibility of a large B-factor and/or a large t-quark mass, one is led to the conclusion that if the present trend of preliminary experimental results on ϵ'/ϵ_0 is confirmed it will be difficult to accommodate together our present understanding of the $\Delta I=1/2$ rule and CP-violation in the standard model. We recall the two preliminary experimental results of the high-precision measurements of ϵ'/ϵ_0 are:

$$\epsilon'/\epsilon_0 = 0.0017 \pm 0.0082 \quad \text{, ref. [57],} \tag{4.133}$$

and

$$\mathrm{Re}\ \epsilon'/\epsilon_0 = -0.0046 \pm 0.0053\ (\text{stat.}) \pm 0.0023\ (\text{syst.}),\ \text{ref.[58].} \tag{4.134}$$

We hope for interesting new developments in this field in the near future.

I began these lectures with a strophe of a poem by Pablo Neruda. Let me finish this incomplete written version with the last strophe of the same poem:

"Asi es la vida, Federico, aqui tienes
las cosas que te puede ofrecer mi amistad
de melancólico varón varonil.
Ya sabes por ti mismo muchas cosas.
Y otras irás sabiendo lentamente".

Acknowledgements

During my lectures at the Gif-Luminy school and at the ELAF school I benefited very much from interesting questions and comments from the participants, which have helped me to prepare the parts I have written here. Discussions with P. Pascual and A. Buras on various topics of the lectures have also been very helpful to me. I wish to thank the organizers of both schools and the members of the theory group of the Max-Planck-Institute in Munich for their kind hospitality.

References

Textbooks

An excellent introduction to the modern theory of Particle Physics is:

T.1 T.D. Lee, Particle Physics and Introduction to Field Theory, Contemporary Concepts in Physics, Vol. 1; Harwood Acad. Publ. (1981).

A useful introduction to the topics discussed in these lectures is the book of Okun (unfortunately the English version is very expensive ...):

T.2 L.B. Okun, Leptons and Quarks, North-Holland Pub. Comp. Amsterdam (1982).

As a short introduction to QED and QCD some readers may find useful my incomplete lecture notes:

T.3 E. de Rafael, Lectures on Quantum Electrodynamics, UAB-FT-D-1 (1977).

T.4 E. de Rafael, Quantum Chromodynamics at a Theoretical Framework of the Hadronic Interactions, Ecole d'Eté de Physique des Particules. Gif-sur-Yvette, 1978; Springer L.N.P. 118 (1980).

A larger coverage of QCD can be found in the recent books:

T.5 F.J. Yndurain, Quantum Chromodynamics, Springer Verlag (1983).

T.6 P. Pascual and R. Tarrach, QCD: Renormalization for the Practitioner, Springer-Verlag L.N.P. 194 (1984).

For consultation on many topics in Quantum Field Theory, we recommend:

T.7 C. Itzykson and J.B. Zuber, Quantum Field Theory, Mc Graw-Hill, New York (1980).

A good coverage of modern phenomenology can be found in:

T.8 C. Quigg, Gauge Theories of the Strong, Weak, and Electromagnetic Interactions, Benjamin Cummings (1983).

Other textbooks in gauge theories with useful information are:

T.9 J.C. Taylor, Gauge Theories of Weak Interactions, Cambridge Univ. Press (1976).

T.10 P. Ramond, Field Theory, A Modern Primer, Benjamin Cummings Pub. Company Inc., Reading Mass. (1981).

T.11 M. Chaichian and Nelipa, Introduction to Gauge Field Theories, Springer-Verlag (1984).

Review Articles

R.1 J. Kim, P. Langacker, M. Levine, and H.H. Williams, Rev. Mod. Phys. 53, 211 (1981).

R.2 P. Musset and J.P. Vialle, Phys. Rep. Vol. 39, No. 1 (1978).

R.3 M. Davier, Dubrovnik Conf. on Phenomenology of Unified Theories, 1983.

R.4 M. Aguilar-Benitez et al., Tables of Particle Properties, Rev. Mod. Phys. 56, No. 2, Part II, April 1984.

R.5 J. Calmet, S. Narison, M. Perrottet and E. de Rafael, Rev. Mod. Phys. 49, No. 1, 1977.

R.6 E. Abers and B.W. Lee, Phys. Rep. 9C, 1 (1973).

R.7 M.A.B. Bég and A. Sirlin, Ann. Rev. of Nucl. Sci. 24, 379 (1974).

R.8 M.A.B. Bég and A. Sirlin, Phys. Rep. Vol. 88, No. 1 (1982).

R.9 W. Marciano and H. Pagels, Phys. Rep. 36c, 137 (1978).

R.10 A. Mueller, Phys. Rep. 73c, 237 (1981).

R.11 A. Petermann, Phys. Rep. 53, 157 (1979).

R.12 A.J. Buras, Rev. Mod. Phys. 52, 199 (1980).

R.13 G. Alterelli, Phys. Rep. Vol. 81, No. 1 (1982).

R.14 J. Gasser and H. Leutwyler, Phys. Rep. 87, 77 (1982).

R.15 M. Cruetz, Quarks, Gluons and Lattices, Cambridge U.P. (1983).

R.16 K. Kleinknecht, Ann. Rev. Nucl. Sci., 26, 1 (1976).

R.17 H. Fritzsch and P. Minkowski, Phys. Rep. 73, 67 (1981).

R.18 R.D.C. Miller and H.I. Kellar, Phys. Rep. 106, No. 4,5 (1984).

R.19 L.L. Chau, Phys. Rep. 95, 1 (1983).

R.20 R. Rückl, Weak Decays of Heavy Flavour, to be published in Physics Repts (1985).

Section I

[1] C.N. Yang and R.L. Mills, Phys. Rev. 96, 191 (1954).

[2a] S.L. Glashow, Nucl. Phys. 22, 579 (1961).

[2b] S. Weinberg, Phys. Rev. Lett., 19, 1264 (1967).

[2c] A. Salam, Proc. Eigth Nobel Symp. (John Wiley, New York, 1968)

[2d] A. Salam and J.C. Ward, Phys. Lett. 13, 168 (1964).

[3a] P.W. Higgs, Phys. Lett. 12, 132 (1964); Phys. Rev. Lett., 13, 508 (1964); and Phys. Rev. 145, 1156 (1966).

[3b] T.W.B. Kibble, Phys. Rev. 155, 1554 (1967).

[3c] F. Englert and R. Brout, Phys. Rev. Lett. 13, 321 (1964).

[3d] G.S. Guralnik, C.R. Hagen and T.W.B. Kibble, Phys. Rev. Lett. 13, 585 (1964).

[4] S.L. Glashow, J. Iliopoulos, and L. Maiani, Phys. Rev. D2, 1285 (1970).

[5] M. Kobayashi and K. Maskawa, Progr. Theor. Phys. 49, 652 (1973).

[6a] J.J. Aubert et al., Phys. Rev. Lett. 33, 1404 (1974).

[6b] J.E. Augustin et al., Phys. Rev. Lett. 33, 1406 (1974).

[7] C. Bouchiat, J. Iliopoulos, and Ph. Meyer, Phys. Lett. 38B, 519 (1972).

[8] M.L. Perl et al., Phys. Rev. Lett. 35, 1489 (1975); Phys. Lett. 63B, 466 (1976).

[9a] S.W. Herb et al., Phys. Lett. 76B, 243 (1978).

[9b] C.W. Darden et al., Phys. Lett. 76B, 243 (1978).

[10] UAI Collaboration Preliminary Result

[11a] D. Ross and M. Veltman, Nucl. Phys. B95, 135 (1975).

[11b] B.W. Lee, Proc. XVI Inter. Conf. on High Energy Physics, eds. J.D. Jackson and A. Roberts, FNL, Batavia, IL (1972), Vol. IV, p. 266.

[12a] G. Arnison et al., Phys. Lett. 122B, 103 (1983); ibid. 126B, 398 (1983).

[12b] M. Banner et al., Phys. Lett. 122B, 476 (1983).

[13] W.J. Marciano and A. Sirlin, Phys. Rev. 329, 945 (1984).

[14] G. t' Hooft, Nucl. Phys. B22, 173 (1971); ibid. B35, 167 (1971).

[15] M. Veltman, Proc. of the 6th International Symposium on Electron and Photon Interactions at High Energies, eds. H. Rollnik and W. Pfeil, Amsterdam, North-Holland Pub. Co. (1974).

[16a] H.D. Politzer, Phys. Rev. Lett. 30, 1346 (1973).

[16b] D.J. Gross and F. Wilczek, Phys. Rev. Lett. 30, 1343 (1973).

[17] M.A. Shifman, A.I. Vainshtein and V.I. Zakharov, Nucl. Phys. B147, 385, 448, 519 (1979).

[18] E. de Rafael, Some Comments on QCD Sum Rules. Proc. of the Topical Conf. on Phenomenology of Unified Theories, Dubrovnic (1983).

[19] S. Narison, Chiral Symmetry Breaking and the Light Meson Systems. Proc. of the XXII International Conference in High Energy Physics, Leipzig (RDA) 1984.

[20] W.A. Bardeen, R. Gastmans, and B.E. Lautrup, Nucl. Phys. B46, 319 (1972).

[21] T. Kinoshita, B. Nizic, and Y. Okamoto, Preprint CLNS-84/606.

[22] L. Maiani, All you need to know about the Higgs boson, Proc. of the 1979 Gif Ecole d'Eté de Physique des Particules, IN2P3.

[23] G. Veneziano, CP and U(1) problems, and their relationship to QCD, Proc. of the eighteen rencontre de Moriond, Vol. 2, p. 299, ed. J. Tran Thanh Van 1983.

Section II

[1] M. Bourquin et al., Z. Phys. C - Particles and Fields 12, 307 (1982); 21, 1, 17, 27 (1982).

[2] D.H. Wilkinson, in Nuclear Physics with Heavy Ions and Mesons, eds. R. Balian, M. Rho and G. Ripka, North Holland, Amsterdam (1978).

[3] A. Sirlin, Nucl. Phys. B196, 83 (1984); Phys. Rev. D22 (1980).

[4] H. Leutwyler and M. Roos, CERN Preprint TH 3830 (1984).

[5] K. Kleinknecht and B. Renk, Z. Phys. C. - Particles and Fields 20, 71 (1983).

[6] H. Abramowicz et al., Z. Phys. C - Particles and Fields 15, 19 (1982).

[7] C. Bourrely, B. Machet and E. de Rafael, Nucl. Phys. B189, 157 (1981).

[8] N.S. Lockyer et al., Phys. Rev. Lett. 51, 1316 (1983).

[9] E. Fernandez et al., Phys. Rev. Lett. 51, 1022 (1983).

[10] CLEO Collaboration, J. Chauveau, talk given at the 7th Int. Conf. Exp. Meson Spectroscopy, B.NL April 1983.

[11] CUSB Collaboration, Phys. Lett. 130B, 444 (1983).

[12] J.L. Cortes, X.Y. Pham, A. Tounsci, Phys. Rev. D25, 188 (1982).

[13] D.E. Klem et al., Phys. Rev. Lett. 53, 1873 (1984).

Section III

[1] K.G. Wilson, Phys. Rev. 179, 1499 (1969).

[2] M.K. Gaillard and B.W. Lee, Phys. Rev. Lett 33, 108 (1974).

[3] G. Altarelli and L. Maiani, Phys. Lett. 52B, 351 (1974).

[4] M.A. Shifman, A.I. Vainshtein, and V.I. Zakharov, JETP Lett. 22, 55 (1975).

[5] M.A. Shifman, A.I. Vainshtein, and Y.I. Zakharov, Nucl. Phys. B120, 316 (1977).

[6] M.B. Wise and E. Witten, Phys. Rev. 20D, 1216 (1979).

[7] F.S. Gilman and M.B. Wise, Phys. Rev. 20D, 2392 (1979).

[8] B. Guberina and R.D. Peccei, Nucl. Phys. B163, 289 (1980).

[9] E. Witten, Nucl. Phys. B104, 445 (1976).

[10] G. Altarelli, G. Curci, G. Martinelli, and S. Petrarca, Phys. Lett. 99B, 141 (1981); Nucl. Phys. (1981).

[11] W. Zimmermann, Local Operator Products and Renormalization in Quantum Field Theory, Brandeis Lectures 1970, Vol. 1.

[12] A.J. Buras and W. Słominski, Preprint MPI-PAE/PTh 68/84 (1984).

[13] T.N. Pham, Phys. Lett. 145B, 113 (1984).

[14] H. Galić, B. Guberina, I. Picek, D. Tadić and J. Trampetić, FZKAAA 12(3) 149 (1980).

[15] J.F. Donoghue, E. Golowich, W.A. Ponce and B.R. Holstein, Phys. Rev. 21D, 186 (1980).

[16] J.F. Donoghue, E. Golowich and B.R. Holstein, Phys. Rev. D23, 1213 (1981).

[17] P. Colić, D. Tadić and J. Trampetić, Phys. Rev. D26, 2286 (1982).

[18] F.J. Gilman and J.S. Hagelin, Phys. Lett. 126B, 111 (1983).

[19] P.H. Ginsparg and M.B. Wise, Phys. Lett. 127B, 265 (1983).

[20] N. Bilić and B. Guberina, Preprint RBI-TP-6/84 (1984).

Section IV

[1] M. Gell-Mann and A. Pais, Phys. Rev. 97, 1387 (1955).

[2] T.D. Lee, R. Oehme and C.N. Yang, Phys. Rev. 106, 340 (1957).

[3] R.G. Sachs, Ann. Phys. NY 22, 239 (1963).

[4] J.M. Christenson, J. Cronin, W.L. Fitch and R. Turlay, Phys. Rev. Lett. 13, 138 (1964).

[5] T.T. Wu and C.N. Yang, Phys. Rev. Lett. 13, 380 (1964).

[6] J.S. Bell and J. Steinberger, Proc. Int. Conf. Elem. Part., Oxford (1965).

[7] T.D. Lee and C.S. Wu, Ann. Rev. Nucl. Sci. 16, 511 (1966).

[8] J. Steinberger, K Decay and CP Violation, CERN lectures, CERN 70-1 (1970).

[9] B.R. Martin and E. de Rafael, Phys. Rev. 162, 1453 (1967).

[10] O. Nachtmann and E. de Rafael, On Radiative Corrections to the $\Delta I=1/2$ Rule in $K_S \to \pi^+\pi^-$ and $K_S \to \pi^0\pi^0$ decays, CERN-Preprint TH-1031 (1969) (unpublished).

[11] N.N. Biswas et al., Phys. Rev. Lett. 47, 1378 (1981).

[12] L. Wolfenstein, Phys. Lett. 13, 562 (1964).

[13] M.K. Gaillard and B.W. Lee, Phys. Rev. D10, 897 (1974).

[14] A.J. Buras, Phys. Rev. Lett. 46, 1354 (1981).

[15] T. Inami and C.S. Lim, Prog. Theor. Phys. 65, 297 (1981).

[16] M. Veltman, Nucl. Phys. B123, 89 (1977).

[17] A.I. Vainshtein et al., Yod. Fiz. 23, 1024 (1976) (Sov. J. Nucl. Phys. 23, 540 (1976).

[18] V.A. Novikov et al., Phys. Rev. 16D, 223 (1977).

[19] E. Witten, Nucl. Phys. B122, 109 (1977).

[20] F.J. Gilman and M.B. Wise, Phys. Lett. 93B, 129 (1980).

[21] F.J. Gilman and M.B. Wise, Phys. Rev. 27D, 1128 (1982).

[22] R.E. Schrock and S.B. Treiman, Phys. Rev. 19D, 2148 (1979).

[23] C. Bourrely, B. Machet and E. de Rafael, Nucl. Phys. B189, 157 (1981).

[24] J. Trampetić, Phys. Rev. D27, 1565 (1983).

[25] L. Wolfenstein, Nucl. Phys. B160, 501 (1979).

[26] J.F. Donoghue, E. Golowich and B.R. Holstein, Phys. Lett. 119B, 412 (1982).

[27] P. Ginsparg and M.B. Wise, Phys. Lett. 127B, 265 (1983).

[28] J. Gasser and H. Leutwyler, CERN Preprints TH.3689 (1983) and TH.3798 (1984).

[29] N. Cabibbo, G. Martinelli and R. Petronzio, CERN Preprint TH.3774 (1984).

[30] P. Colić, B. Guberina, D. Tadić and T. Trampetić, Nucl. Phys. B221, 141 (1983).

[31] F. Wagner, MPI-Preprint MPI-PAE/PTh 89/83 (1983).

[32] P. Cea, G. Nardulli and G. Preparata, CPT-Preprint P. 1626 (1984).

[33] B. Guberina, B. Machet and E. de Rafael, Phys. Lett. 128B, 269 (1983).

[34] J.B. Conway, Functions of one complex variable, Graduate text in Mathematics (Springer, Berlin (1973)).

[35] M.A. Shifman, A.I. Vainshtein and V.I. Zakharov, Nucl. Phys. B147, 385 (1979).

[36] R. Omnès, Nuovo Cim. 8, 316 (1958).

[37] B.R. Martin, D. Morgan and G. Shaw, Pion-pion interactions in particle physics, Academic Press, London (1976).

[38] W. Hoogland et al., Nucl. Phys. B69, 266 (1974).

[39] L. Wolfenstein, Nucl. Phys. B160, 501 (1979).

[40] C. Hill, Phys. Lett. 97B, 275 (1980).

[41] D. Hochberg and R.G. Sachs, Phys. Rev. D27, 6061 (1983).

[42] J.F. Donoghue, E. Golowich and B.R. Holstein, Phys. Lett. 135B, 481 (1984).

[43] G. Ecker, Vienna Preprint, UWThPh-21-1984.

[44] I.I. Bigi and A.I. Sanda, Aachen Preprint PIThA 84/14 (June 84) revised.

[45] M.R. Pennington, Durham Preprint DTP/84/73, August 1984.

[46] R. Jacob and R.G. Sachs, Phys. Rev. 121, 350 (1961).

[47] T. Truong, Phys. Rev. Lett. 17, 1102 (1968).

[48] C. Itzykson et al., Nuovo Cim. Supp. 5, 978 (1967).

[49] J. Cronin, Phys. Rev. 161, 1483 (1967).

[50] S. Weinberg, Physica 96A, 327 (1979).

[51] P. Ginsparg, S.L. Glashow and M. Wise, Phys. Rev. Lett. 50, 1415 (1983).

[52] J.F. Donoghue and B.R. Holstein, UMHEP-Preprint 188 (1983).

[53] J.S. Hagelin, MIU-THP-Preprint-84/010 (1984).

[54] R.A. Bertlmann, G. Launer and E. de Rafael, Preprint CPT-84/P.1586 (to be published in the Nucl. Phys. B) 1984.

[55] A.J. Buras, Preprint MPI-PAE/PTh 46/84 (1984).

[56] F. Gilman and J. Hagelin, Phys. Lett. 133B, 443 (1983).

[57] J.K. Black et al., (Yale-BNL Collaboration) preliminary result (1984).

[58] Chicago-FNAL-Saclay Collaboration, preliminary result (1984).

Index of Contributors

Fluctuations and Sensitivity in Nonequilibrium Systems

Proceedings of an International Conference, University of Texas, Austin, Texas, March 12–16, 1984

Editors: **W. Horsthemke, D. K. Kondepudi**

1984. 108 figures. IX, 273 pages. (Springer Proceedings in Physics, Volume 1)
ISBN-13: 978-3-642-93291-5

Contents: Basic Theory. - Pattern Formation and Selection. - Bistable Systems. - Response to Stochastic and Periodic Forcing. - Noise and Deterministic Chaos. - Sensitivity in Nonequilibrium Systems. - Contributed Papers and Posters. - Index of Contributors.

H. Haken

Synergetics

An Introduction
Nonequilibrium Phase Transitions and Self-Organization in Physics, Chemistry, and Biology

3rd revised and enlaraged edition. 1983. 161 figures. XIV, 371 pages. (Springer Series in Synergetics, Volume 1)
ISBN-13: 978-3-642-93291-5

Contents: Goal. - Probability. - Information. - Chance. - Necessity. - Chance and Necessity. - Self-Organization. - Physical Systems. - Chemical and Biochemical Systems. - Applications to Biology. - Sociology and Economics. - Chaos. - Some Historical Remarks and Outlook. - References, Further Reading, and Comments. - Subject Index.

Chaos and Statistical Methods

Proceedings of the Sixth Kyoto Summer Institute, Kyoto, Japan September 12–15, 1983

Editor: **Y. Kuramoto**

1984. 123 figures. XI, 273 pages. (Springer Series in Synergetics, Volume 24). ISBN-13: 978-3-642-93291-5

Contents: General Concepts. - Fractals in Dynamical and Stochastic Systems. - Onset of Chaos. - One Dimensional Mappings. - Bifurcations and Normal Forms. - Solition Systems. - Fluid Dynamics. - Chemical and Optical Systems. - Anomalous Fluctuations. - Index of Contributors.

Springer-Verlag
Berlin
Heidelberg
New York
Tokyo

Dynamical Systems and Chaos

Proceedings of the Sitges Conference on Statistical Mechanics, Sitges, Barcelona/Spain, September 5–11, 1982

Editor: **L. Garrido**

1983. XIV, 298 pages. (Lecture Notes in Physics, Volume 179). ISBN-13: 978-3-642-93291-5

Unified Theories of Elementary Particles

Critical Assessment and Prospects

Proceedings of the Heisenberg Symposium, Held in München, July 16–21, 1981

Editors: **P. Breitenlohner, H. P. Dürr**

1982. VI, 217 pages. (Lecture Notes in Physics, Volume 160). ISBN-13: 978-3-642-93291-5

Statistical Models and Turbulence

Proceedings of a Symposium Held at the University of California, San Diego (La Jolla), July 15–21, 1971

Editors: **M. Rosenblatt, C. Van Atta**

Reprint. 1975. 129 figures. VIII, 492 pages. (Lecture Notes in Physics, Volume 12). ISBN-13: 978-3-642-93291-5

Dissipative Systems in Quantum Optics

Resonance Fluorescence, Optical Bistability, Superfluorescence

Editor: **R. Bonifacio**
With contributions by numerous experts

1982. 60 figures. XI, 151 pages. (Topics in Current Physics, Volume 27). ISBN-13: 978-3-642-93291-5

Solitons

Editors: **R. K. Bullough, P. G. Caudrey**
With contributions by numerous experts

1980. 20 figures. XVIII, 389 pages. (Topics in Current Physics, Volume 17). ISBN-13: 978-3-642-93291-5

Springer-Verlag
Berlin
Heidelberg
New York
Tokyo

GPSR Compliance
The European Union's (EU) General Product Safety Regulation (GPSR) is a set of rules that requires consumer products to be safe and our obligations to ensure this.

If you have any concerns about our products, you can contact us on

ProductSafety@springernature.com

In case Publisher is established outside the EU, the EU authorized representative is:

Springer Nature Customer Service Center GmbH
Europaplatz 3
69115 Heidelberg, Germany

www.ingramcontent.com/pod-product-compliance
Lightning Source LLC
Chambersburg PA
CBHW061153240926
51846CB00006B/1003
* 9 7 8 3 6 4 2 9 3 2 9 1 5 *